Handbook of Research on Interdisciplinarity Between Science and Mathematics in Education

Bento Cavadas
Lusófona University, Portugal & Polytechnic Institute of Santarém, Portugal

Neusa Branco
Polytechnic Institute of Santarém, Portugal & CIAC, Algarve University, Portugal

A volume in the Advances in Educational Technologies and Instructional Design (AETID) Book Series

IGI Global
PUBLISHER of TIMELY KNOWLEDGE

Published in the United States of America by
IGI Global
Information Science Reference (an imprint of IGI Global)
701 E. Chocolate Avenue
Hershey PA, USA 17033
Tel: 717-533-8845
Fax: 717-533-8661
E-mail: cust@igi-global.com
Web site: http://www.igi-global.com

Library of Congress Cataloging-in-Publication Data

Names: Cavadas, Bento, 1979- editor. | Branco, Neusa, 1979- editor.
Title: Handbook of research on interdisciplinarity between science and mathematics in education / Bento Cavadas, and Neusa Branco, editors.
Description: Hershey PA : Information Science Reference, [2023] | Includes bibliographical references and index. | Summary: "The goal of this book is to provide a compilation of current research and good practices about interdisciplinarity between science and mathematics in education, especially for the professional development of educators who want to explore different levels of connections between science and mathematics"-- Provided by publisher.
Identifiers: LCCN 2022039872 (print) | LCCN 2022039873 (ebook) | ISBN 9781668457658 (hardcover) | ISBN 9781668457665 (ebook)
Subjects: LCSH: Science--Study and teaching. | Mathematics--Study and teaching. | Interdisciplinary approach in education.
Classification: LCC Q181.A2 I58 2023 (print) | LCC Q181.A2 (ebook) | DDC 507.1--dc23/eng20221025
LC record available at https://lccn.loc.gov/2022039872
LC ebook record available at https://lccn.loc.gov/2022039873

This book is published in the IGI Global book series Advances in Educational Technologies and Instructional Design (AETID) (ISSN: 2326-8905; eISSN: 2326-8913)

British Cataloguing in Publication Data
A Cataloguing in Publication record for this book is available from the British Library.

For electronic access to this publication, please contact: eresources@igi-global.com.

Advances in Educational Technologies and Instructional Design (AETID) Book Series

Lawrence A. Tomei
Robert Morris University, USA

ISSN:2326-8905
EISSN:2326-8913

Mission

Education has undergone, and continues to undergo, immense changes in the way it is enacted and distributed to both child and adult learners. In modern education, the traditional classroom learning experience has evolved to include technological resources and to provide online classroom opportunities to students of all ages regardless of their geographical locations. From distance education, Massive-Open-Online-Courses (MOOCs), and electronic tablets in the classroom, technology is now an integral part of learning and is also affecting the way educators communicate information to students.

The **Advances in Educational Technologies & Instructional Design (AETID) Book Series** explores new research and theories for facilitating learning and improving educational performance utilizing technological processes and resources. The series examines technologies that can be integrated into K-12 classrooms to improve skills and learning abilities in all subjects including STEM education and language learning. Additionally, it studies the emergence of fully online classrooms for young and adult learners alike, and the communication and accountability challenges that can arise. Trending topics that are covered include adaptive learning, game-based learning, virtual school environments, and social media effects. School administrators, educators, academicians, researchers, and students will find this series to be an excellent resource for the effective design and implementation of learning technologies in their classes.

Coverage

- Higher Education Technologies
- Adaptive Learning
- Instructional Design
- Web 2.0 and Education
- Collaboration Tools
- K-12 Educational Technologies
- Hybrid Learning
- E-Learning
- Social Media Effects on Education
- Instructional Design Models

IGI Global is currently accepting manuscripts for publication within this series. To submit a proposal for a volume in this series, please contact our Acquisition Editors at Acquisitions@igi-global.com or visit: http://www.igi-global.com/publish/.

The Advances in Educational Technologies and Instructional Design (AETID) Book Series (ISSN 2326-8905) is published by IGI Global, 701 E. Chocolate Avenue, Hershey, PA 17033-1240, USA, www.igi-global.com. This series is composed of titles available for purchase individually; each title is edited to be contextually exclusive from any other title within the series. For pricing and ordering information please visit http://www.igi-global.com/book-series/advances-educational-technologies-instructional-design/73678. Postmaster: Send all address changes to above address.

Titles in this Series

For a list of additional titles in this series, please visit: http://www.igi-global.com/book-series/advances-educational-technologies-instructional-design/73678

New Approaches to the Investigation of Language Teaching and Literature
Aitor Garcés-Manzanera (University of Murcia, Spain) and María Encarnacion Carrillo García (University of Murcia, Spain)
Information Science Reference • © 2023 • 325pp • H/C (ISBN: 9781668460207) • US $215.00

Handbook of Research on Facilitating Collaborative Learning Through Digital Content and Learning Technologies
Jared Keengwe (University of North Dakota, USA)
Information Science Reference • © 2023 • 495pp • H/C (ISBN: 9781668457092) • US $270.00

Handbook of Research on Advancing Teaching and Teacher Education in the Context of a Virtual Age
Aaron Samuel Zimmerman (Texas Tech University, USA)
Information Science Reference • © 2023 • 412pp • H/C (ISBN: 9781668484074) • US $270.00

Handbook of Research on Interdisciplinary Preparation for Equitable Special Education
Dena D. Slanda (University of Central Florida, USA) and Lindsey Pike (University of Central Florida, USA)
Information Science Reference • © 2023 • 657pp • H/C (ISBN: 9781668464380) • US $270.00

Shaping the Future of Online Learning Education in the Metaverse
Gürhan Durak (Balıkesir University, Turkey) and Serkan Cankaya (Izmir Democracy University, Turkey)
Information Science Reference • © 2023 • 350pp • H/C (ISBN: 9781668465134) • US $215.00

Digitalization, New Media, and Education for Sustainable Development
Lars Keller (University of Innsbruck, Austria) Gerd Michelsen (Leuphana University, Germany) Martin Dür (University of Innsbruck, Austria) Syamsul Bachri (Universitas Negeri Malang, Indonesia) and Michaela Zint (University of Michigan, USA)
Information Science Reference • © 2023 • 335pp • H/C (ISBN: 9781799850335) • US $225.00

Assessment Methods and Success Factors for Digital Education and New Media
Francisco Vicente Cipolla Ficarra (Latin Association of Human-Computer Interaction, Spain & International Association of Interactive Communication, Italy)
Information Science Reference • © 2023 • 411pp • H/C (ISBN: 9781799887218) • US $215.00

IGI Global
PUBLISHER of TIMELY KNOWLEDGE

701 East Chocolate Avenue, Hershey, PA 17033, USA
Tel: 717-533-8845 x100 • Fax: 717-533-8661
E-Mail: cust@igi-global.com • www.igi-global.com

List of Contributors

Table of Contents

Section 2
Connections Between Science and Mathematics on Educational Practices

Section 3
Integration of Science and Mathematics in Teacher Education

Detailed Table of Contents

Many arguments have been advanced for the integration of mathematics and science education including economic arguments, integration being logical, being engaging, increasing transfer, increasing conceptual learning, and that the real world is interdisciplinary in nature. This chapter explores some of the issues and contradictions with each of these arguments drawing on Bernstein's theory of boundaries. An example of integration in the policy sphere (STEM policy in England) is first discussed, and some of the tensions arising are explored. Crossing the boundary is more challenging than is often implied in discussions of integration with issues of epistemology, status, and language needing to be addressed. Further, integration may not yield the expected benefits and could even decrease conceptual learning in the disciplines. The author argues that the policy context should be considered when advocating integration and that careful consideration be given as to whether integration is genuinely the most appropriate solution to identified educational issues.

Museums and science centers play a leading role in non-formal education, inspiring visitors and raising social, political, and environmental awareness. These are powerful educational institutions, as they gather the infrastructure and competencies to provide opportunities to address topics related to science, technology, society, and arts, among others, through a non-traditional lens. Moreover, their projects and activities can be tailored to meet the needs and expectations of both families and schools, which makes them flexible engagement platforms for many audiences. In this chapter, the authors present an example of how the Natural History and Science Museum of the University of Porto is exploring the links between nature and mathematics in its educational programs. Taking advantage of this authentic context, an interdisciplinary educational activity was designed and implemented as part of the museum's holiday program. In it, 6- to 12-year-old children are challenged to observe and creatively interact with collections and exhibitions.

This study aims to examine the interdisciplinary connection between physics and mathematics according to the secondary education physics curriculum in Turkey and to identify the mathematical associations in the objectives (outcomes) and content (subject list) of the secondary education physics curriculum in Turkey through the secondary education mathematics curriculum. The study was designed with an analytical research model and carried out within the frame of document analysis. Data for the study were obtained from the 2018 secondary education physics (Year 9, 10, 11, and 12) and mathematics (Year 9, 10, 11, and 12) curricula in Turkey. Data in the study, which were considered as documents, were analysed by content analysis. At the end of the study, it was determined that the secondary education physics curriculum contains mathematical associations at every class level; however, generally speaking, these associations are not up to par.

Evolutionary biology is a core discipline in science. Thus knowledge of it is key to understanding other branches of biology. However, there are many authors who have reported difficulties in the teaching-learning processes of this content across several educational levels. In this chapter, the authors delve into pre- and in-service Spanish teacher perceptions about the cross-curricular nature of evolutionary biology by means of semi-structured interviews. All interviewees consider that biological evolution is a unifying content and so has to be approached in the classroom. Thus, treating this content as a central axis for biology could be the way towards an effective educational proposal for biological evolution, but how to reach and design it is still complex. With this in mind, this chapter describes a sequence of activities for secondary students aimed at addressing some learning challenges about biological evolution while considering its cross-curricular nature and its Spanish curricular framework.

This chapter reports the views from 39 students, averaging 14 years of age, in a secondary school in Singapore towards the use of a mathematics-based STEM project in Design and Technology (D&T) lessons to demonstrate their creativity using scientific principles. An instrument in the form of a questionnaire was designed to gather the students' views. This instrument infused elements from Guilford's and Torrance's frameworks for creativity (fluency, flexibility, originality, and elaboration) and indicators of product creativity (novelty, resolution, and style) through Besemer's creative product analysis model. Findings from this classroom-based research highlight that students found the project useful in promoting their interest, confidence, and creativity. This study contributes a way of strengthening the links between D&T, mathematics, and science in a feasible manner in school practices.

The main objective of this work has been to assess the potential cognitive and emotional benefits of several practical experiences integrating STEM areas. A sample of 13 students in the 6th grade of primary school was accessed. Several questionnaires were designed, depending on the contents to be dealt with, and were applied at different times in order to ascertain the evolution of the variables under study. The data from the post-tests applied after the workshops confirm that the practical and integrated sessions carried out with the students improve the level of knowledge, even finding statistically significant differences (Sig.< 0.05) with respect to the initial values. The emotional results show that following an active and integrated methodology has benefited the promotion of positive emotions in the participants.

Teaching is made by providing students with theoretical lessons and giving them the basic concepts of a subject and potentiating further development. Although this is an important step, student learning differs, and they have difficulty in transposing their theoretical knowledge into real-world problems. To help them make a bridge between the two realities, practical activities are essential. However, due to class restrictions, such as time and number of students, the activities may fall far from the expected results. In this matter, the University of Minho has been introducing different methods of teaching, interacting, and captivating students through learning-by-doing approaches. In light of these experiences, the objective of this work is to provide an overview of how with simple activities the correlation between science and mathematics can be demonstrated to high school students and also to undergraduate students. These types of activities open their horizons and promote their interest and willingness to continue studying and learning actively.

This work aims to understand how the use of problem-based learning (PBL), from an interdisciplinary point of view, can enhance the integration of concepts in different disciplines. In addition, the study strives to verify whether the use of computer programming can enhance this integration. In this sense, an interdisciplinary inquiry learning space (ILS) was implemented in the physics module of some lectures of undergraduate engineering degrees. For this study, the authors analyzed the learning evolution of two groups: the first with the application of ILS without a computer programming component and the second with the same ILS but with a computer programming component in MATLAB©. The ILS was organized around Newtonian mechanics, focused on the problem of planetary orbits and the possibility of orbiting around a black hole. From the observation conducted and the results of a survey, it is concluded that this type of approach can enhance the learning of physics, not only by increasing the motivation to learn, but also by the use of computer programming.

Collaboration between teachers is essential for professional development. This work presents two collaborative teacher cycles of inquiry, with the aim of characterizing the dynamics between four higher education teachers involved in a collaborative experience related to the creation, implementation, and evaluation of interdisciplinary inquiry learning spaces (ILS). Written communications between the participants, field notes, the ILS produced, ILS assessment rubrics, and meetings summary were used as data sources. Initially, teachers met to organize the interdisciplinary work, and two subteams were formed, each one with one math teacher and one science teacher. Collaborative work is evident in each subteam and between both teams on dialogue, decision making, action, and evaluation components. The interdisciplinary collaboration had a strong focus on mathematical and science content knowledge as well as pedagogical content knowledge of the ILS. Furthermore, it promoted the reflection on teachers' own practice and their effects on student learning.

The Malthus principle, which determines that the natural populations grow at a faster rate than the resources of which they depend, was a crucial concept for Darwin to discover natural selection. A previous study shows that an educational activity between biology and mathematics that explores the Malthus principle, implemented with elementary school students, effectively promoted their understanding of natural selection. In the chapter, the authors analyse the impacts of this activity in students' abilities to solve mathematical problems involving geometric progressions. For that, they applied a pre- and post-test approach in two target groups. The answers were ranked and coded by levels of performance. The analysis consisted in comparing order statistics (boxplots) and by the application of Wilcoxon signed rank tests. The results show that the educational activity impacted significatively the student ability to solve mathematical problems involving geometric progressions.

Primary education curricula highlight the need for integrating STEM subjects in the classroom. Thus, preservice teachers' training programs must involve subject integration. Students' personal value to this integration is a crucial component of their motivation, influencing their outcomes. However, practitioners lack validated instruments to monitor student task-value changes to specific subject combinations. Using math in a biological context exemplifies STEM subject integration. This chapter shows the adaptation of the Math-Biology Values Instrument to preservice teachers, a tool to measure task values of using math in biology. Structural and construct validities were obtained from a sample of Spanish and an independent multicultural sample of preservice teachers, respectively. Multigroup confirmatory factor analysis supported partial measurement invariance across sexes, backgrounds, and countries. The instrument provides enough validity to monitor task-value in preservice teachers.

In recent years, there is a growing interest in educational approaches that include the integration of STEAM disciplines. However, to contribute to increase students' interest in these subjects from an early age, it is crucial to prepare primary teachers to carry out an integrated STEAM approach in schools. This study is aimed to demonstrate the potential of engaging primary pre-service teachers in a STEAM program within a science and a mathematics course. The results of the study emerged from the analysis of the participants' responses to a pre-post questionnaire, STEAM lesson plans they have developed, and reflective writing assignments they carried out at the end of the STEAM program. Findings show that primary pre-service teachers' attitudes toward a STEAM-integrated approach evolved positively through developing lesson plans. The results revealed benefits and challenges of planning STEAM activities according to the participants and their confidence to implement a STEAM-integrated approach in the context of practice. This work was supported by Portuguese Foundation for Science and Technology, I.P., Grant/Award Number UIDP/04748/2020.

Foreword

Since the launch of the Russian satellite Sputnik in 1957, the US has competitively pursued the advancement of science and technology, leading to numerous innovations such as the first mobile phone, the first space shuttle launch, and the Macintosh® computer. By the 1990s, the proliferation and significance of science, technology, engineering, and mathematics (STEM) were recognized nationally by educational institutes. As increased globalization leveled the economic playing field over time, the importance of remaining competitive in a global market became increasingly apparent. However, a 2005 US report, "Rise Above the Gathering Storm," highlighted a severe disparity in STEM proficiency among American students compared to those in other countries. This report emphasized the critical need for an American focus on STEM education. The National Science Foundation initially used the acronym SMET to collectively refer to the fields of Math, Science, Engineering, and Technology. In the early 2000s, the more widely known acronym STEM was first coined.

Since the acronym's inception, STEM has become globally established as a novel, innovative, and interdisciplinary instructional approach and area of educational research. While some researchers prefer to view STEM as a collection of independent disciplines, others view STEM as the interdisciplinary connection between two or more individual STEM subjects. In the 21st century, as the need for modernization, creativity, and problem-solving skills increases, there is a global interest in STEM+C (STEM plus computing) education. More importantly, the interdisciplinary connection between science and mathematics carries a significant role as both disciplines convey the techniques of math and science and are, therefore, the sub-core STEM subjects.

The notion of connecting individual ideas and concepts from varying subjects to form an integrated interdisciplinary curriculum has gained traction in recent years. Within the real-world community of STEM practitioners, it is uncommon for STEM subjects to be applied independently from one another. Rather, they are often interdependently connected. Mathematics and scientific domains significantly complement each other and are often applied in collaboration in the working world. Within the classroom, an integrated mathematics and science curriculum offers students broader contextual knowledge of these interrelated disciplines, reinforcing their authentic understandings of the two and how they connect. It also equips learners with inquiry-based skills that can be applied across both subjects, like asking questions and planning and carrying out investigations. In addition, an integrated understanding of mathematics and science can potentially boost students' interest in pursuing them as a set of disciplinary career paths, as they are made aware of the 'bigger picture'.

As such, a comprehensive understanding of science and mathematics gives students the foundation to make sense of the world around them through a logic-grounded lens, thus increasing their motivation to learn. It also provides learners with the problem-solving skills necessary to address complex and multifaceted issues they might face in the workforce. In addition, the broader perspective an integrated approach envisions may allow for increased innovation and creativity in students. From an educator's perspective, an integrated understanding of both subjects puts teachers in a better position to impart context-grounded content and skills in the respective fields in which they specialize. However, this requires pre-service and in-service teachers to have adequate training and understanding of both subjects.

The natural overlaps between the mathematics and science disciplines have long been observed. Thus, a need for an interdisciplinary approach to the pedagogy and understanding of these subjects has been identified. Exploring the two disciplines in tandem also allows us to explore the overlaps between them to a greater extent, furthering our comprehension of both subjects as a whole and creating more avenues for research. Approaches to teaching an integrated mathematics and science curriculum have been explored from elementary to university education levels with varying efficacy. Still, the consensus is that possessing a multi-faceted and interconnected understanding of both subjects is largely beneficial. As such, it is important that curriculum, teaching practices, and educational policies reflect the intersections of mathematics and science.

In line with this effort, this book covers pre-service teacher training in integrating mathematics and science, interdisciplinary connections between mathematics and science, specifically physics and biology, and interdisciplinary teaching approaches and curricula. Some chapters examine existing curriculum and their impacts, while others investigate the perceptions of students and teachers on interdisciplinary pedagogy. Interesting topics addressed are how mathematics is employed in environmental issues and biology, as well as discussions about simple machines, computer programming, and STEM policy. Both quantitative and qualitative studies from around the world are included, expanding the scope of the analysis. The discussions will benefit educators, policymakers, researchers, and curriculum developers as it aids in the understanding of existing interdisciplinary STEM pedagogy and its impacts, hence providing insight into how they might continue to better the efficacy of the education system.

Ibrahim H. Yeter
National Institute of Education, Nanyang Technological University, Singapore

Preface

INTRODUCTION

Interdisciplinarity is a desired goal in education and many efforts are being made to create connections between science and mathematics, in different educational levels (McHugh et al., 2018; Ríordáin et al., 2016). In interdisciplinary practices the perspectives of different disciplines are integrated to create a deeper understanding of the problems under study (Greef et al., 2017). Many benefits are attributed to the integration of science and mathematics due to their complementary nature, such as that mathematics and science engage and better motivate the students, form together a more powerful problem-solving tool and are best learned interconnectedly and together (Aguirre-Munõz et al., 2022; Berlin & White, 1994; Ríordáin et al., 2016).

Although working in an interdisciplinary approach is a long time pursued, it continues being a difficult goal of science and mathematics education (Koirala & Bowman, 2003). Zhang et al. (2015) suggested that it can be easy for teachers to find connections between the content of mathematics and science but can be more difficult to identify the overlapping skills and dispositions students will need in those contexts. This can be achieved considering Bossé et al. (2010) idea that mathematics and science are learned in the same way considering both the extensive similarities and the little dissimilarity between the mathematics process standards and the 5 Es of science (Engage, Explore, Explain, Elaborate, Evaluate). These strong connections are also present in the similarity of research topics across science and mathematics education, such as curriculum, policy, and assessment, learning and instruction, and teaching and teacher education (Park et al., 2021). Inclusively, Park et al. (2021) noticed that science education and mathematics education researchers pose similar types of research questions in their research work. These strong connections are the reason that led many North American universities to provide graduate programs in "science and mathematics education" jointly rather than separately (Park et al., 2021). However, other researchers, such as Wong and Dillon (2019) stated that science is dependent on mathematics, but the reverse is not true. Therefore, this asymmetric dependency turns difficult for mathematics and science to work together for common benefit. Though, changing the disciplinary framework to an interdisciplinary one continues to be the subject of multiple debates and controversies (Raynaut, 2018). That's the reason because many models were proposed to deal with science and mathematics integration. Davidson et al. (1995) proposed five different types of science and mathematics integration:

- Discipline specific integration involves activities that include two or more branches of mathematics or science.

- Content specific integration involves the integration of mathematics and science content and learning objectives.
- Process integration is the conflation of aspects of the scientific process, such as observation, controlling variables and evaluating hypotheses, with mathematical skills, such as problem solving, reasoning and communication.
- Methodological integration occurs when students research issues of science and mathematics using common strategies, such as discovery, inquiry, and the learning cycle.
- Thematic integration is based on a theme which becomes the medium with which science and mathematics interact.

While overarching topics, contexts or problems are a useful way to structure curriculum and pedagogical approaches are an important consideration, Millar (2020) warned there is an absence in terms of discussing the specifics of what knowledge students need to know and to what depth. Maybe this is the reason that led in the past to the creation of the continuum models, such as the ones put forward by Lonning and DeFranco (1997) or Huntley (1998), which suggested three levels of integration: absence, partial or full integration. These levels of integration could be expressed through different connections between science and mathematics in the educational policies, curriculum, or in the science and mathematics teachers' educational practices, when they create common educational resources and deliver instruction in co-teaching, for example. In Ng (2019) perspective, the interdisciplinary approach can be based on achieving a higher level of integration across objects by focusing on a shared concept and at the same time trying to overcome any separation of knowledge and skills within the disciplines.

Some studies show that in-service teachers are generally positive and supportive towards integration of mathematics and science in the classroom (Leszczynski, 2014). In-service teachers seem aware of the possible benefit for the students of teaching science and mathematics in an integrated manner because it provides more authentic learning environments where real-life situations are introduced more naturally (Lee et al., 2013). This idea is also shared by Bossé et al. (2010) who stated that science enables authentic context for mathematics, and using mathematics, students can establish better relationships. Interdisciplinary practices lead to interdisciplinary learning, which occurs when students explore knowledge by relating it to aspects of their environment (Wiles & Bondi, 2007). For that aim, Bossé et al. (2010) argued that teachers should develop rich science and mathematics learning activities because this will strengthen the students' learning experience in each field. However, mathematics teachers seem to perceive integration as two separate lessons taught together, whether science teachers perceive integration to be a seamless blend of two disciplines (Leszczyinski, 2014). This difference is related with their past experiences with mathematics and science connections (Leszczyinski, 2014), which could result from inadequate integration models. The semantic meaning of the same terms to mathematics and science, such as "modeling" could also prejudice integration if a common ground is not achieved (Reynante et al., 2020).

Therefore, it is important that in-service teachers have the possibility to participate in professional development activities about science and mathematics integration. For example, in-service teachers' participation in a STEM education program in Turkey proved to positively affect teachers' views of STEM education and they were more confident in preparing teaching plans for integrating STEM subjects (Altan & Ercan, 2016). The authors noted that the in-service teachers reinforced their perception that STEM education is needed to improve creativity, problem solving, inquiry and technology-utilization skills and to develop positive attitudes towards these disciplines. In fact, the appearance of STEM education brought to curriculum integration more popularity (Czerniak & Johnson, 2014). However, the level of the content knowledge of the subject matter in which teachers are not the experts can affect their confidence during the creation of integrated science and mathematics lessons (Lee et al., 2013). Another integration difficulty is the struggle that teachers face to maintain focus on the subject content when exploring integrated activities (Dare et al., 2014). Teachers are also concerned that integration could lead to a loss of the scientific and mathematical content covered in common projects (Wong & Dillon, 2020) and the exclusion of some practices (Reynante et al., 2020). Evaluating learning that results from the integration of science and mathematics is another challenge for teachers. Teachers need assistance to create clear goals to assess specific content knowledge in integrated activities (Dare et al., 2014). Furthermore, teachers have some concerns about the structured curriculum they use, and they reported that flexibility in the curriculum is necessary to integrate different disciplines (Lehman et al., 2014). Additionally, a successful model of integration requires significant degrees of teamwork and collaboration among teachers (Lee et al., 2013; Geiger et al., 2018), administrative support for teachers having time to collaborate on the development of integrated lessons (Aguirre-Munõz et al., 2022) and the creation of effective communities of practice (Frade et al., 2009; Lehman et al., 2014). Research also identifies that teachers experience difficulties in planning and teaching integrated lessons (Czerniack & Johnson, 2014; Widjaja et al., 2019) and lack the time for planning collaboratively (Widjaja et al., 2019; Frade et al., 2009). In fact, creating integrated tasks is a big challenge and requires more time than specific tasks for just one subject, as stated by Brown and Bogiages (2019). McComas and Burgin (2020) also warned that "thinking and working outside the disciplinary boxes is often quite profitable in solving real-world problems but doing so is likely not the best way to help learners gain important foundational understanding" (p. 826).

The challenges of science and mathematics integration are also identified in teacher education (An, 2017). The literature suggests that an effective and integrated science and mathematics education could help to improve preservice teacher education (Czerniak & Johnson, 2014), and provide opportunities for less fragmented knowledge and more relevant and stimulating learning experiences (Frykholm & Glasson, 2005). It is known that preservice teachers value integration (Berlin & White, 2010; Koirala & Bowman, 2003), however student teachers consider that interdisciplinary work is important but report little exposure to it during their teacher education (Biseth et al., 2022). Therefore, for a successful integration in their practices, preservice teachers must experience integrated methods in their teacher education courses (Koirala & Bowman, 2003; McHugh et al. 2018). This is the reason why You (2017) alerted to one of the most important problems of integrated education: "If nobody teaches students how to integrate the different disciplines, students are prevented from discovering and creating links between relevant subject matter" (p.73). There is also evidence that courses focused on science and mathematics integration methods can develop their interdisciplinary science inquiry (Chowdary et al., 2014) and interdisciplinary pedagogical content knowledge (e.g. An, 2017; An & Tillman, 2018). These training courses could also reduce preservice teachers' concern and anxiety when teaching science (Mellado et al., 2004).

Although traditional activities of teaching still prevail, teacher educators are in general positive to interdisciplinary work and understand its importance to both the individual and society (Biseth et al., 2022). Geiger et al. (2018) reported that an interdisciplinary collaboration between mathematicians, scientists, and mathematics and science educators in an online module development on mathematical modeling for preservice teachers provided a rich boundary crossing experience. The participants highlighted they were especially pleased with the way mathematical and science disciplinary knowledge was presented in authentic modeling contexts (Geiger et al., 2018). In fact, the authors identified that there are opportunities for learning at intersections between members of different communities who value and create distinct types of knowledge.

Teachers' education programs related with STEM education need to focus on interdisciplinary themes and pedagogical approaches that encompass a broader spectrum of content, methodologies, and practices (Ng, 2019). DeLuca et al. (2015) suggested that an integrated approach contributes to preservice teachers' broader conceptions of integration, stronger relationships, greater emphasis on pedagogy and enhanced efficiency in teaching. Nevertheless, for integration being effective, teacher education programs should include deeper understanding of content across STEM and strategies for collaboration and teamwork to make integrated instruction time more efficient and less difficult to manage (Berlin & White, 2012). Kim and Bolger (2017) showed that preservice teachers found several positive aspects of having the opportunity to engage in planning STEAM lessons within their science methods course, in addition to potential building of confidence. However, most preservice teachers perceive integration of subjects as difficult and costly (Leszczynski, 2014). Creating plans for integrating curriculum and teaching students requires a lot of hard work and carries a high cost of time for preservice teachers (Kim & Bolger, 2017). Other barriers to integration are preservice teachers' perceptions of the difficulties and challenges of integration (Berlin & White, 2010), and confidence about teaching mathematics and science (Baxter et al. 2014; Weinberg & McMeeking, 2017). Difficulties on content knowledge (Frykholm & Glasson, 2005; You, 2017) and weak content knowledge integration (You, 2017) also are relevant aspects that can condition integration. You (2017) highlighted that if teachers have insufficient content knowledge in the specific disciplines, they will also lack the ability to integrate the concepts.

Other challenges that integration of science and mathematics faces are deeply rooted academic traditions in teacher training (Samson, 2014). Institutions are commonly divided into distinct departments or academic fields (Mason, 1996), which results in science teachers still being trained and certified to teach separate disciplines (Lederman & Lederman, 2013; You, 2017). This problem is extended to other educational levels because disciplines are divided into different subjects since primary education, moving away from a global view of science (Stone, 2014).

The contemporary world has many problems to solve, whose complexity needs knowledge from different subjects and a set of competences that must be integrated to create innovative and effective solutions for those problems. For instance, Honey et al. (2014) and McComas and Burgin (2020) highlighted the importance of using knowledge and skills from multiple subjects to solve complex phenomena or situations. For example, about the complexity of sustainability problems, Bianchi et al. (2022) suggested that should be developed the competences of systems thinking, critical thinking and problem framing. Ng (2019) also argued the interdisciplinary approach, with a focus on STEM practices, can provide students with a solid content knowledge that enables them to create and propose innovative solutions to the problems of the society and the world. Therefore, science and mathematics interdisciplinary thinking could contribute to solving some of the major global problems, such as climate change, energy crisis, loss of biodiversity, food production, new diseases, and loss of freshwater. It is in the school that interdisciplinary thinking can start being developed from the first ages through problem-solving activities, for example. In fact, one goal of science and mathematics integration is to foster students' ability to use skills from both areas in problem-solving activities (Furner & Kumar, 2007; McHugh et al., 2018).

So, the dissemination of current research and good practices about interdisciplinarity between science and mathematics in education – the goal of this book – could support educators who wish to integrate science and mathematics in their practices. The book also presents pedagogical experiences that can be useful for science and mathematics teachers, teacher educators and preservice teachers. This book can be useful for in-service teachers of all levels, preservice teachers, teacher educators, university professors in the STEM areas, collaborators of science museums, master's and doctoral students in education, and other stakeholders in education that want to know more about interdisciplinary between science and mathematics in education

CHAPTER OVERVIEW

The book *Handbook of Research on Interdisciplinarity Between Science and Mathematics in Education* is organized into three sections with a total of 15 chapters.

Section 1: Policies and Approaches of Interdisciplinarity Between Science and Mathematics

In Chapter 1, "Problematizing Integration in Policy and Practice," the author discusses arguments advanced for integration and explores issues and contradictions with these arguments, using Bernstein's theory of boundaries. The author explores some of the issues with the arguments for curriculum integration of mathematics and science. An example of STEM policy in England is also presented. The author highlights the importance of the political context when advocating integration and the analysis that should be made of the role of integration in the identified educational issues.

In Chapter 2, "Connections Between Nature and Mathematics: The Fibonacci's Sequence in the Natural History and Science Museum of the University of Porto," the authors present an interdisciplinary approach between science and mathematics in the educational programs of a Natural History and Science Museum. An activity to explore the Fibonacci's sequence is presented. The participants, 6-12 years old children, identified a set of structures that could serve as natural examples of the Fibonacci's sequence, interacting with the collections and exhibitions in the botanical garden and museum. By bringing together insights from diverse areas, such as mathematics, natural sciences and arts, the museum visitors were given the chance to witness how the boundaries between science and mathematics are mixed in nature.

In Chapter 3, "An Examination of the Interdisciplinary Approaches Between Physics and Mathematics According to Secondary Education Physics Curriculum: The Case of Turkey," the authors found associations between the physics and mathematical curriculum of secondary education (years 9 to 12), but highlighted that these associations are not on a par. The authors identify the need to take into consideration the connection between physics and mathematics in the curriculum effectively. For that aim, synchronization between the outcomes of physics and the outcomes of mathematics classes is needed.

In Chapter 4, "Cross-Curricular Nature of Evolutionary Biology: A Call, Perceptions, and Teaching Approaches," the author explores connections between evolutionary biology and other disciplines, such as mathematics, technology, and natural science. The author, departing from the cross-curricular nature and the Spanish curriculum of biological evolution, presents a sequence of activities for secondary school students. A specific activity - *SHAPE MATTERS!* – is detailed in this chapter. Students work with real datasets and participate in the same practices used by researchers and scientists to become informed about the key role of science in solving real problems.

Section 2: Connections Between Science and Mathematics on Educational Practices

In Chapter 5, "Views From Singapore Students Towards the Use of a Mathematics-Based STEM Project to Demonstrate Their Creativity Using Scientific Principles," the author presents the views of 39 students from a secondary school in Singapore about the use of a mathematics-based STEM project to demonstrate their creativity using scientific principles. Findings show that students recognize that the project was useful in promoting their interest, confidence, and creativity by coming up with design variations of a mathematics-based kit through a variety of scientific principles.

In Chapter 6, "Design and Implementation of Practical Workshops for Teaching STEM Content: Analysis of Cognitive and Emotional Variables," the authors present several practical experiences related to the integration of STEM areas. Those activities were applied with 13 students of the 6th grade of primary school to assess their potential cognitive and emotional benefits. The results show the benefit of practical work for improving the knowledge and promoting positive emotions of students.

In Chapter 7, "Different Teaching Methods: A Step for Student Motivation, the authors present several activities related to engineering problems done by high school students and undergraduate students during outreach events in a Portuguese University. Those activities aim to proportionate different knowledge transfer methodologies to highlight the connections between science and mathematics. The students had the opportunity to work with research teams from the University, experience the importance of research in the development of improved solutions to real-life problems and the role of mathematics in engineering problems as well.

In Chapter 8, "Integrated Physics Learning Using an Interdisciplinary Inquiry Learning Space: An Exploratory Study Using Computer Programming," the authors present an analysis of problem-based learning (PBL) and computer programming usage to improve integration of different disciplines of undergraduate engineering degrees. The authors created and implemented an interdisciplinary inquiry learning space (ILS) about planetary orbits with two groups, whereas only the second used a computer programming component in MATLAB©. Most students identify mathematical concepts related with the interdisciplinarity between Physics and Mathematics proposed by ILS. Also, the use of computer programming can enhance the learning of Physics.

In Chapter 9, "Science and Mathematics Teacher Collaboration in Higher Education: A Pedagogical Experience Using Inquiry Learning Spaces," the authors present an educational experience based on collaborative teacher cycles of inquiry. The authors characterize the dynamics between four higher education teachers that collaborate to create, implement, and evaluate two interdisciplinary inquiry learning spaces (ILS) using the four components of the collaborative teacher cycle of inquiry as a framework.

In Chapter 10, "Malthus' Principle in the Teaching of Evolution as an Integrative Context of Science and Mathematics in Elementary Education," the authors present an educational activity that explores the Malthus Principle, developed, and implemented with elementary school students. The authors analyze the impacts of this biological context in students' abilities to solve mathematical problems involving geometric progressions.

Section 3: Integration of Science and Mathematics in Teacher Education

In Chapter 11, "Integrating Science and Mathematics in Teacher Education Programs," the author focuses on science and mathematics integration models and integration of science and mathematics on teacher education. The author proposed a model of an integrated course focused on the usage of the engineering design process as a bridge to effectively connect science and mathematics. The overall aim of the course is to provide a rich environment to enhance pre-service teachers' content knowledge in science and mathematics and contribute to the development of their pedagogical content knowledge for teaching an integrated curriculum.

In Chapter 12, "Mathematics in Environmental Issues: Research-Based Interdisciplinary Practices in Preservice Teacher Education," the authors present and assess an implementation of a learning sequence on environmental and mathematics education on initial teacher education. The authors noticed that the preservice teachers who participated in the study were able to design, implement, and assess environmental and mathematics interdisciplinary educational activities, based on situated environmental issues and real-world problems.

In Chapter 13, "Adaptation of the Math-Biology Values Instrument to Preservice Teachers: German, Portuguese, and Spanish Versions," the authors present the adaptation of the Math–Biology Values Instrument (MBVI) to monitor changes in the subjective value that preservice teachers from Germany, Portugal, and Spain ascribe to using mathematics in biology. The study concludes that although the interest, utility and cost values attributed to biology and mathematics by preservice teachers are relevant for their performance in integrated tasks, the success in math-biology integration depends on students completing individual tasks, daily and weekly.

In Chapter 14, "Primary Pre-Service Teachers' Changing Perspectives About Integrated STEAM Approach," the authors present the potential of engaging primary preservice teachers in a STEAM program within a science and a mathematics course. Considering the work proposed, the results show that primary pre-service teachers developed positive attitudes toward STEAM integrated approach. An improvement in the confidence of preservice teachers to implement an integrated STEAM approach in the context of practice was noticed.

In Chapter 15, "How Did the Egyptians Transport the Stone Blocks of the Pyramids? Design of a STEM Practice to Address the Simple Machines in Initial Teacher Training," the authors present the design and reflection about an educational intervention focused on the STEM approach to address simple machines with a sample of preservice primary teachers. Preservice teachers created an inclined plane with a pulley reusing different materials and carried out some tests with each model, collected, analyzed data and performed complementary tasks. Although the preservice teachers were motivated and worked as a team, a lack of their mathematical reasoning and a poor understanding of the conceptual meaning of some STEM areas was noticed.

FINAL REMARKS

Handbook of Research on Interdisciplinarity Between Science and Mathematics in Education presents well-grounded empirical research about interdisciplinarity between science and mathematics focusing educational policies, curriculum, or in the science and mathematics teachers' educational practices. Several approaches and practices of integration between science and mathematics are explored from primary to higher education, on different chapters. STEM practices which mobilized different types of content knowledge and teaching practices were also presented in this book and proportionate interesting insights that can be used by other teachers.

For teacher and teacher educators, this book discusses different topics and approaches involving mathematics and sciences that can bring relevant insights to improve the pedagogical content knowledge about interdisciplinary practices. Moreover, a set of theoretical reflections, research designs and research instruments for the study of interdisciplinarity practices are present in the chapters, contributing for further informing the researchers who want to contribute further research on this topic.

Bento Cavadas
Lusófona University, Portugal & Polytechnic Institute of Santarém, Portugal

Neusa Branco
Polytechnic Institute of Santarém, Portugal & CIAC, Algarve University, Portugal

REFERENCES

Aguirre-Muñoz, Z., Dang, B., & Garro, E. S. L. (2022). Impact of Integrated Science and Mathematics Instruction on Middle School Science and Mathematics Achievement. In M. Kalogiannakis & M. Ampartzaki (Eds.), *Advances in Research in STEM Education* (pp. 1–19). IntechOpen. doi:10.5772/intechopen.104082

Altan, E. B., & Ercan, S. (2016). STEM education program for science teachers: Perceptions and competencies. *Journal of Turkish Science Education*, *12*, 103–117.

An, S. A. (2017). Preservice teachers' knowledge of interdisciplinary pedagogy: The case of elementary mathematics–science integrated lessons. *ZDM Mathematics Education*, *49*(2), 237–248. doi:10.100711858-016-0821-9

An, S. A., & Tillman, D. A. (2018). Preservice teachers' pedagogical use of "Gerrymandering" to integrate social studies and mathematics. *Journal of Mathematics Education*, *11*(3), 33–53.

Baxter, J. A., Ruzicka, A., Beghetto, R. A., & Livelybrooks, D. (2014). Professional development strategically connecting mathematics and science: The impact on teachers' confidence and practice. *School Science and Mathematics*, *114*(3), 102–113. doi:10.1111sm.12060

Berlin, D., & White, A. L. (2012). A longitudinal look at attitudes and perceptions related to the integration of mathematics, science and technology education. *School Science and Mathematics*, *112*(1), 20–30. doi:10.1111/j.1949-8594.2011.00111.x

Berlin, D. F., & White, A. L. (1994). The Berlin-White integration science and mathematics model. *School Science and Mathematics*, *94*(1), 12–14. doi:10.1111/j.1949-8594.1994.tb12280.x

Berlin, D. F., & White, A. L. (2010). Preservice mathematics and science teachers in an integrated teacher preparation program for grades 7-12: A 3-year study of attitudes and perceptions related to integration. *International Journal of Science and Mathematics Education*, *8*(1), 97–115. doi:10.100710763-009-9164-0

Bianchi, G., Pisiotis, U., & Cabrera Giraldez, M. (2022). GreenComp – The European sustainability competence framework. In EUR 30955 EN. Publications Office of the European Union.

Biseth, H., Svenkerud, S. W., Magerøy, S. M., & Rubilar, K. H. (2022). Relevant transformative teacher education for future generations. *Frontiers in Education*, *7*(806495), 806495. doi:10.3389/feduc.2022.806495

Bossé, M. J., Lee, T. D., Swinson, M., & Faulconer, J. (2010). The NCTM Process Standards and the Five Es of Science: Connecting Math and Science. *School Science and Mathematics*, *110*(5), 262–276. doi:10.1111/j.1949-8594.2010.00033.x

Brown, R. E., & Bogiages, C. A. (2019). Professional development through STEM integration: How early career math and science teachers respond to experiencing integrated STEM tasks. *International Journal of Science and Mathematics Education*, *17*(1), 111–128. doi:10.100710763-017-9863-x

Chowdhary, B., Liu, X., Yerrick, R., Smith, E., & Grant, B. (2014). Examining science teachers' development of interdisciplinary science inquiry pedagogical knowledge and practices. *Journal of Science Teacher Education*, *25*(8), 865–884. doi:10.100710972-014-9405-0

Czerniak, C. M., & Johnson, C. C. (2014). Interdisciplinary science teaching. In S. K. Abell & N. G. Lederman (Eds.), *Handbook of Research on Science Education* (2nd ed., pp. 395–411). Routledge.

Dare, E. A., Ellis, J. A., & Roehrig, G. H. (2014). Driven by beliefs: Understanding challenges physical science teachers face when integrating engineering and physics. *Journal of Pre-College Engineering Education Research*, *4*(2), 47–61. doi:10.7771/2157-9288.1098

Davidson, D. M., Miller, K. W., & Metheny, D. L. (1995). What does integration of Science and Mathematics really mean? *School Science and Mathematics*, *95*(5), 226–230. doi:10.1111/j.1949-8594.1995.tb15771.x

de Greef, L., Post, G., Vink, C., & Wenting, L. (2017). *Design Interdisciplinary Education. A practical handbook for university teachers*. Amsterdam University Press. doi:10.5117/9789462984769

DeLuca, C., Ogden, H., & Pero, E. (2015). Reconceptualising Elementary Preservice Teacher Education: Examining an Integrated-Curriculum Approach. *New Educator*, *11*(3), 227–250. doi:10.1080/1547688X.2014.960986

Frade, C., Winbourne, P., & Braga, S. M. (2009). A mathematics-science community of practice: Reconceptualising transfer in terms of crossing boundaries. *For the Learning of Mathematics*, *29*, 14–22.

Frykholm, J. A., & Glasson, G. E. (2005). Connecting science and mathematics instruction: Pedagogical context knowledge for teachers. *School Science and Mathematics*, *105*(3), 127–14. doi:10.1111/j.1949-8594.2005.tb18047.x

Furner, J. M., & Kumar, D. D. (2007). The mathematics and science integration argument: A stand for teacher education. *Eurasia Journal of Mathematics, Science and Technology Education*, *3*(3), 185–189. doi:10.12973/ejmste/75397

Geiger, V., Mulligan, J., Date-Huxtable, L., Ahlip, R., Jones, D. H., May, E. J., Rylands, L., & Wright, I. (2018). An interdisciplinary approach to designing online learning: Fostering pre-service mathematics teachers' capabilities in mathematical modelling. *ZDM Mathematics Education*, *50*(1-2), 217–232. doi:10.100711858-018-0920-x

Honey, M., Pearson, G., & Schweingruber, A. (2014). *STEM integration in K-12. Status, Prospects, and an agenda for research*. National Academies Press.

Huntley, M. A. (1998). Design and implementation of a framework for defining integrated mathematics and science education. *School Science and Mathematics*, *98*(6), 320–327. doi:10.1111/j.1949-8594.1998.tb17427.x

Kim, D., & Bolger, M. (2017). Analysis of Korean elementary pre-service teachers' changing attitudes about integrated STEAM pedagogy through developing lesson plans. *International Journal of Science and Mathematics Education*, *15*(4), 587–605. doi:10.100710763-015-9709-3

Koirala, H. P., & Bowman, J. K. (2003). Preparing middle level preservice teachers to integrate mathematics and science: Problems and possibilities. *School Science and Mathematics*, *103*(3), 145–154. doi:10.1111/j.1949-8594.2003.tb18231.x

Lederman, N. G., & Lederman, J. S. (2013). Is it STEM or "S & M" that We Truly Love? *Journal of Science Teacher Education*, *24*(8), 1237–1240. doi:10.100710972-013-9370-z

Lee, M. M., Chauvot, J. B., Vowell, J., Culpepper, S. M., & Plankis, B. J. (2013). Stepping into iSMART: Understanding science–mathematics integration for middle school science and mathematics teachers. *School Science and Mathematics*, *113*(4), 159–169. doi:10.1111sm.12015

Lehman, J. D., Kim, W., & Harris, C. (2014). Collaborations in a community of practice working to integrate engineering design in elementary science education. *Journal of STEM Education: Innovations and Research*, *15*(3), 21–28.

Leszczynski, E. (2014). *The Study of Middle School Mathematics and Science Teachers' Practices, Perceptions, and Attitudes Related to Mathematics and Science Integration* [Unpublished doctoral dissertation]. University of Montclair, Canada.

Lonning, R. A., & DeFranco, T. C. (1997). Integration of science and mathematics: A theoretical model. *School Science and Mathematics*, *97*(4), 212–215. doi:10.1111/j.1949-8594.1997.tb17369.x

Mason, T. C. (1996). Integrated curricula: Potential and problems. *Journal of Teacher Education*, *47*(4), 263–270. doi:10.1177/0022487196474004

McComas, W. F., & Burgin, S. F. (2020). A Critique of "STEM" education. *Science & Education*, *29*(4), 805–829. doi:10.100711191-020-00138-2

McHugh, L., Kelly, A. M., & Burghardt, M. D. (2018). Professional development for a middle school mathematics-infused science curriculum. *Journal of Science Teacher Education*, *29*(8), 804–828. doi:10.1080/1046560X.2018.1514825

Mellado, V., Borrachero, A. B., Brígido, M., Melo, L. V., Dávila, M. A., Cañada, F., Conde, M. C., Costillo, E., Cubero, J., Esteban, R., Martínez-Borreguero, G., Ruiz, C., Sánchez, J., Garritz, A., Mellado, L., Vázquez, B., Jiménez, R., & Bermejo, M. L. (2014). Las emociones en la enseñanza de las ciencias. *Enseñanza de las Ciencias*, *32*(3), 11–36. doi:10.5565/rev/ensciencias.1478

Millar, V. (2020). Trends, issues and possibilities for an interdisciplinary STEM Curriculum. *Science and Education*, *29*(4), 929–948. doi:10.100711191-020-00144-4

Ng, B. S. (2019). *Exploring STEM competences for the 21st century. Current and Critical Issues in Curriculum, Learning and Assessment, 30.*

Park, W., Kim, D., & Kang, D. Y. (2021). Research trends in science and mathematics education in South Korea 2014-2018: A cross-disciplinary analysis of publications in selected local journals. *Asia-Pacific Science Education*, *7*(2), 280–308. doi:10.1163/23641177-bja10029

Raynaut, C. (2018). Paradoxos e ambiguidades na ideia de interdisciplinaridade. *Desenvolvimento e Meio Ambiente*, *47*, 13–48. doi:10.5380/dma.v47i0.62428

Reynante, B. M., Selbach-Allen, M. E., & Pimentel, D. R. (2020). Exploring the promises and perils of integrated STEM through disciplinary practices and epistemologies. *Science & Education*, *29*(4), 785–903. doi:10.100711191-020-00121-x

Ríordáin, M. N., Johnston, J., & Walshe, G. (2016). Making mathematics and science integration happen: Key aspects of practice. *International Journal of Mathematical Education in Science and Technology*, *47*(2), 233–255. doi:10.1080/0020739X.2015.1078001

Samson, G. (2014). From writing to doing: The challenges of implementing integration (and interdisciplinarity) in the teaching of mathematics, sciences, and technology. *Canadian Journal of Science. Mathematics and Technology Education*, *14*(4), 346–358. doi:10.1080/14926156.2014.964883

Stone, E. M. (2014). Guiding students to develop an understanding of scientific inquiry: A science skills approach to instruction and assessment. *CBE Life Sciences Education*, *13*(1), 90–101. doi:10.1187/cbe-12-11-0198 PMID:24591508

Weinberg, A., & McMeeking, L. (2017). Toward meaningful interdisciplinary education: High school teachers' views of mathematics and science integration: Meaningful interdisciplinary education. *School Science and Mathematics*, *117*(5), 204–213. doi:10.1111sm.12224

Widjaja, W., Hubber, P., & Aranda, G. (2019). Potential and challenges in integrating science and mathematics in the classroom through real-world problems: A case of implementing an interdisciplinary approach to STEM. In Y. S. Hsu & Y. F. Yeh (Eds.), *Asia-Pacific STEM teaching practices* (pp. 157–171). Springer. doi:10.1007/978-981-15-0768-7_10

Wiles, J., & Bondi, J. (2007). *Curriculum development. A guide to practice* (7th ed.). Pearson Education.

Wong, V., & Dillon, J. (2019). 'Voodoo maths', asymmetric dependency and maths blame: Why collaboration between school science and mathematics teachers is so rare. *International Journal of Science Education*, *41*(6), 782–802. doi:10.1080/09500693.2019.1579945

Wong, V., & Dillon, J. (2020). Crossing the boundaries: Collaborations between mathematics and science departments in English secondary (high) schools. *Research in Science & Technological Education*, *38*(4), 396–416. doi:10.1080/02635143.2019.1636024

You, H. S. (2017). Why teach science with an interdisciplinary approach: History, trends, and conceptual frameworks. *Journal of Education and Learning*, *6*(4), 66–77. doi:10.5539/jel.v6n4p66

Zhang, D., Orrill, C., & Campbell, T. (2015). Using the mixture Rasch model to explore knowledge resources students invoke in mathematics and science assessments. *School Science and Mathematics*, *115*(7), 356–365. doi:10.1111sm.12135

Acknowledgment

The editors thank to all contributors of this book, especially the authors, reviewers, members of the Editorial Advisory Board and assistant development editors.

Editorial Advisory Board

Section 1

Policies and Approaches of Interdisciplinarity Between Science and Mathematics

Chapter 1
Problematising Integration in Policy and Practice

Victoria Wong
https://orcid.org/0000-0002-1735-0045
University of Exeter, UK

ABSTRACT

Many arguments have been advanced for the integration of mathematics and science education including economic arguments, integration being logical, being engaging, increasing transfer, increasing conceptual learning, and that the real world is interdisciplinary in nature. This chapter explores some of the issues and contradictions with each of these arguments drawing on Bernstein's theory of boundaries. An example of integration in the policy sphere (STEM policy in England) is first discussed, and some of the tensions arising are explored. Crossing the boundary is more challenging than is often implied in discussions of integration with issues of epistemology, status, and language needing to be addressed. Further, integration may not yield the expected benefits and could even decrease conceptual learning in the disciplines. The author argues that the policy context should be considered when advocating integration and that careful consideration be given as to whether integration is genuinely the most appropriate solution to identified educational issues.

INTRODUCTION

It has been claimed that mathematics is essential to science. This is not an uncontested notion, and the nature of the mathematics required for different branches of science varies considerably, but there is little disagreement that at least some mathematics is necessary to do science. Partly due to this dependency there have been calls for mathematics and science education to work more closely together, often described as integration.

Defining integration is problematic. Berlin and White (2012) note a plethora of terms in the literature which are used to refer to what they would describe as integration: integration; connection; co-operation; co-ordination; correlated; cross-disciplinary; fused; interactions; interdependent; interdisciplinary; interrelated; linked; multidisciplinary; trans-disciplinary; unified. In the UK the term cross-curricular is also

DOI: 10.4018/978-1-6684-5765-8.ch001

used. For some authors integration means the traditional discipline boundaries are blurred or even lost, while for others integration retains boundaries but stresses the interactions between disciplines (Lederman & Niess, 1998). The lack of clarity means that while 'integration' might suggest mathematics and science are taught in the same classroom by the same teacher at the same time, a variety of other meanings are evident in the literature, with authors not always articulating exactly what they mean by any of the terms above. To make matters more challenging, not all science educators refer to 'mathematics', with some instead using the terms 'numeracy' or 'quantitative skills.' The use of neither of these terms is well-delineated. In this chapter I am using the term integration to mean crossing the boundary (the meaning of boundary is discussed below) and bringing mathematics and science teaching together in some way; describing or theorising the exact nature of that integration I will leave to others.

Among the anglophone nations mathematics-science integration is increasingly discussed as STEM (science, technology, engineering and mathematics) education. However, internationally STEM rarely means a programme of technology education (Williams, 2011) and in the UK neither does it include engineering at school level (Wong et al, 2016). McComas and Burgin (2020) similarly note the tendency to use STEM to refer to science and mathematics, ignoring the T and the E. Thus, while some of the literature that I will draw on in problematising the notion of mathematics-science integration focuses on STEM integration, rather than specifically mathematics and science integration, in practice these are often almost identical.

I begin the chapter with a brief discussion of Bernstein's (2000) theory of boundaries. I then turn to an example of integration in a policy context – the STEM policies and initiatives in England. This is an unusual example as although the mathematics and science education communities worked closely together in formulating policy during this period, there was no call for science and mathematics teachers in schools to do the same; in fact government policy at this time strongly promoted clearly delineated school subjects, particularly at secondary level. As the government policy which I am discussing relates to secondary (11-16) education, that is the focus of this chapter. The arguments may be somewhat different were primary education in view. I then consider six reasons which are given for integration. I will discuss some of the problems and contradictions in these arguments.

BACKGROUND THEORY: BOUNDARIES

In this chapter I will draw on the theories of boundaries (Bernstein, 2000) to explore some of the issues with the arguments for curriculum integration of mathematics and science.

Bernstein argues that 'A can only be A if it can separate itself from B' (2000, p. 6). To achieve such separation requires boundaries around the separate entities, in this case around school subjects or departments. These boundaries have a significant impact on the relationship between mathematics and science, in schools and in the policy sphere (Wong, 2018). Bernstein argues that boundaries are created by, and closely associated with, the operation of power. He argues that for categories (subjects) to be differently specialised from each other they need a space in which to develop their unique identity. This space is not within the subject itself, but in the distance which separates one subject from another, which he terms *classification*. In order to be differently specialised 'they must have a space in which to develop their unique identity, an identity with its own internal rules and special voice' (p. 6).

In other words, the identity of one subject is reliant on it being different from, separated from, insulated from another subject. Science can only be science if it can effectively insulate itself from mathematics;

mathematics can only be mathematics if it can effectively insulate itself from science. If the insulation between the categories is broken, then it can become impossible to tell where one ends and the other begins and they are in danger of losing their unique identity.

Newman argues that these borders and boundaries 'are [...] there to be crossed. From the moment they are established, there are always groups who have an interest in finding ways to move beyond the barrier (2003, p. 14). There is, indeed, a long history of authors and researchers calling for the integration of science and mathematics (Berlin & Lee, 2005). However, Newman cautions that 'crossing the border does not always bring the expected benefits' (2003, p. 14).

Bernstein (2000) suggests that when classification changes from strong to weak or vice versa, as is the case when departments or subjects collaborate or integrate, we should always ask whose interests are served.

The subjects of mathematics and science in schools are insulated in many ways including by differences in discourse and language, by specialised teachers, by specialised teaching spaces and support staff (technicians), the nature of the curriculum, the asymmetric dependency between mathematics and science, and blaming the mathematics department for students' difficulties with mathematics in science (Wong & Dillon, 2020).

INTEGRATION IN A POLICY CONTEXT: STEM EDUCATION IN ENGLAND

In this section I will briefly discuss a specific example of mathematics-science integration in the policy sphere: the STEM strategies in England in the first decade of this century.

A two-phase qualitative study was carried out with an analysis of government documents followed by interviews with 21 long-standing and acknowledged key contributors to the science and/or mathematics education discourse in England, including 11 science educators, 8 mathematics educators, an engineering educator and a retired civil servant. The aim was to try to understand the purposes and underlying philosophies of STEM policy in England in the period 2002-2014. For further details of the methods see Wong et al (2016).

Policy is often forgotten in education research, but policy provides the context in which schools operate (Maguire, Ball, & Braun, 2012) and therefore a better understanding of policy leads to better understanding of practice in schools.

The Economic Argument

Calls for a greater focus on science and mathematics education often arise from concerns about science more broadly (Wong et al, 2016). In the early 2000s, the UK government was concerned that the supply of high-quality scientists and engineers should not be a constraint on future research and development (R&D) and innovation performance, due to the high importance such activity had for the economy. They commissioned a review by Sir Gareth Roberts (2002), at the heart of which was a fear that the country would become less economically competitive unless changes were made. The actions advised included many directly related to mathematics and science education and led to a series of STEM initiatives. The economic drivers behind education policy were widely acknowledged for the STEM initiatives; the majority of those interviewed stated that the reasons why STEM is promoted and was strongly funded by government are economic at heart.

Mathematics and the sciences were the subjects deemed most significant and received the most funding. A key focus and performance indicator for STEM in England was the number of students choosing physical science and mathematics subjects post-16. Compulsory education ends at 16 in England and for the majority of students there are no mandatory subjects after that age. Consequently, numbers of students studying mathematics in upper secondary education are low compared to the OECD average (Hodgen et al, 2010). There are a number of different courses on offer for post-16 students, but the focus of the STEM initiatives was to increase the numbers choosing mathematics and science A-levels, the most widely followed pre-university qualification. This preoccupation with A-level numbers was noted by the majority of the participants.

Even within science, biology was seen as a lower priority. At least part of the suggested reason for this was that the numbers studying biological sciences at both post-16 and university level were considerably higher than in the other science subjects. However, one participant pointed out that 'biology is actually becoming more and more and more important in economic terms as well as general educational terms' (quoted in Wong et al, 2016, p. 2359) such that bioscience could be funded more highly in future. In other words, the higher status of the physical sciences and mathematics are in part about concerns around numbers of students but also about perceived economic importance. A key difference between policy in England and in some other nations is that these skills were deemed to be developed through the disciplinary study of mathematics and science, particularly physics. There was no push for interdisciplinary teaching, indeed it was actively discouraged.

Diversity and High Status Students

The focus on A-level numbers also ensured a focus on only some students, as in England progression onto A-levels is only possible with high enough grades in the GCSE exams taken at age 16. For example, one science educator noted that:

From the point of view of the government [STEM] is about macro-economic policy. So it is driven by a thesis that says for an advanced country like this one our economic future lies in high-tech advanced technology, using the products of science, and in order to do that you need a technologically educated workforce. And therefore you need good people with qualifications in science and mathematics. (Quoted in Wong, 2018, p. 197)

The identification of the need for 'good people' shows that this set of policies is aimed at those who will be likely to get high grades in science and mathematics qualifications, not at all students. For these 'good people' to be recognised requires ranking and ordering; such recognition is based on national curriculum tests and is highly unlikely to be socially just (Apple, 1993; Archer et al, 2016). The perceived need for a supply of people (and there may not actually be a shortage of qualified people (Smith & White, 2019)) to be part of a technologically educated workforce becomes a target to increase the numbers of students studying STEM subjects in post-compulsory education and thus a driver in science and mathematics education policy. When asked about the aims, purpose and achievements of the STEM initiatives, three-quarters of the respondents referred to an increase in numbers of students choosing to study A-level sciences and mathematics and, subsequently, STEM subjects at undergraduate level. Therefore, the additional financial resources from the national STEM initiatives went to high status subjects to help them attract high status students. According to Goodson, 'the close connection between academic status and resources is a fundamental feature of our educational system' (1995, p. 173), but it is not a hallmark of a system which is socially just.

Not all interviewees found the emphasis on high-status students comfortable, with both science and mathematics educators expressing reservations about both this focus and the curricula changes subsequently introduced.

Indeed, in the early STEM documentation it is possible to see an awareness of the problem of a lack of diversity in STEM. For example, the Roberts' (2002) review recommended that action was required to improve both the number of girls and women and to address the under-representation of ethnic minorities in STEM. In the 2006 STEM programme report (Department for Education and Skills) 1 of the 17 recommended actions relates to diversity. However, the later documentation makes no mention of diversity. While virtually all interviewees mentioned improving the numbers of young people choosing to continue to study STEM subjects at some level, only two mentioned the need to increase the diversity of those people by increasing numbers of women and ethnic minorities studying STEM subjects, suggesting that any early aims to increase diversity were not operationalised, at least by 2014 (Wong et al, 2016). Since then, however, the government has part-funded research by the Institute of Physics into improving the gender balance in physics A-level; for more than 30 years only a fifth of those taking A-level physics have been female (Institute of Physics, 2017).

Risk, Benefit, and Asymmetric Dependency

England has had a national curriculum since the late 1980s and this curriculum has always consisted of discrete subjects. The possibilities of coordination were ignored in the creation of the original curriculum (Orton & Roper, 2000) with the development of the mathematics and science curriculum operating in separate channels (Wong, 2018). However, while it may appear obvious to a science educator that there should be links to the mathematics curriulcum, mathematics links to far more subjects than just to science. One of the mathematics participants described this as: 'your Venn diagrams look different' (quoted in Wong, 2018, p. 217).

Many participants who were mathematics educators were concerned that for mathematics to be seen as prioritising links to science could be perceived as privileging some reasons for studying mathematics over others. Perhaps worryingly for the science education community, some mathematics educators suggested that for mathematics to be part of STEM could be seen as ignoring the needs of the majority of school students as science could be presumed to be for only the highest-achieving students. Given the focus of the STEM initiatives on high attaining students, this could be seen as a valid concern. As there is no part of science which does not fit within STEM, these tensions were not present for science educators in the same way (Wong et al, 2016). Additionally, mathematics educators were concerned that mathematics could get lost in STEM as it was seen to have a science focus.

Wong and Dillon (2019) argue that as science is dependent on mathematics but mathematics is not dependent on science there is an asymmetric dependency between the two subjects. This asymmetry of dependency, is rarely, if at all, discussed in the education literature and yet it is critical in understanding the relationship between the disciplines. The asymmetry of dependency means that when science and mathematics work together more closely – in Bernstein's (2000) terms, the classification weakens – there will tend to be greater benefits for science from such collaboration. This insight comes directly from asking, as Bernstein (2000) suggested, who benefits from the change in classification.

Thus in the STEM policy context, perhaps unsurprisingly, the economic argument was dominant. Despite the bringing together of science and mathematics under the same umbrella policy – STEM – there were very limited calls for science and mathematics teachers in schools to collaborate or integrate their teaching. Indeed, mathematics educators were wary about close involvement with STEM, seeing risks for mathematics education in close association with science.

REASONS TO INTEGRATE

In this section I will discuss six arguments which are advanced for integration. Although the boundaries are challenging to cross, Newman (2003) argues that as soon as boundaries are established there are groups who want to find ways to move beyond the barrier. In the case of mathematics and science integration there are several reasons given for doing so. These include: the workforce argument (e.g. Millar, 2020); that there is overlap such that integration is logical (e.g. Frade et al, 2009); that it will lead to an increase in engagement (e.g. Venville et al, 2002); that it will increase transfer (e.g. Honey et al, 2014); that it will increase conceptual learning in the disciplines (e.g. Reynante et al, 2020) and that the real world is interdisciplinary in nature (e.g. Venville, 2002). I will consider some of the problems with each of these arguments.

The Workforce or Economic Argument

This argument is based on the idea that in bringing mathematics and science teaching together, students will be able to develop desirable skills. This may be framed as skills required for the national interest (e.g. Wong et al, 2016) or to increase students' employability (e.g. Honey et al, 2014). The economic argument is favoured by many governments, at least in the English-speaking west (Millar, 2020). Millar (2020) argues that this is an instrumental argument for interdisciplinarity, where other arguments are more conceptual.

Defining what skills students learn through integrated teaching is not straightforward. In subjects learning is evaluated in terms of conceptual development, whereas the outcomes in integrated curricula are often poorly defined, as Venville et al (2002) acknowledge. As a result, they argue that integrated curricula do not fit well with the expectation in many countries that the school curriculum should be academically orientated towards traditional disciplines, and should emphasise written work, individual study and examinable aspects of the syllabus. In an era when clearly defined outcomes against arguably narrow subject criteria are the most highly prized in education, it is perhaps easy to see why, if integrated teaching has fuzzier aims, it struggles to gain much traction. Given how strongly established the disciplines are in schools, they propose that integrated approaches should not attempt to ignore the traditional subjects. Gao et al (2020) reviewed 49 empirical research articles focusing on interdisciplinary STEM in secondary and college education. They found that most assessemnts focused on disciplinary knowledge or affective domains, even when the program aimed to improve students' interdisciplinary skills. Thus assessment did not align with aims. When it is difficult to define the skills and thus to assess them one has to question whether what students are expected to learn is actually worthwhile.

Economic arguments for education, including mathematics-science integration, are not without their critics. For example, McComas and Burgin (2020) offer a strong critique of the workforce argument for integration and suggest that there are far fewer jobs in STEM than there are graduates. The economic or workforce argument goes beyond mathematics and science, with the prevailing political ideology being that education must serve the economy (Hill, 2013). Apple, in a paper discussing the purpose of national curricula, identifies a similar focus on industry and the economy in education in the US, identifying 'growing pressure to make the perceived needs of business and industry into the primary goals of the school' (1993, p. 4). Young suggests we need to see the curriculum not as 'an instrument for achieving goals such as 'contributing to the economy' [...] but as intrinsic to why we have schools at all' (2014, p. 91). Thus, whether you accept either the economic or the skills arguments for integration will be related to your beliefs about the fundamental aims and purpose of education.

There Is Overlap in Subject Content

Partly this argument is a time-saving measure: in a crowded curriculum integration could lead to fewer learning standards (Reynante et al, 2020). Some authors, such as Pang and Good (2000) and Zhang et al. (2015) suggest that it should be relatively easy for teachers to find points of overlap in the content of mathematics and science. However, this is naïve for several reasons. Finding connections in another subject takes teachers beyond their knowledge base (Millar, 2020). To find points of overlap presupposes that teachers are able to see connections between science and mathematics, when to do so would require content knowledge of both subjects and an understanding of the connections within and between them. Having such knowledge and appreciating the connections is recognised as being challenging and evidence of expert practice in only one of the subjects (Turner & Rowland, 2011); seeing it across two subjects when the teacher probably teaches in only one is unlikely to be easy. Frade et al (2009) found that in practice it did indeed take a considerable investment of time for two experienced teachers to understand one another's disciplinary knowledge in sufficient depth to find overlaps in the mathematics and science curriculum. In a study of six schools with some level of integration between mathematics and science, Wong and Dillon (2020) found that all described finding points of overlap to be challenging. Such epistemological boundaries in knowledge present significant challenges for all involved in interdisciplinary projects (Millar, 2020).

However, the boundaries between mathematics and science teachers go beyond the epistemological. Bernstein (2000) suggests that where there is strong classification, teaching staff are tied to their departments and that their identity is bound up with that of the department. He gives two reasons for this phenomenon: firstly, 'the department is symbolic of their category' and, secondly, promotion comes by appropriate activities in the department (p. 10). Ball (1987) observed in practice that for many secondary teachers their subject, and their commitment to it, is a key part of their overall commitment to teaching. Being a *subject* teacher is central to the satisfactions that they achieve from their work.

Teacher identification with the subject goes beyond the enthusiasm for what they teach. Siskin noticed that secondary teachers of different departments speak quite different languages and:

demonstrate the distinctive vocabularies, logics, and concerns of their subject specialities in subject-specific ways [...] these are more than simply idiosyncratic appearances of technical jargon; rather the discipline's language and epistemology are interwoven in the ways teachers – as subject-matter specialists – conceptualise their world, their roles within it, and the nature of knowledge, teaching, and learning [...]

Even when teachers do not directly reference the subject matter, disciplinary background reveals itself in the choice of words, the structure of their arguments, or the goals they hold. (Siskin, 1994, pp. 152-153)

Ball argues that there are differences between subjects in their views about the broader purposes of education, about how children learn and about the classroom responsibilities of the teacher and that these 'complexes of epistemological, pedagogical and education values and assumptions constitute, in each case, a subject sub-culture' (1987, p. 41). The same was found to be true in the USA (Grossman & Stodolsky, 1995).

Furthermore, there is perhaps a difference between requiring a mathematics teacher to teach science and a science teacher to teach mathematics – as might be required for integrative practice. Western democracies prize logical-mathematical thinking (Crombie, 1994) more highly than other types of intelligence (Caprara & Cervone, 2000). When this type of thinking is so valued, it could be difficult for a science teacher to admit to a limited knowledge of mathematics. Lack of knowledge of science could be seen as a lack of knowledge of science 'facts' which could be an easier thing to admit to as it does not require logical-mathematical thinking. Mathematics teachers frequently do not have a strong science background so there would be no expectation or loss of face in admitting that they do not have scientific knowledge. Science teachers are expected to be highly mathematical, even if that expectation is unrealistic, and thus admitting to struggling with mathematics could also challenge teachers' identity, perhaps that of teaching an elite subject, and status.

The consequences of all these differences are explained by Bernstein who argues that when subjects are strongly classified:

the staff cannot relate to each other in terms of their intrinsic function, which is the reproduction of pedagogic discourse. Where the lines of communication between staff are established by a system of this kind, there will be weak relations between staff with respect to pedagogic discourse, as each is differently specialised. (2000, p. 10)

In other words, differences in language, discourse, culture and status between departments result in breaks in communication leading to weak relations between staff from different departments. Epistemological boundaries are significant, but they are not the only cause of the communication difficulties between mathematics and science teachers. When calling for science and mathematics teachers to integrate their teaching, the significant barriers that exist between them should at least be acknowledged.

Integration Increases Engagement

Engagement is a slippery concept which is hard to define (Godec et al, 2018). Fredericks et al (2004) identify three types of engagement: affective-emotional including attitudes and interest; cognitive including persistence and motivation; and behavioural including taking part in activities. Godec et al (2018) highlight that when considering engagement it is important to pay attention to the opportunities students have to engage, including how those who do not meet school expectations may find it harder to engage with science in the context of school science lessons. The same argument can also be made for mathematics education. They call for a focus on greater inclusion and social justice as part of any discussion of engagement. This is a particular issue for science and mathematics as what counts as knowledge in these subjects has often been dictated by a western, white, masculine identity, which can marginalise minorities and women (Reynante et al, 2020). Integration offers possibilities to disrupt the status quo and value different ways of knowing. However, integrated teaching would usually be undertaken by the existing teachers of science and mathematics. To truly increase diversity and inclusion would require that teachers change their beliefs about who can hold valid knowledge and, furthermore, communicate this change to the other students in the class. Godec et al. (2018) demonstrate that teachers can make such changes in their practice within the context of teaching science in the existing curriculum. Therefore, while the aim of increasing engagement is a valid one, perhaps integration of science and mathematics teaching is not the most appropriate response to a justified concern. Indeed, Holmlund et al. (2018) found that integration did not necessarily lead to a focus on inclusion of all students and argue that when equity is not explicitly addressed it is usually overlooked. It would thus seem that if an increase in social justice within the classroom is what is desired then professional development which supports teachers in changing their practice to be more inclusive might be the more successful approach.

Furthermore, expecting that integration in and of itself will lead to an increase in engagement is rather simplistic. Wong and Dillon (2020) note that teachers reported that problems with behaviour increased during some integrated mathematics-science projects, suggesting a decrease in engagement. Issues with behaviour management were likewise noted as a barrier to integrated teaching by Czerniak and Johnson (2014). There are, however, few studies which specifically focus on student engagement in integrated contexts. An exception to this is a mixed methods study by Struyf et al (2019) of how student engagement varies in different STEM learning environments. They found that an integrated STEM learning environment had an initially positive impact on student engagement, compared to lessons in individual subjects. However, they suggest that this increase in engagement can be explained by a higher degree of student-centred teaching, rather than by the integrated nature of that teaching. They do, however, suggest that an integrated approach 'seems to facilitate the teachers' implementation of a [...] student centred approach' (ibid, p1402). They sound a note of caution, however, that teachers' attitudes towards integrated teaching can influence their instructional practice, a finding which is echoed by Wong and Dillon (2020).

In other words, integrated teaching can indeed increase engagement, particularly in the hands of an enthusiastic teacher who is a good classroom practitioner. However, it may not be the integrated approach in and of itself which is the reason for that increase in engagement.

Integration Increases Transfer

In the educational literature, transfer is the use or application in one subject of knowledge learned in another. It is a contested idea with authors expressing a wide range of views as to what it is, whether it can be promoted through teaching and even whether it exists at all (Wong & Dillon, 2019). The argument for transfer is almost always that integrated teaching will improve students' use of mathematics in science; there is rarely a call for transfer of knowledge in the other direction (Wong, 2018).

Transfer as a lens to explore the relationship between school science and mathematics is problematic for a number of reasons. Expectation of transfer by science teachers (as reported by, for example, Turşucu et al., 2017 and Goldsworthy et al., 1999) leads to a deficit view of both students and mathematics teachers when the expected transfer is not seen (Wong & Dillon, 2019). Furthermore, Redish and Kuo (2015) in the context of undergraduate physics and Wong (2017) in the context of school graphing practices, argue that there are differences in how mathematics is used in mathematics itself and in science, which make straightforward transfer from mathematics to science unlikely.

The expectation of transfer is understandable; both subjects require students to manipulate numbers and use graphs, for example. However, Redish (2017) argues that in physics equations are linked to physical systems and that this adds information about how they should be interpreted. In the simplest terms, it is not possible to add two numbers if they have different units; one cannot add length to mass. Scientists use physical knowledge when applying mathematics to physical systems, which might involve knowing the limits to the numbers which can be put into an equation. For example, in the speed equation (speed = distance ÷ time) none of the values would be negative as speed, distance and time are scalar quantities which are always positive values. There is no mathematical reason why a negative number could not be introduced into the equation, but there are reasons based on physical knowledge. However, in the similar equation for velocity (velocity = displacement ÷ time) displacement and velocity are vector quantities and can be negative. Redish (2017) argues that physicists use physical knowledge in doing mathematics as much as they use mathematics in doing physics. Redish and Kuo argue that part of the 'acculturation of a physics student is learning to interpret the math physically, not to only focus on mathematical structure and manipulations' (2015, p. 567). This acculturation would not necessarily be emphasised in an integrated course and neither would it be taught in a mathematics class ready to be transferred into science.

However, the main difficulty with the argument that integration increases transfer between the subjects is that there is no evidence that it does so (Honey et al, 2014). This does not necessarily mean that transfer is not promoted by integration as transfer is difficult to evidence. It does mean that authors should be wary of using the general notion of transfer as an argument for greater integration between science and mathematics.

Integration Increases Conceptual Learning in the Subjects

Mathematics and science each contain distinct knowledge, specialized practices and particular skills (Reynante et al, 2020). Lederman and Niess (1998) argue that in mathematics and science there are different epistemologies, in other words different ways of knowing and building knowledge, with the nature of science and mathematics fundamentally different. The use and importance of empirical evidence and data is a key area where the domains diverge.

Mathematics does not rely on the external world to support or falsify knowledge but on other reference points such as the importance of logic. Science does refer to the external world. The ultimate arbiter of knowledge is empirical (derived from experience) observation. Lederman and Niess argue that because of these fundamental differences, the disciplinary boundaries between the disciplines should be maintained even when they are being integrated. However, they express concern that much integrated work emphasises inquiry and problem solving. They argue that what counts as data upon which decisions can be made is fundamentally different in the two disciplines and thus attempting to combine them equally will result in a hybrid version of inquiry which is 'confused and chaotic' because 'different disciplines view the world and how one comes to know it differently' (p. 283). Instead, rather than trying to dissolve the disciplines and create hybrids we should focus on interactions between them.

Reynante et al (2020), however, argue that the clear distinctions between the disciplines could be changing. Statistics, data science and applied mathematics are considered to be part of mathematics in many classrooms even while there is an on-going debate as to whether they should properly be considered mathematics. They suggest that as these mathematical sciences increase in use it could be argued that the epistemology of mathematics now includes 'elements of epiricism' (p. 788) which blurs the boundary between mathematics and the sciences.

In spite of the possible blurring of boundaries, key differences in the conceptual ideas in the subjects remain. Wong and Dillon (2020), in a study of four schools carrying out integrated project based teaching, reported that teachers expressed concerns about the scientific and mathematical content which could be covered in joint projects. In other words, prioritising integration led to a loss of subject-specific content and learning. The longer the project, the more this issue was felt to be problematic. Williams et al. (2016), argue that there is nothing obvious to exchange when science and mathematics teachers work together; no clear benefit to both sides. In consequence it is difficult to define project outcomes which are meaningful and useful to both subjects.

This finding chimes with work by Becker and Park (2011) who found negative effect sizes for mathematics in some studies into integrated teaching of STEM, alongside often positive effect sizes for science. Reynante et al (2020) suggest mathematical learning could be reduced as there is a lower likelihood of students taking up abstract ideas when interdisciplinary learning tends to be focused on real world contexts at the expense of thinking about abstract relational ideas.

Indeed, in a wide ranging review of the literature on interdisicplinary STEM teaching, including mathematics-science integration, Honey et al. (2014) suggest that integrated approaches could impede learning by placing excessive demands on resource limited cognitive processes such as working memory and attention. Like Lederman and Neiss (1998), they express concern that integrated approaches could obscure important differences between disciplines about how knowledge is constructed and revised. They argue that there is limited research evdience as to 'whether more explicit connections or integration across the disciplines significantly improves student learning, retention, achievement, or other valued outcomes' (Honey et al, 2014, p. 22). Nine years on, evidence for integrated teaching improving student disciplinary learning remains limited.

Real World Challenges Are Interdisciplinary

In the real world there will, of course, always be problems which require solutions which transcend disciplinary boundaries and require an interdisciplinary apporach (McComas and Burgin, 2020). For example, during the recent Covid-19 crisis the UK government had advisory committees of people with expertise in a range of disciplines including medicine, microbiology, statistics, behavour sciences, public health and epidemiology (Scientific Advisory Group for Emergencies, 2022). The nature and complexity of the problem demanded that the detailed knowledge of the practices and content of each discipline be brought together in the search for viable solutions to the evolving emergency. No one person or discipline had all the knowledge and expertise required; that an interdisiciplinary approach is required to meet such challenges is recognised in the composition of such an advisory group. Indeed, Millar argues that 'while much of the discussion of interdisciplinarity assumes that it is inherently different to disciplinarity, an alternative school of thought poses that interdisciplinarity presupposes disciplinarity such that one is integral to the other' (2020, p. 939). However, the need for a range of expertise in solving real world challenges, described by McComas and Burgin as 'thinking and working outside the disciplinary boxes' (2020, p. 826), does not mean that working on such problems is the best way to help learners gain foundational knowledge and understanding within each discipline.

Perhaps as a result of the focus on real world contexts, a common approach to integrated projects is to start with a science context to which mathematics is added (Wong, 2018, Frykholm and Glassom,

2005 and Pang and Good, 2000). Such a science-first approach can lead to concerns from mathmeatics educators that the mathematics and mathematics learning can be lost in the excitement of the real world context (Wong and Dillon, 2019). The science-first approach may help to explain Becker and Park's (2011) finding that when mathematics is taught in an integrated context with science, mathematics attainment can be lower than when it is taught separately. Nonetheless, Struyf et al (2019) found that students find the use of authentic real world problems engaging.

CONCLUSION

In this chapter I have discussed some of the policy discourse that emerged from the bringing together of mathematics and science in England's STEM policy initiatives. I have then explored many of the arguments given for integrating mathematics and science and have examined some of the problems with each argument.

Many of the arguments for integration in schools can be seen in the policy context, albeit often with different foci. For example, the workforce argument, or at least the economic argument, is seen in England's STEM policy. However, it is expressed as a focus on student numbers in physics and mathematics rather than specific skills. Physics and mathematics as they stand are perceived to develop skills that are required by the country. This perhaps reflects the status and value western democracies place on logical-mathematical thinking (Caprara & Cervone, 2000). It also reflects the emphasis in STEM policy on what the country was perceived to need, rather than what might be of benefit to individual students in their future lives.

Desire for an increase in transfer was seen in the STEM government data, although it was not strongly present. Most discussions of transfer frame it as from mathematics to science but the belief among some UK policy experts was that an increase in mathematics in science would lead to greater achievement in mathematics assessments (Wong, 2019). It was, in any case, not an argument for greater integration in schools.

The aim of an increase in engagement is evident in both the literature calling for an increase in integration and the STEM policy documentation from England. The focus on engagement from the policy documentation is strongly on numbers of high status students ('good people') choosing to study science. In the focus on high status students, calls for an increase in diversity are lost or diluted and the STEM initiatives become, in effect, a call for more of the same type of student rather than an attempt to attract students from a wider range of backgrounds.

Neither the 'overlap so integration is logical' nor the 'real-world challenges are interdisciplinary' arguments are seen in England's policy documentation and are rarely present in the interview data. This reflects the government focus on a curriculum made up of discrete subjects (Young, 2011), with the knowledge that students are expected to know and that schools are expected to teach laid down in detail in the national curriculum. The most recent curriculum, being developed as the research was carried out, is a deliberate and determined move (Young, 2011) away from a curriculum based on the competencies or skills which are often the basis of interdisciplinary projects.

Wong and Dillon (2020), in a study of mathematics-science collaboration in schools, found that for teachers to engage in integrated (or cross-departmental) practice required the endorsement of, or even promotion by, school leaders, a finding echoed by Struyf (2019). The leaders are influenced, in turn, by the national context in which they operate. Where neither input regulation (such as national curricula) nor output regulation (such as measurement of school performance by attainment data, external inspections and audits) (Priestley et al, 2021) value integrative practice it makes it hard for school leaders to support it. Indeed, Wong and Dillon (2020) found that the collaborative practice was both rare and fragile, collapsing rapidly when support was removed. The balance of risks and benefits for individual schools and teachers in engaging in integrated teaching will be strongly influenced by the external context in which they operate and which educational outcomes are valued within that context. Fensham (2009) argues that the political and cultural context is often ignored in science education research, as is the interplay between stakeholders in school and beyond who determine the way the curriculum is enacted. Thus, the opportunities for teachers to engage in integrated practice will be constrained by school leaders. Decisions by school leaders about the balance between the risks and possible benefits of integration will be strongly influenced by the policy context in which they operate. Therefore, any promotion of integrated practice should take into account the policy context in which schools operate.

Given the difficulties with the arguments for integration that I have outlined, should science and mathematics educators persist with integration and is it a worthwhile educational aim? The answer to this question will depend on the beliefs held about the aims of education, the policy context, the amount or proportion of time that will be spent on integrated teaching and, in particular, what issue the integration is intending to address. There are many and significant barriers for science and mathematics teachers to cross in order to integrate their teaching. Careful consideration should be given to the problem that integration is attempting to solve and whether integration is likely to be the most effective approach or response.

REFERENCES

Apple, M. (1993). The politics of official knowledge: Does a national curriculum make sense? *Discourse (Abingdon), 14*(1), 1–16. doi:10.1080/0159630930140101

Archer, L., Moote, J., Francis, B., DeWitt, J., & Yeomans, L. (2016). Stratifying science: A Bourdieusian analysis of student views and experiences of school selective practices in relation to 'Triple Science' at Key Stage 4 in England. *Research Papers in Education*, 1–21.

Ball, S. (1987). *The Micropolitics of the School*. Routledge.

Becker, K., & Park, K. (2011). Effects of integrative approaches among science, technology, engineering, and mathematics (STEM) subjects on students' learning: A preliminary meta-analysis. *Journal of STEM Education: Innovations and Research, 12*, 23–37.

Berlin, D., & Lee, H. (2005). Integrating Science and Mathematics Education: Historical Analysis. *School Science and Mathematics, 105*(1), 15–24. doi:10.1111/j.1949-8594.2005.tb18032.x

Berlin, D., & White, A. L. (2012). A longitudinal look at attitudes and perceptions related to the integration of mathematics, science and technology education. *School Science and Mathematics, 112*(1), 20–30. doi:10.1111/j.1949-8594.2011.00111.x

Bernstein, B. (2000). *Pedagogy, symbolic control and identity* (2nd ed.). Rowman & Littlefield.

Caprara, G. V., & Cervone, D. (2000). *Personality: determinants, dynamics, and potentials*. Cambridge University Press. doi:10.1017/CBO9780511812767

Crombie, A. C. (1994). *Styles of scientific thinking in the European tradition*. Duckworth.

Czerniak, C., & Johnson, C. (2014). Interdisciplinary science teaching. In N. Lederman & S. Abell (Eds.), *Handbook of research on science education* (Vol. 2, pp. 395–411). Routledge.

Department for Education and Skills. (2006). *The STEM programme report*. HM Government.

Fensham, P. (2009). The link between policy and practice in science education: The role of research. *Science Education, 93*(6), 1076–1095. doi:10.1002ce.20349

Frade, C., Winbourne, P., & Braga, S. M. (2009). A mathematics-science community of practice: Re-conceptualising transfer in terms of crossing boundaries. *For the Learning of Mathematics*, *29*, 14–22.

Fredericks, J., Blumenfeld, P., & Paris, A. (2004). School Engagement: Potential of the Concept, State of the Evidence. *Review of Educational Research*, *74*(1), 59–109. doi:10.3102/00346543074001059

Frykholm, J., & Glassom, G. (2005). Connecting Science and Mathematics Instruction: Pedagogical context knowledge for teachers. *School Science and Mathematics*, *105*(3), 127–141. doi:10.1111/j.1949-8594.2005.tb18047.x

Gao, X., Li, P., Shen, J., & Sun, H. (2020). Reviewing assessment of student learning in interdisciplinary STEM education. *International Journal of STEM Education*, *7*(4), 24. Advance online publication. doi:10.118640594-020-00225-4

Godec, S., King, H., Archer, L., Dawson, E., & Seakins, A. (2018). Examining Student Engagement with Science Through a Bourdieusian Notion of Field. *Science and education*, *27*(5-6), 501–521. doi:10.100711191-018-9988-5

Goldsworthy, A., Watson, R., & Wood-Robinson, V. (1999). *Getting to grips with graphs*. ASE.

Goodson, I. (1995). Becoming a School Subject. In I. Goodson (Ed.), *The making of the curriculum* (pp. 155–177). The Falmer Press.

Grossman, P., & Stodolsky, S. (1995). Content as Context: The Role of School Subjects in Secondary School Teaching. *Educational Researcher*, *14*(5), 5–23. doi:10.3102/0013189X024008005

Hill, M. J. (2013). *The Public Policy Process* (5th ed.). Pearson.

Hodgen, J., Pepper, D. S., & Ruddock, G. (2010). *Is the UK an Outlier? An international comparison of upper secondary mathematics education*. Nuffield Foundation.

Holmlund, T., Lesseig, K., & Slavit, D. (2018). Making sense of "STEM education" in K-12 contexts. *International Journal of STEM Education*, *5*(32), 32. Advance online publication. doi:10.118640594-018-0127-2 PMID:30631722

Honey, M., Pearson, G., & Schweingruber, H. (2014). *STEM Integration in K-12 Education: Status, Prospects, and an Agenda for Research*. National Academies Press. Retrieved from https://www.nap.edu/catalog.php?record_id=18612

Institute of Physics. (2017). *Improving gender balance*. Institute of Physics. Retrieved from https://www.iop.org/sites/default/files/2019-07/IGB-reflections-intervention.pdf

Lederman, N., & Niess, M. (1998). 5 Apples + 4 oranges =? *School Science and Mathematics*, *98*(6), 281–284. doi:10.1111/j.1949-8594.1998.tb17422.x

Maguire, M., Ball, S., & Braun, A. (2012). *How schools do policy: policy enactments in secondary schools*. Routledge.

McComas, W., & Burgin, S. (2020). A Critique of "STEM" education. *Science and Education*, *29*(4), 805–829. doi:10.100711191-020-00138-2

Millar, V. (2020). Trends, Issues and Possibilities for an Interdisciplinary STEM Curriculum. *Science and Education*, *29*(4), 929–948. doi:10.100711191-020-00144-4

Newman, D. (2003). On borders and power: A theoretical framework. *Journal of Borderlands Studies*, *18*(1), 13–25. doi:10.1080/08865655.2003.9695598

Orton, T., & Roper, T. (2000). Science and Mathematics: A relationship in need of counselling? *Studies in Science Education*, *35*(1), 123–153. doi:10.1080/03057260008560157

Pang, J., & Good, R. (2000). A Review of the Integration of Science and Mathematics: Implications for Further Research. *School Science and Mathematics*, *100*(2), 73–82. doi:10.1111/j.1949-8594.2000.tb17239.x

Priestley, M., Philippou, S., Alvunger, D., & Soini, T. (2021). Curriculum making: a conceptual framing. In M. Priestley, D. Alvunger, S. Philippou, & T. Soini (Eds.), *Curriculum making in Europe: policy and practice within and across diverse contexts* (pp. 1–28). Emerald Publishing Limited.

Redish, E. F. (2017). Analysing the competency of mathematical modelling in physics. In Key competences in physics teaching and learning (pp. 25-40). Springer. doi:10.1007/978-3-319-44887-9_3

Redish, E. F., & Kuo, E. (2015). Language of Physics, Language of Math: Disciplinary Culture and Dynamic Epistemology. *Science and education*, *24*(5-6), 561–590. doi:10.100711191-015-9749-7

Reynante, B., Selbach-Allen, M., & Pimentel, D. (2020). Exploring the promises and perils of integrated STEM through disciplinary practices and epistemologies. *Science and Education*, *29*(4), 785–903. doi:10.100711191-020-00121-x

Scientific Advisory Group for Emergencies. (2022). *Coronavirus (COVID-19)*. Retrieved from gov.uk: https://www.gov.uk/government/organisations/scientific-advisory-group-for-emergencies

Siskin, L. S. (1994). *Realms of Knowledge: Academic Departments in Secondary Schools*. The Falmer Press.

Smith, E., & White, P. (2019). Where do all the STEM Graduates Go? Higher Education, the Labour Market and Career Trajectories in the UK. *Journal of Science Education and Technology*, *28*(1), 26–40. doi:10.100710956-018-9741-5

Struyf, A., De Loof, H., Boeve-de Pauw, J., & Van Petegem, P. (2019). Students' engagement in different STEM learning environments: Integrated STEM education as promising practice? *International Journal of Science Education*, *41*(10), 1387–1407. doi:10.1080/09500693.2019.1607983

Turner, F., & Rowland, T. (2011). The knowledge quartet as an organising framework for developing and deepening teachers' mathematics knowledge. In K. Ruthven & T. Rowland (Eds.), *Mathematical knowlege in teaching* (pp. 195–212). Springer. doi:10.1007/978-90-481-9766-8_12

Turşucu, S., Spandaw, J., Flipse, S., & de Vries, M. J. (2017). Teachers' beliefs about improving transfer of algebraic skills from mathematics into physics in senior pre-university education. *International Journal of Science Education*, *39*(5), 587–604. doi:10.1080/09500693.2017.1296981

Venville, G., Wallace, J., Rennie, L., & Malone, J. (2002). Curriculum Integration: Eroding the high ground of science as a school subject. *Studies in Science Education*, *37*(1), 43–83. doi:10.1080/03057260208560177

Williams, J. (2011). STEM Education: Proceed with caution. *Design and Technology Education. International Journal (Toronto, Ont.)*, *16*(1), 26–35.

Williams, J., Roth, W.-M., Swanson, D., Doig, B., Groves, S., Omuvwie, M., . . . Mousoulides, N. (2016). *Interdisciplinary Mathematics Education*. Springer Open. Retrieved from https://link.springer.com/book/10.1007/978-3-030-11066-6

Wong, V. (2017). Variation in graphing practices between mathematics and science: Implications for science teaching. *The School Science Review*, *98*(365), 109–115.

Wong, V. (2018). *The relationship between school science and mathematics education*. King's College London.

Wong, V. (2019). Authenticity, transition and mathematical competence: An exploration of the values and ideology underpinning an increase in the amount of mathematics in the science curriculum in England. *International Journal of Science Education*, *41*(13), 1805–1826. doi:10.1080/09500693.2019.1641249

Wong, V., & Dillon, J. (2019). 'Voodoo maths', asymmetric dependency and maths blame: Why collaboration between school science and mathematics teachers is so rare. *International Journal of Science Education*, *41*(6), 782–802. doi:10.1080/09500693.2019.1579945

Wong, V., & Dillon, J. (2020). Crossing the boundaries: Collaborations between mathematics and science departments in English secondary (high) schools. *Research in Science & Technological Education*, *38*(4), 396–416. doi:10.1080/02635143.2019.1636024

Wong, V., Dillon, J., & King, H. (2016). STEM in England: Meanings and motivations in the policy arena. *International Journal of Science Education*, *38*(15), 2346–2366. doi:10.1080/09500693.2016.1242818

Young, M. (2011). The return to subjects: A sociological perspective on the UK Coalition government's approach to the 14–19 curriculum. *Curriculum Journal*, *22*(2), 265–278. doi:10.1080/09585176.2011.574994

Young, M. (2014). The progressive case for a subject-based curriculum. In M. Young & D. Lambert (Eds.), *R. C., & M. Roberts, Knowledge and the future school: Curriculum and social justice* (pp. 89–110). Bloomsbury Academic.

Zhang, D., Orrill, C., & Campbell, T. (2015). Using the mixture Rasch model to explore knowledge resources students invoke in mathematic and science assessments. *School Science and Mathematics*, *115*(7), 356–365. doi:10.1111sm.12135

ADDITIONAL READING

Reynante, B., Selbach-Allen, M., & Pimentel, D. (2020). Exploring the promises and perils of integrated STEM through disciplinary practices and epistemologies. *Science and Education*, *29*(4), 785–903. doi:10.100711191-020-00121-x

Wong, V., & Dillon, J. (2019). 'Voodoo maths', asymmetric dependency and maths blame: Why collaboration between school science and mathematics teachers is so rare. *International Journal of Science Education*, *41*(6), 782–802. doi:10.1080/09500693.2019.1579945

Wong, V., & Dillon, J. (2020). Crossing the boundaries: Collaborations between mathematics and science departments in English secondary (high) schools. *Research in Science & Technological Education*, *38*(4), 396–416. doi:10.1080/02635143.2019.1636024

Wong, V., Dillon, J., & King, H. (2016). STEM in England: Meanings and motivations in the policy arena. *International Journal of Science Education*, *38*(15), 2346–2366. doi:10.1080/09500693.2016.1242818

KEY TERMS AND DEFINITIONS

A-Levels: Post-16, pre-university qualification taken by students in England and Wales. Usually only three subjects are chosen for study at this stage.

Asymmetric Dependency: Science education is dependent on mathematics, mathematics education is not dependent on science (or at least is less dependent on science), thus the dependency is asymmetric.

Collaboration: In this chapter, I am using this term to mean any form of working together by teachers from mathematics and science disciplinary backgrounds and school departments which focuses on teaching and learning.

Secondary Education: Schooling for 11-18 year olds. In England, education is compulsory up to the age of 16; post-16 a variety of courses are on offer, including A-levels.

STEM: Science, technology, engineering, and mathematics.

Chapter 2

Connections Between Nature and Mathematics:

The Fibonacci's Sequence in the Natural History and Science Museum of the University of Porto

Nuno Teles
https://orcid.org/0000-0003-0431-5731
MHNC-UP, Portugal

Rosário Chaves
MHNC-UP, Portugal

Joana Torres
MHNC-UP, Portugal

Maria João Fonseca
MHNC-UP, Portugal

ABSTRACT

Museums and science centers play a leading role in non-formal education, inspiring visitors and raising social, political, and environmental awareness. These are powerful educational institutions, as they gather the infrastructure and competencies to provide opportunities to address topics related to science, technology, society, and arts, among others, through a non-traditional lens. Moreover, their projects and activities can be tailored to meet the needs and expectations of both families and schools, which makes them flexible engagement platforms for many audiences. In this chapter, the authors present an example of how the Natural History and Science Museum of the University of Porto is exploring the links between nature and mathematics in its educational programs. Taking advantage of this authentic context, an interdisciplinary educational activity was designed and implemented as part of the museum's holiday program. In it, 6- to 12-year-old children are challenged to observe and creatively interact with collections and exhibitions.

DOI: 10.4018/978-1-6684-5765-8.ch002

INTRODUCTION

Philosophy is written in this grand book, which stands continually open before our eyes (I say the 'Universe'), but cannot be understood without first learning to comprehend the language and know the characters as it is written. It is written in mathematical language, and its characters are triangles, circles, and other geometric figures, without which it is impossible to humanely understand a word; without these one is wandering in a dark labyrinth. Galileo (1623) Cited by (Alonso, 2003).

Education is a central process in the life of each of us, taking place as we grow and age and in every environment we encounter. As today's societies face increasingly complex local and global challenges, the solutions required to address this complexity entail scientific, social, cultural, and economic dimensions that are often transversal to the educational, health, and ecological systems (English, L. D., 2009). Therefore, it is of paramount importance to foster collaboration between formal and non-formal educational agents and platforms to promote meaningful and lifelong learning opportunities.

In modern societies, people are increasingly becoming more detached from nature, a phenomenon that can come about as the result of our search for comfort and the modifications we introduce in the natural world, especially in urban environments, to achieve it (Kahn, Weiss, and Harrington, 2018).

During the last decades, there has been a tendency for science and technology museums to restructure their material to make it more efficient from a communicative point of view, as well as to create privileged relationships with institutions of formal education. (Gkouskou, 2021, p. 217). Science centers and museums can be seen as educational institutions that provide tools and activities, frequently addressing pressing issues related to the interaction of science, technology, society, and the arts, many of which can be articulated with formal school programs (Teles & Fonseca, 2019, p. 132). By focusing on the connections that can be established between different areas of knowledge and building an interdisciplinary context, it is possible to prompt the development of scientific reasoning and advance knowledge of key scientific concepts. This had led to the recognition of the educational worth of interdisciplinarity in both formal and non-formal settings (Santaolalla, 2020). One of the major challenges science educators face nowadays pertains to the identification and use of efficient strategies to engage children in problem-solving when complex systems are involved and within interdisciplinary contexts. Studies have shown that a possible way to meet this goal is through the promotion of direct contact and exploration of the natural world (Kahn, Weiss, and Harrington, 2018; Chawla, 2020). These authentic interactions are known to foster an increased understanding of scientific and artistic phenomena. Humans need to feel connected with nature to commit to its protection (Chawla, 2020). Contact with natural environments, both wild and managed, has been identified as a way to connect with nature. Studies have shown that children and adults alike, who express higher levels of connection with nature are more likely to report actions in favor of the environment, besides displaying increased levels of health and well-being, and that these effects are strengthened through the frequency and duration of the contact (Chawla, 2020). Accordingly, opportunities for children to connect with nature are important when aiming to preserve the biosphere, by prompting action toward biodiversity conservation. And in this regard, the use of exploratory and discovery activities has been shown to reinforce and deepen the connection (Barrable, 2019). In particular, the link between science and mathematics has been shown to have considerable educational potential when applied to the development of active learning activities (Tytler, 2021). Museums, botanical gardens, nature centers, and planetariums, for example, can provide complementary learning experiences inspired by contact with nature. How these spaces operate provides room for interdisciplinarity (Seligmann, 2014).

By engaging in creative practices underpinned by interdisciplinarity, these cultural and non-formal education platforms can enhance their ability to scaffold real social change (Seligmann, 2014). In museums, exhibitions are often interdisciplinary and usually designed to appeal to a wide age range of audiences (Hartman, 2015, p. 289), providing a ground from which to sustain consistent educational programming (Hartman, 2015, p. 292).

Having all this in mind, in this study, the authors present the description and outcomes of an educational activity applied in a non-formal learning setting – the Natural History and Science Museum of the University of Porto (MHNC-UP), to promote contact with nature. Acknowledging that science and mathematics involve distinctive but overlapping epistemic processes (Tytler, 2021, p.1946), this practical activity was outlined based on the dialogue between science, mathematics, and arts. The outcomes obtained through its implementation demonstrate the positive impact of interdisciplinarity in non-formal educational practices in the development of meaningful learning and the increase of scientific literacy.

A university museum, like MHNC-UP, is a museum that is built with materials that have a direct connection with educational and research practices throughout time (Carchon, 2021, p.139). The MHNC-UP is a university museum based in Porto, Portugal, which encompasses three units: a core Pole, harboring heritage and recent collections in a wide range of fields, including archaeology, ethnography, biological anthropology, zoology, botany, paleontology, mineralogy, exact sciences and history of science, as well as a document and audiovisual archive, and currently under reconstruction but promoting a dynamic program of temporary exhibitions; the Hall of Biodiversity (GB), open to the public since 2017; and the Botanical Garden of the University of Porto (JBUP).

The MHNC-UP's requalification process is being carried out according to a new museology approach coined by Jorge Wagensberg. The *total museology* philosophy foresees the use of real objects and museographic metaphors and aphorisms to present natural phenomena, by concentrating on reality instead of replacing it. Visitors are urged to question reality and their own conceptions about it, while engaging in active dialogue and conversation, with the objects they observe, with one another, through social interaction, and with themselves. For this purpose, aesthetically loaded installations, appealing interactives, and a wide range of sensory experiments are offered, aimed to spark curiosity and intellectual enjoyment and foster a hands-on, minds-on, and hearts-on engagement (Wagensberg, 2006). This new paradigm is deeply rooted and clearly visible in the GB (Rossi, 2019), but also in the JBUP, a space of science and fruition, bearing important landscape, ecological, literary, and artistic values, which, in addition, houses a living botany collection, settled well in the heart of the city of Porto.

In line with the interdisciplinary vision of the art and science that guides the MHNC-UP's mission and lines of action, the activity "Natureza e Geometria" ("Nature and Geometry") was developed to promote the understanding of biological phenomena through the connection of mathematics, nature, and arts, by encouraging participants to creatively construct representations of natural phenomena, while strategically guiding them towards established understandings about the science.

This activity was prepared to provide participants with time outdoors in an urban garden, as well as in a museum setting, aiming to make them feel comfortable and willing to understand natural phenomena while encouraging them to feel and understand how they can make a positive difference for the environment.

In particular, the activity explored Fibonacci's sequence as a key to identifying patterns in natural and artificial shapes and structures.

Fibonacci's Sequence

Leonardo Bigollo, or as we all know him – Fibonacci (contraction of Bonacci's son) – lived about 1200 years ago in Pisa, Italy. Fibonacci was a mathematician who was interested in introducing the western world to the use of the Indo-Arabic numeral system, the numeration currently used in Europe and other parts of the world, as a replacement for the Roman system, which he perceived to be very laborious even when carrying out simple math operations. Fibonacci studied the rate of development of rabbits, giving rise to the now widely popular succession that bears his name and which can explain various natural phenomena. In his book *Liber Abaci* ("The book of the abacus"), Fibonacci proposed a problem concerning the growth of a colony of rabbits that should be reproduced from the second month of life in such a way that each pair generated a new pair every month. The Fibonacci problem consists in calculating the number of pairs that exist after *n* months. If one starts with a couple in the first month (1), in the second month, one couple is not fertile so you only have the first reproductive couple (1), then change in the third if they reproduced and there are two couples (2), in the fourth, there are three (3), and then successively, achieving the numbers 5, 8, 13, 21, 34, 55, and so on. This leads us to conclude that the law of formation of the terms of Fibonacci's sequence is given by the fact that to obtain the term of the sequence you need to add the last two numbers. This phenomenon can be observed in other living beings like sunflowers (*Helianthus sp.*) and pinecones, as it is very common in the plant world. For example, in pineapples (*Ananas comosus*) the leaves are arranged according to 8 spirals on the right, 13 on the left, and 21 on the vertical axis, that is, consecutive terms of the Fibonacci's sequence. Furthermore, the shape of the sunflower spirals, and not just their number, is related to Fibonacci's sequence and the golden ratio, both related to the golden number, an irracional number (1,618033...), also known as the letter phi (ϕ). The key feature of Fibonacci's sequence is that each term in this sequence is equal to the sum of the two preceding ones (Livio, 2008, p. 97-98). The sequence starts with 0, followed by the number 1. Other numbers follow based on the previously mentioned rule: 0, 1, 1, 2, 3, 5, 8, 13, 21, 34, 55, 89, 144, 233, 377, 610, 987, 1597, 2584, 4181, … This immediately explains the formation of other spirals, their shape, which is logarithmic, like, for instance, the shape of the Nautilus shell (*Nautilus sp.*). The logarithmic spiral shape is abundant in nature, for example in the nautilus shell, the structure of the tobacco virus or the horns of various mammals (Alonso, 2003) are only a few examples of this shape in nature.

MATERIAL AND METHODS

Nature and Geometry: Description of the Activity

"Nature and Geometry" was developed and enacted directly by the MHNC-UP's educational services team, as part of the MHNC-UP's holidays program "Férias com o Museu" ("Holiday with the Museum") in 2020, 2021 and 2022.

It took place over six distinct days, more specifically: 24th July and 21st August 2020; 22nd July, August 19, and December 22, 2021; and July 15, 2022; from 9 am to 5 pm. The activity was developed in the JBUP and the GB, taking advantage of the features of these spaces, and mobilizing a total number of 67 children aged 6 to 12 years old.

The activity began with a welcome to the participants. They were able to introduce themselves as well as to know more about the educators. Then, following a brief introduction about the main activity, the work plan for the day was disclosed. Special attention was paid to the main topics to be addressed. Using a *PowerPoint* slides presentation, the participants were acquainted with Fibonacci's sequence, which would be used to guide their work through eight main tasks.

Task 1: Identifying Fibonacci's Numbers and Shapes in Nature

The participants were asked to observe a *PowerPoint* slides presentation about the Fibonacci's sequence. In it, different patterns based on Fibonacci's sequences that can be observed in natural structures and phenomena were shown.

By observing this spiral and its successive repetition, the participants were able to identify geometric patterns, for instance, the shape associated with the formation of hurricanes (Figure. 1), the curling of chameleons' tails (Figure. 2), snails' shells (Figure. 3), and even the floral patterns of plant species, such as sunflowers (Figure. 4), or even the shape of pinecones (Figure. 5).

Figure 1. Spiral in the formation of hurricanes.
Source: @ Totem Learning

Figure 2. Spiral in the curling of chameleons' tails.
Source: @ Totem Learning

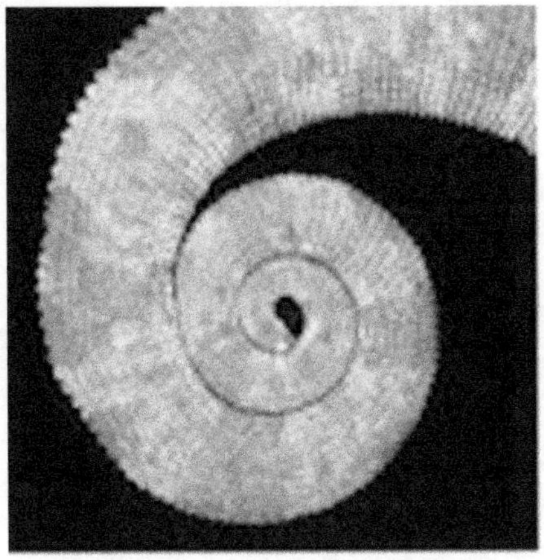

Figure 3. Spiral in snail shells.
Source: @MHNC-UP

Figure 4. Spiral geometric pattern in sunflowers.
Source: @ Totem Learning

Figure 5. Spiral geometric pattern in pinecones.
Source: @bzanchi/Shutterstock

Task 2: Fibonacci's Numbers Sequence

A practical challenge was then launched. Starting the sequence with the first two numbers, 0 and 1, the participants had the opportunity to complete part of the subsequent numbers, resulting from the addition of the two previous numbers. Using the whiteboard, the educator wrote the following numbers. The numbers 0, 1, 1, 2, 3, 5, 8, 13, 21, and 34 came about as the result of the collaborative work carried out. The participants had the chance to perceive that the sequence built by following these steps is infinite, but were instructed that in this specific activity only a few numbers would be used. To fully grasp how the sequence works, they were asked to build successive squares on squared paper. The MHNC-UP's team presented the ideal shape for the construction of this geometric pattern, fitting into the sequence of squares explained in task 3.

Task 3: Square Construction

The third task refers to the construction of an illustrative sequence of squares with sides equal to the numbers of Fibonacci's sequence (Figure 6). For this, they were asked to use a pencil, an eraser, and a ruler. Square side 1, square side 1, square side 2, square side 3, square side 5, square side 8, square side 13, and finally square side 21. After having created these shapes, the participants had to paint all the squares with a permanent marker, so if they mistakenly drew the arch too wide or too narrow, they could start over and not erase the squares. They were asked to write down the number of the radius of the circumference in the middle of the square so that they could follow their mathematical reasoning.

Figure 6. Drawing of the squares Fibonacci's sequence
Source: @ MHNC-UP

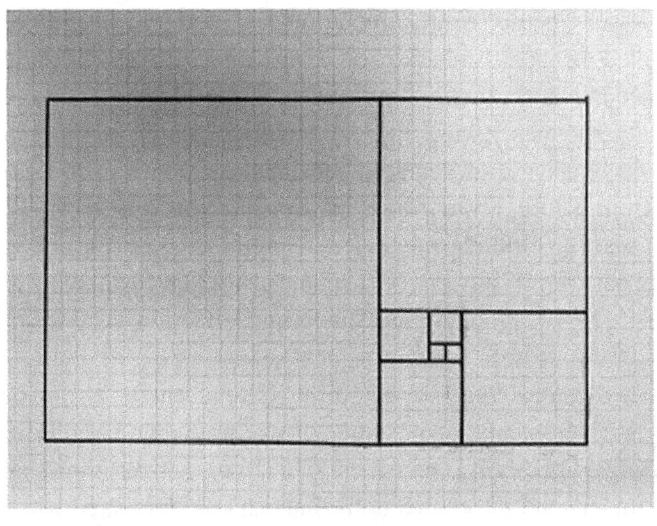

Task 5: The Golden Spiral

After task 4, participants were able to then draw the golden spiral, through guided drawing. They drew 1/4 of the circumference with a radius equal to the side of the square. By drawing all semi-circumferences, and creating successive arches, it was possible to draw the golden spiral (Figure 7). They then pencil painted each of the spirals with their colors of choice, giving wings to their imagination.

Figure 7. Sequence of squares according to the Fibonacci's sequence.
Source: https://super.abril.com.br/wp-content/uploads/2018/07/513f755d9827683068000278fibonacci_spiral_geogebra-svg.png

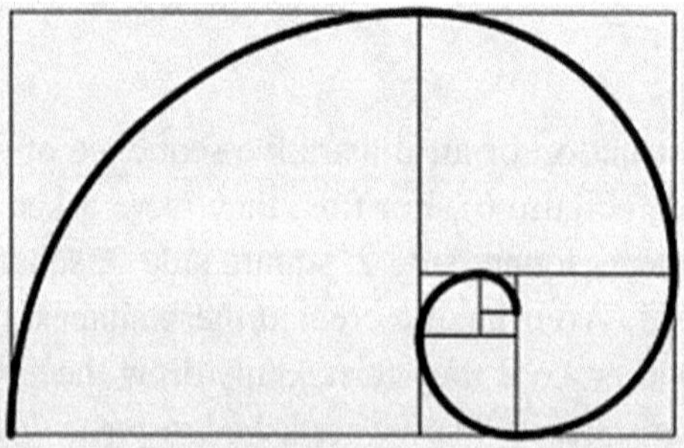

Task 6: Exploring Nature

The next challenge consisted of a visit to the JBUP, in which the participants had to find parts of plants, such as leaves, flowers, petals, fruits, among other structures, which would show the sequence numbers previously worked through. Embarked on a mission to JBUP to search for spiral patterns and find where they took place in nature. In this way, they carefully observed and collected flowers, leaves, floral structures, and fruit structures, taking their time in interacting with these natural components and being instructed on how to engage in rational and sensible collection: getting to know what and how much they can retrieve from nature. From the search for spiral patterns in the JBUP, it was possible to find spirals in succulent plants, floral rearrangements, and pinecones (Figure 8), as well as to observe geometric and numerical patterns in flowers, leaves, and other plant structures associated with phenology and plant development. Besides looking into the living collection, they were also invited to observe different built structures in this space.

They were then asked to illustrate what they saw through representative drawings. By drawing through observation and making measurements, two techniques explored on a very basic level, the participants identified important starting points, such as sequences of numbers and different shapes, sizes, textures, and colors, and were able to understand the connections that can be established with science, math, and art.

Figure 8. An example of the golden spiral pattern in the formation of pinecones.
Source: @ MHNC-UP

Task 7: Exploring the Museum

The next challenge was a visit to the GB. At the main entrance of the GB, the participants were given the opportunity to observe the floor mosaic tiles and built structures (Figure. 9) they had seen in the *PowerPoint* slides presentation, with their abounding floristic and faunal patterns. Therein, it was possible to once again observe shapes, including spirals, inspired by nature, displaying Fibonacci's sequence numbers. Attention was called to a very interesting feature in the study of mathematics and nature – symmetry. The participants discussed the notion of symmetry, which was easily visible in these tiles and the shapes they hold.

After identifying all the structures that could serve as natural examples of Fibonacci's sequence, the participants were able to recognize shapes that were related to its numbers. By bringing together insights from diverse fields, such as mathematics, natural sciences, and arts, they were given the chance to witness how the boundaries between areas of knowledge are blurred when it comes to naturally occurring phenomena.

Figure 9. Fibonacci's sequence numbers in some structures in the mosaic tiles
Source: @ MHNC-UP

Following this session, a visit to the permanent exhibition of the GB was promoted. Particular attention was given to two main rooms of the permanent exhibition, "Diversidade de Formas" ("Diversity of Shapes") & "Biodiversidade que fala Português" ("Biodiversity which speaks Portuguese").

In the GB's module "Diversidade de Formas", the participants were given some notions about the concepts addressed in this module and were given the chance to read the texts on display and, in it, a very special exhibit that addresses the spiral shape, by demonstrating his main function – compacting (Figure 10). There, the participants could see *in loco* the representation of the golden spiral in the shell of *Nautilus sp.* (Figure 11), as well as the application of this shape in everyday life.

Figure 10. Modular proposal of the spiral
Source: @ MHNC-UP

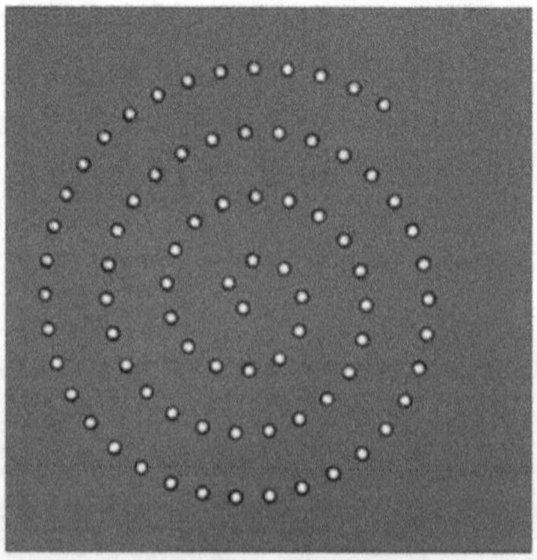

Figure 11. Representation of the golden spiral in the formation of Nautilus sp
Source: @ MHNC-UP

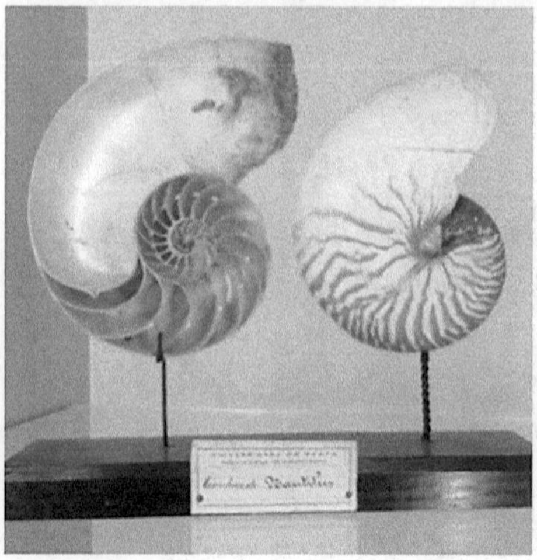

Afterwards, it was time to visit the room "Biodiversidade que fala Português", which depicts a cabinet of curiosities, with its walls full of maps, illustrations, and paintings and its cabinets filled with incredible objects, such as skeletons, minerals, shells and taxidermies of various animals. There, the participants had the chance to once again identify shapes and structures pertaining to Fibonacci's sequence. Together with these general examples, some specific ones pertaining to the GB themselves were shown, revealing Fibonacci's sequence in countless collection objects on display, such as shark teeth forming a triangle - a figure with 3 sides (Figure. 12) and starfish specimens - with 5 sides (Figure. 13), both Fibonacci's numbers.

Figure 12. Fibonacci's sequence numbers in some structures in shark tooth
Source: @ MHNC-UP

Figure 13. Fibonacci's sequence numbers in sea stars
Source: @ MHNC-UP

Task 8: Symmetry in Nature

The search for symmetry guided the participants in the collection of leaves from the JBUP in task 6. Based on the notion of symmetry, showed in mosaic tiles at task 7, they were challenged to prepare and organize a collection of leaves from the plants in the JBUP, based on two criteria: each participant had to collect a leaf of a different shape from the others and each of these leaves had to have different colors or shades. Once collected, with the help of scissors the participants were asked to cut out half of the previously collected leaves they glued the halves on a sheet of white paper size A4 they were able to draw with the pencil the symmetrical shape of the leaf. Finally, they were asked to paint half of the leaf drawn with the corresponding shades of green, yellow, red, or brown. By doing so, the participants were made sensitive to the importance and relevance of all species getting to know more about biodiversity.

After all the tasks were completed, participants took all drawings and collected material home. This way, parents could be interested in questioning participants on how all tasks were made, giving them the opportunity to explain them.

RESEARCH DESIGN AND METHODOLOGY

This was an empirical study in which data were collected using a qualitative assessment approach based on the naturalistic observation of the participants. Naturalistic observation allows understanding if the learning process is occurring based on the participants' behaviors and attitudes toward the subject at hand (Coutinho, 2019), with a very low degree of interference. An important strength of observation methods stems from the fact that they allow to assess the participants' behavior without having to rely on their responses. This is particularly useful when attempting to avoid bias and when the study subjects have difficulties expressing themselves. Allowing participants to be self-aware of their surroundings and able to self-express themselves (Chawla, 2020).

Sample

The study involved a convenience sample of 67 participants between the ages of 6 and 12, with an average age of 8.8, (M = 8.8 years old), including 42 females (62.7%) and 25 males (37.3%).

These participants were selected based on their participation in the MHNC-UP's Holidays Program "Férias com o Museu" (Holiday with the Museum). Permission for the children's participation in this study was obtained from their parents, by email. No personal data, including names or photographs, were used and disclosed for the purpose of this study.

The characterization of the sample is provided in Table 1, Figure 14, and Figure 15.

Table 1. Sample characterization

Demographic Variables		Participants (n=67)	
		n	%
Gender	Female	42	62.7
	Male	25	37.3
Age	Mean	8.8	
	Minimum	6	

Figure 14. Participants' age distribution.
Source: Authors

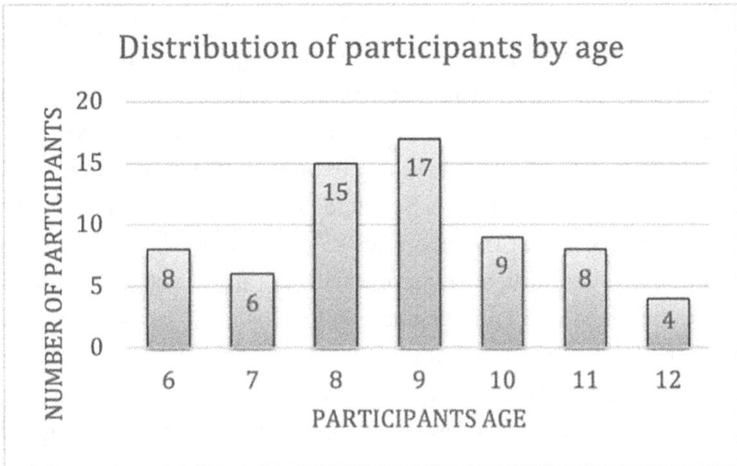

Figure 15. Participants' gender distribution.
Source: Authors

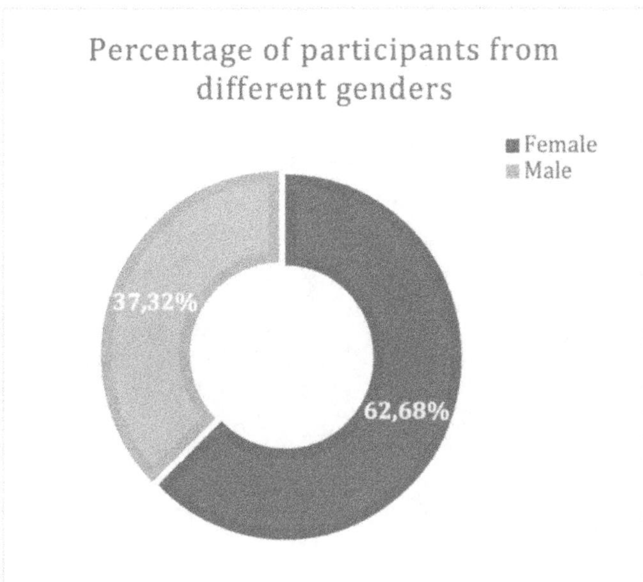

RESULTS AND DISCUSSION

This section highlights some of the main outcomes of the implementation of the activity "Nature and Geometry." It is important to have in mind that the children who took part in this activity had no formal instruction on the core mathematical concepts and processes addressed prior to engaging in the activity.

By the end of the activity, it was possible to observe that not only were they acquainted with Fibonacci's sequence and what it stands for, but they were also able to reproduce it and identify it in various natural and built structures. This way they were able to understand and discuss why and how this activity was developed. Throughout the challenges, it was possible to witness how easily and swiftly the participants identified and addressed symmetry and proportion issues. Throughout history, mathematicians have often called attention to the search for beauty as a relevant criterion that is underpinned by proportions (Giménez, 2009). Having this in mind, the previously explained tasks' results reveal the engagement of the activities and the main subject – Fibonacci's Sequence in nature. Drawings of squares and the golden spiral (Figure. 16) demonstrate some of the examples of works made by participants. In the same way, the representation of Fibonacci's numbers sequence with botanical parts (Figure. 17) allows them to establish the connections between math and science. With this task, participants were able to identify, by themselves and with the help of the museum educators, which parts of the plants they could use to complete the sequence. Some of them, the older ones, searched for the plant parts with larger numbers as 18 and 21, trying to achieve a further number of the sequence. The task regarding symmetry was successful as well, since they had the opportunity to cut, draw and paint half leaves on a sheet of paper (Figure. 18). Since almost all of the tasks were possible for the participants to use their imagination and creativity, the researchers saw the potential of arts in developing these activities.

Figure 16. Drawing of the golden spiral in squares - Fibonacci's sequence
Source: @ MHNC-UP

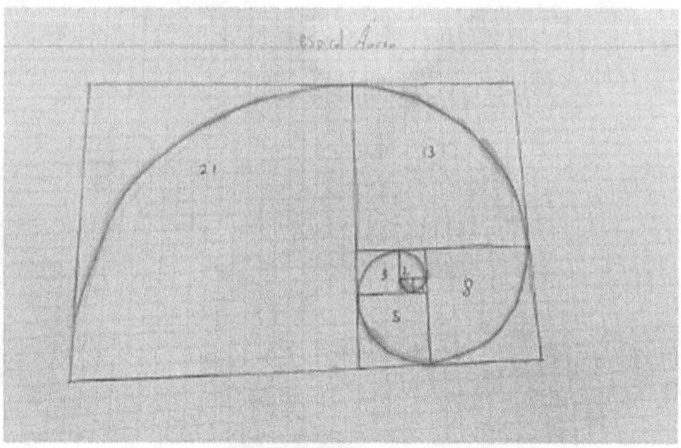

Figure 17. Fibonacci's sequence (1, 2, 3, 5, 8, 13, 21) with botanical parts.
Source: @ MHNC-UP

Figure 18. Representation of the symmetry of leaves developed by participants ages 6 to 12 years old
Source: @ MHNC-UP

At the end of the activity, participants displayed all their work in the activity room almost like an exhibition. Since the activity was developed in MHNC-UP, participants felt like they were responsible to do a good job for parents to see. This way, their parents were able to see the work developed that day, and witness how their children can make artistic drawings inspired by nature and integrating mathematics and science. It was interesting to see how the parents asked their children how they did it. This gave the participants the opportunity to explain, in their own words, simple concepts, such as adding numbers, symmetry, shapes, and colors. It is important to highlight that some of the parents showed an interest in knowing more about the museum and other subjects regarding this activity in the following days. Some of them visited the GB and the JBUP, and after this activity, had their own children explaining some of the tasks they were able to do. Other participants expressed their interest in knowing more about the subject, having the museum educator explain some of the topics about Fibonacci's Sequence and the golden ratio to parents, for future research at home. The understanding of natural phenomena using images, patterns, and drawings, and in direct contact with nature, can lead to enhanced learning in children. One of the best-known aspects of the practical usefulness of mathematics is its great capacity for modeling natural phenomena since the study of these models allows for a better understanding of specific processes (Alonso, 2003). To some extent, the way in which the participants engaged in the tasks at hand, supports these notions, as they were shown to be able to apply the concepts consistently and actively, they learned in subsequent tasks. They were also shown to be increasingly aware and attentive toward the patterns they encountered.

The outcomes obtained therefore suggest that by combining mathematics, biology, and arts it is possible to foster meaningful learning experiences that are engaging and motivating. In addition, based on the participants' feedback and comments along the activity, the outcomes obtained seem to be in line with other studies relating to museum-based interventions, which confirm that participants involved in interdisciplinary experiences have a positive perception of museums as places for learning (Santaolalla, 2020). The learning that took place during this process was enhanced by the close contact with nature and the social interaction that took place when the participants' experiences were shared with one another. More than merely allowing children to play, it was possible to witness the significant learning that arises from their engagement with the natural world (Kahn Jr, 2018).

The children who participated in this activity were invited to carefully observe their surroundings and identify the Fibonacci's sequence within the structure of the plants and animals encountered. By doing so, they were given the opportunity to describe naturally occurring geometric and numerical arrangements using mathematical formulas inscribed in the harmony of living things. Therefore, harmony and beauty were two of the concepts introduced. Harmony was presented as the "balance in the proportions between the different parts of a whole" (Giménez, 2009), and said to be linked with the perception of beauty. Given the aesthetic appeal of nature, didactical approaches aimed at attaining a connection with the natural environment should foster the perception of beauty and engagement with it. To promote the connections foreseen, other fields of understanding had to be taken into account. For example, the origins of the development of an organism, as well as the evolutionary relationship between groups of organisms, were two relevant aspects considered in the communication established with the participants during the activity. Indeed, the language used purposely emphasized the ontogenetic and phylogenetic significance of living beings (Kahn Jr, 2018).

Another aspect worth mentioning refers to one of the main features of the learning sequence implemented, which entailed setting up problems or challenges that engaged participants in the inquiry. By designing tasks that challenged participants to make observations and represent their understandings and findings, direct interaction with nature was even more thoroughly promoted. Participants were introduced to the main subject – Fibonacci's sequence and possible ways of addressing it –, then challenged to discover different representations of this sequence. They were granted time and space for reflection and discussion and encouraged to find new and creative approaches. This way, it was possible to create an interactive and collaborative learning environment and stimulating and enriched experiences through the interplay with the MHNC-UP's team and by establishing interdisciplinary relationships between knowledge (Teles & Fonseca, 2019, p.133).

As younger generations seem to be losing the habit of engaging in a free-ranging exploration of the surrounding natural world (Kahn, 2018), through this activity, an attempt was made to meet what Chawla (2020) refers to as 'socialization of the natural world. Specifically, more than just learning, a sense of well-being and a positive connection with nature was encouraged, by keeping children safe while enabling independent exploration of the MHNC-UP. The participants were encouraged to talk with one another and to report what they observed through the activities. The goal of this approach was for them to feel empowered by their knowledge, mobilizing actions to preserve their surroundings (Chawla, 2020).

FUTURE RESEARCH DIRECTIONS

The findings obtained in this study are promising. However, given its exploratory nature, new and more focused research is needed to fully grasp the impact of the activity described.

When it comes to its experimental design, even though observation has some advantages, it also has some weaknesses. For example, the reasons behind the observed behavior may not be clear and the data analysis may be more time-consuming than other collection methods (Coutinho, 2019). So, future studies built on this one may involve the use of other data collection methods, namely including more structured qualitative analysis and also a quantitative dimension.

The researchers believe it is important to replicate and expand this type of interdisciplinary action, allowing for new pedagogical experiences that are useful for science, arts, and mathematics, among other fields.

An interesting feature of this activity is that it allowed the mobilization of several teams from the MHNC-UP: educators, collection curators, and researchers. By this, future research can be developed using STEM (Science, Technology, Engineering, and Maths) as a cross-curricular strategy. Keeping in mind the benefits that STEM receives from the collaboration of the different teams working in a museum (Seligmann, 2014), this was an important achievement, which can be further explored in future activities developed by bringing together other fields of knowledge, as for instance archaeology, mineralogy, zoology, and botany. Therefore, STEAM approaches can be even more powerful, since the arts are included and a lot of the activities developed integrate arts to acknowledge all kinds of connections between subjects. This can be a suitable strategy to provide more authentic activities, which clearly translate the identity of the MHNC-UP.

A future line of action can include the adaptation and implementation of this activity in another context, formal, non-formal, and informal activities, as part of the MHNC-UP's educational program for schools. By this, teachers can assure that these activities are important to understand subjects at school by the students. This way, we can recognize the benefits of the synergy between the non-formal activities in museums and science centers and the formal education at school. Students can have access to a better learning process and understand the importance of thinking about science from a holistic perspective. This way, we can develop partnerships between schools and museums and improve scientific literacy in society. This can be a particularly interesting strategy having in mind the curricular guidelines that recommend the exploitation of diverse resources and environments, for instance, literature research and visits to gardens, natural parks, museums, and planetariums. This strategy aims to develop a bridge between science education programs at schools with existing scientific literacy programs. The goal is to take advantage of existing opportunities by engaging students at school with experiences in museums and science centers.

CONCLUSION

With the implementation of this non-formal education activity, it was possible to sustain the perception that museums and science centers, as well as botanical gardens, can enhance learning through interdisciplinary educational practice. By sparking curiosity and the connection with nature, these platforms can be used to scaffold initiatives aimed at addressing new and enduring challenges, demands, and implications of non-formal education, since museums can be used as an educational resource (Santaolalla, 2020).

The incorporation of mathematics in a science-driven activity infused with the arts was relatively straightforward. Moreover, the chance to tackle mathematical concepts and tools allowed to enlighten and expand the knowledge of the scientific concepts at play (Tytler & Mulligan, 2021), as suggested by the positive and active engagement of the participants in identifying and replicating patterns, shapes, and sequences.

Museum educators are responsible for creating a professionalized learning space that ensures that learning does indeed take place. Those responsible for educational actions in museums can carry out the evaluation and possibly the restructuring of their educational role. (Gkouskou, 2021, p. 219). These goals for museum learning are widely accepted in the field of museum education (Seligmann, 2014). By taking advantage of the authentic context provided by the MHNC-UP, the educational activity outline based on an interdisciplinary approach was shown to challenge children to look at museum collections (including a living one) and exhibitions in a different, more creative way, and by paying attention to detail and to the different layers of complexity that a single space may hold.

The excitement of the participants during the activity was definitely a key outcome, which reinforces the beneficial impact that contact with nature has on providing children with opportunities for wonder and amazement, which can act as emotional triggers for a lifetime of curiosity and concern for the natural world (Chawla, 2020).

ACKNOWLEDGMENT

The authors acknowledge all the participants and their parents, whose collaboration was pivotal in making this study possible. They also acknowledge the collaboration of all the MHNC-UP's teams who were involved in the development of the activity and the preparation of this article. The study was carried out with the support of PRISC (Portuguese Research Infrastructure of Scientific Collections).

REFERENCES

Alonso, J. C. P. (2003). Las matemáticas en la naturaleza [Mathematics in nature]. *Sigma: revista de matemáticas, 22*, 161-171.

Barrable, A. (2019). Refocusing environmental education in the early years: A brief introduction to a pedagogy for connection. *Education Sciences*, *9*(1), 61. doi:10.3390/educsci9010061

Carchon, R., & Segers, D. (2021). The Use of the Museum Collection for Educational Purposes. In *Historical Scientific Instruments in Contemporary Education* (pp. 139–157). Brill.

Chawla, L. (2020). Childhood nature connection and constructive hope: A review of research on connecting with nature and coping with environmental loss. *People and Nature*, *2*(3), 619–642. doi:10.1002/pan3.10128

Coutinho, C. P. (2019). Metodologia de investigação em ciências sociais e humanas: Teoria e prática [Research methodology in social and human sciences: Theory and practice] (2nd ed.). Almedina.

English, L. D. (2009). Promoting interdisciplinarity through mathematical modelling. *ZDM*, *41*(1), 161–181. doi:10.100711858-008-0106-z

Giménez, J., Abdounur, O. J., Badillo, E., Balbás, S., Corbalán, F., Dos Santos, J. M., & Spinadel, V. W. (2009). *La proporción: arte y matemáticas* [The proportion: Art and mathematics]. Graó.

Gkouskou, E., & Koliopoulos, D. (2021). Describing the educational role of Natural History Museums: An analysis tool for pre-service and in-service teachers. *European Journal of Education Studies*, *8*(1). Advance online publication. doi:10.46827/ejes.v8i1.3526

Hartman, S., & Hines-Bergmeier, J. (2015). Building connections: Strategies to address rurality and accessibility challenges. *Journal of Museum Education*, *40*(3), 288–303. doi:10.1179/1059865015Z.000000000105

Kahn, P. H. Jr, Weiss, T., & Harrington, K. (2018). Modeling child–nature interaction in a nature preschool: A proof of concept. *Frontiers in Psychology*, *835*, 835. Advance online publication. doi:10.3389/fpsyg.2018.00835 PMID:29896143

Livio, M. (2008). *The golden ratio: The story of phi, the world's most astonishing number*. Crown.

Rossi-Linnemann, C., & De Martini, G. (Eds.). (2019). *Art in Science Museums: Towards a Post-Disciplinary Approach*. Routledge. doi:10.4324/9780429491597

Santaolalla, E., Urosa, B., Martín, O., Verde, A., & Díaz, T. (2020). Interdisciplinarity in teacher education: Evaluation of the effectiveness of an educational innovation project. *Sustainability*, *12*(17), 6748. doi:10.3390u12176748

Seligmann, T. (2014). Learning museum. *Journal of Museum Education*, *39*(1), 42–53. doi:10.1080/10598650.2014.11510794

Teles, N., & Fonseca, M. J. (2019). A importância do microscópio ótico na Revolução Científica - Das práticas educacionais à Representação Museológica [The importance of the optical microscope in the Scientific Revolution - From educational practices to Museological Representation]. *História Da Ciência e Ensino: Construindo Interfaces*, *20*, 126–140. doi:10.23925/2178-2911.2019v20espp126-140

Tytler, R., Mulligan, J., Prain, V., White, P., Xu, L., Kirk, M., Nielsen, C., & Speldewinde, C. (2021). An interdisciplinary approach to primary school mathematics and science learning. *International Journal of Science Education*, *43*(12), 1926–1949. doi:10.1080/09500693.2021.1946727

Wagensberg, J. (2006). *CosmoCaixa: The total museum through conversation between architects and museologists* (No. 069.2: 5 COS).

ADDITIONAL READING

Błaszak, M., Rybska, E., Tsivitanidou, O., & Constantinou, C. P. (2019). Botanical Gardens for Productive Interplay between Emotions and Cognition. *Sustainability, 11*(24), 7160. doi:10.3390u11247160

Ivanda, T. K. (2022). The museum-school mentoring model and personalization of education. *Journal of Elementary Education/Revija za Elementarno Izobraževanje, 15.*

Meisner, G. B. (2018). *The golden ratio: The divine beauty of mathematics.* Race Point Publishing.

Mujtaba, T., Lawrence, M., Oliver, M., & Reiss, M. J. (2018). Learning and engagement through natural history museums. *Studies in Science Education, 54*(1), 41–67. doi:10.1080/03057267.2018.1442820 PMID:29951162

Sanders, D. L., Ryken, A. E., & Stewart, K. (2018). Navigating nature, culture and education in contemporary botanic gardens. *Environmental Education Research, 24*(8), 1077–1084. doi:10.1080/1350 4622.2018.1477122

Schneider, M. (1995). *A Beginner's Guide to Constructing the Universe: Mathematical Archetypes of Nature, Art, and Science.* Harper Perennia.

Young, S., Eadie, T., Suda, L., & Church, A. (2022). LEARN: Essential elements of museum education programs for young children. *Curator (New York, N.Y.), 65*(1), 209–223. doi:10.1111/cura.12456

KEY TERMS AND DEFINITIONS

Biodiversity: Biological diversity among and within species of living beings in a given environment.

Fibonacci's Sequence: A sequence starting with 0 and followed by 1, in which each subsequent number is the sum of the two proceeding ones.

Formal Education: Structured education system that runs from primary school to university, and includes specialized programs for vocational, technical, and professional training.

Holistic Perspective: Viewpoint characterized by the belief that the parts of something are intimately interconnected and explicable only by reference to the whole.

Interdisciplinarity: Any academic or scientific study that draws on the expertise of more than one discipline.

Non-Formal Education: Programs and processes of personal and social education designed to improve a range of skills and competences outside the formal educational curriculum.

Scientific Literacy: An individual's understanding and command of scientific concepts, phenomena, and processes.

Sustainability: Meeting present needs without compromising the ability of future generations to meet their own needs.

Symmetry: The quality of exactly similar parts facing each other or around an axis.

Chapter 3
An Examination of the Interdisciplinary Connections Between Physics and Mathematics According to Secondary Education Physics Curriculum:
The Case of Turkey

İpek Derman
https://orcid.org/0000-0002-2024-2619
Hacettepe University, Turkey

Sevim Bezen
https://orcid.org/0000-0002-0304-5314
Hacettepe University, Turkey

ABSTRACT

This study aims to examine the interdisciplinary connection between physics and mathematics according to the secondary education physics curriculum in Turkey and to identify the mathematical associations in the objectives (outcomes) and content (subject list) of the secondary education physics curriculum in Turkey through the secondary education mathematics curriculum. The study was designed with an analytical research model and carried out within the frame of document analysis. Data for the study were obtained from the 2018 secondary education physics (Year 9, 10, 11, and 12) and mathematics (Year 9, 10, 11, and 12) curricula in Turkey. Data in the study, which were considered as documents, were analysed by content analysis. At the end of the study, it was determined that the secondary education physics curriculum contains mathematical associations at every class level; however, generally speaking, these associations are not up to par.

DOI: 10.4018/978-1-6684-5765-8.ch003

INTRODUCTION

The amount of life skills preschool children employ is awe-inspiring. Children learn quickly to meet their natural needs or satisfy their curiosity during this period. For instance, a child building a sandcastle may not be able to answer questions about the ratio of water and sand to be used, what angle the castle walls should be constructed, and how much water pressure the castle walls will be exposed to when they fill it with water. However, they have cognitive, sensory, and psychomotor learning while building the sandcastle (Johnston, 2005). When such learning is transferred to a school environment, the knowledge used in making the sandcastle becomes physical formulae, mathematical calculations, and chemical equations, independent of one another. Regarded as such, teaching concepts, principles, and theories under disciplines such as mathematics, physics, and geography independently of one another makes learning difficult. An interdisciplinary connection alleviates this problem for students to learn better. For instance, it has been suggested that there should be a multidisciplinary connection between mathematics and physics because they are compatible. For example, it is argued that physics and mathematics should adopt an interdisciplinary connection because these two courses are compatible. However, it is also underlined that this theoretical compatibility has not been reflected in either curriculum in a balanced manner (Park et al., 2021). Incorporating real-life problems into the teaching process is said to be the way to prevent this problem (Johnston et al., 2020). In this respect, it is argued that to support the natural process of learning, learning at school should be interdisciplinary; to provide students with a better learning experience, especially the multidisciplinary connection in science should be presented via a learning environment that exceeds the boundaries of the subject (Broggy et al., 2017). For instance, in many north American universities, graduate programs are offered in "science and mathematics education" not as separate entities but together (Park et al., 2021). STEM is a teaching method that integrates content and skills related to science, technology, engineering, and mathematics. In this respect, STEM in teaching science is a way of supporting students in developing 21st-century skills, which they need to sustain a quality life in the complex modern world (Johnson et al., 2020). In this integration in which engineering is a context, the aim is for students to make sense of real-life events using technology and various models, visuals, and symbols, using mathematical and scientific concepts (Martin-Paez & Aguilera, 2019). As such, to ensure STEM integration in a teaching program, first, there is a need for a structure onto which various teaching strategies could be incorporated, different combinations of disciplines comprising STEM could be included, and one field could be emphasized more than the other (National Academy of Sciences, 2014). In other words, the relationship between mathematics and physics, two of the significant disciplines of STEM, should be reflected in schools via curricula. This integration should be examined in the literature (Bahar et al., 2018). Positive sciences are a fundamental branch of science that aims to understand science's nature. To understand the nature of science through positive sciences, students should have a good grasp of mathematical forms and representations (Osborne, 2014). Moreover, the support of specific skills within mathematics such as data analysis and interpretation, probable and statistical thinking, and the use of mathematical tools provided for science should not be disregarded. In addition, it should be kept in mind that mathematical knowledge is essential in ensuring calculative thinking, the symbolic representation of variables and relations, and the prediction of conclusions. In this respect, mathematics is not only a tool for data analysis for science but also a way of thinking and reasoning (Wong, 2018).

THE RELATIONSHIP BETWEEN PHYSICS AND MATHEMATICS

The relationship between physics and mathematics is as old as the hills. Indeed, the relations posited in Aristotle's *Physics* and *Posterior Analytics* had determined the frame of discussion for the nature of mixed sciences until the 17th century. Writings of Newton and Huygens in the 17th century are accepted as the beginning of mathematics in physics. Throughout history, theories point to one common conclusion: the formulation of physics is mathematics (Gingras, 2001; Pospiech, 2015). This historical connectedness between physics and mathematics is twofold. Firstly, there is a need for a consistent frame and compound theories to define, explain, guess, and understand the analogies between the physical objects in the real world and the structures in the world of ideas. In this sense, physics has become inseparable from mathematical modeling (Ceuppens, 2019; Gifford & Finkelstein, 2021; Pospiech, 2015). In other words, mathematics is a language through which physics theories are constructed by using symbols and operations to make sense of equations and representations in physics (Buick, 2007; Mumthas & Abdulla, 2019; Polverini, 2021). Therefore, students should learn mathematics and the vernacular used in physics. Secondly, sound mathematical knowledge and skill are necessary to solve physics problems successfully. This means having a good comprehension of mathematics and the ability to connect physics and mathematics. This is a tricky thing to do for students of all levels. The literature argues that students' knowledge in physics is not independent of their ability in mathematics; it maintains that it contains their ability to relate these two fields and blend the information they receive from both areas (Ceuppens, 2019) because the primary aim of physics is to define nature through the help of such physical concepts as force, energy, heat, etc., and to generate physical laws, models, and theories. Later, these contribute to explaining physical problems or phenomena faced daily. These aims are based on evaluating experiments and using mathematical tools and structures inductively and deductively. These two foundations are closely related. For instance, predicting, evaluating, and interpreting experiments cannot be conceived without using mathematics (Pospiech, 2015). In this respect, as mentioned frequently in advanced science and engineering classes and this article, "real world" problem necessitates mathematics skills. The relationship between physics and mathematics brings together multiple mental models and contributes to students' carrier-related skills. Consequently, mathematics is not merely an arithmetical tool for physics; it helps shape and conceptualizes the model (Polverini, 2021). However, as mentioned before, even though mathematics is the backbone of physics, learning difficulties are still rampant among students today because mathematical principles and operations are not used in physics (Hansson et al., 2021; Mumthas & Abdulla, 2019). In this respect, the close and dependent relationship between physics and mathematical thought and tools cannot be reflected in schools or education policies (Wong, 2018). At this point, it is necessary to focus on the quality of teaching materials, the flexibility of the curriculum about interdisciplinarity, the structure of the content, and the level of knowledge of teachers in non-area learning when it comes to drawing interdisciplinary relations (Broggy et al., 2017; Johnston et al., 2020; Miron & Staicu, 2010; Treacy, 2021).

BACKGROUND

Recent arguments suggest that for students to compete in a globalized world, the pedagogy should be reinforced by integrating mathematics and physics (McHugh et al., 2018). The first step of this integration should be through curricula because while teachers in Turkey are aware of the close relationship between physics and mathematics, they fall short in incorporating this relationship into their classes, and teaching programs should include interdisciplinary connections to alleviate deficiencies in teachers (Başkan et al., 2010; Berlin & White, 2012; Furner & Kumar, 2007; McHugh et al., 2018; Wong & Dillon, 2020); parallel to these, teacher training and in-service professional development programs should be re-designed to address this change (Aguirre-Munoz et al., 2022; Davison et al., 1995; Mumthas & Abdulla, 2019). Curricula that ensure the integration of mathematics and physics based on an interdisciplinary connection are argued to support students' meaningful learning (Aguirre-Munoz et al., 2022; Ceuppens, 2019; De Ataide & Greca, 2013; Dehipawala et al., 2014; McHugh et al., 2018; Treach, 2021; Tytler et al., 2021; Winarno et al., 2020) and thus to improve their problem-solving skills (De Ataide & Greca, 2013; Dehipawala et al., 2014; Treacy, 2021). It is also underlined that students would be more self-confident while learning (Aguirre-Munoz et al., 2022; McHugh et al., 2018), and they could be more actively involved in the learning process due to a problem-solving method based on critical and creative reasoning and collaboration in classes where an interdisciplinary connection is adopted (Cavadas et al., 2022; Ceuppens, 2019; Tytler et al., 2021). For instance, a study carried out in Finland argues that the interdisciplinary approach, which was considered to be a pioneer in the country, has resulted in a profound change in the education system and it has made it easier for students to sort problems (Bastos, 2017). Similarly, a study in the United States indicates that more than one discipline plays a role in solving problems in a world of advanced technology and meaningful learning is ensured when multiple disciplines are taught about one another (Next Generation Science Standards [NGSS], 2013).

Moreover, a multidisciplinary connection can steer students towards multiple viewpoints from different disciplines, sub-branches, and fields of expertise; indeed, an interdisciplinary connection will pioneer a novel process in education (Broggy et al., 2017; Treach, 2021). Moreover, teachers would cooperate with colleagues from different fields, and thus their collaboration would be time-efficient in carrying out the curriculum (Berlin & White, 2012; Bezen et al., 2018; Broggy et al., 2017; Johnston et al., 2020; Miron & Staicu, 2010; Wong & Dillon, 2020). Nevertheless, teachers may inevitably have trouble identifying common points (Wong & Dillon, 2020) because the rigid boundaries between the physics and mathematics curricula and their unilateral relationship would make collaboration difficult (Wong, 2018). In this respect, teachers have a huge responsibility in this process. Aware of the importance of the relationship between mathematics and physics, they are expected to be cognizant of the path they would like to follow when applying the interdisciplinary connection (Başkan et al., 2010; Cura & Ercan Yalman, 2019; Retnawati et al., 2018). Thus, it is believed that the findings of this study would guide policymakers, academics who train teacher candidates, and teachers in determining the relationship between physics and mathematics according to the secondary education physics curriculum and lay bare the need for redesigning the curriculum based on an interdisciplinary perspective.

Purpose

When the curricula of countries successful in science, especially in international evaluations, such as the United States, the UK, Singapore, Japan, Estonia, Finland, Hong Kong (China), Taiwan (China), South Korea, and Russia are examined, one can see a close relationship between physics and mathematics and they have the common understanding that they should "prepare individuals for real life" (Derman, 2019). Their curricula make sure the interdisciplinary relationship between mathematics and physics is well understood, and the skills nurturing each other in both fields are acquired by the students. This is reflected in the Turkish secondary education physics curriculum under the title "Mathematical proficiencies and basic proficiencies in science/technology" as the improvement of mathematical thinking to solve the problems faced in daily life. In other words, the Turkish secondary education physics curriculum emphasizes a process, activity, and knowledge based on excellent arithmetical skills. It can be argued that there are skills concerning mathematical proficiency and the different uses of rational and spatial thinking and formulae, models, constructs, graphics, tables, and mathematical presentations (Ministry of Education, [MofEd], 2018a).

Although the Turkish secondary education physics curriculum frequently refers to mathematics-related outcomes under the title "Turkey Proficiencies Framework," the number of studies in the national literature examining them in terms of purpose (outcomes) and content (subject list) is limited. In this respect, this study focuses on the interdisciplinary connections between physics and mathematics regarding objective (products) and content (subject list) to address this gap in the literature. It aims to identify the mathematical associations in the goal (outcomes) and content (subject list) elements in Turkey's secondary education physics curriculum. Analyzing the existing programs to strengthen the interdisciplinary connection between physics and mathematics would be a crucial starting point in designing future programs. To this end, this study seeks to answer the problem question, "What is the interdisciplinary connection between physics and mathematics in the objective and content of the secondary education physics curriculum in Turkey?" It is believed that findings about which mathematical subjects are employed to reach the objectives in the secondary education physics curriculum would contribute to the updating and development of studies of physics and mathematics curricula. Curricula in which interdisciplinary connections are strengthened would be an important guideline for physics and mathematics teachers. Moreover, emphasizing the interdisciplinary connection between physics and mathematics in curricula would also be an important finding for the physics and mathematics education programs in Turkey and the world. As such, it is hoped that one could train teachers who have internalized the interdisciplinary connection between these two fields and who are ready to work collaboratively.

METHOD

Design

Since this study aimed to examine the interdisciplinary connections between mathematics and physics in the objective and content of the secondary education physics curriculum in Turkey, which has been in use since 2019 in Turkey, and to identify the mathematical associations through the secondary education mathematics curriculum, it was designed in the analytical research model. Documents and texts were analyzed by an analytical research model, which does not fit into either qualitative or quantitative research. In other words, the study focused on document analysis through an analytical research model. In this sense, data obtained from the objective (outcomes) and content (subject list) of the secondary education physics curriculum were interpreted via the contexts in the secondary education mathematics curriculum (Ersoy, 2015; McMillan, 2004).

Data Source

Data of this study are the secondary education physics and mathematics curricula (Years 9, 10, 11, and 12) in use since it was published and ratified by the Directory of the Board of Education in 2018

(MofEd, 2018a; MofEd, 2018b). Electronic versions of these curricula were accessed via the Directory of the Board of Education's official website. Only the Turkish versions of these curricula are available on the mentioned website.

Data Analysis

Data obtained as documents were analyzed by content analysis. Content analysis was done in five steps (Figure 1).

Figure 1. Analysis process

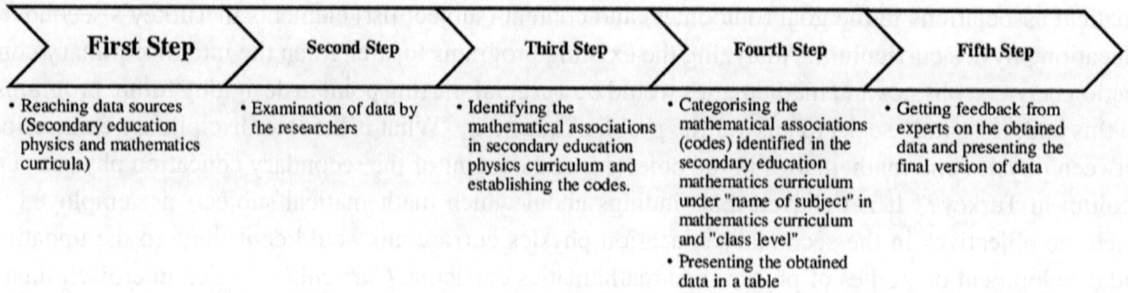

As can be seen in Figure 1, the secondary education physics and mathematics curricula were accessed via the official website in the first step. In the second step, the researchers examined data in detail: the secondary education physics and the secondary education mathematics curriculum. In the third step, mathematical associations in the curriculum were identified on two separate occasions based on the objectives (outcomes) and content (subject list) in the secondary education physics curriculum. For instance, the outcome *"Volume calculations are made for regularly-shaped objects such as cubes, rectangular prisms, cylinders, and spheres, and objects with irregular shapes"* under the subject "matter and density" was considered to be an outcome that requires mathematical association. Mathematical associations identified as such constituted the codes of this study. Then, in the fourth step, researchers examined the secondary education mathematics curriculum, and the codes identified in the previous step were gathered under the categories titled subject and class level of the subject. Codes and categories were then turned into a table by the researchers based on the views and suggestions of experts, who are educators in physics and mathematics. Data took its final shape based on expert opinions; in the last step, they were presented in detail in the findings section (Bezen et al., 2018; Creswell, 2014).

Validity and Reliability Studies

To ensure validity, data obtained in the study were reported in detail, and direct quotations from the secondary education physics and mathematics curricula were presented in the findings section of the survey. Moreover, data obtained from the secondary education physics and mathematics curricula were presented to two experts, who were independent coders; the experts repeated it at separate times to form the codes and the categories; then these codes and categories were checked; finally, findings of the study were analyzed to ensure data validity. To ensure reliability, experts in physics and mathematics education examined the codes; similarities and differences between the views of the experts and researchers were identified. Then, the experts and researchers sat together to discuss the differences between their examinations until a 100% consensus was reached. As such, research reliability was established (Yıldırım & Şimşek, 2016).

FINDINGS

The study aimed to identify the mathematical associations in the objective (outcomes) and content (subject list) elements of the Turkish secondary education physics curriculum. To this end, this section focused on the mathematical associations in the objectives and contents of the Year 9, 10, 11, and 12 physics curricula.

Mathematical Associations in the Objective and Contents of the Secondary Education Year 9 Physics Curriculum

In the study, the objectives and content of secondary education Year 9 physics and mathematics curricula were examined under the categories of "name of the subject,", "number of outcomes," "number of outcomes related to mathematics," "the subject of mathematics related to the outcome," and "class level." Findings about Year 9 were explained in detail in Table 1.

Table 1. Mathematical associations in the objective and contents of the secondary education Year 9 physics curriculum

Secondary Education Physics Curriculum				Secondary Education Mathematics Curriculum	
Year 9 Physics Course Units	**Name of the Subject**	**Number of Outcomes**	**Number of Outcomes Related to Mathematics**	**The Subject of Mathematics Related to the Outcome**	**Class Level**
Introduction to Physics	Importance of Physics as a branch of science	1	-	-	-
	Application Areas of Physics	1	-	-	-
	Classification of Physical Qualities	1	1	Counting and Probability	Year 10
	Scientific Research Centres	1	-	-	-
Matter and its Characteristics	Matter and Density	2	1	Solid Geometry	Year 10
	Durability	1	-	-	-
	Adhesion and Coherence	1			
Motion and Force	Motion	6	2	Data	Year 9
	Force	1	-	-	-
	Newton's Law of Motion	3	1	Data Conversions	Year 9 Year 12
	Frictional Force	1	1	Equations and Disequilibrium Conversions	Year 9 Year 12
Energy	Work, Energy, and Power	2	1	Equations and Disequilibrium Conversions	Year 9 Year 12
	Mechanical Energy	1	1	Equations and Disequilibrium Conversions	Year 9 Year 12
	Preservation of Energy and Energy Conversions	2	-	-	-
	Yield	2	-	-	-
	Energy Resources	1	-	-	-
Heat and Temperature	Heat and Temperature	5	1	Equations and Disequilibrium Conversions	Year 9 Year 12
	Change of State	1	1	Conversion	Year 12
	Thermal Equilibrium	1	1	Equations and Disequilibrium	Year 9
	Ways of Energy Conduction and Speed of Energy Conduction	5	1	Equations and Disequilibrium	Year 9
	Expansion	1	-	-	-

Continued on following page

Table 1. Continued

Secondary Education Physics Curriculum				Secondary Education Mathematics Curriculum	
Year 9 Physics Course Units	**Name of the Subject**	**Number of Outcomes**	**Number of Outcomes Related to Mathematics**	**The Subject of Mathematics Related to the Outcome**	**Class Level**
Electrostatics	Electric charges	4	2	Equations and Disequilibrium Conversions	Year 9 Year 12
Total		**44**	**14**	***(Level of association 31,81%)***	

Table 1 shows that each unit of secondary education Year 9 physics curriculum includes a mathematical association. At the same time, mathematical associations were made with counting and probability, showing data on graphics, conversions, equations, and disequilibrium in the objective and contents of the secondary education Year 9 physics curriculum. It was determined that 14 of the 44 outcome expressions in the secondary education Year 9 physics curriculum were associated with mathematics (31,81%). For instance, the subject "counting and probability" was associated with the outcome "*One makes sure students do the adding up in vectors in a single dimension*" in the secondary education Year 9 physics curriculum. Related to solid geometry, the outcome in the curriculum is as follows: "*One does volume calculations for regularly shaped objects, namely, cubes, rectangular parallelepiped, cylinders, spheres, and irregularly-shaped objects.*" The outcome is as follows: "*Students are made sure to collect data by experimenting or through simulations, draw position-time and speed-time graphs, interpret them and convert between the drawn graphs.*" The outcome concerning conversions is as follows: "*Mathematical model between net force, velocity, and mass are presented*" and "*Mathematical models of displacement kinetic energy, gravitational potential energy, and flexibility potential energy.*" Finally, the equations and disequilibrium outcomes are: "*Students are taught to determine the relationship between the change of temperature and heat and thermal equilibrium*" and "*Students are made sure to determine the relationship between variables and to make deductions from data. Obtained from experiments or simulations.*" In the study, it was determined that some of the mathematical skills present in the objectives and content of the secondary education Year 9 physics curriculum are present in higher classes in the secondary education mathematics curriculum.

Mathematical Associations in the Objective and Contents of the Secondary Education Year 10 Physics Curriculum

In the study, the objectives and content of the secondary education Year 10 physics curriculum were examined under the categories of "name of the subject,", "number of outcomes," "number of outcomes related to mathematics," "the subject of mathematics related to the outcome," and "class level." Findings about Year 10 were explained in detail in Table 2.

Table 2. Mathematical associations in the objective and contents of the secondary education Year 10 physics curriculum

Secondary Education Physics Curriculum				Secondary Education Mathematics Curriculum	
Year 10 Physics Course Units	**Name of the Subject**	**Number of Outcomes**	**Number of Outcomes Related to Mathematics**	**The Subject of Mathematics Related to the Outcome**	**Class Level**
Electrics and Magnetism	Electric Current, Potential Difference, and Resistance	2	1	Equations and Disequilibrium Conversions	Year 9 Year 12
	Electric Circuits	4	1	Data Equations and Disequilibrium Conversions	Year 9 Year 12
	Magnet and Magnetic Field	1	-	-	-
	Current and Magnetic Field	2	-	-	-
Pressure and Ascending Force	Pressure	2	1	Equations and Disequilibrium Conversions	Year9 Year 12
	Ascending Force	2	1	Data Conversions	Year 9 Year 12
Waves	Waves	2	1	Conversions	Year 12
	Spring Wave	2	-	-	-
	Water Wave	4	-	-	-
	Sound Wave	2	-	-	-
	Seismic Wave	2	-	-	-
Optics	Illumination	2	1	Equations and Disequilibrium Conversions	Year 9 Year 12
	Shadow	1	-	-	-
	Reflection	1	-	-	-
	Plane Mirror	1	-	-	-
	Spherical Mirror	2	-	-	-
	Fraction	3	2	Equations and Disequilibrium Data Conversions	Year 9 Year 12
	Optics	2	-	-	-
	Prism	1	-	-	-
	Colour	1	-	-	-
Total		**39**	**8**	***(Level of association 20,51%)***	

Table 2 shows that the secondary education Year 10 physics curriculum has mathematical associations in every unit. It was determined that mathematical associations were made in the secondary education Year 10 physics curriculum's objective and contents with data and equations and disequilibrium. It was also determined that 8 of the 39 outcomes in the curriculum were associated with mathematics (20,51%). For instance, regarding conversions, the secondary education Year 10 physics curriculum has the following outcomes: "*Mathematical models related to solid and hydrostatic pressure as well as pressure force are presented*" and "*Mathematical models between the concepts of progress speed of the wave, wavelength, and frequency are presented.*" The outcomes related to data are: "*The magnitudes of the weight of the object and the ascending force during swimming, floating, and sinking are compared,*" and "*Students are made sure to experiment and draw the path light follows and to make associations with daily life phenomena.*" Finally, the outcomes related to the subject of equations and disequilibrium are as follows: "*It is underlined that refractive index is a dependent variable related to the average speed of light in the environment and its speed in space,*" "*Students are made sure to determine the relationship between variables by utilizing experiments and simulations and to draw the mathematical model,*" and "*Experiments and simulations establish the relationship between illumination intensity, light intensity, and light flow.*" It was determined in the study that some of the mathematical skills present in the objectives and content of the secondary education Year 10 physics curriculum are given in higher classes in the secondary education mathematics curriculum.

Mathematical Associations in the Objective and Contents of the Secondary Education Year 11 Physics Curriculum

In the study, the objectives and content of secondary education Year 11 physics thematics curriculum were examined under the categories of "name of the subject,", "number of outcomes," "number of outcomes related to mathematics," "the subject of mathematics related to the outcome," and "class level." Findings about Year 11 were explained in detail in Table 3.

Table 3. Mathematical associations in the objective and contents of the secondary education Year 11 physics curriculum

Secondary Education Physics Curriculum				Secondary Education Mathematics Curriculum	
Year 11 Physics Course Units	**Name of the Subject**	**Number of Outcomes**	**Number of Outcomes Related to Mathematics**	**The Subject of Mathematics Related to the Outcome**	**Class Level**
Force and Motion	Vectors	4	2	Trigonometry	Year 11
	Relative Motion	3	1	Counting and Probability	Year 10
	Newton's Laws of Motion	2	2	Counting and Probability	Year 10
	Constant Acceleration Motion in One Dimension	6	5	Data Conversions	Year 9 Year 12
	Motion in Two Dimensions	2	1	Equations and Disequilibrium	Year 9
	Energy and Motion	3	3	Functions	Year 10
	Impulse and Linear Momentum	4	3	Functions Conversions	Year 10 Year 12
	Torque	3	1	Equations and Disequilibrium	Year 9
	Equilibrium and Equilibrium Conditions	3	1	Equations and Disequilibrium	Year 9
	Simple Machines	3	1	Equations and Disequilibrium	Year 9
Electricity and Magnetism	Electrical Force and Electric Field	3	1	Equations and Disequilibrium	Year 9
	Electrical Potential	3	2	Equations and Disequilibrium	Year 9
	Uniform Electric Field and Capacitance	6	1	Conversions	Year 12
	Magnetism and Electromagnetic Induction	11	6	Equations and Disequilibrium Conversions	Year 9 Year 12
	Alternative Current	4	-	-	-
	Transformators	2	-	-	-
Total		**62**	**30**	***(Level of association 48,38%)***	

Table 3 shows a mathematical association in every secondary education Year 11 physics curriculum unit. Associations were done in the objective and contents of the secondary education Year 11 physics curriculum with trigonometry, counting, probability, data, functions, conversions, equations, and disequilibrium. It was determined that 30 of the 62 objectives in the curriculum contain a mathematical association (48,33%). For example, the outcome related to trigonometry in the curriculum is "*By giving them the co-sinus theorem, students are ensured to calculate the magnitude of the complex vector.*" Concerning counting and probability, the outcomes in the curriculum are "*Students make calculations about relative motion*" and "*Students are made sure to calculate the magnitude of the friction force by determining its direction on the lateral, vertical, and inclined plane.*" The outcome related to data is "*Students are ensured to draw and interpret position-time, speed-time, and velocity-time graphs related to constant acceleration motion; they are ensured to convert graphs to one another.*" The function outcome is "*Students make work calculations utilizing force-road graphs.*" Outcomes related to equations and disequilibrium are "*Students are ensured to make mathematical calculations related to efficiency in simple machines*" and "*Students make calculations about the electric force and electric field in point charges.*" Finally, outcomes associated with conversions are as follows: "*Students are made sure to obtain the mathematical model about capacitance*" and "*Students are given the mathematical model of limit velocity.*" It was determined in the study that some of the mathematical skills present in the objectives and content of the secondary education Year 11 physics curriculum are present in higher classes in the secondary education Year 12 mathematics curriculum.

Mathematical Associations in the Objective and Contents of the Secondary Education Year 12 Physics Curriculum

Finally in the study, the objectives and content of the secondary education Year 12 physics curriculum were examined under the categories of "name of the subject,", "number of outcomes," "number of outcomes related to mathematics," "the subject of mathematics related to the outcome," and "class level." Findings about Year 11 were explained in detail in Table 4.

Table 4. Mathematical associations in the objective and contents of the secondary education Year 12 physics curriculum

Secondary Education Physics Curriculum				Secondary Education Mathematics Curriculum	
Year 12 Physics Course Units	**Name of the Subject**	**Number of Outcomes**	**Number of Outcomes Related to Mathematics**	**The Subject of Mathematics Related to the Outcome**	**Class Level**
Cyclical Motion	Uniform Circular Motion	4	4	Equations and Disequilibrium Triangles Circle and Cycle Trigonometry Conversions	Year 9 Year 11 Year 12
	Rotatory Translational Motion	3	-	-	-
	Angular Momentum	4	2	Equations and Disequilibrium	Year 9
	Gravitational Force	3	2	Equations and Disequilibriums Conversions	Year 9 Year 12
	Kepler Laws	1	-	-	-
Simple Harmonic Motion	Simple Harmonic Motion	5	5	Equations and Disequilibriums Functions Conversions	Year 9 Year 10 Year 12
Wave Mechanics	Diffraction in Waves, Interference, and Doppler Event	6	-	-	-
	Electromagnetic Waves	2	-	-	-

Continued on following page

Table 4. Continued

Secondary Education Physics Curriculum				Secondary Education Mathematics Curriculum	
Year 12 Physics Course Units	**Name of the Subject**	**Number of Outcomes**	**Number of Outcomes Related to Mathematics**	**The Subject of Mathematics Related to the Outcome**	**Class Level**
Introduction to Atomic Physics and Radioactivity	Historical Development of the Concept of Atom	3	-	-	-
	Big Bang and the Formation of the Universe	4	-	-	-
	Radioactivity	4	-	-	-
Modern Physics	Special Relativity	4	-	-	-
	Introduction to Quantum Physics	1	1	Functions	Year 10
	Photoelectric Event	6	3	Data Equations and Disequilibrium	Year 9
	Compton Scattering and De Broglie Wavelength	4	1	Equations and Disequilibrium	Year 9
Applications of Modern Physics in Technology	Screening Technologies	2	-	-	-
	Semi-Conductor Technology	5	-	-	-
	Super Conductors	2	-	-	-
	Nanotechnology	3	-	-	-
	Laser Beams	2	-	-	-
Total		**68**	**18**	***(Level of association 26,47%)***	

Table 4 shows that three of the six units in the secondary education Year 12 physics curriculum (*Cyclical Motion, Simple Harmonic Motion, Modern Physics*) contain mathematical associations; the remaining three units (*Wave Mechanics, Introduction to Atomic Physics and Radioactivity,* and *Applications of Modern Physics in Technology*) do not have any mathematical associations. It was determined that mathematical associations were made in the subjects of Triangles, Equations and Disequilibrium, Cycle and Circle, Trigonometry, Function, Data, and Conversions in the objective and contents of the secondary education Year 12 physics curriculum. 18 of the 68 accurate expressions were associated with mathematics (26,47%). For instance, the outcome related to triangles in the curriculum is as follows: "*Students are made sure to find out the intensity of acceleration and show its direction by drawing the linear velocity vector on the circle for the uniform circular motion.*" For the equations and disequilibrium, the outcomes are "*By experiments and simulations; students are made sure to determine the relationship between variables centripetal force is dependent upon*" and "*Students explain angular momentum by relating it to linear momentum.*" About cycles and circles, the outcome is "*Students draw the free-body diagrams of objects which have uniform circular motion at the horizontal and vertical plane.*" The work related to trigonometry is "*Students make the necessary calculations for the safe turning conditions of cars on vertical, horizontal, and inclined surfaces.*" The function outcome is: "*Students are made to draw and interpret position-time graphs by conducting experiments or simulations.*" The work related to data is "*Students draw the maximum kinetic energy-frequency graph for different metals.*" Finally, the outcomes related to conversions are "*By experiments and simulations, students are made sure to determine the relationship between variables centripetal force is dependent upon,*" "*Mathematical model is presented,*" and "*Gravitational force is mentioned. A mathematical model is presented.*"

DISCUSSION

This study aimed to examine the interdisciplinary connections between physics and mathematics according to the secondary education physics curriculum and to identify the mathematical associations in the physics curriculum with the help of the secondary education mathematics curriculum. To this end, it was concluded that the years 9, 10, 11, and 12 levels of the secondary education physics curriculum all contain mathematical associations. It was determined that there is 31,81% mathematical association in the outcomes of the Year 9 physics curriculum, 20,51% in Year 10, 48,38% in Year 11, and 26,47% in Year 12. Based on this, it can be argued that the level of association of the outcomes the secondary education physics curriculum sets to achieve at every class level with the subjects in the secondary education mathematics curriculum is 32,86%. It was also determined that associations were drawn at every class level of the secondary education physics curriculum when it comes to the following subjects in the secondary education mathematics curriculum: "Data," "Equations and Disequilibrium," "Counting and Probability," and "Conversions;" mathematical associations were made about "Solid Geometry" at Year 9 and Year 10 levels; mathematical associations were made about "Functions" and "Trigonometry" at Year 11 and Year 12 levels, and mathematical associations were made about "Triangles" and "Cycle and Circle" only at Year 12 level. When these associations were examined in terms of the unit titles in mathematics courses, it was seen that the variety of mathematics subjects related to the outcomes in Year 9 and Year 10 physics courses (Data, Equations and Disequilibrium, Counting and Probability, Solid Geometry, Conversions) is less than those needed in Year 11 and Year 12. The highest number of associations is at the Year 12 level. In this regard, it is believed that students will benefit more from mathematics subjects in Year 11 and Year 12 physics classes. Moreover, when the associations between the secondary education physics curriculum and secondary education mathematics curriculum were examined in proportion, it was seen that the highest relation level was established at Year 11 level. A noteworthy finding is that the relation level increases at Year 11. At the same time, it decreases at Year 12 because it is believed that mathematical association can be established at a lower level in the first years of physics classes in parallel to the level of prior mathematical knowledge; it is also believed that as students' level of mathematical knowledge increases so should the number of mathematical associations in physics classes (Retnawati et al., 2018).

Consequently, it was determined in the study that the number of mathematical associations in the physics curriculum in Turkey is not sufficient. In other words, the outcomes of the secondary education physics curriculum and the outcomes of the secondary education mathematics curriculum are not synchronized when one looks at the mathematical associations in the outcomes of secondary education physics in terms of the class level in secondary education mathematics courses. Put more clearly, it was determined that to achieve the outcomes in Year 9 physics course, students should achieve the Year 10 and Year 12 outcomes in mathematics; to achieve the outcomes of the Year 10 physics course, they should achieve the Year 12 outcomes in mathematics; and to achieve the outcomes of Year 11 physics course, they would need the Year 12 outcomes of mathematics courses. In other words, it was determined that students had to employ certain mathematical knowledge for the first time in physics classes before they learn it in mathematics classes first. Results in the literature also indicate that the lack of synchronization between the physics and mathematics curricula is the basis for the difficulty of transferring mathematical knowledge during secondary education physics classes. Several studies also show that this disjointedness results in physics courses being regarded as difficult. It is also stated that this problem is related more to the gaps in students' pre-requisite learning, which requires mathematical knowledge and skill, than to the difficulty of secondary education physics courses (Ayvacı & Bebek, 2018; Basson, 2002; Duit et al., 2007; Mumthas & Abdulla, 2019; Pektaş et al., 2022; Pietrocola, 2008; Retnawati et al., 2018). As such, students prior knowledge of secondary education mathematics is effective in their success in secondary education physics classes (Başkan et al., 2010; Buick, 2007; McHugh et al., 2018; Retnawati et al., 2018; Surma, 2020; Vallo & Valovicova, 2019). Moreover, when the secondary education mathematics course does not support the secondary education physics curriculum, students come face to face with mathematical operations for the first time in physics courses, and this becomes a burden and extra difficulty for the teacher who is supposed to teach them (Mumthas & Abdulla, 2019; Retnawati et al., 2018; Treacy, 2021). In the literature, this difficulty is regarded by the physics teachers as a barrier in front of meaningful learning as well as a waste of time because physics teachers experience difficulties both since they must cover topics that are supposed to be covered by a mathematics teacher and due to ineffective use of time. In the literature, these problems teachers face are cited as inevitably reflected in students (Treacy, 2021; Xu et al., 2021). In this respect, basic mathematical proficiencies to be utilized in physics courses should be taught earlier before students are introduced to physics. At the same time, the relationship between physics and mathematics, two of the STEM disciplines, should be reflected in schools through curricula, real-world problems should be frequently included within teaching, and this integration should be prioritized (Bahar et al., 2018). In other words, both curricula should be redesigned and re-structured, keeping the pre-requisite knowledge necessary in both fields (Buick, 2007; Meltzer, 2002; Mumthas & Abdulla, 2019; Retnawati et al., 2018).

CONCLUSION AND RECOMMENDATIONS

This study revealed that mathematical associations are present in every class level in the secondary education physics curriculum; however, these associations are not at the required level. Another important result obtained in the study is that there is no synchronization between the outcomes of physics and the outcomes of mathematics classes. This makes students regard physics as a tough course; moreover, it causes physics teachers to waste time in a large portion of their classes teaching students the prerequisite mathematical information. This finding made it clear that better to manage students' learning processes in physics classes, improvement efforts of secondary education mathematics and physics curricula should be carried out with experts from both disciplines coming together because the integration of these two curricula and making sure STEM integration is complete are critical to the sustainability in education (Mumthas & Abdulla, 2019). At the same time, while discipline-based and content-heavy curricula focus on theory, the application should have a healthy part, and real-life events should be reflected in the school experience through an interdisciplinary connection. In other words, holistic curricula containing application-based critical education cases should be designed with a multidisciplinary connection (Haatainen et al., 2021; Miron & Staicu, 2010). However, at this point, it should be kept in mind that mathematics should be included in physics courses as an algebraic tool and a reasoning process serving students' advanced thinking skills. As such, one could help train individuals with science literacy and contribute to training generations who could produce technology (Davison et al., 1995; Johnston et al., 2020; Tytler et al., 2021). In addition, it should be kept in mind that teachers tend to repeat the education models they experienced when they were students themselves (Romero Ariza et al., 2021). In this respect, it could be suggested that innovative teacher training programs and teaching materials should be designed to ensure that new curricula take into consideration the connection between physics and mathematics (Aguirre-Munoz et al., 2022; Tytler et al., 2021). Moreover, in-service training could be provided for teachers so that programs with a strengthened connection between physics and mathematics as such could be effectively put into practice. In these strengthening and renovation studies, teachers' proficiency perceptions and attitudes toward interdisciplinary teachers should not be disregarded (Haatainen et al., 2021). Interdisciplinary problem-solving hours run by teachers from both disciplines could be added to the curriculum in addition to the already existing physics and mathematics classes. "Good practice" examples from various countries in which interdisciplinary applications are run effectively could be identified, digital education content could be created to disseminate them.

LIMITATIONS AND DIRECTIONS FOR FUTURE STUDIES

This study was limited to the secondary education physics and mathematics curricula prepared by the Republic of Turkey, Ministry of Education Board of Education. It is believed that to obtain more in-depth information about the relationship between physics and mathematics courses, one could have a more longitudinal analysis starting with the primary education science curriculum and primary school mathematics curriculum. Moreover, the study's findings could be comparatively studied internationally. In addition, it is thought that such variables as students' cognitive levels, class conditions, and teachers' fields of expertise during the curriculum's application could influence the study's findings (Winarno et al., 2020). In this respect, classes in which curricula are put into practice could be observed, and future studies could be conducted to determine to what extent the application reflects the official curriculum. Likewise, textbooks are essential means of the application of curricula. It is believed that it could contribute to the literature examining secondary education physics and mathematics textbooks used in schools to see to what extent the associations in the curriculum are reflected in these books (Ekmen & Bakar, 2019).

REFERENCES

Aguirre-Munoz, Z., Dang, B., & Garro, E. S. (2022). *Impact of integrated science and mathematics instruction on middle school science and mathematics achievement. In Advances in research in STEM education*. IntechOpen.

Ayvacı, H. Ş., & Bebek, G. (2018). A study on evaluation of the problems faced with teaching physics. *Kastamonu Education Journal, 26*(1), 1–10.

Bahar, M., Yener, D., Yılmaz, M., Emen, H., & Gürer, F. (2018). The changes of standards in the 2018 science curriculum and STEM integration. *Bolu Abant İzzet Baysal University Journal of Faculty of Education, 18*(2), 702–735.

Başkan, Z., Alev, N., & Karal, I. S. (2010). Physics and mathematics teachers' ideas about topics that could be related or integrated. *Procedia: Social and Behavioral Sciences, 2*(2), 1558–1562. doi:10.1016/j.sbspro.2010.03.235

Basson, I. (2002). Physics and mathematics as interrelated fields of thought development using acceleration as an example. *International Journal of Mathematical Education in Science and Technology, 33*(5), 679–690. doi:10.1080/00207390210146023

Bastos, R. M. B. (2017). The surprising success of the Finnish educational system in a global scenario of commodified education. *Revista Brasileira de Educação, 22*, 802–825. doi:10.15901413-24782017227040

Berlin, D. F., & White, A. L. (2012). A longitudinal look at attitudes and perceptions related to the integration of mathematics, science, and technology education. *School Science and Mathematics, 112*(1), 20–30. doi:10.1111/j.1949-8594.2011.00111.x

Bezen, S., Başal, C., Aykutlu, I., Seçken, N., & Bayrak, C. (2018). An ınterdisciplinary look at the comparative examination of physics and chemistry coursebooks. *Journal of Qualitative Research in Education, 6*(3), 267–283. doi:10.14689/issn.2148-2624.1.6c3s13m

Broggy, J., O'Reilly, J., & Erduran, S. (2017). Interdisciplinarity and science education. In *Science Education* (pp. 81–90). Brill Sense. doi:10.1007/978-94-6300-749-8_6

Buick, J. M. (2007). Investigating the correlation between mathematical pre-knowledge and learning gains in service physics. *European Journal of Physics, 28*(6), 1073–1080. doi:10.1088/0143-0807/28/6/004

Cavadas, B., Rezio, S., Nogueira, J. R., & Branco, N. (2022). A framework and a research design proposal to identify preservice teachers' integration performance of science and mathematics. *Canadian Journal of Science. Mathematics and Technology Education, 22*(1), 101–129. doi:10.100742330-022-00198-2

Ceuppens, S. (2019). *Learning of physics and mathematics concepts in an integrated STEM curriculum*. KU Leuven Faculty of Science.

Creswell, J. W. (2014). *Research design: Qualitative, quantitative and mixed methods approaches* (4th ed.). Sage.

Cura, G., & Ercan Yalman, F. (2019). Investigating science teacher candidates' use of interdisciplinary instructional approach. *Online Science Education Journal, 4*(2), 131–153.

Davison, D. M., Miller, K. W., & Metheny, D. L. (1995). What does integration of science and mathematics really mean? *School Science and Mathematics, 95*(5), 226–230. doi:10.1111/j.1949-8594.1995.tb15771.x

De Ataíde, A. R. P., & Greca, I. M. (2013). Epistemic views of the relationship between physics and mathematics: Its influence on the approach of undergraduate students to problem solving. *Science & Education, 22*(6), 1405–1421. doi:10.100711191-012-9492-2

Dehipawala, S., Shekoyan, V., & Yao, H. (2014). Using mathematics review to enhance problem solving skills in general physics classes. In *Proceedings of the 2014 Zone 1 Conference of the American Society for Engineering Education* (pp. 1- 4). IEEE. 10.1109/ASEEZone1.2014.6820631

Derman, İ. (2019). *The level of association of science course with daily life* [Unpublished doctoral dissertation]. Hacettepe University, Ankara, Turkey.

Duit, R., Schecker, H., Höttecke, D., & Niedderer, H. (2007). *Teaching physics handbook of research on science education*. Routledge.

Ekmen, C., & Bakar, E. (2019). Digital competencies in the curriculum and texsbooks of primary education. *Milli Eğitim Dergisi*, *48*(221), 5–35.

Ersoy, A. (2015). Investigation of Phd students' initial qualitative research experiences via their diaries. *Pegem Journal of Education and Instruction*, *5*(5), 549–568.

Furner, J. M., & Kumar, D. D. (2007). The mathematics and science integration argument: A stand for teacher education. *Eurasia Journal of Mathematics, Science and Technology Education*, *3*(3), 185–189. doi:10.12973/ejmste/75397

Gifford, J. D., & Finkelstein, N. D. (2021). Applying a mathematical sense-making framework to student work and its potential for curriculum design. *Physical Review. Physics Education Research*, *17*(1), 010138-1–27. doi:10.1103/PhysRevPhysEducRes.17.010138

Gingras, Y. (2001). What did mathematics do to physics? *History of Science*, *39*(4), 383–416. doi:10.1177/007327530103900401

Haatainen, O., Turkka, J., & Aksela, M. (2021). Science teachers' perceptions and self-efficacy beliefs related to integrated science education. *Education Sciences*, *11*(272), 1–20. doi:10.3390/educsci11060272

Hansson, L., Hansson, Ö., Juter, K., & Redfors, A. (2021). Curriculum emphases, mathematics and teaching practices: Swedish upper-secondary physics teachers' views. *International Journal of Science and Mathematics Education*, *19*(3), 499–515. doi:10.100710763-020-10078-6

Johnston, J. (2005). *Early explorations in science*. Mcgraw-Hill Education.

Johnston, J., Walshe, G., & Riordain, M. (2020). Supporting key aspect of practice in making mathematics explicit in science lessons. *International Journal of Science and Mathematics Education*, *18*(7), 1399–1417. doi:10.100710763-019-10016-1

Martin-Paez, T., Aguilera, D., Perales-Palacios, F. J., & Vílchez-González, J. M. (2019). What are we talking about when we talk about STEM education? A review of literature. *Science Education*, *103*(4), 799–822. doi:10.1002ce.21522

McHugh, L., Kelly, A. M., & Burghardt, M. D. (2018). Professional development for a middle school mathematics-infused science curriculum. *Journal of Science Teacher Education*, *29*(8), 804–828. doi: 10.1080/1046560X.2018.1514825

McMillan, J. H. (2004). *Educational research: Fundamentals for the consumer* (4th ed.). Pearson Education, Inc.

Meltzer, D. E. (2002). The relationship between mathematics preparation and conceptual learning gains in physics: A possible "hidden variable" in diagnostic pretest scores. *American Journal of Physics, 70*(12), 1259–1268. doi:10.1119/1.1514215

Ministry of National Education (MEB). (2018a). *Ortaöğretim fizik dersi öğretim programı* [Secondary education physics curriculum]. Talim ve Terbiye Kurulu Başkanlığı. http://mufredat.meb.gov.tr/Dosyalar/201812103112910-ortaöğretim_fizik_son.pdf

Ministry of National Education (MEB). (2018b). *Ortaöğretim matematik dersi öğretim programı* [Secondary education mathematics curriculum]. Talim ve Terbiye Kurulu Başkanlığı. http://mufredat.meb.gov.tr/Dosyalar/201821102727101-OGM%20MATEMATİK%20PRG%2020.01.2018.pdf

Miron, C., & Staicu, I. (2010). The impact of interdisciplinarity on the physics-mathematics scientific education in high schools. *Romanian Reports in Physics, 62*(4), 907–917.

Mumthas, N. S., & Abdulla, S. U. (2019). Substandard Performance in Mathematical Problem Solving in Physics among Higher Secondary School Students in Kerala—An Investigation on Teacher Perceptions and Student Difficulties. *Online Submission, 7*(1), 31–39. doi:10.15415/iie.2019.71005

National Academy of Sciences. (2014). *STEM integration in K-12 education: Status, prospects, and an agenda for research.* National Academies Press.

Next Generation Science Standards (NGSS). (2013). http://www.nextgenscience.org

Osborne, J. (2014). Teaching scientific practices: Meeting the challenge of change. *Journal of Science Teacher Education, 25*(2), 177–196. doi:10.100710972-014-9384-1

Park, W., Kim, D., & Kang, D. Y. (2021). Research trends in science and mathematics education in South Korea 2014-2018: A cross-disciplinary analysis of publications in selected local journals. *Asia-Pasific Science Education, 7*(2), 280–308. doi:10.1163/23641177-bja10029

Pektaş, M., Sadak, M., İncikabı, L., & Kayabaşı, B. (2022). Examining the mathematics content of science textbooks: Connected textbooks. *Mehmet Akif Ersoy University Journal of Education Faculty, 61*, 98–121.

Pietrocola, M. (2008). Mathematics as structural language of physical thought. *Connecting Research in Physics Education with Teacher Education, 18*(2), 193–205.

Polverini, G. (2021). *Exploring interdisciplinarity between physics and mathematics: the design of a linguistic and an epistemological tool for analysing texts about the parabolic motion* [Unpublished master's thesis]. Alma Mater Studiorum-University of Bologna, Italy.

Pospiech, G. (2015). Interplay of mathematics and physics in physics education. *International Symposium of Mathematics and its Connections to the Arts and Sciences* (s. 36-43). University of Education Schwäbisch Gmünd.

Retnawati, H., Arlinwibowo, J., Wulandari, N. F., & Pradani, R. G. (2018). Teachers' difficulties and strategies in physics teaching and learning that applying mathematics. *Journal of Baltic Science Education, 17*(1), 120–135. doi:10.33225/jbse/18.17.120

Romero Ariza, M., Armenteros, A. Q., & Castro, A. E. (2021). Promoting critical thinking through mathematics and science teacher education: The case of argumentation and graphs interpretation about climate change. *European Journal of Teacher Education*, 1–19. doi:10.1080/02619768.2021.1961736

Surma, K. (2020). *Does the mathematics level affect student success in high school physics?* School of Education Student Capstone Theses and Dissertations. https://digitalcommons.hamline.edu/hse_all/4510

Treacy, P. (2021). A conceptual framework for integrating mathematics and science in the secondary classroom. *SN Social Sciences*, *1*(6), 1–17. doi:10.100743545-021-00166-x

Tytler, R., Mulligan, J., Prain, V., White, P., Xu, L., Kirk, M., Nielsen, C., & Speldewinde, C. (2021). An interdisciplinary approach to primary school mathematics and science learning. *International Journal of Science Education*, *43*(12), 1926–1949. doi:10.1080/09500693.2021.1946727

Vallo, D., & Valovicova, L. (2019). Interdisciplinary relations of mathematics and physics in curriculum of conic sections. *AIP Conference Proceedings*, 1–7. doi:10.1063/1.5124779

Winarno, N., Rusdiana, D., Riandi, R., Susilowati, E., & Afifah, R. (2020). Implementation of integrated science curriculum: A critical review of the literature. *Journal for the Education of the Gifted*, *8*(2), 795–817.

Wong, V., & Dillon, J. (2020). Crossing the boundaries: Collaborations between mathematics and science departments in English secondary (high) schools. *Research in Science & Technological Education*, *38*(4), 396–416. doi:10.1080/02635143.2019.1636024

Wong, V. J. (2018). *The relationship between school science and mathematics education* [Unpublished doctoral dissertation]. King's College London. King's Research Portal: https://kclpure.kcl.ac.uk/portal/

Xu, L., Prain, V., & Speldewinde, C. (2021). Challenges in designing and assessing student interdisciplinary learning of optics using a representation construction approach. *International Journal of Science Education*, *43*(6), 844–867. doi:10.1080/09500693.2021.1889070

Yıldırım, A., & Şimşek, H. (2016). *Sosyal bilimlerde nitel araştırma yöntemleri* [Qualitative research methods in the social sciences]. Seçkin.

ADDITIONAL READING

Eli, M., Munkebye, Scheie, E., Gabrielsen, A., Jordet, A., Misund, S., Nergård, T., & Øyehaug, A. B. (2020). Interdisciplinary primary school curriculum units for sustainable development. *Environmental Education Research*, *26*(6), 795–811. doi:10.1080/13504622.2020.1750568

King, N. C., & Hambrook, G. (2021). Examining the impact of pre-university qualifications on success in interdisciplinary science. *Journal of Further and Higher Education*, *45*(9), 1192–1205. doi:10.1080/0309877X.2020.1854695

Oprisan, S. A. (2022). Interdisciplinary curriculum for computational neuroscience at primarily undergraduate institutions. *Journal of Computational Science*, *61*, 101642. doi:10.1016/j.jocs.2022.101642

Quezada-Espinoza, M., Dominguez, A., & Zavala, G. (2021). Academic and professional relevance of physics: Comparing perceptions of engineering students from Mexico and Chile. *International Journal of Engineering Education*, *37*(5), 1174–1185.

Slomka, J. M. (2019). Analysing the nature of ministry guidelines for developing interdisciplinary science courses (grades 11–12) in Ontario (Canada). *Canadian Journal of Science. Mathematics and Technology Education*, *19*(4), 367–386. doi:10.100742330-019-00062-w

Suceta Zulueta, L., Chivás Tito, Y., & Delfino Ferreira, A. (2018). The interdisciplinary in the teaching learning process of the subject Scholar Physics' Fundamentals I. *Foro Educ*, *30*, 85–100.

Vieyra, R., & Himmelsbach, J. (2022). Teachers' disciplinary boundedness in the ımplementation of ıntegrated computational modeling in physics. *Journal of Science Education and Technology*, *31*(2), 153–165. doi:10.100710956-021-09938-9

Zúñiga-Meléndez, A., Durán-Apuy, A., Chavarría-Vásquez, J., Gamboa-Araya, R., Carballo-Arce, A. F., Vargas-González, X., ... Torres-Salas, I. (2020). Diagnosis of training needs of teachers of biology, chemistry, physics, and mathematics, in disciplinary, pedagogical areas, and use of technologies to promote scientific thinking skills. *Revista Electrónica Educare*, *24*(3), 469–497.

KEY TERMS AND DEFINITIONS

Curriculum: It is an education plan that explains how subjects in the relevant disciplines will be taught and the order in which topics will be introduced.

Interdisciplinary Connection: It denotes a connection that supports meaningful learning by integrating concepts or subjects from two or more disciplines.

Mathematical Association: It is a connection that ensures the formation of new information by increasing the interaction with mathematical concepts.

Mathematics: Mathematics is a branch of science that examines the connection between numbers and qualities through deductive reasoning.

Physics: Physics is a branch of science that focuses on phenomena that could be mathematically defined and examined through experimentation.

STEM Integration: It is a teaching idea that combines different disciplines and presents various teaching methods from the actual world into the problem's solution.

Teacher: It denotes the person who carries out teaching activities in education environments in a planned and organized manner.

Chapter 4
Cross-Curricular Nature of Evolutionary Biology: A Call, Perceptions, and Teaching Approaches

María Martín-Peciña
https://orcid.org/0000-0001-5304-5737
University of Jaén, Spain

ABSTRACT

Evolutionary biology is a core discipline in science. Thus knowledge of it is key to understanding other branches of biology. However, there are many authors who have reported difficulties in the teaching-learning processes of this content across several educational levels. In this chapter, the authors delve into pre- and in-service Spanish teacher perceptions about the cross-curricular nature of evolutionary biology by means of semi-structured interviews. All interviewees consider that biological evolution is a unifying content and so has to be approached in the classroom. Thus, treating this content as a central axis for biology could be the way towards an effective educational proposal for biological evolution, but how to reach and design it is still complex. With this in mind, this chapter describes a sequence of activities for secondary students aimed at addressing some learning challenges about biological evolution while considering its cross-curricular nature and its Spanish curricular framework.

INTRODUCTION

Evolutionary Biology is a comprehensive discipline in science that is aimed at explaining the history of life in this planet. Understanding the Theory of Biological Evolution is therefore key to the different branches of Biology and other fields of knowledge. However, there are many authors who have reported difficulties in the teaching-learning processes, among others: Van Dijk & Reydon (2010), Tamayo (2010), González-Galli & Meinardi (2017). Recently, Gefaell et al. (2020) have identified a poor knowledge about the Theory of Evolution through the KEE questionnaire (Knowledge of Evolution Exam;

DOI: 10.4018/978-1-6684-5765-8.ch004

Moore et al., 2009) in undergraduate students belonging to different degrees, including samples from the Bachelor of Biology Degree in Spain, despite the high acceptance of this theory among students as a result of the MATE questionnaire (Measure of Acceptance of the Theory of Evolution; Rutledge & Warden, 1999). Similar results have been reported by Kuschmierz et al. (2021) in a cross-country study at European level. This fact suggests a review of the teaching strategies of Evolutionary Biology from the pre-university levels

In this sense, the acceptance of the Theory of Evolution has also been evaluated in Secondary Education students (Martínez-Hernández et al., 2017) finding a high acceptance in last-course students, with respect to those of previous ones, in schools located in Almería (Spain). In fact, the level of acceptance of the Theory of Evolution in these students was not far from that reported for undergraduate students in Spain (Gefaell et al., 2020) and was higher than that of Biology students from other countries (Turkey: Deniz & Sahin, 2016; United States: Barnes et al., 2019). However, the information regarding the knowledge in Secondary Education students about the Theory of Evolution is less satisfactory, as mentioned before, and even a low understanding of the Theory of Evolution has been also described in Secondary Education teachers (Yates & Marek, 2015 – USA; To et at., 2017 – UK; Hermann et al., 2020 – USA). Acosta-Pérez (2014) reported difficulties in High School students when distinguishing between facts, theories, evidence and results of Biological Evolution. Gallego & Muñoz (2015) also observed a significant percentage of students who showed Lamarckian interpretations of the Theory of Evolution both in Secondary Education and High School.

One of the problems underpinning this situation could lie in a lack of knowledge among teachers about the factors that would improve the students' understanding about the Theory of Evolution. For instance, the study by Gefaell et al. (2020), mentioned above, suggests a positive impact of the hours dedicated to Evolutionary Biology to understand the Theory of Evolution in undergraduate students. There are also data supporting that teaching Genetics prior to Evolutionary Biology improves the understanding of the latter, but not necessarily its acceptance (Mead et al., 2017). Some time ago, Jensen & Finley (1996) detected a slight improvement in the understanding of the Theory of Evolution after the use of peer problem-solving with a historically enriched curriculum as teaching strategy compared to traditional approaches. Therefore, there are indicators that point to a relevant role of the teaching-learning processes to improve the understanding and acceptance of the Theory of Biological Evolution.

A teaching sequence can be defined as the orderly implementation of learning processes focused on the achievement of a more scientific and technical development of the pedagogy behind the content taught (Bingham & Davis, 2012). Therefore, the design of teaching sequences has been revealed as a useful tool to improve teaching-learning processes when applied to several disciplines (Ogborn, 2010). Many authors have designed and implemented this approach as an effective teaching instruments in different contexts, such as in Physics (Viiri & Savinainen, 2008), for the concept of energy (Opitz, 2016), to teach about ecosystems (Yeo & Lee, 2016) or human nutrition (Cabello-Garrido et al., 2018). In the area of Evolutionary Biology, many efforts are also being made to develop teaching sequences that improve students' understanding of that matter. In this regard, the proposals of Hagman, Olander & Wallin (2003) stand out, implemented in Swedish High School students specialized in Science and Technology. These authors carried out strategies based on discussions between students, guided by the teacher, to reflect on their own ideas and those of their classmates about the Theory of Evolution in different biological contexts. However, no more than 43% of the students used the Theory of Evolution consistently in the questionnaires filled out after an educational intervention applying that teaching sequence (Wallin, 2008).

In fact, one key challenge in the design of a generalist teaching sequence towards the domain the Theory of Evolution could be the wide range of individual learning pathways identified among 13 years-old students by Zabel & Gropengiesser (2011). Evolutionary Biology drinks from and nourishes different disciplines, which must be reflected in their teaching-learning processes. In fact, Evolutionary Biology itself can be understood as a compendium of interconnected disciplines and subdisciplines that explain the natural processes of Biological Evolution through various scientific models (Álvarez-Pérez et al., 2016). Moreover, Soler (2002) presented a modified scheme of Futuyma et al. (1998) showing a structure of Evolutionary Biology and its relations with other fields of knowledge. Considering all these ideas together, this chapter will revolve around the cross-curricular nature of Evolutionary Biology and the teaching approaches to deal with this feature focusing on the interplay between mathematics, technology and natural science.

BACKGROUND

Biological Evolution explains the history of life on Earth and, as such, it is a discipline that has aroused the interest of society since long ago, touching on unappealable questions about our own human nature:

- *Where do we come from?* or
- *How did we get here?*

This genuine curiosity about the evolution of species has been reported since the first approaches to understand the diversity of life on Earth made by philosophers such as Empedocles (484-424 BC) in 400 B.C. asserting that species could arise from natural causes guided by unify and diversify forces (Caetano-Anollés & Janko, 2022). For his part, Aristotle (384-322 BC) in his *scala naturae* established a hierarchical classification of species corseted according to a divine order and defending a fixist idea of biodiversity (Aristotle, 343 B. C./1910). This fixism was accepted and argued by successive thinkers such as the Stoics and the same thesis was recovered and integrated during the Middle Ages in the prevailing doctrine in the West under creationism views (Solinas, 2015). On the other hand, in the first millennium after Christ the Arab zoologist Al-Jahiz (776 -868) in The Book of Animals raises an evolutionary idea of species based on the conditions of their environment (Malik et al., 2017). However, it was not until the 18th-19th centuries with the appearance of the Linnaean taxonomy (Linnaeus, 1758), as well as advances in Geology and the interpretation of fossils, with figures such as Cuvier (1769-1832), Hutton (1726-1797) or Lyell (1797-1875), when ideas about the evolution of life began to materialize in scientific knowledge of a certain scope (Moore, 1984). Cuvier's catastrophism with his idea of immutable species that appeared and disappeared due to natural catastrophes (the last of them the Universal Flood) was confronted with Lyell's uniformitarianism, according to which all the processes that have taken place historically on Earth could be explained by the same causes acting at the present moment (King, 1877). This last idea implies a gradualism and balance in the processes of erosion and sedimentation, which could also apply to the extinction of species and the appearance of new ones according to the fossil record.

In this historical context, Lamarck (1744-1829) developed a hypothesis about Biological Evolution in which he formulated that species would be transformed as a result of adaptation to the environment through a slow and gradual process (Lamarck, 1809). According to Lamarck, the organs are strengthened with use and atrophy/eliminated with disuse, which is transmitted by inheritance through generations. This idea necessarily implies the inheritance of acquired characters and the function as force of organ creation, but no experimental evidence has been found to support these assumptions, so lamarckian hypothesis has been discarded as an evolutionary mechanism (Koonin & Wolf, 2009), although it continues to wander in the minds of a large part of society (González-Galli & Meinardi, 2015, 2017). Charles Darwin (1809-1882) and Alfred Russel Wallace (1823-1913) formulated in parallel the hypothesis about one of the mechanisms by which species evolve and that explains the history of life on Earth. The expeditions on the HMS Beagle of the former and the studies on Indonesian biodiversity of the latter, together with Lyell's Principles of Geology, Tomas Malthus' (1766-1834) theses on the dynamics of populations in terms of resources availability, and the evolution of species proposed by Lamarck, was what composed the breeding ground for the development of the Theory of Evolution by Natural Selection that was initially proposed by Darwin and Wallace (1858) and further developed by Darwin in his work "On the Origin of Species by Means of Natural Selection" (Darwin, 1859). The structure of the Theory of Evolution by Natural Selection is configured around three pillars: descent with modification, inheritance of characters from generation to generation, and differential survival and reproduction among individuals in a population (Lewontin, 1970). In this sense, in a population that presents variability, those individuals with higher fitness (greater capacity to survive and reproduce) will pass on their characteristics to future generations, which will lead, over time, to the predominance of these features in the population.

This theory proposed by Darwin and Wallace, also called darwinism, was complemented with the continuous advances that took place during the 20th century, in disciplines such as Molecular Genetics, Geology or Systematics since the rediscovery of Mendel's laws (1822-1884) in 1900. These findings gave rise to the known as the Synthetic Theory of Evolution or Modern Evolutionary Synthesis; sometimes called neo-darwinism; although this term was coined earlier and does not cover many aspects of the Synthesis (Mayr, 1984); by figures such as Julian Huxley (1887-1975), Theodosius Dobzhansky (1900-1975), Ernst Mayr (1904-2005), George G. Simpson (1902-1984), Bernhard Rensch (1900-1990), and George Ledyard Stebbins (1906-2000) (Kutschera, 2011). Taking into account the curricular content of the Secondary Education in Spain, the Modern Synthesis is the main content to be treated in this educational level (BOE, 2014). However, it is observed that special reference is made in the curriculum to the historical development about the Biological Evolution knowledge, something that does not happen with other contents.

Science is a body of knowledge based on an inquiry nature, constantly exposed to criticism and reformulation of ideas to propose the best explanatory models for natural phenomena at a certain historical moment. For this reason the Extended Evolutionary Synthesis has recently been suggested with the purpose of including new concepts, patterns and evolutionary mechanisms, thus recognizing that they are more enriched than the pioneers of Evolutionary Biology understood them (Pigliucci & Müller, 2010). These new findings, coming from the hand of progress in scientific knowledge, are totally complementary to the Theory of Evolution by Natural Selection and Modern Synthesis. González-Galli & Meinardi (2013) compiled some of these new models as follows:

- **Punctuated equilibrium:** Darwinism calls for an evolutionary gradualism guiding the change processes of natural species that was initially retained by the Modern Synthesis (Koonin, 2009). However, Niles Eldredge and Stephen Gould (1941-2002) proposed the theory of interrupted equilibria (or punctuated) to explain the evident discontinuities in the fossil record that opposes to the hypothesis that explains this scarcity due to the difficulty in finding fossils and the low probability of finding transitional forms towards new species due to their low abundance in nature

(Eldredge & Gould, 1972). These authors argued that changes to new species originate through a relatively rapid process of phenotypic change during the time of speciation (punctuation) and, once formed, are virtually unchanged (stasis). Therefore, the speed of evolutionary change would not be uniform. Far from what many authors controversially mention, equating punctuality with saltationism (Cano-Espinosa, 2015; Amorós-Hernández, 2018), this theory could not replace evolution by natural selection since it indicates that these stages follow one another relatively quickly during speciation but does not add or confront the same causes of adaptation proposed by darwinism (Hancock et al., 2021). In addition, the experimental evidence supporting punctuated equilibrium is scarce although some authors have recently discussed their findings in the light of punctuated equilibrium based on statistical models to empirically explore it (Pennell et al., 2014).

- **Serial endosymbiosis:** This theory formulated by Lynn Margulis (1938-2011) establishes the origin of the eukaryotic cell in a mutualistic symbiotic relationship with another prokaryotic cell that would have incorporated the first as organelles, in which different cellular functions would be carried out (Margulis, 1981). As a result of this idea, some authors consider endosymbiosis as an evolutionary mechanism in itself and different from natural selection if the latter is understood exclusively in terms of competition (O'Malley, 2015). However, this would not necessarily be the case since the endosymbiotic process would be articulated as a way of generating variability that would later be selected (or not), which would be compatible with the Theory of Evolution by Natural Selection (Blackstone, 2016). The same would apply to horizontal gene transfer, a process closely linked to endosymbiosis (Debortoly et al., 2016).
- **Neutralist Theory of Evolution:** The biologist and mathematician Mooto Kimura (1924-1994), after his studies using mathematical and statistical models on the frequency of mutations, found that most of them do not have beneficial or deleterious effects on carrier individuals, instead they are neutral and therefore not likely to be selected, thus granting a high randomness to the evolutionary processes and a key role to the genetic drift (Kimura, 1968). However, this theory would be framed within molecular evolution and it would not conflict with the Theory of Evolution by Natural Selection that concerns the adaptation to the environment.
- **Evolutionary Developmental Biology (Eco-Devo):** This discipline, arising from the discovery of HOX genes (Bateson, 1894), addresses the connection between evolution and the development process. Hall (2012) explains some of the goals of Eco-Devo including questions focus on exploring why some phenotypes, especially in terms of morphology, are very frequent and others never appear in nature, how individual development plans constrain and bias evolution of the appearance of evolutionary innovations that imply radical changes in development patterns and the birth of new groups of organisms. In this sense, it was found that small changes in developmental gene networks (homologous genes, evolutionarily conserved) cause large morphological changes, which would not be in line with the prevailing gradual thinking of darwinism (Hagolani et al., 2021). However, these views would not be confronted since natural selection would be the process driving the frequency changes of those traits in the populations (Brakefield, 2011).
- **Epigenetic inheritance:** This subject deals with the study of phenotypic variations that are not based on variations in the DNA sequence itself, but rather on chemical modifications of it, often unstable and reversible, which are transmitted to future generations (Ashe et al., 2021). This premise has often been misunderstood to vindicate lamarckian ideas about evolution (Loison, 2018). However, it could not be understood as an alternative to natural selection since it is indifferent to the physical substrate of inheritance (nucleotide sequence or changes in methylation patterns),

the only thing that the definition requires is that there is heritable variation and associated differences in reproduction potential (Haig, 2007). In addition, the sites that can be chemically modified to alter DNA expression (gene regulation) and the effects of these changes can also be under natural selection (Ashe et al., 2021). When Darwin published his work "On the Origin of Species by Means of Natural Selection" in 1859, the structure of DNA and its significance as a molecule carrying hereditary information had not yet been described. The DNA molecule was discovered in 1869 by Friedrich Miescher (1844-1895) and its structure was not published until 1953 by James Watson (1928) and Francis Crick (1916-2004) thanks to the discoveries of Rosalin Franklin (1920-1958). As this branch of science advances, it will be possible to understand how to configure and string together these new ideas about the evolution of life on Earth.

This historical landscape regarding Evolutionary Biology shows the birth and development of the discipline from the findings and reflections of professionals from different branches of knowledge: geologists, paleontologists, biologists, mathematicians, computer scientists, historians or sociologists. Therefore, Evolutionary Biology cannot be understood far from an interdisciplinary perspective. In fact, some authors have argued recently that this could be the way to facilitate the understanding of the evolution of living beings (Sá-Pinto et al., 2021) and that too focused teaching-learning approaches in genetics and exclusively related to the modern synthesis could hinder a holistic and integrative understanding of Biological Evolution (Hanisch & Eirdosh, 2020).

A literature review regarding mental models on Biological Evolution shows the existence of preponderant alternative ideas on the subject among students and teachers as those below mentioned. Knowing these alternative ideas is a requirement from the didactic perspective to design effective educational proposals. Embracing educational constructivism, it is necessary for students to be aware of and overcome alternative conceptions in order to rework knowledge about a subject in a solid way (Ausubel, 2000). In the field of didactics about Biological Evolution, many authors have described alternative conceptions around some concepts reflecting resistant and intuitive thought models that, frequently, represent historical ideas that have been surpassed in Evolutionary Biology (Ha & Nehm, 2014).

In this way, plenty of authors have described lamarckian conceptions regarding the evolution of living beings such as the inheritance of acquired characters, the use/disuse of organs or biological change guided by necessity (Gallego & Muñoz, 2015; González-Galli & Meinardi, 2015, 2017). In turn, also a lack of knowledge about the variability between individuals of the same species has been reported (Andersson & Wallin, 2006; Prinou et al., 2008) and the confusion between individual, population or species, observing the individual rather than the population as the unit of selection on evolutionary processes (Bardapurkar, 2008; Evans, 2010). On the other hand, there is a distorted idea of geological time, due, among other causes, to the abstraction that this concept requires (James & Clark, 2006; Decker et al., 2007). The history of life on Earth is often understood as a short period of time in which biological changes and the appearance of new species occur rapidly (Libarkin et al., 2007). Related to a lamarckian perspective, Biological Evolution is also often understood as a process with a determined purpose (Kampourakis & Zogza, 2008; Gallego & Muñoz, 2015) and guided towards a goal (teleological thought), even mutations would be directed, which leads to think that the human being is the summit of the evolutionary process under a completely anthropocentric prism (Richard et al., 2017; Arenson & Coley, 2018). In addition, considering the polysemy of the everyday context in which the concept of evolution moves, Biological Evolution can be understood as a belief or a hypothesis instead of as the scientifically proven fact that it is (Kampourakis, 2013; Chaves-Mejía, 2016). This idea is reinforced by socio-religious arguments (De Souza et al., 2010; Glaze & Goldston, 2015) as well as by the lack of knowledge about the scientific method leading to the confusion between terms such as proof, fact, hypothesis or theory that are often misunderstood in the scientific field and contaminated by their use and meaning in different contexts (Van Dijk & Reydon, 2010; Acosta-Pérez, 2014).

These erroneous alternative ideas about Biological Evolution have been reported at all educational levels:

- Creationism and lamarckian conceptions (Cañal, 2009; Ceballos et al., 2017) as well as a lack of knowledge about human evolution (Ceballos et al., 2017) are common among students from Early Childhood and Primary Education. For a review of early concepts on evolution in pre-adolescents see Bruckermann et al. (2021).
- In the case of Secondary Education and Baccalaureate students, all the alternative ideas indicated above have been described (Acosta-Pérez, 2014; Yates & Marek, 2015; González-Galli & Meinardi, 2015).
- At university levels, lamarckian explanations about evolution persist as well as a sense of purpose and perfectionism in evolutionary processes (Cunningham & Wescott, 2009; Gregory & Ellis, 2009; Pazza et al., 2010; Pinxten et al., 2020; Gefaell et al., 2020).
- Among teachers, misconceptions about the nature of science, teleological ideas or the need as the reason for the appearance of adaptations have been reported (Primary Education: Prinou et al., 2011; Pre-service teachers: Sánchez et al., 2017; Torkar & Šorgo, 2020; Karataş, 2020; Secondary and High School Education: Rutledge & Warden, 1999; Tidon & Lewontin, 2004; Nehm & Schonfeld, 2007; Buchan, 2019).
- Among the cultured population not specialized in science, lamarckian responses to evolutionary questions were found regardless the highest educational level reached (Fernández & San José, 2007).

Recently, Kuschmierz et al. (2020) have compiled a large number of studies on the acceptance and knowledge about the Theory of Evolution in which some of the cases mentioned here are reported as well as comparisons between results obtained through different methodological approaches and at a cross-country level. The authors show that, despite being largely accepted among European students, Biological Evolution is still poorly understood in this population so it would be advisable to invest focused educational efforts to face this issue broadly reported in the scientific literature.

INTERDISCIPLINARY APPROACHES TO ADDRESS BIOLOGICAL EVOLUTION

Analysis of the Spanish Curriculum for Secondary Education

The curriculum framing this study is regulated by the Royal Decree 1105/2014 of December 26 (BOE 1105/2014) which establishes and develops the management and curriculum of Compulsory Secondary Education (ESO) and Baccalaureate (Bacc) in Spain. The elements of the Spanish curriculum (blocks of contents, topics addressed and evaluation criteria) that are indicated from now on have been translated into English. This educational research and the sequence of activities proposed in this chapter is aimed at the subject of Biology and Geology in the 4th year of ESO where the contents included within the Block

1 - "The evolution of life" (Annex I, pg. 205, BOE 1105/2014) are programmed, see Table 1. However, throughout the chapter, references will be made to content related to other blocks and subjects with the intention of treating the Theory of Evolution as a cross-curricular content, understood as an integrating and holistic discipline. Furthermore, for an thorough tracking of the explicit presence of Evolutionary Biology across the Spanish curricula at all the compulsory educational levels see also Bugallo-Rodríguez & Vázquez-Ben (2020).

In accordance with the Royal Decree 1105/2014 which establishes the basic curriculum of Compulsory Secondary Education and Baccalaureate, the content related to the Theory of Evolution is explained for the first time in the first block of the 4th year of ESO called "The evolution of life" (Annex I, pg. 205, BOE 1105/2014). However, only 4 out of the 19 evaluation criteria associated with this block of contents (numbers 1 and 8 could also be included, see Table 1) would have a direct and explicit relationship with the Theory of Evolution, in fact, the word "evolution" appears exclusively in two of them. At this point, one might think of *What happens to the remaining criteria included in the block? Do they have something to do with The evolution of life?* Indeed, they are strongly connected to this discipline, since they are directly linked to cell division, heredity and Genetics. This could be interpreted as the lack of a tangential approach in the curriculum through evaluation criteria considering the relationship between these topics with Evolutionary Biology. For instance, as shown in Table 1, the evaluation criteria number 4., "Formulate the main processes that take place in mitosis and meiosis and review their meaning and biological importance", makes direct reference to variation in natural populations while the evaluation criteria 5, 6 and 7 (concerning nucleic acids, genetic information and the genetic code) can be directly linked to homology and common ancestry. In addition, the evaluation criteria 11, related to hereditary diseases, could address non-adaptative variations while other criteria such as numbers 12, 13, 14 and 15 could be the way to delve into artificial selection.

As indicated above, taking into account the current education law, it is in the 4th year of ESO when the discipline of Evolutionary Biology is addressed. However, in the Secondary Education and Baccalaureate curriculum there are mentions of Biological Evolution in the subject of "Biology and Geology" programmed for other grades but also in other subject and disciplines covered over all the Secondary Education and Baccalaureate courses. Thus, content related to Biological Evolution can be found in the core subject (compulsory or optional depending on the course) of Biology and Geology of all Secondary Education years (except in 2nd course of ESO in which this subject does not exist) and in the two courses of the Baccalaureate in Sciences and Technology (during the first one within the subject of Biology and Geology and for the second one in the subject of Biology). In particular, apart from the thematic block dedicated to "The evolution of life" in 4th ESO, additional curricular topics that could be directly addressed from an evolutionary perspective appears in the following thematic blocks (Annex I, pg. 204-222, BOE 1105/2014):

- 1st of ESO: "Biodiversity on the planet".
- 3rd of ESO: "People and health. Health promotion" and "Earth relief and its evolution".
- 4th of ESO: "The dynamics of the Earth" and "Ecology and environment".
- 1st of Baccalaureate: "Cellular organization", "Biodiversity", "Plants: their functions and adaptations to the environment", "Animals: their functions and adaptations to the environment» and "History of the Earth".
- 2nd of Baccalaureate: "The living cell: Morphology, structure, cell physiology" and "Genetics and Evolution".

Furthermore, there are other subjects in the Spanish curriculum, apart from Biology and Geology, that include content directly connected to Evolutionary Biology. This is the case of certain contents addressed in the subject of Geography and History (1st of ESO), such as those dedicated to human evolution and the hominization process (Annex I, pg. 300, BOE 1105/2014), being the moment in which students are exposed to content about Biological Evolution for the first time during Secondary Education. In the subject Scientific Culture (4th of ESO and 1st of Bacc) there is content dedicated specifically to Biological Evolution (Annex II, pg. 465 and 467, BOE 1105/2014). Furthermore, the subject of Geology (2nd of Bacc) addresses content concerning the Geological and Biological Evolution of the Earth from the Archaic eon to the present within the block "Geological Time and Historical Geology" putting special emphasis on the interpretation of the fossil record (Annex I, pg. 313, BOE 1105/2014). In that sense, it should be noted that the first allusion to the term fossil does not appear in the ESO and Baccalaureate curriculum until the second block of contents of 4th ESO "The dynamics of the Earth" and it does so in an evaluation criterion in reference to the application of fossils to dating studies not to Biological Evolution itself ("Recognize and date geological eons, eras, and periods, using knowledge from fossil guides" in Annex I, pg. 212, BOE 1105/2014). Bearing this in mind, a question arises about how students can understand the existence of fossils as proof of evolution if the concept of fossilization has not been dealt prior to approaching the evolution of life and if the first mention of the term in educational law refers to the application of fossils for dating purposes without going through the concept of fossil itself and with no explicit mention to Evolutionary Biology.

On the other hand, there are several contents in a variety of subjects throughout Secondary Education and Baccalaureate that could present a link with Biological Evolution, being more diffuse than in the cases indicated previously, but which also deserve to be noted. The subject of Economics (4th of ESO; Annex I, pg. 242, BOE 1105/2014) addresses the topic "The scarcity, the choice and the allocation of resources. The opportunity cost." that would be aligned with the idea of the economist Thomas Malthus about resources as a limiting and regulating factor of the dynamics of human populations, one of the scaffolds on which Charles Darwin supported his Theory on the Origin of Species through Natural Selection. Special mention deserves the case of the subject of Philosophy (4th ESO, 1st and 2nd of Bacc, History of Philosophy in this last year), although apparently far from Biological Evolution, it includes an evaluation criterion related to "Writing reflection and group discussion about the possible impact of the genetic inheritance and what has been acquired on the formation of personality", and several contents around the origin of societies and culture as an adaptation of the human being. In fact, within the block of contents about Philosophy of Science, "The hypothetico-deductive method", "The philosophical implications of evolution" or "The nature-culture dialectic in the process of anthropogenesis" appear as topics with a clear fit in an Evolutionary Biology context (Annex I, pg. 254, BOE 1105/2014).

Moreover, most of the scientific subjects included in the Secondary Education and Baccalaureate curriculum, such as Mathematics (during all the years comprising the Secondary Education and Baccalaureate) or Physics and Chemistry (2nd, 3rd and 4th of ESO as well as both Baccalaureate years), provide crucial instruments and skills to understand the scientific knowledge in which, of course, the Theory of Evolution is framed. Thus, in these subjects there are contents such as "Scientific activity. The scientific method: its stages", "design of simulations and the elaboration of predictions on diverse mathematical situations" (key to understand the effect of natural selection, migration, mutations or genetic drift in the natural populations of species) or "the preparation of reports and documents on the processes carried out and the results and conclusions obtained" (Annex I, pg. 258 and 391, BOE 1105/2014). On the other hand, the subject of Language and Literature deals with the communicative skills necessary to understand any type of text and to express different ideas concretely and effectively, including information and argumentation in science (Annex I, pg. 357 to 359, BOE 1105/2014). These skills are strictly necessary to facilitate the assimilation and manage the multiplicity of concepts in Evolutionary Biology as well as to build arguments including the proofs that support the Biological Evolution of the species (Asterhan & Schwarz, 2007).

Finally, a wide range of subjects assigned to both the scientific and the humanities areas of knowledge present a syllabus in which certain aspects of human evolution and its closest aspects to biological, social and cultural anthropology are cited. As an example, in Technology (2nd, 3rd and 4th of ESO) there is "Understanding technological evolution throughout history" as content (Annex I, pg. 453, BOE 1105/2014), the subjects Initiation to Entrepreneurial and Business Activity (1st, 2nd and 3rd of ESO) aims to develop competence about "Take risks, be innovative, have persuasion, negotiation and strategic thinking skills" (Annex II, pg. 505, BOE 1105/2014), even in Performing Arts and Dance (4th year of ESO) and Fundamentals of Art (1st and 2nd year of Bacc) content is included on "Critically expose personal opinion regarding the role of performing arts and its importance as a means of expressing feelings, emotions, ideas and sensations" (Annex II, pg. 461, BOE 1105/2014) or "Cave art: painting and sculpture" (Annex I, pg. 280, BOE 1105/2014) respectively. Also, in History of Spain (2nd of Bacc) appears "Prehistory: the evolution from the Paleolithic to the Neolithic" (Annex I, pg. 322, BOE 1105/2014). These topics are suitable to approach the importance of technology for human Biological Evolution (Bejan, 2020), the relevant role of dance as a driver of sexual selection (Shogren et al., 2022) or the evolutionary origin of the Arts (Dissanayake, 2017).

Moreover, in the Ethical Values subject (1st and 2nd of ESO) there is one evaluation criteria that has a lot to say about some problems that the knowledge and acceptance of the Theory of Evolution has faced throughout history that are strongly linked to the understanding of the Nature of Science (Athanasiou et al., 2012; Cofré et al., 2018). This evaluation criteria states "Recognize that, at present, there are cases in which scientific research is not neutral, but is determined by political, economic, etc. interests through the analysis of the idea of progress and its wrong interpretation, when the objectives that are sought do not respect an ethical code based on the Universal Declaration of Human Rights" (Annex II, pg. 541, BOE 1105/2014).

Therefore, the current educational law itself mirrors the cross-curricular and interdisciplinary nature of Evolutionary Biology. That is why a real integrated and coordinated teaching approach for Evolutionary Biology should be considered to prevent students from confusion and the misunderstanding of some concepts depending on the way they were addressed in previous courses and subjects. In fact, students face the content on the Theory of Evolution relatively late in Secondary Education (4th year within the Biology and Geology subject) and this could pose an added challenge to its holistic understanding specially taking into account that in that year Biology and Geology is an optional subject.

Table 1. Contents, evaluation criteria and evaluable learning standards of BOE 1105/2014 for Block 1 of the Biology and Geology subject of the 4th year of ESO. CCL: linguistic communication; CMCT: mathematical competence and basic competences in science and technology; CD: digital competence; CPAA: learning to learn; CSC: social and civic competences; SIEE: sense of initiative and entrepreneurial spirit; CEC: cultural awareness and expressions. Note that the contents, evaluation criteria and learning standards highlighted in bold type directly address the Theory of Evolution. Information from BOE 1105/2014 has been translated here into English

Block 1. The Evolution of Life.	
Contents	The cell. Cellular cycle. nucleic acids. DNA and molecular genetics. DNA replication process. Gene concept. Expression of genetic information. Genetic code. **Mutations**. relations with evolution. The inheritance and transmission of characters. Introduction and development of Mendel's Laws. Chromosomal basis of Mendel's laws. Applications of Mendel's laws. Genetic Engineering: techniques and applications. Biotechnology. Bioethics. **Origin and evolution of living beings. Hypothesis about the origin of life on Earth. Theories of evolution. The fact and mechanisms of evolution. Human evolution: hominization process.**
Evaluation criteria	**1. Determine the analogies and differences in the structure of prokaryotic and eukaryotic cells, interpreting the evolutionary relationships between them. CMCT.** 2. Identify the cell nucleus and its organization according to the phases of the cell cycle through direct or indirect observation. CMCT. 3. Compare the structure of chromosomes and chromatin. CMCT. 4. Formulate the main processes that take place in mitosis and meiosis and review their meaning and biological importance. CMCT. 5. Compare the types and composition of nucleic acids, relating them to their function. CMCT. 6. Relate DNA replication with the conservation of genetic information. CMCT. 7. Understand how genetic information is expressed, using the genetic code. CMCT. **8. Assess the role of mutations in genetic diversity, understanding the relationship between mutation and evolution. CMCT.** 9. Formulate the basic principles of Mendelian Genetics, applying the laws of inheritance in solving simple problems. CMCT. 10. Differentiate the inheritance of sex and that linked to sex, establishing the relationship between them. CMCT. 11. Know some hereditary diseases, their prevention and social scope. CMCT, CSC, CYEC. 12. Identify the techniques of Genetic Engineering: recombinant DNA and PCR. CMCT. 13. Understand the cloning process. CMCT. 14. Recognize the applications of Genetic Engineering: GMOs (genetically modified organisms). CMCT. 15. Assess the applications of recombinant DNA technology in agriculture, livestock, the environment and health. CMCT, CSC, CYEC. **16. Know the evidence of evolution. Compare Lamarckism, Darwinism, and Neo-Darwinism. CMCT.** **17. Understand the mechanisms of evolution, highlighting the importance of mutation and selection. Analyze the debate between gradualism, saltationism and neutralism. CMCT, CAA.** **18. Interpret phylogenetic trees, including the human. CMCT, CAA.** **19. Describe hominization. CCL, CMCT.**

Continued on following page

Table 1. Continued

Block 1. The Evolution of Life.	
Evaluable learning standards	1.1. Compares the prokaryotic and eukaryotic cell, the animal and the plant, recognizing the function of the cellular organelles and the relationship between morphology and function. 2.1. Distinguishes the different components of the nucleus and their function according to the different stages of the cell cycle. 3.1. Recognize the parts of a chromosome using it to build a karyotype. 4.1. Recognizes the phases of mitosis and meiosis, differentiating both processes and distinguishing their biological meaning. 5.1. Distinguish between different nucleic acids and list their components. 6.1. Recognizes the role of DNA as a carrier of genetic information, relating it to the concept of gene. 7.1. Illustrates the mechanisms of gene expression through the genetic code. 8.1. Recognizes and explains what mutations and their types consist of. 9.1. Recognizes the basic principles of Mendelian Genetics, solving practical problems of crosses with one or two characters. 10.1. Solve practical problems on the inheritance of sex and sex-linked inheritance. 11.1. Identifies the most frequent hereditary diseases and their social scope. 12.1. Difference working techniques in genetic engineering. 13.1. Describes animal cloning techniques, distinguishing therapeutic and reproductive cloning. 14.1. Analyzes the ethical, social and environmental implications of Genetic Engineering. 15.1. Critically interprets the consequences of current advances in the field of biotechnology. **16.1. Distinguishes the differentiating characteristics between Lamarckism, Darwinism and Neo-Darwinism** **17.1. Establishes the relationship between genetic variability, adaptation and natural selection.** **18.1. Interpret phylogenetic trees.** **19.1. Recognize and describe the phases of hominization.**

Teaching-learning Difficulties Concerning Evolutionary Biology

The alternative ideas that have been exposed about the Theory of Biological Evolution are the product of certain difficulties or obstacles, both conceptual and socio-emotional, that the content about Biological Evolution presents in terms of its understanding and acceptance. Some of these difficulties are associated with the intrinsic inclination of people to understand the construction of their knowledge based on a causal understanding of natural phenomena that would be directed towards a purpose, which has been called "common sense of teleology" and "linear causal reasoning" (Kelemen & Rosset, 2009; González-Galli & Meinardi, 2015). This explains why alternative ideas are resistant to change and persist even after some educational interventions (Maskiewicz & Lineback, 2013). Evolutionary Biology is particularly sensitive to this learning process that must be attended in order to avoid reinforcing the idea of evolutionary change always directed by a cause towards a goal. Thus, it is common to hear among students (and teachers) expressions such as "birds developed wings to fly" that may strength students' teleological misconceptions. In addition, learning by "individual-centered reasoning" (González-Galli & Meinardi, 2015) can also be observed, which favors the acquisition of ideas based on the importance of the individual in evolutionary processes, forgetting the population as a unit of selection, and ultimately led to an anthropocentric view of biological evolution. Another obstacle on which several alternative ideas about Biological Evolution are articulated is the existentialism by which it is understood that different categories are equal and the limits between them are strictly defined (Gelman & Rhodes, 2012).

In addition, there are teaching-learning challenges associated with the subject itself, the understanding of which would require significant intellectual maturity as well as a good capacity for abstraction (Lawson & Thomson, 1988). The very cross-curricular nature of Evolutionary Biology could support, as a consequence, one of the most important epistemological obstacles for its understanding. That is the management of multiple concepts with a great variety of different meanings in other academic and colloquial contexts (polysemic concepts) that lead to imprecision when teaching them and, therefore, confusion in its learning (Van Dijk & Reydon, 2010; Acosta-Pérez, 2014). In this sense, Chaves-Mejía (2016) proposes a review of these terms through a conceptual profile for the continuous training of teachers. In that study, there are some examples about how the concept of evolution has several context-dependent meanings: progress of a situation (everyday use, socioeconomic, social and human sciences), change or gradual transformation of something (every day and academic use) or changes towards a defined state (non-biological science such as Astronomy, Physics or Chemistry). Kampourakis (2013) reviews the concept of adaptation in the same way. In this case, the use of the term "adaptation" in colloquial language, in the media or in other sciences can lead to misunderstanding their meaning under the evolutionary umbrella. Furthermore, as several authors reported the lack of a historical perspective on the Theory of Evolution as well as on the very Nature of Science (Athanasiou et al., 2012; Cofré et al., 2018), assuming uncertainty and overcoming concepts based on scientific evidence, has been a challenge to understand and an opportunity to mistrust Biological Evolution ("evolution is just a theory", note here the polysemy of the therm "theory" within every day and scientific contexts).

Many authors point to the lack of teachers' training in Biological Evolution as a teaching-learning obstacle for this discipline (Berkman et al., 2008; Sanders & Ngxola, 2009; Smith, 2010; Sickel & Friedrichsen, 2013; Hawley & Sinatra, 2019) as well as the contextual fragmentation in the didactics of the subject in which initial training programs, pedagogical guides or reliable teaching materials/ resources available to teachers are scarce and heterogeneous (Friedrichsen et al., 2016). In fact, the usual sources of knowledge of teachers about Evolutionary Biology are TV documentaries and general culture (Ceballos et al., 2017). In spite of their power to massively reach people, these channels appear to have been ineffective to promote science education among society in a rigorous manner and based on research evidence considering the low scientific literacy level reported in Spain by Mellizo-Soto & Romero (2015). After its apparent simplicity, the complexity of the evolutionary concepts can lead to confusion and even cause stress among the teachers (Griffith & Brem, 2004), which percolates in the students and hinders the learning of the subject (Castro, 2007; Araujo & Cantó, 2015).

Finally, some religious beliefs have also shown to be an obstacle to the acceptance of the Theory of Biological Evolution. This is the case of those who support a creationist perspective of life on Earth as it currently exists, unsustainable from the evolutionary perspective. In fact, in several studies a negative relationship between religious faith and the acceptance of evolution is observed both among students (Eder et al., 2011; Betti et al., 2020) and teachers (Athanasiou et al., 2016; Deniz & Sahin, 2016). However, in the United States (McCain & Kampourakis, 2018) and in Europe (Beniermann, 2019), religious faith alone was not a predictor of rejection of evolution, and a large percentage of religious believers participating in those studies accepted evolution.

Studies on acceptance and understanding of the Theory of Evolution have not yielded conclusive results on the relationship between both concepts as it is reviewed by Kuschmierz et al., 2020. In most of them, a significant correlation between them has not been found or it has been shown to be weak. However, these authors also observed that there are cases in which a high acceptance of the Theory of Evolution is detected along with a low understanding of it, which suggests that a deep knowledge of Biological Evolution is not a requirement for its acceptance. In any case, a review of the concept "acceptance of Evolution" by Peñaloza (2016) indicates that understanding this concept is key to teach and learn Biological Evolution.

This diversity of factors from wide range origins, which often appear combined, hinders the learning of Evolutionary Biology and the concepts of the evolutionary field, which are designated as "problematic nuclei or nodes" for the teaching-learning processes of the Theory of Biological Evolution (Hernández et al., 2009).

Examples of Teaching Approaches Found in the Literature to Address Evolutionary Biology in the Classroom

A large part of the didactic proposals for addressing Evolutionary Biology found in the scientific literature are intended to overcome didactic difficulties through metacognition about the obstacles associated with that content, which have previously been commented (Mariano-Pérez et al., 2018; Bernard-Cobo, 2019). In any case, the textbook continues to be the primary teaching resource for teachers as they perceive it as a reliable source of information for students while reducing the effort invested in teaching planning (Jiménez-Aleixandre, 1994; Abimbola & Baba, 1996; Ottevanger, 2001). These potential strengths of textbooks are especially notable in the case of teachers who show little training in the subject or take the textbook as a curricular update for them (Verspoor & Wu, 1990; Ottevanger, 2001). However, several studies about the convenience of the use of textbooks to teach about Biological Evolution have reported shortcomings associated with the language and evolutionary terms used, the contents, the activities or the diagrams included (Rosenthal, 1985; Jiménez-Aleixandre, 1994; Stern, 2004; Decker et al., 2007; Tshuma & Sanders, 2015; Sanders & Makotsa, 2016). As a particular case, even though there is a good representation of cladograms in textbooks, their correct interpretation is not explained and they could reinforce alternative ideas in Evolutionary Biology (Catley & Novick, 2008), although a slight improvement is seen in the textbooks used currently in the United States compared to previous years (Schramm et al., 2021). Furthermore, the use of non-cladogenic diagrams in textbooks also leads to confusion about evolutionary concepts (Catley et al., 2010). On the other hand, gaps in coverage of human evolution have been reported in a wide sample of textbooks (United States: Skoog, 2005; several European and African countries: Quessada et al., 2008; Muslim countries: Asghar et al., 2014).

These shortcomings described for textbooks regarding Biological Evolution find their greatest problem if teachers hold misconceptions about the subject (Prinou et al., 2008; Molefe, 2013; Yates & Marek, 2015) and cannot fulfill their role as detectors and filters of information or errors associated with textbooks (Abimbola & Baba, 1996) so these are inevitably transferred to students.

Fortunately, the development and growing interest about Science Education has given rise to the approach of several alternative or complementary didactic proposals to the use of the textbook as the main teaching resource for Evolutionary Biology. As detailed in Appendix I - Table 3, there are many teaching approaches that value the historical vision of the Theory of Biological Evolution and the knowledge of the nature of science itself as an effective teaching-learning proposal in Evolutionary Biology (Torreblanca et al., 2011; Nelson et al., 2019). However, the most widely used teaching approach for this content has been, without a doubt, the use of simulations, which have yielded good results even in elementary grades (Sá-Pinto et al., 2021) but some mathematical and computational notions are usually required for the experience to be completely fruitful (Appleton & Mackie, 2019; McGuire et al., 2022). Recently, there has been an increase in the number of works that bet on inquiry and project-based learning to address Biological Evolution, which promotes contextualized learning based on real practical cases (Desantis, 2009; Cooper et al., 2019). Laboratory practices specifically designed to address evolutionary concepts are also included in this kind of teaching proposals. Other approaches are based on the use and interpretation of phylogenetic trees as a tool for understanding Biological Evolution (Rosa & Tricarico, 2016; Kong et al., 2017). In addition, argumentation, gamification, contextualized narration, the flipped classroom, case studies or the use of analogies have been key teaching strategies in other educational interventions (Gómez-Moreno, 2020; Sanabria-Totaitive & Arango-Martínez, 2021; see Table 3 from the Appendix I for more references).

Although some scientific papers, PhD, Master's or Bachelor's thesis present an introduction of concepts and contents concerning the Extended Evolutionary Synthesis, it is still being questioned whether it is appropriate to incorporate them within a teaching frame about Biological Evolution (Folguera & González-Galli, 2012; Franco-Goyena, 2019). In this sense, Zudaire & Napal-Fraile (2020) affirm that it is possible to successfully introduce knowledge about Epigenetics in the secondary school curriculum, but special attention must be paid to the key concepts in the initial courses of this educational stage to avoid confusion. The teaching of Evo-Devo has also been studied in this regard and has been proposed as an opportunity to address the teaching of Biological Evolution from an interdisciplinary perspective (Chaves-Mejía, 2010; Love, 2013). In the same way, the debate between the role of serial endosymbiosis and natural selection is proposed as a way to face a traditionally biased teaching of Evolutionary Biology (Chaves-Mejía, 2012).

All this wide range of teaching proposals for Evolutionary Biology gives a good account of the heterogeneity of the discipline, which is one of the reasons why it is complex to extrapolate results between different studies as well as to assess the real scope of the different teaching-learning approaches applied. In addition, most of these approaches are articulated considering students as their main target, but they should not forget the low knowledge about Biological Evolution and about the pedagogical content knowledge that has been reported among teachers, which is a crucial aspect to improve the teaching-learning processes for any discipline.

Finally, there are not too many proposals aimed at approaching the teaching of Evolutionary Biology from a holistic and integrating perspective within the Biology curriculum and other subjects, despite the suitability of this kind didactic approach for these particular contents (Hanisch & Eirdosh, 2020). Some teaching proposals in this sense are limited to the use of Evolution and the history of life as an introductory framework for the Biology subject (Alles, 2001; Burke da Silva, 2012), to the presentation of conceptual maps (Apodaca et al., 2019), the link between Biological Evolution and Medicine or Microbiology (Cooper et al., 2019) but also to explore transdisciplinarity between Biological Evolution and Mathematics (Sá-Pinto et al., 2021) (see Table 3 from Appendix 1 for more references).

Perceptions From Spanish Teachers about the Cross-curricular Nature of Evolutionary Biology

Following the interview structure that can be found at https://doi.org/10.5281/zenodo.5649365 (Martín-Peciña & Rams, 2021) as a research instrument, 6 pre-service (PT) and 3 in-service teachers (IT) were interviewed about the four thematic blocks: teaching methodology (9 questions), didactic resources (10 and 9 questions for IT and PT respectively), teaching-learning difficulties (11 and 12 questions for IT and PT respectively) and cross-curricular nature of Biological Evolution (8 questions). The overall results from these interviews can be found at Martín-Peciña & Rams (2022) while the present chapter focuses on the perception of interviewed teachers about the cross-curricular nature of Evolutionary Biology and its integrative teaching approach in the classroom.

Participating pre-service teachers were students of the Master Degree for Teachers of Secondary Education (specialized in Biology and Geology) from the University of Granada during the academic course 2020/2021 when the interviews were performed while in-service ones were teaching Biology and Geology in the Secondary Education level at public schools located in Granada (Spain). Google Meet was used to carry out interviews that were recorded and later translated for their analysis. Participants gave their signed consents to voluntarily participate in this study.

In this regard, in-service teachers consider Biological Evolution as a unifying content, they agree that it should be treated as such in the classroom and they try to do so. However, they indicate few concrete proposals to treat Evolutionary Biology in an integrating way. Accordingly, pre-service teachers also maintain that Evolutionary Biology is a cross-curricular matter but they do not believe that this integrative nature is being considered by in-service teachers when addressing these contents in the classroom while they would do that in their future practice:

"I think I would do it, I would prepare the classes to be able to connect contents, it requires effort" (PT1)

"Students are asked a lot to know how to relate content when it is not done, the teacher himself is not doing it" (PT2)

"Topic pass, topic forgotten" (PT6)

In addition, in-service teachers find a lot of affinity between Evolutionary Biology and the subjects of Scientific Culture, Information and Communications Technologies, Geography and History and Mathematics (Figure 1, right section). These results were very similar in the case of pre-service teachers, adding the subjects such as Philosophy and Social Changes and Gender (see Figure 1, left section). However, while the total amount of in-service teachers interviewed here declared that Mathematics is a subject strongly connected with Evolutionary Biology (they valued the affinity between both disciplines as High-4), pre-service teachers did not perceive such a high degree of affinity (Figure 1). As reported in the literature, although pre-service teachers show a more positive attitude toward integration (Pang & Good, 2000), they also encounter more obstacles to design integration proposals or when trying to transfer them to the classroom, being less knowledgeable about integration that in-service ones (Lehman, 1994; West et al., 2006; Siew et al., 2015). In fact, Steflitsch (2022) reviewed some studies showing that Mathematics pre-service teachers could suffer the "practice or transition shock" after entering the teaching profession which could hide them to detect innovation opportunities. In this case, in-service teaching experience must allowed them to detect the power of Mathematics as a solving tool or a matter directly related to Biological Evolution as shown in Figure 1 which brings closer this topic to a STEM context.

"There is transdisciplinarity with other subjects such as Mathematics, speaking of the years ago when different events occurred, such as the appearance of the first hominid", "With mathematics, as I have commented before when making graphic representations" or "I have seen in some centers life-size hominid murals that are very nice." (IT2)

Figure 1. Teachers' perception of the affinity between Biological Evolution and each school subject included in the Spanish curriculum for Secondary Education based on semi-structured interview including questions answered with discrete qualitative values (none, low, medium and high affinity).

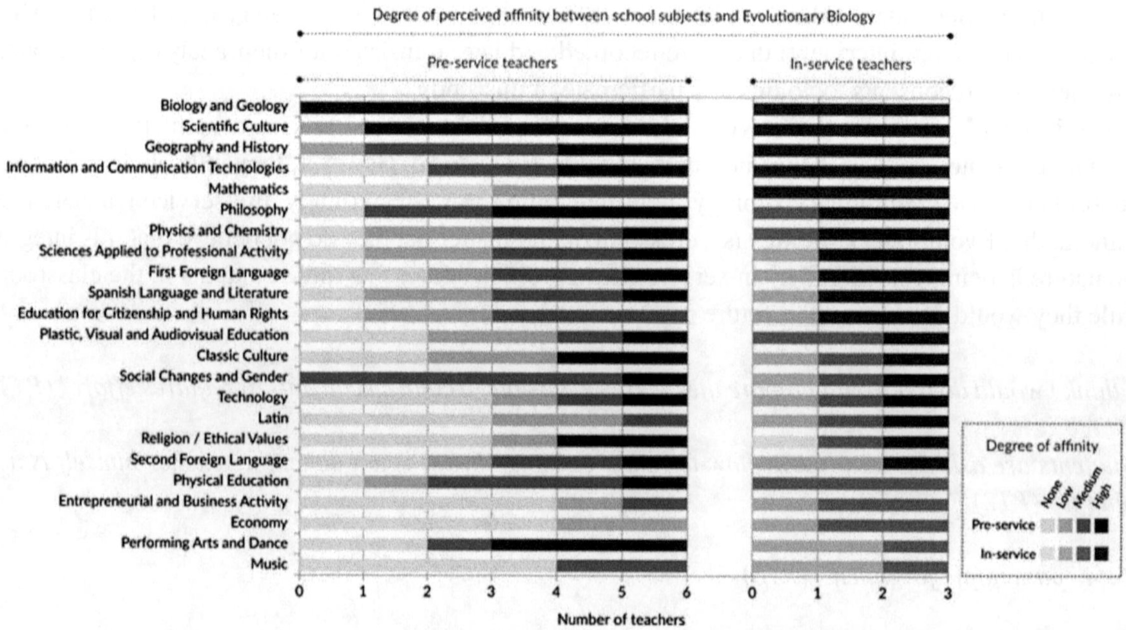

Despite this agreement of teachers about the cross-curricular nature of Evolutionary Biology, they were not able to provide specific proposals to deal with biological evolution from an integration approach with other subjects (question 4.6.). Concerning these proposals, IT1 indicates that a scheme could be made at the end of the course gathering the knowledge acquired under the evolutionary prism to offer students a broad perspective of the matter that helps its compression. IT2 indicates that a life-size mural of the different hominid species could be built to bring together human evolution and arts. Lastly, IT3 proposes to carry out a practice with shark teeth to show the reduction in size of the species over time, but the interdisciplinary here should be more evident, since it is not mentioned in what aspect of the practice it would be revealed. In relation to the question about possible didactic proposals that pre-service teachers thought were adequate to deal with Biological Evolution in a transversal way, the interviewees answered hesitantly and indicated that it is complex for them to propose such a proposal in a short time.

"You can explain Geology with Evolution...as they can complement each other...how to do it is a bit complicated...I don't know how but I see a clear relationship." (PT1)

"I don't know how to come up with an specific proposal like this, it would be to give it a lot of thought." (PT2)

"Using family trees but I don't know... it is difficult." (PT3)

"Right now I can't think of any, I think it can be related to many contents" (PT4)

"I can't think of something specific like that...from the beginning...actually since elementary grades." (PT5)

"Well, I don't know...when a teacher is giving a lesson asking the students about...but a didactic proposal, I don't know, I'm very bad at that." (PT6)

Some answers focused on linking plate tectonics and other geological content with Biological Evolution (PT1), others include dedicating part of each topic to an evolutionary reflection (PT2 and PT6), using family trees, the example of the reduction in size of sharks or the fragmentation of Pangaea (PT3), addressing the origin of the eukaryotic cell from an evolutionary perspective (PT4), finally PT5 recommends dealing with the subject of evolution from primary education with small references during school hours.

In addition, in-service teachers highlighted the great difficulty that exists in bringing cross-curricular teaching proposals to the classroom and reported some of the most frequent obstacles that prevent these proposals from becoming a reality such as the high workload and time constrains they suffer, the lack of motivation in part of the students or low interest about innovative teaching methodologies among teachers, emphasizing on several occasions that there is a problem of coordination with the rest of the teachers responsible for other subjects to carry out interdisciplinary teaching approaches.

"You have to coordinate with other teachers.", "It depends on the rest of the teachers.", "Sometimes I try it other times there is a lack of motivation from the students or interest from the rest of the teachers" (IT2)

"I consider that perspective but I know that other colleges don't" (IT3)

Therefore, this finding could suggest that teachers should be provided with more training to detect connections between different subjects serving as a common ground for teaching them through an cross-curricular approach. Moreover, it would be also necessary to provide and support teachers with professional development courses that prepare them to overcome these challenges allowing the design and implementation of integration proposals for teaching Evolutionary Biology that fit into the real context of the classroom and the school as frequently suggested in the literature (Hanisch & Eirdosh, 2020; Hartelt et al., 2022).

SOLUTIONS AND RECOMMENDATIONS

The analysis of the Spanish curriculum shows that it is possible to treat certain contents of Evolutionary Biology from different disciplines following an interdisciplinary and integrated teaching approach. In addition, the interviewed teachers confirm that the nature of Biological Evolution makes it necessary to treat this content from a cross-curricular perspective in order to achieve a deep understanding of the subject. On the other hand, the high abstraction of the evolutionary contents represents an obstacle for students preventing a deep and enriched learning, so it is essential to bring Evolutionary Biology to contexts with which students can feel identified, enhancing their motivation to learn these contents, and thus facilitating its understanding (Brandt et al., 2022). Therefore, it is necessary to propose activities that comply with these premises and one of them based on the construction of phylogenetic trees is proposed below to treat Biological Evolution from an interdisciplinary approach. In fact, the interplay between Mathematics and Evolutionary Biology through Bioinformatics seems to be a powerful strategy to learn about phylogenetic trees and Biological Evolution (Jungck & Weisstein, 2013).

The construction of phylogenetic trees is a method to solve problems within the framework of Biological Evolution but that can be contextualized in real situations closer to the students' life. For example, they are used to find out the relationship between species, the origin of zoonotic diseases or the changes/extinctions of populations or species that can be related to certain environmental variations, major geological events or geographic redistribution (Wiley, 2010) but also they help us to understand the human cultural evolution (Tehrani, 2013). In addition, the interpretation of phylogenetic trees is very useful for students to overcome alternative ideas about evolutionary concepts (Meisel, 2010) and this proposal tries to do so (details for the preparation of the activity in Table 2).

Table 2. Teaching details about the sequence of activities SHAPE MATTERS!!

Learning goals (LG)	1. Assess the importance of phylogenetic trees as a representation of the evolutionary history of species. 2. Know databases and computer applications to build phylogenies. 3. Build and interpret phylogenetic trees. 4. Understand the use of phylogenetic markers. 5. Represent species differences and similarities using Mathematics.
Contents - Block 1. The evolution of life (4th ESO)	The fact and mechanisms of evolution.
Evaluation criteria - Block 1. The evolution of life (4th ESO)	18. Interpret phylogenetic trees, including the human phylogeny.
Related contents from Biology subject	1st of ESO - Block 3. Biodiversity on planet Earth. 1st of Bacc - Block 1. Living beings: composition and function; Block 4. Biodiversity. 2nd of Bacc - Block 3. Genetics and Evolution.
Related contents from other school subjects	Scientific culture: • 4th of ESO - Block 2. The Universe. (Annex II, pg. 465, BOE 1105/2014). • 1st of Bacc - Block 2. The Earth and life. (Annex II, pg. 467, BOE 1105/2014). Philosophy: • 1st of Bacc - Block 3. Knowledge. (Annex I, pg. 252, BOE 1105/2014). Mathematics: • 1st, 2nd, 3rd and 4th of ESO, 1st and 2nd Bacc - Block 1. Processes, methods and attitudes in mathematics. (Annex I, pg. 382, 386, 391, 395, 399, 404, 409, 414, 419, BOE 1105/2014). Physics and Chemistry: • 2nd, 3rd and 4th of ESO, 1st and 2nd of Bacc - Block 1. Scientific activity. (Annex I, pg. 258, 263, 268, 273, 448, BOE 1105/2014). Spanish Language and Literature • 2nd, 3rd and 4th of ESO, 1st and 2nd Bacc - Block 2. Written communication: reading and writing. (Annex I, pg. 361, 367, 372, 376, BOE 1105/2014). Information and communication technologies • 4th of ESO - Block 2. Computers, operating systems and networks; Block 3. Organization, design and production of digital information. (Annex II, pg. 532, BOE 1105/2014).
Materials	Computers with internet connection Screen and projector Blackboard, notebook/class diary and pen Guidelines including the steps to perform this practice
Room	Computer rooms
Duration	5 hours

The sequence of activities presented here is called SHAPE MATTERS!!. The title of this task refers to the importance of avoiding understanding evolution in a linear way instead of perceiving it as a branching process represented by phylogenetic trees, which is a frequent alternative conception setting an obstacle for learning about Biological Evolution (Hartelt et al., 2022). The task was reviewed, apart from the designer – the chaper's author, by a Biology teacher at Secondary Education level and by an Evolutionary Biology researcher who made some improvement suggestions about providing teachers and/or students with additional information concerning the links to download the precise sequence to perform the practice. The sequence of activities was tried in a class of 1° Baccalaureate in Granada for design optimization purposes but its impact on students has not been assessed using a research evaluation concept yet.

SHAPE MATTERS!! starts from a discussion around the memes included in Figure 2a. That discussion will focus on the importance of phylogenies to understand the evolution of species away from a linear process, thus the following questions will be raised by way of brainstorming and critical reflection *What is the meaning of these memes?* and *How do we interpret them in the light of Evolutionary Biology?*.

Then, students will start a computational practice to build phylogenetic trees from open-access online resources following the guidelines described in https://doi.org/10.5281/zenodo.5946898 (Martin-Peciña, 2022). During this practice, students will have the chance to explore how a database is made (LG 2 in Table 2), to manage and download data from online repositories in several formats (LG 2 in Table 2), to make their own comparisons between different species features (LG 1 and 5 Table 2), to know about cooperative work (LG 1 to 4 in Table 2) and to realize the power of phylogenetic for understanding the evolution of species (LG 1, 3 and 4 in Table 2). Therefore, this practice allows to approach contents from different subjects as Mathematics or Information and Communication Technologies when using digital resources as bioinformatics tools, analyzing and interpreting data, building distance matrix, evaluating parsimonious ways to align sequences or exploring statistical concepts.

First of all, students will need to download the DNA sequences indicated in the guidelines belonging to a mitochondrial gene (COII) and a nuclear gene (18S) for each of the species targeted in the practice (*Drosophila melanogaster*, *Arabidopsis thaliana*, *Homo sapiens*, *Mus musculus* and *Locusta migratoria*) using the GenBank repository (https://www.ncbi.nlm.nih.gov/genbank/). This gene selection allows to explore the presence of mitochondrial DNA as an evidence of endosymbiosis and its implications when used as phylogenetic marker due to its common maternal inheritance, among other factors, compared to nuclear DNA (Rubinoff & Holland, 2005). GenBank is an open-access repository hosting freely downloadable molecular data from thousands of species including DNA sequences which could really motivate students as they notice that DNA sequences appearing in textbooks are real data that researchers around the globe use in their experiments. In addition, during this sequence search they will be able to visualize how a gene is represented in bioinformatics as well as the collaborative work in science that exists behind the construction of public databases (Figure 2c).

Then, they will align the DNA sequences of all species for each gene separately (one alignment for the mitochondrial gene and one for the nuclear gene) and use the information to build a phylogenetic tree using each gene as marker (software for sequence alignment: https://www.ebi.ac.uk/Tools/msa/clustalo/). The alignment will show mutations and homologies between the DNA sequences of the studied species allowing for visualizing comparisons between species and different algorithms for alignments regarding different statistical approaches and likelihood methods (Figure 2c).

After that, students will build two phylogenetic trees from the alignment of both genes using an online tool (http://www.phylogeny.fr/one_task.cgi?task_type=tnt) and the expected results can be found in the Figure 3a and 3b.

Moreover, SHAPE MATTERS! comprises an activity to build a phylogenetic tree using the information about the genome size of each species. This activity was designed to provoke a cognitive conflict based on the C-value paradox and the inappropriateness of genome size as a phylogenetic marker. The resulting phylogenetic tree in this case will be different from the previous ones and will not represent the evolutionary history of the species. For that purpose, students will need to retrieve the data regarding genome size of each species from online repositories (Animals: http://www.genomesize.com/, Plants: https://cvalues.science.kew.org/search). Then, they will elaborate a distance matrix by simply subtracting the values of the genome size between different species so that they can get a matrix as the one shown in Figure 2b. Finally, they will use that matrix to built a phylogenetic tree using the online software http://www.trex.uqam.ca/index.php?action=trex that should result in a tree topology similar to the one appearing in Figure 3c.

Figure 2. Examples of resources, results and supporting material including in SHAPE MATTERS!!

a)

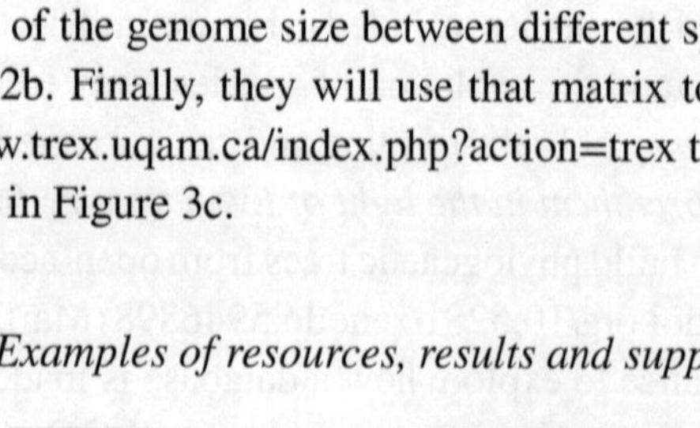

b)

	Lmig	Hsap	Dmel	Mmus	Atha
Lmig	0.000	2.370	5.810	2.760	5.680
Hsap	2.370	0.000	3.440	0.390	3.310
Dmel	5.810	3.440	0.000	3.050	0.130
Mmus	2.760	0.390	3.050	0.000	2.920
Atha	5.680	3.310	0.130	2.920	0.000

c)

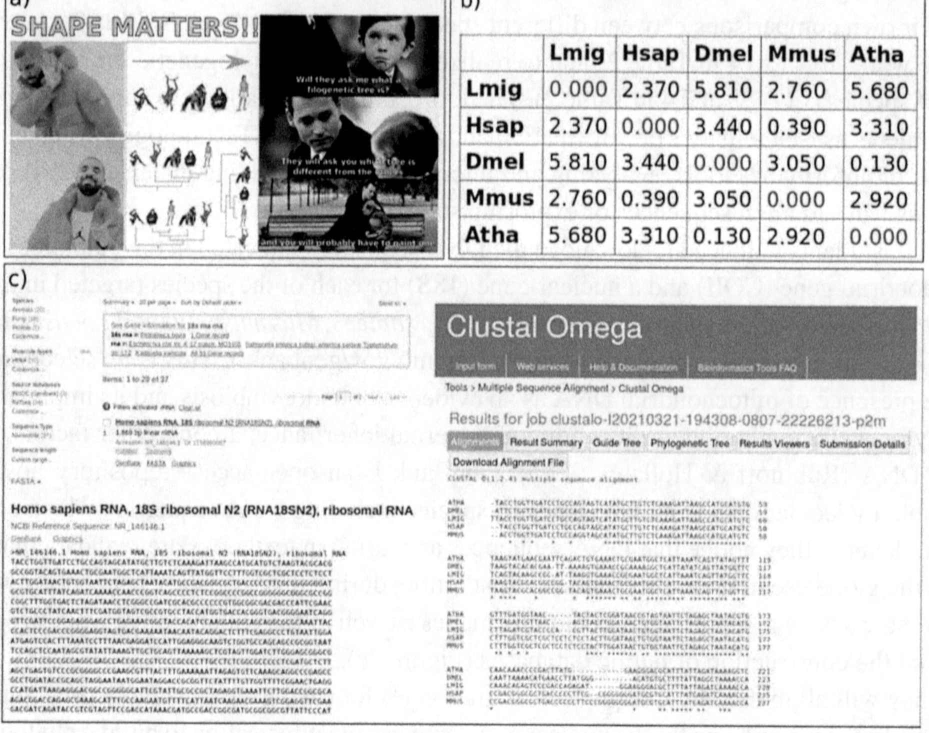

Once the three phylogenetic trees (using as markers, a mitochondrial gene, a nuclear gene and the genome size) have been built, the results will be discussed in the big group interpreting and arguing about agreement and disagreements found about tree topology and their evolutionary information (considering swapping the nodes around). Then, each student will prepare a report following a scientific paper format guided by the questions included in the guidelines document for this practice at https://doi.org/10.5281/zenodo.5946898 (Martin-Peciña, 2022). Although the entire protocol is designed to be done by students, the Appendix 2 of this chapter contains supporting information for teachers such as the precise links to download each DNA sequence, the entry format required by the alignment software and the format of the distance matrix that should be used as entry data for phylogenetic reconstruction. This information could be provided also to students as a way to adapt the task to lower educational levels and in case students can not complete the activities by their own.

Figure 3. The three resultant phylogenetic trees built using the guidelines described in SHAPE MATTERS!! a) Using DNA sequences from a nuclear gene (18S), b) Using DNA sequences from a mitochondrial gene (COII), c) From a distance matrix using the genome size as phylogenetic marker (GS)

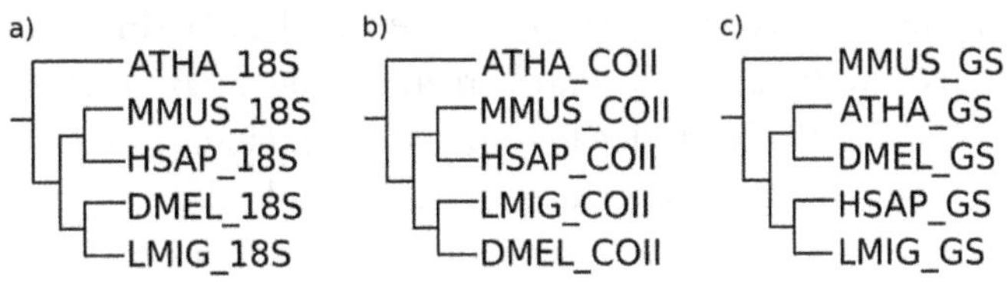

One of the most interesting results of this practice is that students may be aware that certain characters do not contain evolutionary information and can lead to confusion if they are used to build phylogenetic trees, as is the case of genome size (tree from Figure 3c). This particular case is known as the C-value paradox and refers to the non-existence of a relationship between the amount of DNA content, the content of genes encoded by that genome, the complexity of the species or their evolutionary history (Brookfield, 2000). That is the reason why in Figure 3c *H. sapiens* appears to be closer to the grasshopper *L. migratoria* than to the mammal *M. musculus*. Exploring the C-value paradox through phylogenetic trees will be useful to rethink about the concept of species complexity and the human as the pinnacle of evolution (Moran, 2007). Therefore, the process of design thinking to plan this kind of activities using phylogenies is key to highlighting the power of phylogenetic trees to provoke cognitive conflicts and their usefulness to overcome misconceptions such as the supremacy of the human, the confusion between a gene tree and a species tree, assuming the uncertainty of science and raise awareness about respect for all forms of life.

FUTURE RESEARCH DIRECTIONS AND CONCLUSION

The cross-curricular nature of Biological Evolution as a backbone in Biology and other sciences is reinforced according to the perception of teachers interviewed in the frame of this study. However, some teachers expressed the existing difficulties when implementing interdisciplinary proposals for Evolutionary Biology since it requires the coordination of teaching staff and the investment of an effort to which they are often reluctant due to the great workload and lack of time to deal with all their responsibilities during the academic year. As Rosales-López (2015) stated: "to avoid the possible rejection of the teaching staff to implement cross-curricular teaching proposals in the classroom, it should be noted that, far from adding new blocks of content to be dealt with, an integrative proposal should be brought to the classroom with the reconfiguration of already existing content in a qualitative way". Therefore, although the approach to these interdisciplinary approaches is not exempt from limitations, to promote effective coordination between the teaching staff and the latter with the students and their environment will facilitate the process and its final scope. It is then crucial to coordinate the educational community to implement fruitful proposals that, due to their cross-curricular nature, run the risk of being conceived as isolated plots in a disintegrated landscape distorting their unifying features and diluting its teaching potential.

The sequence of activities proposed here (discussion about evolutionary concepts guided by memes, exploration of bioinformatics databases, sequences alignment or building of phylogenetic trees) make up a bigger task (the SHAPE MATTER!! proposal) that intends to approach Biological Evolution from interdisciplinarity considering Biological Evolution as an unifying subject. By handling real datasets and participating in the same practices used by researchers and scientist, students will become aware of the key role of science for solving real problems, which is completely aligned with the STEM approach. Along these lines, in addition to the evolutionary contents directly related to this proposal (Table 2), the sequence connects with many other issues and skills. For example, SHAPE MATTERS!! links directly to the block "Abilities, skills and strategies. Scientific methodology." (Biology and Geology – 1st and 3rd of ESO; Annex I, pg. 205, BOE 1105/2014) due to its practical and inquiry-based nature. In addition, the sequence of activities presented here is expected to touch multiple skills required to face the demands of the current society that are based on the new Spanish educational law (LOMLOE, 2020) and that are developed for the Secondary Education in BOE 217/2020 as eight key competences (Annex I, pg. 25-32, BOE 217/2020) that could be addressed through this proposal as follows:

- **Competence in linguistic communication (CCL):** the proposed task could be helpful to develop communication skills through the discussion around memes or the elaboration of a writing report formatted as a scientific paper (Stewart et al., 2016; Paul, 2020).
- **Multilingual competence (CP):** the digital resources explored in these activities are in English as it is the broadly used language in scientific context. Thus, students could develop "use different languages, oral or signed, in an appropriate and effective way for learning and communication" as the Spanish curriculum states (Annex I, pg. 27, BOE 217/2020).
- **Competence in Mathematics, Science, Technology and Engineering (STEM):** mathematics and technology knowledge is essential to perform these research-based activities since students must manage several databases, online resources and computer tools but also learn about models, statistics and elaborate a distance matrix. Then, this proposal could help students to "understand the world using scientific methods, mathematical thinking and representation, technology and engineering methods" (Annex I, pg. 28, BOE 217/2020).
- **Digital competence (CD):** the use of digital information is key to the successful performance of these activities that are mainly focused on bioinformatics. Therefore, students would "safe, healthy, sustainable, critical and responsible use digital technologies for learning" (Annex I, pg. 29, BOE 217/2020).
- **Personal, social and learning to learn competence (CPSAA):** the design of SHAPE MATTERS!! could allow to learn from cognitive conflicts (as the C-value paradox) that emerge from the results of the practice helped by a guided reflection around them. In fact, cognitive conflicts can be useful for Cognitive Acceleration and metacognition (Beach et al., 2020).
- **Citizen competence (CC):** the discussion around the C-value paradox found as an outcome of these activities could motivate students to reduce anthropocentric views (the human as the perfect species having the larger genome) and to promote respectful attitudes and sustainable behaviors

leading to a environmental awareness and the conservation of biodiversity (Brookfield, 2020; Moran, 2007).

- **Entrepreneurial competence (CE):** as these activities offer students the use of freely accessible databases built thanks to the cooperation of researchers, they will be encouraged to autonomously go deep into them without limits since those resources are available to everyone and students could also contribute enriching their content, now or in the future. In fact, some authors have reported the effectiveness of working with databases to develop entrepreneurial mindset in students (Stanchev, 2015).
- **Competence in cultural awareness and expressions (CCEC):** SHAPE MATTERS!! starts with the use of memes as a cultural expression emerged from the nowadays digital world that is close to students and with which they can feel comfortable and identified (Shifman, 2014; Shchukina et al., 2020).

The search for solutions to the growing environmental and social problems cannot and should not be left out of the educational umbrella since the creation of a competent environmental citizenry is necessary to meet the challenges established in the European Green Deal (European Commission. The European Green Deal. 2020). This environmental competence, developed within the educational system must be approached in a contextualized way and including both the individuals themselves and the community, in order to achieve a significant impact on problem solving (Romero-Ariza et al., 2022; see initiatives as the MOST project https://icse.eu/international-projects/most/). Evolutionary Biology understood from a cross-curricular and integrating angle could help to understand important key challenges for the European Green Deal, such as the loss of biodiversity or the destruction of habitats (Caroll et al., 2014; Gilbert, 2020; Matthews et al., 2020) promoting environmental competence through education and scientific literacy (Hadjichambis & Paraskeva-Hadjichambi, 2020). This places Evolutionary Biology in an ideal position to address sustainability in the classroom, but for this, the synergies between Evolutionary Biology, the great current challenges of the planet and the educational curriculum must be clearly established. In addition, it is necessary to design programs for teachers training that consider the unifying and cross-curricular nature of Evolutionary Biology, and other disciplines, within the STEM context since they are key agents for the improvement of students' learning. In fact, there is no other way apart from an integrating teaching approach to build the path towards a didactic paradigm based on the acquisition of the required skills to solve major current problems that society faces.

ACKNOWLEDGMENT

This research was supported by an introductory research grant for master students of the University of Granada [5.B Program – BOJA nº 30, 13th of February, 2020]. The author's chapter acknowledges S. Rams for improvement feedback to some parts of this work. The author would like to express her deepest gratitude to the interviewed teachers for their kindness and availability but also for their highly valuable perspectives and strong commitment to research-based education. Finally, the feedback from the chapter's reviewers, in particular from reviewer 2, has been highly appreciated and it has really enriched and improved the first version of the manuscript.

REFERENCES

Abimbola, I. O., & Baba, S. (1996). Misconceptions & Alternative Conceptions in Science Textbooks: The Role of Teachers as Filters. *The American Biology Teacher*, *58*(1), 14–19. doi:10.2307/4450067

Acosta-Pérez, R. M. (2014). *Evidencias de la evolución biológica: aprendiendo evolución desde la metodología científica* [Tesis de licenciatura, Universidad Nacional Autónoma de México]. Repositorio de la Dirección General de Bibliotecas y Servicios Digitales de Información. https://repositorio.unam.mx/contenidos/238472

Ahuja, A. (2021). Backyard evolutionary biology: Investigating local flowers brings learning to life. *Ecology and Evolution*, *11*(8), 3459–3463. doi:10.1002/ece3.7199 PMID:33898002

Aivelo, T., & Uitto, A. (2016). Digital gaming for evolutionary biology learning: The case study of parasite race, an augmented reality location-based game. *LUMAT: International Journal on Math, Science and Technology Education*, *4*(1), 1–26. doi:10.31129/LUMAT.4.1.3

Alles, D. L. (2001). Using Evolution as the Framework for Teaching Biology. *The American Biology Teacher*, *63*(1), 20–23. doi:10.2307/4451025

Alonso, M., Galagovsky, L. R., Stella, C., & Yankilevich, M. L. (2005). Una simulación con planilla de cálculo para el aprendizaje de mecanismos evolutivos a partir del equilibrio de Hardy-Weinberg. *Enseñanza de las ciencias,* 1-6.

Álvarez-Pérez, E., & Ruiz, R. (2016). Proposal for Teaching Evolutionary Biology: A bridge between research and educational practice. *Journal of Biological Education*, *50*(2), 123–146. doi:10.1080/00219266.2015.1007887

Amorós-Hernández, M. (2018). Teoría de la evolución y modelización pedagógica: Una perspectiva ecológica. *Temps d'Educacio*, *55*, 171–190.

Andersson, B., & Wallin, A. (2006). On developing content-oriented theories taking biological evolution as an example. *International Journal of Science Education, Abingdon*, *28*(6), 673–695. doi:10.1080/09500690500498385

Apodaca, M. J., McInerney, J. D., Sala, O. E., Katinas, L., & Crisci, J. V. (2019). A Concept Map of Evolutionary Biology to Promote Meaningful Learning in Biology. *The American Biology Teacher*, *81*(2), 79–87. doi:10.1525/abt.2019.81.2.79

Appleton, L., & Mackie, J. (2019). Using Digital Organism Evolutionary Software in the Classroom. *The American Biology Teacher*, *81*(1), 12–17. doi:10.1525/abt.2019.81.1.12

Araujo, R., & Cantó, J. (2015). El árbol de la vida, un juego para la superación de los obstáculos al aprendizaje de la evolución biológica. *Escritos sobre la Biología y su enseñanza,* *8*(15), 59-71.

Arenson, M., & Coley, J. D. (2018). Anthropocentric by Default? Attribution of Familiar and Novel Properties to Living Things. *Cognitive Science*, *42*(1), 253–285. doi:10.1111/cogs.12501 PMID:28481420

Arias, I. A., & Sánchez, F. M. (2016). *La transversalidad de las temáticas de evolución biológica en el PCLB como posibilidad para la configuración del conocimiento profesional del profesor en Ciencias* [tesis de licenciatura, Universidad Pedagógica Nacional]. Repositorio Institucional UPN: http://hdl.handle.net/20.500.12209/1766

Aristotle. (1910). *Historia Animalium* (W. Ogle, Trans.). Clarendon Press. doi:10.5962/bhl.title.147382

Asghar, A., Hameed, S., & Farahani, N. K. (2014). Evolution in Biology Textbooks: A Comparative Analysis of 5 Muslim Countries. *Religious Education (Chicago, Ill.)*, *41*(1), 1–15. doi:10.1080/15507394.2014.855081

Ashe, A., Colot, V., & Oldroyd, B. P. (2021). How does epigenetics influence the course of evolution?. *Philosophical Transactions of the Royal Society B, 376*(1826), 20200111.

Asterhan, C. S. C., & Schwarz, B. B. (2007). The effects of monological and dialogical argumentation on concept learning in evolutionary theory. *Journal of Educational Psychology*, *99*(3), 626–639. doi:10.1037/0022-0663.99.3.626

Athanasiou, K., Katakos, E., & Papadopoulou, P. (2012). Conceptual ecology of evolution acceptance among Greek education students: The contribution of knowledge increase. *Journal of Biological Education*, *46*(4), 234–241. doi:10.1080/00219266.2012.716780

Athanasiou, K., Katakos, E., & Papadopoulou, P. (2016). Acceptance of evolution as one of the factors structuring the conceptual ecology of the evolution theory of Greek secondary school teachers. *Evol Educ Outreach*, *9*(1), 7. doi:10.118612052-016-0058-7

Ausubel, D. (2000). *Adquisición y retención del conocimiento: Una perspectiva cognitiva* (G. Sánchez, Trans.). Paidós.

Bardapurkar, A. (2008). Do students see the "selection" in organic evolution?: A critical review of the causal structure of student explanations. *Evolution (New York)*, *1*(3), 299–305. doi:10.100712052-008-0048-5

Barnes, M. E., Dunlop, H. M., Holt, E. A., Zheng, Y., & Brownell, S. E. (2019). Different evolution acceptance instruments lead to different research findings. *Evolution (New York)*, *12*(1), 4. doi:10.118612052-019-0096-z

Bateson, W. (1894). *Materials for the study of variation.* McMillan and Co.

Beach, P. T., Anderson, R. C., Jacovidis, J. N., & Chadwick, K. L. (2020). *Making the abstract explicit: The role of metacognition in teaching and learning*. Organisation du Baccalauréat International.

Bejan, A. (2020). Human evolution is biological & technological evolution. *Bio Systems*, *195*, 104156. doi:10.1016/j.biosystems.2020.104156 PMID:32492467

Beniermann, A. (2019). *Evolution-Von Akzeptanz und Zweifeln, Empirische Studien über Einstellungen zu Evolution und Bewusstsein* [Evolution – of acceptance and doubts, Empirical studies on attitudes towards evolution and the evolution of the human mind]. Springer Fachmedien.

Berkman, M. B., Pacheco, J. S., & Plutzer, E. (2008). Evolution and Creationism in America's Classrooms: A National Portrait. *PLoS Biology*, *6*(5), e124. doi:10.1371/journal.pbio.0060124 PMID:18494560

Bernard-Cobo, J. (2019). *Errores y Malinterpretaciones De La Teoría De La Evolución* [tesis de maestría, Universidad de Jaén]. Repositorio institucional: https://hdl.handle.net/10953.1/11448

Betti, L., Shaw, P., & Behrends, V. (2020). Acceptance of biological evolution by first-year life sciences University Students. *Science & Education*, *29*(2), 395–409. doi:10.100711191-020-00110-0

Bingham, C. B., & Davis, J. P. (2012). Learning Sequences: Their Existence, Effect, and Evolution. *Academy of Management Journal*, *55*(3), 611–641. doi:10.5465/amj.2009.0331

Bixler, A. (2007). Teaching Evolution with the Aid of Science Fiction. *The American Biology Teacher*, *69*(6), 337–340. doi:10.1662/0002-7685(2007)69[337:TEWTAO]2.0.CO;2

Blackstone, N. W. (2016). An Evolutionary Framework for Understanding the Origin of Eukaryotes. *Biology (Basel)*, *5*(2), 18. doi:10.3390/biology5020018 PMID:27128953

BOE. (2014). Real Decreto 1105/2014, de 26 de diciembre, por el que se establece el currículo básico de la Educación Secundaria Obligatoria y del Bachillerato. Boletín Oficial del Estado, 3 de enero de 2015, 3, 169-546.

BOE. (2020). Real Decreto 217/2022, de 29 de marzo, por el que se establece la ordenación y las enseñanzas mínimas de la Educación Secundaria Obligatoria. Boletín Oficial del Estado, 30 de marzo de 2022, 76, 1-198.

Brakefield, P. M. (2011). Evo-devo and accounting for Darwin's endless forms. *Philosophical Transactions of the Royal Society of London. Series B, Biological Sciences*, *366*(1574), 2069–2075. doi:10.1098/rstb.2011.0007 PMID:21690125

Brandt, M., Groom, Q., Magro, A., Misevic, D., Narraway, C. L., Bruckermann, T., . . . Jenkins, T. (2022). Promoting scientific literacy in evolution through citizen science. *Proceedings of the Royal Society B*, *289*(1980), 20221077. 10.1098/rspb.2022.1077

Brewer, M. S., & Gardner, G. E. (2013). Teaching Evolution through the Hardy-Weinberg Principle: A Real-Time, Active-Learning Exercise Using Classroom Response Devices. *The American Biology Teacher*, *75*(7), 476–479. doi:10.1525/abt.2013.75.7.6

Brookfield, J. F. Y. (2000). Genomic sequencing: The complexity conundrum. *Current Biology*, *10*(14), R514–R515. doi:10.1016/S0960-9822(00)00581-9 PMID:10898997

Brown, S. A., Ronfard, S., & Kelemen, D. (2020). Teaching natural selection in early elementary classrooms: Can a storybook intervention reduce teleological misunderstandings? *Evolution (New York)*, *13*(1), 12. doi:10.118612052-020-00127-7

Bruckermann, T., Fiedler, D., & Harms, U. (2021). Identifying precursory concepts in evolution during early childhood – a systematic literature review. *Studies in Science Education*, *57*(1), 85–127. doi:10.1080/03057267.2020.1792678

Buchan, L. (2019). *Development and evaluation of resources for teaching evolution in primary schools* [tesis de doctorado, University of Batch]. Repositorio institucional https://purehost.bath.ac.uk/ws/portalfiles/portal/194268489/Loredana_Buchan_PhD_thesis_PURE_pdf_Redacted.pdf

Bugallo-Rodríguez, A., & Vázquez-Ben, L. (2020). Estado actual de la teoría de la evolución en la normativa educativa curricular española. *eVOLUCIÓN. Revista de la Sociedad Española de Biología Evolutiva*, *14*(11), 47–59.

Burke da Silva, K. (2012). Evolution-centered teaching of biology. *Annual Review of Genomics and Human Genetics*, *13*(1), 363–380. doi:10.1146/annurev-genom-090711-163749 PMID:22559325

Burks, R. L., & Boles, L. C. (2007). Evolution of the Chocolate Bar: A Creative Approach to Teaching Phylogenetic Relationships Within Evolutionary Biology. *The American Biology Teacher*, *69*(4), 229–237. doi:10.1662/0002-7685(2007)69[229:EOTCBA]2.0.CO;2

Burmeister, A. R., & Smith, J. J. (2016). Evolution across the Curriculum: Microbiology. *Journal of Microbiology & Biology Education*, *17*(2), 252–260. doi:10.1128/jmbe.v17i2.988 PMID:27158306

Cabello-Garrido, A., España-Ramos, E., & Blanco-López, Á. (2018). Developing a Human Nutrition Learning Progression. *International Journal of Science and Mathematics Education*, *16*(7), 1269–1289. doi:10.100710763-017-9838-y

Caetano-Anollés, G., & Janko, R. (2022). The rise of hierarchy and modularity in biological networks explained by empedocles' double tale ~2,400 years before darwin and systems biology. *Frontiers in Genetics*, *13*, 973233. doi:10.3389/fgene.2022.973233 PMID:36061206

Campos, R., & Sá-Pinto, X. (2013). Early evolution of evolutionary thinking: Teaching biological evolution in elementary schools. *Evolution (New York)*, *6*(25), 1–13. doi:10.1186/1936-6434-6-25

Cañal, P. (2009). Acerca de la enseñanza sobre la evolución biológica en la escuela infantil y primaria. *Alambique*, *62*, 75–91.

Cano-Espinosa, D. (2015). La Evolución biológica en la Biofilosofía de F. J. Ayala. *Pensamiento. Revista De Investigación E Información Filosófica, 71*(269), 1015-1054. . doi:10.14422/pen.v71.i269.y2015.001

Carroll, S. P., Jørgensen, P. S., Kinnison, M. T., Bergstrom, C. T., Denison, R. F., Gluckman, P., Smith, T. B., Strauss, S. Y., & Tabashnik, B. E. (2014). Applying evolutionary biology to address global challenges. *Science*, *346*(6207), 1245993. Advance online publication. doi:10.1126cience.1245993 PMID:25213376

Castro, L. (2007). Docencia y evolución: La biología evolutiva de la enseñanza secundaria. *eVOLUCIÓN. Revista de la Sociedad Española de Biología Evolutiva*, *2*(1), 63–66.

Castro-Moreno, J.A. (2012). La Biología como ciencia histórica: el caso de la evolución biológica. *Biografía: Escritos sobre la Biología y su Enseñanza, 5*(9), 19-37.

Catley, K. M., & Novick, L. R. (2008). Seeing the Wood for the Trees: An Analysis of Evolutionary Diagrams in Biology Textbooks. *Bioscience*, *58*(10), 976–987. doi:10.1641/B581011

Catley, K. M., Novick, L. R., & Shade, C. K. (2010). Interpreting evolutionary diagrams: When topology and process conflict. *Journal of Research in Science Teaching*, *47*(7), 861–882. doi:10.1002/tea.20384

Ceballos, M., Vílchez, J. E., & Escobar, T. (2017). La enseñanza de la Evolución en Primaria. Opinión del profesorado y exploración de ideas inadecuadas en los niños. *Boletín de la Real Sociedad Española de Historia Natural Sección Aula. Museos y Colecciones*, *4*, 55–68.

Chaves-Mejía, G. A. (2010). ¿Enseñamos Evo-Devo? Implicaciones de las nuevas hipótesis sobre evolución biológica y su incidencia en la escuela. *Bio-grafía*, *3*(4), 155–166. doi:10.17227/20271034.vol.3num.4bio-grafia155.166

Chaves-Mejía, G. A. (2016, July-December). La enseñanza y el aprendizaje de la evolución biológica (eb) con la perspectiva teórica del perfil conceptual: Implicaciones en la formación continua del profesorado. *Bio-grafía*, *9*(17). Advance online publication. doi:10.17227/20271034.vol.9num.17bio-grafia109.117

Chaves-Mejía, G. A. (2017). La enseñanza de la evolución biológica desde la historia y la filosofía de la biología: aportes a la formación continua del profesorado. *Tecné, Episteme y Didaxis: TED. Número Extraordinario.* Memorias, Séptimo Congreso Internacional sobre Formación de Profesores de Ciencias, 12 al 14 de octubre de 2016, Bogotá.

Chavez-Mejía, G. A. (2012). ¿Es la endosimbiosis seriada una alternativa a la teoría sintética de la evolución? Implicaciones epistemológicas y didácticas. *Bio-grafía*, *5*(9), 67–79. doi:10.17227/20271034.9biografia67.79

Chuang-Tzu. (2005). *Obra completa.* Traducción de Cristóbal Serra Simó. Edicions Cort.

Cofré, H. L., Santibáñez, D. P., Jiménez, J. P., Spotorno, A., Carmona, F., Navarrete, K., & Vergara, C. A. (2018). The effect of teaching the nature of science on students' acceptance and understanding of evolution: Myth or reality? *Journal of Biological Education*, *52*(3), 248–261. doi:10.1080/00219266.2017.1326968

Cook, K. (2009). A Suggested Project-Based Evolution Unit for High School: Teaching Content through Application. *The American Biology Teacher*, *71*(2), 95–98. doi:10.1662/005.071.0209

Cooper, V. S., Warren, T. M., Matela, A. M., Handwork, M., & Scarponi, S. (2019). EvolvingSTEM: A microbial evolution-in-action curriculum that enhances learning of evolutionary biology and biotechnology. *Evolution (New York)*, *12*(1), 12. doi:10.118612052-019-0103-4 PMID:32647555

Cunningham, D. L., & Wescott, D. J. (2009). Still More "Fancy" and "Myth" than "Fact" in Students' Conceptions of Evolution. *Evolution (New York)*, *2*(3), 505–517. doi:10.100712052-009-0123-6

Darwin, C. (1859). *On the Origin of Species by Means of Natural Selection, or the Preservation of Favoured Races in the Struggle for Life*. John Murray. doi:10.5962/bhl.title.82303

Darwin, C., & Wallace, A. R. (1858). On the tendency of species to form varieties; and on the perpetuation of varieties and species by natural means of selection. *Journal of the Proceedings of the Linnean Society Zoology*, *20*, 45–62.

De Fraga, F. (2018). Towards an Evolutionary Perspective in Teaching and Popularizing Microbiology. *Journal of Microbiology & Biology Education, 19*(1), 19.1.45. . doi:10.1128/jmbe.v19i1.1531

De Souza, R. F., De Carvalho, M., Matsuo, T., & Zaia, D. A. M. (2010). Study on the opinion of university students about the themes of the origin of Universe and evolution of life. *International Journal of Astrobiology*, *9*(2), 109–117. doi:10.1017/S1473550410000054

Debortoli, N., Li, X., Eyres, I., Fontaneto, D., Hespeels, B., Tang, C. Q., Flot, J. F., & Van Doninck, K. (2016). Genetic Exchange among Bdelloid Rotifers Is More Likely Due to Horizontal Gene Transfer Than to Meiotic Sex. *Current Biology*, *26*(6), 723–732. doi:10.1016/j.cub.2016.01.031 PMID:26948882

Decker, T., Summers, G., & Barrow, L. (2007). The treatment of geologic time & the history of life on Earth in high school biology textbooks. *The American Biology Teacher*, *69*(7), 401–405. doi:10.1662/0002-7685(2007)69[401:TTOGTT]2.0.CO;2

Deniz, H., & Sahin, E. A. (2016). Exploring the factors related to acceptance of evolutionary theory among Turkish preservice biology teachers and the relationship between acceptance and teaching preference. *The Electronic Journal of Science Education*, *20*(4), 21–42.

Desantis, L. R. G. (2009). Teaching Evolution through Inquiry-Based Lessons of Uncontroversial Science. *The American Biology Teacher*, *71*(2), 106–111. doi:10.2307/27669381

Dissanayake, E. (2017). Ethology, Interpersonal Neurobiology, and Play: Insights into the Evolutionary Origin of the Arts. *American Journal of Play*, *9*(2), 143–168.

Downie, J. R. (2004). Evolution in Health and Disease: The Role of Evolutionary Biology in the Medical Curriculum. *Bioscience Education*, *4*(1), 1–18. doi:10.3108/beej.2004.04000004

Eder, E., Turic, K., Milasowszky, N., Van Adzin, K., & Hergovich, A. (2011). The relationships between paranormal belief, creationism, intelligent design and evolution at secondary schools in Vienna (Austria). *Science & Education*, *20*(5-6), 517–534. doi:10.100711191-010-9327-y

Eldredge, N., & Gould, S. J. (1972). Punctuated equilibria: an alternative to phyletic gradualism. *Models in Paleobiology*, 82-115.

Emmons, N., Lees, K., & Kelemen, D. (2017). Young children's near and far transfer of the basic theory of natural selection: An analogical storybook intervention. *Journal of Research in Science Teaching*, *55*(3), 321–347. doi:10.1002/tea.21421

Evans, M. (2010). Conceptual change and evolutionary biology: a developmental analysis. In S. Vosniadou (Ed.), *International handbook of research in conceptual change* (pp. 263–294). Routledge.

Fernández, J. J., & Sanjosé, V. (2007). Permanencia de ideas alternativas sobre Evolución de las Especies en la población culta no especializada. *Didáctica de las Ciencias Experimentales y Sociales*, *21*, 129–149.

Folguera, G., & Gonzalez-Galli L. (2012). La extensión de la síntesis evolutiva y los alcances sobre la enseñanza de la teoría de la evolución. *Biografia: Escritos sobre biología y su enseñanza,* *5*(9), 4-18.

Franco-Goyena, J. (2019). *La evolución biológica en secundaria: contribución a la incorporación de nuevos contenidos y metodologías* [tesis de máster, Universidad Pública de Navarra]. Recuperado de: https://hdl.handle.net/2454/33655

Friedrichsen, P. J., Linke, N., & Barnett, E. (2016). Biology teachers' professional development needs for teaching evolution. *Science Educator*, *25*(1), 51–61.

Futuyma, D. J. (1998). *Biología Evolutiva* (3rd ed.). Sunderland Associates.

Gallego, A., & Muñoz, A. (2015). Análisis de las hipótesis evolutivas en alumnos de Educación Secundaria y Bachillerato. *Revista Electrónica de Enseñanza de las Ciencias*, *14*(1), 35–54.

García de la Torre, E., Sequeiros, L., & Pedrinaci, E. (1995). Tectónica de placas y evolución biológica: Construcción de un paradigma e implicaciones didácticas. *Enseñanza de las Ciencias de la Tierra*, *3*(1), 14–22.

García-Penas, A. (2019). *"Simios contra el reloj": mecánicas de juego y role-playing como herramienta en la enseñanza de la Evolución Biológica* [tesis de maestría, Universidad de Zaragoza]. Repositorio institucional: https://zaguan.unizar.es/record/86683

Gefaell, J., Prieto, T., Abdelaziz, M., Álvarez, I., Antón, J., Arroyo, J., Bella, J. L., Botella, M., Bugallo, A., Claramonte, V., Gijón, J., Lizarte, E., Maroto, R. M., Megías, M., Milá, B., Ramón, C., Vila, M., & Rolán-Alvarez, E. (2020). Acceptance and knowledge of evolutionary theory among third-year university students in Spain. *PLoS One*, *15*(9), e0238345. doi:10.1371/journal.pone.0238345 PMID:32881957

Gelman, S. A., & Rhodes, M. (2012). "Two-Thousand Years of Stasis": How Psychological Essentialism Impedes Evolutionary Understa. In *Evolution Challenges: Integrating Research and Practice in Teaching and Learning about Evolution*. Oxford University Press. doi:10.1093/acprof:oso/9780199730421.003.0001

Gendron, R. P. (2000). The Classification & Evolution of Caminalcules. *The American Biology Teacher*, *62*(8), 570–576. doi:10.2307/4450980

Gilbert, S. F. (2020). Evolutionary developmental biology and sustainability: A biology of resilience. *Evolution & Development*, *12366*. Advance online publication. doi:10.1111/ede.12366 PMID:33400344

Glaze, A. L., & Goldston, M. J. (2015). U.S. science teaching and learning of evolution: A critical review of the literature 2000-2014. *Science Education*, *99*(3), 500–518. doi:10.1002ce.21158

Gómez-Moreno, D. K. (2020). *Enseñanza de la evolución biológica a través de historietas conceptuales contextualizadas* [tesis de maestría, Universidad Distrital Francisco José de Caldas]. Repositorio institucional: http://hdl.handle.net/11349/24454

González-Galli, L. (2017). Darwin y los naturalistas del Plata: una propuesta para la utilización de la historia y la sociología de la ciencia en la enseñanza de la teoría de la evolución. Historia de la ciencia e investigación en formación de profesores de ciencia, 34-59.

González-Galli, L., & Meinardi, E. (2013). ¿Está en crisis el darwinismo? Los nuevos modelos de la biología evolutiva y sus implicaciones didácticas. *Didáctica de las Ciencias Experimentales y Sociales*, *27*(27), 219–234. doi:10.7203/dces.27.2458

González-Galli, L., & Meinardi, S. (2015). Obstáculos para el aprendizaje del modelo de evolución por selección natural, en estudiantes de escuela secundaria de la Argentina. *Ciência & Educação (Bauru)*, *21*(1), 101–122. doi:10.1590/1516-731320150010007

González-Galli, L., & Meinardi, S. (2017). Obstáculos para el aprendizaje del modelo de evolución por selección natural, en estudiantes universitarios de biología. *Revista Eureka sobre Enseñanza y Divulgación de las Ciencias, 14*(3), 435-449. . doi:10.25267/Rev_Eureka_ensen_divulg_cienc.2017.v14.i3.03

Gould, S. (2004). *La Estructura de la Teoría Evolutiva*. Tusquets.

Gregory, T. R. (2008). Understanding Evolutionary Trees. *Evolution (New York), 1*(2), 121–137. doi:10.100712052-008-0035-x

Gregory, T. R., & Ellis, C. A. J. (2009). Conceptions of Evolution among Science Graduate Students. *Bioscience, 59*(9), 792–799. doi:10.1525/bio.2009.59.9.11

Griffith, J. A., & Brem, S. K. (2004). Teaching evolutionary biology: Pressures, stress, and coping. *Journal of Research in Science Teaching, 41*(8), 791–809. doi:10.1002/tea.20027

Ha, M., & Nehm, R. H. (2014). Darwin's difficulties and student's struggles with trait loss: Cognitive-historical parallelisms in evolutionary explanation. *Science & Education, 23*(5), 1051–1074. doi:10.100711191-013-9626-1

Hadjichambis, A. C., & Paraskeva-Hadjichambi, D. (2020). Education for Environmental Citizenship: The Pedagogical Approach. *Environmental Discourses in Science Education, 237*, 237–261. Advance online publication. doi:10.1007/978-3-030-20249-1_15

Hagman, M., Olander, C., & Wallin, A. (2003). Research-based teaching about biological evolution. In J. Lewis, A. Margo & L. Simonneaux (Eds.), *Biology education for the real world. Student - Teacher – Citizen. Proceedings of the IV Conference of European Researchers in Didactic of Biology (ERIDOB)* (pp. 105-119). Ecole national de formation agronomique.

Hagolani, P. F., Zimm, R., Vroomans, R., & Salazar-Ciudad, I. (2021). On the evolution and development of morphological complexity: A view from gene regulatory networks. *PLoS Computational Biology, 17*(2), e1008570. doi:10.1371/journal.pcbi.1008570 PMID:33626036

Haig, D. (2007). Weismann Rules! OK? Epigenetics and the Lamarckian temptation. *Biology & Philosophy, 22*(3), 415–428. doi:10.100710539-006-9033-y

Hall, B. K. (2012). Evolutionary Developmental Biology (Evo-Devo): Past, Present, and Future. *Evolution (New York), 5*(2), 184–193. doi:10.100712052-012-0418-x

Hancock, Z. B., Lehmberg, E. S., & Bradburd, G. S. (2021). Neo-darwinism still haunts evolutionary theory: A modern perspective on Charlesworth, Lande, and Slatkin (1982). *Evolution, 75*(6), 1244–1255. doi:10.1111/evo.14268 PMID:33999415

Hanisch, S., & Eirdosh, D. (2020). Educational potential of teaching evolution as an interdisciplinary science. *Evolution (New York), 13*(1), 25. doi:10.118612052-020-00138-4

Hartelt, T., Martens, H., & Minkley, N. (2022). Teachers' ability to diagnose and deal with alternative student conceptions of evolution. *Science Education, 106*(3), 706–738. doi:10.1002ce.21705

Hawley, P. H., & Sinatra, G. M. (2019). Declawing the dinosaurs in the science classroom: Reducing Christian teachers' anxiety and increasing their efficacy for teaching evolution. *Journal of Research in Science Teaching*, *56*(4), 375–401. doi:10.1002/tea.21479

Hermann, R. S., Shane, J. W., Meadows, L., & Binns, I. C. (2020). Understanding of Evolution Law among K–12 Public School Teachers. *The American Biology Teacher*, *82*(2), 86–92. doi:10.1525/abt.2020.82.2.86

Hernández, R. M. C., Álvarez Pérez, E., & Ruíz Gutiérrez, R. (2009). La selección natural: Aprendizaje de un paradigma. *Teorema*, *28*(2), 107–121.

James, P., & Clark, I. (2006). Overcoming geological misconceptions. *Planet*, *17*(1), 10–13. doi:10.11120/plan.2006.00170010

Jensen, M. S., & Finley, F. N. (1996). Changes in students' understanding of evolution resulting from different curricular and instructional strategies. *Journal of Research in Science Teaching*, *33*(8), 879–900. doi:10.1002/(SICI)1098-2736(199610)33:8<879::AID-TEA4>3.0.CO;2-T

Jiménez-Aleixandre, M. P. (1994). Teaching evolution and natural selection: A look at textbooks and teachers. *Journal of Research in Science Teaching*, *31*(5), 519–535. doi:10.1002/tea.3660310507

Julius, M. L., & Schoenfuss, H. L. (2006). Phylogenetic reconstruction as a broadly applicable teaching tool in the biology classroom: The value of data in estimating likely answers. *Journal of College Science Teaching*, *35*(7), 40–45.

Jung, J. Y., Lim, Y., & Huh, M. K. (2017). Flipped classroom pedagogy enhances student satisfaction and validated motivated strategies in Evolution Biology classrooms. *International Journal of Academic Research and Reflection*, *5*(6), 1–7.

Jungck, J. R., & Weisstein, A. E. (2013). Mathematics and evolutionary biology make bioinformatics education comprehensible. *Briefings in Bioinformatics*, *14*(5), 599–609. doi:10.1093/bib/bbt046 PMID:23821621

Kampourakis, K. (2013). Teaching About Adaptation: Why Evolutionary History Matters. *Science & Education*, *22*(2), 173–188. doi:10.100711191-011-9363-2

Kampourakis, K., & Zogza, V. (2008). Students' intuitive explanations of the causes of homologies and adaptations. *Science & Education*, *17*(1), 27–47. doi:10.100711191-007-9075-9

Karataş, A. (2020). Preservice Science Teachers' Misconceptions About Evolution. *Journal of Education and Training Studies*, *8*(2), 38–46. doi:10.11114/jets.v8i2.4690

Kelemen, D., Emmons, N. A., Schillaci, R. S., & Ganea, P. A. (2014). Young children can be taught basic natural selection using a picture storybook intervention. *Psychological Science*, *25*(4), 893–902. doi:10.1177/0956797613516009 PMID:24503874

Kelemen, D., & Rosset, E. (2009). The human function compunction: Teleological explanation in adults. *Cognition*, *111*(1), 138–143. doi:10.1016/j.cognition.2009.01.001 PMID:19200537

Kerfeld, C. A., & Scott, K. M. (2011). Using BLAST to teach "E-value-tionary" concepts. *PLoS Biology, 9*(2), e1001014. doi:10.1371/journal.pbio.1001014 PMID:21304918

Kimura, M. (1968). Evolutionary rate at the molecular level. *Nature, 217*(5129), 624–626. doi:10.1038/217624a0 PMID:5637732

King, C. (1877). Catastrophism and evolution. *American Naturalist, 11*(8), 449–470. doi:10.1086/271929

Kong, Y., Thawani, A., Anderson, T., & Pelaez, N. (2017). A Model of the Use of Evolutionary Trees (MUET) to Inform K-14 Biology Education. *The American Biology Teacher, 79*(2), 81–90. doi:10.1525/abt.2017.79.2.81

Koonin, E. V. (2009). Towards a postmodern synthesis of evolutionary biology. *Cell Cycle (Georgetown, Tex.), 8*(6), 799–800. doi:10.4161/cc.8.6.8187 PMID:19242109

Koonin, E. V., & Wolf, Y. I. (2009). Is evolution Darwinian or/and Lamarckian? *Biology Direct, 4*(1), 42. doi:10.1186/1745-6150-4-42 PMID:19906303

Kuschmierz, P., Meneganzin, A., Pinxten, R., Pievani, T., Cvetković, D., Mavrikaki, E., Graf, D., & Beniermann, A. (2020). Towards common ground in measuring acceptance of evolution and knowledge about evolution across Europe: A systematic review of the state of research. *Evolution (New York), 13*(1), 18. doi:10.118612052-020-00132-w

Kutschera, U. (2011). From the scala naturae to the symbiogenetic and dynamic tree of life. *Biology Direct, 6*(1), 33. doi:10.1186/1745-6150-6-33 PMID:21714937

Lamarck, J. B. (1809). Philosophie Zoologique, ou Exposition des Considérations relatives à l'Histoire naturelle des Animaux. Dentu et 'l Auteur.

Latham, L. G. II, & Scully, E. P. (2008). Critters! A Realistic Simulation for Teaching Evolutionary Biology. *The American Biology Teacher, 70*(1), 30–33. doi:10.1662/0002-7685(2008)70[30:CARSFT]2.0.CO;2

Lawson, A. E., & Thompson, L. D. (1988). Formal reasoning ability and misconceptions concerning genetics and natural selection. *Journal of Research in Science Teaching, 25*(9), 733–746. doi:10.1002/tea.3660250904

Lehman, J. R. (1994). Integrating science and mathematics: Perceptions of preservice and practicing elementary teachers. *School Science and Mathematics, 94*(2), 58–64. doi:10.1111/j.1949-8594.1994.tb12293.x

Lewontin, R. C. (1970). The Units of Selection. *Annual Review of Ecology and Systematics, 1*(1), 1–18. doi:10.1146/annurev.es.01.110170.000245

Libarkin, J. C., Kurdziel, J. P., & Anderson, S. W. (2007). College student conceptions of geological time and the disconnect between ordering and scale. *Journal of Geoscience Education, 55*(5), 413–422. doi:10.5408/1089-9995-55.5.413

Linnæus, C. (1758). Systema naturæ per regna tria naturæ, secundum classes, ordines, genera, species, cum characteribus, differentiis, synonymis, locis. Tomus I. Editio decima, reformata.

Loison, L. (2018). Lamarckism and epigenetic inheritance: A clarification. *Biology & Philosophy, 33*(3), 1–17.

LOMLOE. (2020). Ley Orgánica 3/2020, de 29 de diciembre de 2020, por la que se modifica la Ley Orgánica 2/2006, de 3 de mayo, de Educación. BOE núm. 340, de 30 de diciembre de 2020, páginas 122868 a 122953.

Love, A. C. (2013). Interdisciplinary Lessons for the Teaching of Biology from the Practice of Evo-Devo. *Science & Education, 22*(2), 255–278. doi:10.100711191-011-9351-6

Lozada-Guzmán, M. M. (2016). *Implementación De Una Secuencia De Enseñanza-Aprendizaje (SEA) En Teoría De Evolución Biológica: Aproximación A La Transformación De Las Concepciones De Naturaleza De La Ciencia De Profesores En Formación* [tesis de maestría, Universidad de Tolima]. Repositorio institucional: http://repository.ut.edu.co/handle/001/1737/statistics

Malik, A. H., Ziermann, J. M., & Diogo, R. (2018). An untold story in biology: The historical continuity of evolutionary ideas of Muslim scholars from the 8th century to Darwin's time. *Journal of Biological Education, 52*(1), 3–17. doi:10.1080/00219266.2016.1268190

Malthus, T. R. (1798). *An essay on the principle of population*. Oxford World's Classics reprint.

Marcelos, M. F., & Nagem, R. (2012). Use of the "Tree" Analogy in Evolution Teaching by Biology Teachers. *Science & Education, 21*(4), 507–541. doi:10.100711191-011-9370-3

Margulis, L. (1981). *Symbiosis in Cell evolution*. Freeman.

Mariano-Pérez, G., Gómez Galindo, A. A., & González Galli, L. (2018). Enseñanza de la evolución: fundamentos para el diseño de una propuesta didáctica basada en la modelización y la metacognición sobre los obstáculos epistemológicos. *Revista Eureka sobre Enseñanza y Divulgación de las Ciencias, 15*(2), 210101-210113.

Martín-PeciñaM. (2022). SHAPE MATTERS! Una práctica de filogenias moleculares con datos reales orientada a superar concepciones alternativas sobre la Evolución Biológica en 4º de ESO. Zenodo. doi:10.5281/zenodo.5946898

Martín-PeciñaM.RamsS. (2021). Instrumento de toma de datos del Trabajo Fin de Máster «Secuencia de Actividades para Biología Evolutiva a partir de los Resultados de una Investigación Educativa sobre la percepción didáctica de la materia en el profesorado (SABE y RIE)» del Máster Universitario en Profesorado de Enseñanza Secundaria Obligatoria y Bachillerato, Formación Profesional y Enseñanzas de Idiomas de la Universidad de Granada (España). Zenodo. doi:10.5281/zenodo.5649365

Martín-Peciña, M., & Rams, S. (2022). Didáctica de la Teoría de la Evolución Biológica desde la perspectiva del profesorado de educación secundaria en activo y en formación inicial. In Investigación educativa ante los actuales retos migratorios (pp. 1173–1185). Dykinson.

Martínez-Hernández, F., Mendoza-Fernández, A. J., Salmerón-Sánchez, E., Mota-Poveda, J. F., & Garzón, A. (2017). Sobre el grado de aceptación de la teoría evolutiva de los alumnos de Enseñanza Secundaria Obligatoria. *Enseñanza de las ciencias: revista de investigación y experiencias didácticas,* 1281-1288. https://www.raco.cat/index.php/Ensenanza/article/view/335260

Maskiewicz, A. C., & Lineback, J. E. (2013). Misconceptions are "so yesterday!". *CBE Life Sciences Education, 12*(3), 352–356. doi:10.1187/cbe.13-01-0014 PMID:24006383

Matthews, B., Jokela, J., Narwani, A., Räsänen, K., Pomati, F., Altermatt, F., Spaak, P., Robinson, C. T., & Vorburger, C. (2020). On biological evolution and environmental solutions. *The Science of the Total Environment, 724*, 138194. doi:10.1016/j.scitotenv.2020.138194 PMID:32251887

Mayr, E. (1984). What is Darwinism Today? *PSA: Proceedings of the Biennial Meeting of the Philosophy of Science Association, 1984*(2), 145–156. https://www.jstor.org/stable/192502

McCain, K., & Kampourakis, K. (2018). Which question do polls about evolution and belief really ask, and why does it matter? *Public Understanding of Science (Bristol, England), 27*(1), 2–10. doi:10.1177/0963662516642726 PMID:27117489

McComas, W. F. (1991). Resources for Teaching Evolutionary Biology Labs: An Analysis. *The American Biology Teacher, 53*(4), 205–209. doi:10.2307/4449270

McGuire, R. M., Hayashi, K. T., Yan, X., Caritá Vaz, M., Cinoğlu, D., Cowen, M. C., Martínez-Blancas, A., Sullivan, L. L., Vazquez-Morales, S., & Kandlikar, G. S. (2022). EcoEvoApps: Interactive apps for theoretical models in ecology and evolutionary biology. *Ecology and Evolution, 12*(12), e9556. doi:10.1002/ece3.9556 PMID:36479028

Mead, R., Hejmadi, M., & Hurst, L. D. (2017). Teaching genetics prior to teaching evolution improves evolution understanding but not acceptance. *PLoS Biology, 15*(5), e2002255. doi:10.1371/journal.pbio.2002255 PMID:28542179

Medina-Talero, N. A. (2017). La enseñanza de la evolución biológica y otros conceptos asociados por medio de la estrategia didáctica "The Caminalcules". *Tecné, Episteme y Didaxis: TED. Número Extraordinario.* Memorias, Séptimo Congreso Internacional sobre Formación de Profesores de Ciencias, 12 al 14 de octubre de 2016, Bogotá.

Meisel, R. P. (2010). Teaching Tree-Thinking to Undergraduate Biology Students. *Evolution, 3*(4), 621–628. doi:10.100712052-010-0254-9 PMID:21572571

Mellizo-Soto, M., & Romero, M. (2015). Alfabetización científica. ¿Qué ha cambiado en la última década? In *Percepción Social de la Ciencia y la Tecnología* (pp. 71–98). FECYT, Fundación Española para la Ciencia y la Tecnología.

Mengascini, A., & Menegaz, A. (2005). "El juego de las mariposas". Propuesta didáctica para el tratamiento del cambio biológico. *Revista Eureka sobre Enseñanza y Divulgación de las Ciencias, 2*(3), 403-415.

Miglino, O., Rubinacci, F., Pagliarini, L., & Lund, H. H. (2004). Using artificial life to teach evolutionary biology. *Cognitive Processing, 5*(2), 123–129. doi:10.100710339-004-0009-z

Molefe, L. (2013). *The pedagogical content knowledge of Life Sciences teachers teaching evolution for the first time* [tesis de maestría, University of the Witwatersrand]. Repositorio institucional: http://hdl.handle.net10539/13385.

Molina-Martínez, A. L. (2015). Propuesta didáctica para la enseñanza-aprendizaje de "la historia y evolución de la vida en la tierra" dirigida a estudiantes de grado octavo utilizando como estrategia el juego [tesis de mestría, Universidad Nacional de Colombia]. Repositorio institucional: https://repositorio.unal.edu.co/handle/unal/78048

Moore, J. A. (1984). Science as a way of knowing—Evolutionary biology. *American Zoologist*, *24*(2), 467–534. doi:10.1093/icb/24.2.467

Moore, R., Cotner, S., & Bates, A. (2009). The influence of religion and high school biology courses on students' knowledge of evolution when they enter college. *The Journal of Effective Teaching*, *9*(2), 4–12.

Moran, L. A. (2007). *The deflated ego problem*. https://sandwalk.blogspot.com/2007/05/deflated-ego-problem.h
tml

Morón-Monge, H., Morón-Monge, M. C., Wamba-Aguado, A. M., & Jiménez-Pérez, R. (2012). Una propuesta metodológica para la enseñanza de la biología y geología en la educación secundaria. *Revista de Educación en Biología*, *2*, 58–68.

Nadelson, L., Culp, R., Bunn, S., Burkhart, R., Shetlar, R., Nixon, K., & Waldron, J. (2009). Teaching evolution concepts to early elementary school students. *Evolution (New York)*, *4*(3), 267–274. doi:10.100712052-009-0148-x

Nehm, R. H., & Schonfeld, I. S. (2007). Does Increasing Biology Teacher Knowledge of Evolution and the Nature of Science Lead to Greater Preference for the Teaching of Evolution in Schools? *Journal of Science Teacher Education*, *18*(5), 699–723. doi:10.100710972-007-9062-7

Nelson, C. E., Scharmann, L. C., Beard, J., & Flammer, L. I. (2019). The nature of science as a foundation for fostering a better understanding of evolution. *Evolution (New York)*, *12*(1), 6. doi:10.118612052-019-0100-7

Nesse, R. M., & Natterson-Horowitz, B. (2019). Evolutionary Medicine – A Great Way to Teach Biology. *The American Biology Teacher*, *81*(8), 533. doi:10.1525/abt.2019.81.8.533

O'Malley, M. A. (2015). Endosymbiosis and its implications for evolutionary theory. *Proceedings of the National Academy of Sciences of the United States of America*, *112*(33), 10270–10277. doi:10.1073/pnas.1421389112 PMID:25883268

Ogborn, J. (2010). Curriculum development as practical activity. En K. Kortland & K. Klaassen (Eds.), Designing Theory-Based Teaching-Learning Sequences for Science Education (pp. 69-78). Utrecht: CDBeta Press – Freudenthal Institute for science and mathematics education (FIsme), Utrecht University – FIsme series on Research in Science Education.

Opitz, S. T. (2016). *Students' Progressing Understanding Of The Energy Concept. An Analysis Of Learning In Biological And Cross- Disciplinary Contexts* [tesis de doctorado, Christian-Albrechts University At Kiel]. Repositorio insitucional https://nbn-resolving.org/urn:nbn:de:gbv:8-diss-190054

Ottevanger, W. (2001). *Teacher support materials as a catalyst for science curriculum implementation Namibia* [tesis de doctorado, University of Twente]. Repositorio institucional https://research.utwente.nl/en/publications/teacher-support-materials-as-a-catalyst-for-science-curriculum-im

Pang, J., & Good, R. (2020). A review of the integration of science and mathematics: Implications for further research. *School Science and Mathematics*, *100*(2), 73–82. doi:10.1111/j.1949-8594.2000.tb17239.x

Paul, A. (2020). Memes as means for understanding interpersonal communication: A formative assignment. *Communication Teacher*, *34*(4), 346–354. doi:10.1080/17404622.2019.1681484

Pazza, R., Penteado, P. R., & Kavalco, K. F. (2010). Misconceptions About Evolution in Brazilian Freshmen Students. *Evolution (New York)*, *3*(1), 107–113. doi:10.100712052-009-0187-3

Peñaloza, G. (2016). La Importancia De La Demarcación: Una Revisión Del Concepto De "Aceptación De La Evolución". *Bio-grafía Escritos sobre la biología y su enseñanza*, *9*(17), 17-28. . doi:10.17227/20271034.vol.9num.17bio-grafia17.28

Pennell, M. W., Harmon, L. J., & Uyeda, J. C. (2014). Is there room for punctuated equilibrium in macroevolution? *Trends in Ecology & Evolution*, *29*(1), 23–32. doi:10.1016/j.tree.2013.07.004 PMID:23978567

Pierce, A. A., & de Man, T. (2019). Antibiotic resistant pathogen outbreak investigation: An interdisciplinary module to teach fundamentals of evolutionary biology. *Journal of Biological Education*, *53*(2), 150–156. doi:10.1080/00219266.2018.1447003 PMID:31073246

Pigliucci, M., & Muller, G. B. (Eds.). (2010). *Evolution–the extended synthesis*. MIT Press. doi:10.7551/mitpress/9780262513678.001.0001

Pinxten, R., Vandervieren, E., & Janssenswillen, P. (2020). Does integrating natural selection throughout upper secondary biology education result in a better understanding? A cross-national comparison between Flanders, Belgium and the Netherlands. *International Journal of Science Education*, *42*(10), 1609–1634. doi:10.1080/09500693.2020.1773005

Pobiner, B. (2012). Use human examples to teach evolution. *The American Biology Teacher*, *74*(2), 71–72. doi:10.1525/abt.2012.74.2.2

Pobiner, B., Beardsley, P. M., Bertka, C. M., & Watson, W. A. (2018). Using human case studies to teach evolution in high school A.P. biology classrooms. *Evolution (New York)*, *11*(1), 3. doi:10.118612052-018-0077-7

Prinou, L., Halkia, L., & Skordoulis, C. (2011). The inability of Primary School to introduce children to the Theory of Biological Evolution. *Evolution (New York)*, *4*(2), 275–285. doi:10.100712052-011-0323-8

Prinou, L., Halkia, L., & Skordoulis, G. (2008). What conceptions do Greek school students form about biological evolution? *Evolution (New York)*, *1*(3), 312–317. doi:10.100712052-008-0051-x

Quessada, M. P., Clément, P., Oerke, B., & Valente, A. (2008). Human evolution in science textbooks: A survey in eighteen countries. *Science Education International*, *19*(2), 147–162.

Quintana-Pinto, A. L. (2018). Modelo de argumentación de toulmin: Una propuesta didáctica para el aprendizaje de la herencia biológica. *Paradigma, 39*(1), 11-22. http://revistaparadigma.online/ojs/index.php/paradigma/article/view/639/636

Ramírez-Olaya, L. C. J. (2014). Juego de Simulación para enseñar evolución por selección natural a estudiantes de noveno grado. *Revista Perspectivas Educativas*, *7*(1), 115–133.

Ramírez-Olaya, L. C. J. (2017). Juego De Simulación Como Propuesta Didáctica Para Abordar El Concepto Evolución Biológica. *Bio-grafía*, *83*, 92. doi:10.17227/20271034.vol.0num.0bio-grafia83.92

Ratcliff, W. C., Raney, A., Westreich, S., & Cotner, S. (2014). A Novel Laboratory Activity for Teaching about the Evolution of Multicellularity. *The American Biology Teacher*, *76*(2), 81–87. doi:10.1525/abt.2014.76.2.3

Richard, M., Coley, J. D., & Tanner, K. D. (2017). Investigating Undergraduate Students' Use of Intuitive Reasoning and Evolutionary Knowledge in Explanations of Antibiotic Resistance. *CBE Life Sciences Education*, *16*(3), ar55. doi:10.1187/cbe.16-11-0317 PMID:28821540

Romero-Ariza, M., Gallego, A. M. A., Quesada, A., & Martín-Peciña, M. (2022). MOST: alfabetización científica y educación para la sostenibilidad a través de Proyectos de Escuela Abierta. In A. Benarroch (Ed.), 30 Encuentros Internacionales de Didáctica de las Ciencias Experimentales. La enseñanza de las ciencias en un entorno intercultural (pp. 311-316). Universidad de Granada.

Rosa, S. M., & Tricarico, H. (2016). Uso de árboles evolutivos para contextualizar científicamente la enseñanza de la biodiversidad vegetal. *Revista Eureka sobre Enseñanza y Divulgación de las Ciencias, 13*(2), 384-394. . doi:10.25267/Rev_Eureka_ensen_divulg_cienc.2016.v13.i2.11

Rosales-López, C. (2015). Evolución y desarrollo actual de los Temas Transversales: Posibilidades y límites. *Foro de Educación*, *13*(18), 143–160. doi:10.14516/fde.2015.013.018.008

Rosenthal, D. B. (1985). Evolution in high school biology textbooks: 1963-1983. *Science Education*, *69*(5), 637–648. doi:10.1002ce.3730690506

Rubinoff, D., & Holland, B. S. (2005). Between two extremes: Mitochondrial DNA is neither the panacea nor the nemesis of phylogenetic and taxonomic inference. *Systematic Biology*, *54*(6), 952–961. doi:10.1080/10635150500234674 PMID:16385775

Rutledge, M. L., & Warden, M. A. (1999). The Development and Validation of the Measure of Acceptance of the Theory of Evolution Instrument. *School Science and Mathematics*, *99*(1), 13–18. doi:10.1111/j.1949-8594.1999.tb17441.x

Sá-Pinto, X., Pinto, A., Ribeiro, J., Sarmento, I., Pessoa, P., Rodrigues, L. R., Vázquez-Ben, L., Mavrikaki, E., & Bernardino Lopes, J. (2021). Following Darwin's footsteps: Evaluating the impact of an activity designed for elementary school students to link historically important evolution key concepts on their understanding of natural selection. *Ecology and Evolution*, *11*(18), 12236–12250. doi:10.1002/ece3.7849 PMID:34594496

Sanabria-Totaitive, I. A., & Arango-Martínez, A. V. (2021). La analogía como estrategia en la enseñanza de la evolución biológica. *Praxis (Bern)*, *17*(1), 11–26. doi:10.21676/23897856.3312

Sánchez, J. Conde, M., & Zapata, V. (2017). Concepciones alternativas sobre evolución. Un estudio en futuros maestros. *Enseñanza de las Ciencias, Número Extraordinario*, 2219-2224.

Sanders, M., & Makotsa, D. (2016). The possible influence of curriculum statements and textbooks on misconceptions: The case of evolution. *Education as Change*, *20*(1), 1–23. doi:10.17159/1947-9417/2015/555

Sanders, M., & Ngxola, N. (2009). Addressing teachers' concerns about teaching evolution. *Journal of Biological Education*, *43*(3), 121–128. doi:10.1080/00219266.2009.9656166

Sandvik, H. (2008). Tree thinking cannot taken for granted: challenges for teaching phylogenetics. *Theory in biosciences = Theorie in den Biowissenschaften, 127*(1), 45–51. . doi:10.1007/s12064-008-0022-3

Schramm, T., Jose, A., & Schmiemann, P. (2021). Seeing the Woods for the Trees Again: Analyzing Evolutionary Diagrams in German and US University-Level Textbooks. *Education in Science*, *11*(8), 367. doi:10.3390/educsci11080367

Shchukina, I., Kudinova, V., Kudinova, O., Kondratenko, N., & Shashkova, S. (2022). Using internet memes to develop students' motivation and cross-cultural competence at the lessons of English. *INTED2022 Proceedings,* 7364-7369.

Shifman, L. (2014). *Memes in digital culture*. MIT Press.

Shogren, E. H., Anciães, M., Barske, J., Cestari, C., DuVal, E. H., Gaiotti, M. G., Johnson, E. I., Kimball, R. T., Marini, M. A., Ryder, T. B., Scholer, M. N., Ungvári, J., White, S. A., & Boyle, W. A. (2022). Dancing drives evolution of sexual size dimorphism in manakins. *Proceedings of the Royal Society B*, *289*(1974), 20212540. 10.1098/rspb.2021.2540

Sickel, A. J., & Friedrichsen, P. (2013). Examining the evolution education literature with a focus on teachers: Major findings, goals for teacher preparation, and directions for future research. *Evolution (New York)*, *6*(1), 23. doi:10.1186/1936-6434-6-23

Siew, N. M., Amir, N., & Chong, C. L. (2015). The perceptions of pre-service and in-service teachers regarding a project-based STEM approach to teaching science. *SpringerPlus*, *4*(1), 8. doi:10.1186/2193-1801-4-8 PMID:25674494

Skoog, G. (2005). The coverage of human evolution in high school biology textbooks in the 20th century and in current state science standards. *Science & Education*, *14*(3-5), 395–422. doi:10.100711191-004-5611-z

Smith, J. J., Johnson, W. R., Lark, A. M., Mead, L. S., Wiser, M. J., & Pennock, R. T. (2016). An Avida-ED digital evolution curriculum for undergraduate biology. *Evolution (New York)*, *9*(1), 9. doi:10.118612052-016-0060-0

Smith, M. U. (2010). Current status of research in teaching and learning evolution: I. philosophical/epistemological issues. *Science & Education*, *19*(6-8), 523–538. doi:10.100711191-009-9215-5

Soderberg, P., & Price, F. (2003). An examination of problem-based teaching and learning in population genetics and evolution using EVOLVE, a computer simulation. *International Journal of Science Education*, *25*(1), 35–55. doi:10.1080/09500690110095285

Soler, M. (2002). *Evolución: la base de la biología*. Proyecto Sur.

Solinas, M. (2015). *From Aristotle's Teleology to Darwin's Genealogy: The stamp of inutility*. Springer.

Stanchev, P. (2015). Teaching the Entrepreneurial Mindset in Database Class. *11th Annual International Conference on Computer Science and Education in Computer Science (CSECS).*

Steflitsch, D. (2022). Math teachers' visions of an ideal math class: What do they tell about bringing innovations into the classroom? *In Twelfth Congress of the European Society for Research in Mathematics Education (CERME12).*

Stern, L. (2004). Effective assessment: Probing students' understanding of natural selection. *Journal of Biological Education, 39*(1), 12–17. doi:10.1080/00219266.2004.9655948

Stewart, A. F., Williams, A. L., Lofgreen, J. E., Edgar, L. J., Hoch, L. B., & Dicks, A. P. (2016). Chemistry writing instruction and training: Implementing a comprehensive approach to improving student communication skills. *Journal of Chemical Education, 93*(1), 86–92. doi:10.1021/acs.jchemed.5b00373

Stewart, J. (2010). Modeling for Understanding in Science Education (Project MUSE): Involving High School Students in Evolutionary Biology with Realistic Problems and Causal Models. *Biology International: Darwin 200 – Evolution in Action. 47,* 78-90.

Tamayo, M. (2010). Dificultades en la enseñanza de la evolución biológica. *eVOLUCIÓN. Revista de la Sociedad Española de Biología Evolutiva, 5*(2), 23–27.

Tehrani, J. J. (2013). The phylogeny of little red riding hood. *PLoS One, 8*(11), e78871. doi:10.1371/journal.pone.0078871 PMID:24236061

Tidon, R., & Lewontin, R. C. (2004). Teaching evolutionary biology. *Genetics and Molecular Biology, 27*(1), 124–131. doi:10.1590/S1415-47572004000100021

To, C., Tenenbaum, H. R., & Hogh, H. (2017). Secondary school students' reasoning about evolution. *Journal of Research in Science Teaching, 54*(2), 247–273. doi:10.1002/tea.21347

Torkar, G., & Šorgo, A. (2020). Evolutionary content knowledge, religiosity and educational background of Slovene preschool and primary school pre-service teachers. *Eurasia Journal of Mathematics, Science and Technology Education, 16*(7), 1–13. doi:10.29333/ejmste/7991

Torreblanca, M., Merino, G., & De Longhi, A. L. (2011). El enfoque histórico contextualizado como facilitador de la enseñanza de los mecanismos evolutivos. La utilidad de un enfoque histórico integrado a un encuadre CTS en la enseñanza de los mecanismos micro y macroevolutivos. *Revista De Educación En Biología, 14*(1), 56–59.

Toth, E. E., Brem, S. K., & Erdos, G. (2009). "Virtual Inquiry": Teaching Molecular Aspects of Evolutionary Biology Through Computer-Based Inquiry. *Evolution (New York), 2*(4), 679–687. doi:10.100712052-009-0163-y

Tshuma, T., & Sanders, M. (2015). Textbooks as a possible influence on unscientific ideas about evolution. *Journal of Biological Education, 49*(4), 354–369. doi:10.1080/00219266.2014.967274

Van Dijk, E., & Reydon, T. (2010). A conceptual analysis of evolutionary theory for teacher education. *Science and Education, 19*(6-8), 655–677. doi:10.100711191-009-9190-x

Verspoor, A., & Wu, K. (1990). *Textbooks and educational development.* Educational and Employment Division, Population and Human Resources Department, The World Bank.

Vicente-Romero, R. (2016). *Trabajar ideas alternativas en el proceso de enseñanza-aprendizaje de la evolución a través de juegos: "Survival & Evolution"* [tesis de maestría, Universidad de Zaragoza]. Repositorio institucional https://zaguan.unizar.es/record/58753

Viiri, J., & Savinainen, A. (2008). Teaching-learning sequences: A comparison of learning demand analysis and educational reconstruction. *Latin-American Journal of Physics Education, 2*(2), 80–86.

Vondrasek, J. R., Janis, A., & Taylor, D. R. (2004). Evolution Kills: A Web Resource for Instructors of Evolutionary Biology. *Bioscene: Journal of College Biology Teaching, 30*(2), 3–8.

Wallin, A. (2008). One year after teaching - How consistent are students in using the scientific theory of biological evolution by natural selection? In M. Hammann, M. Reiss, C. Boulter & S. D. Tunnicliffe (Eds.), *Biology in Context – Learning and teaching for the twenty-first century. Proceedings of the VI Conference of European Researchers in Didactic of Biology (ERIDOB)* (pp. 52-63). University of London.

Watson, J. D., & Crick, F. (1953). Molecular Structure of Nucleic Acids: A Structure for Deoxyribose Nucleic Acid. *Nature, 171*(4356), 737–738. doi:10.1038/171737a0 PMID:13054692

West, S. S., Vasquez-Mireles, S., & Coker, C. (2006). Mathematics and/or science education: Separate or integrate. *Journal of Mathematical Sciences and Mathematics Education, 1*(2), 11–18.

Wiley, E. O. (2010). Why Trees Are Important. *Evo Edu Outreach, 3*(4), 499–505. doi:10.100712052-010-0279-0

Yates, T., & Marek, E. (2015). A Study Identifying Biological Evolution-Related Misconceptions Held by Prebiology High School Students. *Creative Education, 6*(8), 811–834. doi:10.4236/ce.2015.68085

Yeo, C., & Lee, H. (2016). Developing and Assessing a Learning Progression for the Ecosystem. *Journal of The Korean Association For Science Education, 36*(1), 29–43. doi:10.14697/jkase.2016.36.1.0029

Yoho, R. A., & Vanmali, B. H. (2016). Controversy in Biology Classrooms-Citizen Science Approaches to Evolution and Applications to Climate Change Discussions. *Journal of Microbiology & Biology Education, 17*(1), 110–114. doi:10.1128/jmbe.v17i1.1026 PMID:27047604

Zabel, J., & Gropengiesser, H. (2011). Learning progress in evolution theory: Climbing a ladder or roaming a landscape? *Journal of Biological Education, 45*(3), 143–149. doi:10.1080/00219266.2011.586714

Zudaire, I., & Napal-Fraile, M. (2020). Exploring the Conceptual Challenges of Integrating Epigenetics in Secondary-Level Science Teaching. *Research in Science Education.* Advance online publication. doi:10.100711165-019-09899-5

KEY TERMS AND DEFINITIONS

Baccalaureate: The optional educational level in Spain that comprises two academic courses prior to university where students ages range from 17 to 18 years old.

Cross-Curricular: Related to different disciplines.

Distance Matrix: Table showing pairwise comparison between each of the components that usually appear in row and column headers.

DNA (Deoxyribonucleic Acid): A polymeric biomolecule that carries the genetic and inheritable information for the development of an organism.

Evolutionary Biology: Discipline that studies the evolution over time of all the organisms that inhabit the planet.

Fitness: The ability to reproduce and give rise to fertile offspring.

Homology: Similarities that are informative in evolutionary terms as they derived from the existence of common ancestors.

Misconception: Alternative ideas and pre-concepts that are intuitive and resistant to change.

Natural Selection: Central evolutionary process of Darwinism that explains the evolution of species by differential survival and reproduction among the individuals that make up a population.

Phylogenetic Tree: Model of representation showing the evolutionary relationships between organisms.

Secondary Education: The compulsory educational level in Spain that comprise four academic courses where students' ages ranged from 12 to 16 years old.

STEM: Term used for the integration of Science, Technology, Engineering and Mathematics as a broad and unifying discipline in teaching and educational policy contexts.

APPENDIX 1

Table 3. Approaches and pedagogical resources used in the teaching of Biological Evolution

Teaching Approach	Reference	Description
History and nature of science	Torreblanca et al., 2011	Contextualized historical approach as a facilitator of the teaching of evolutionary mechanisms.
	Castro-Moreno, 2012	Biological Evolution to treat Biology as a historical science.
	Lozada-Guzmán, 2016	Conceptions about the nature of science in pre-service teachers.
	Chaves-Mejía, 2017	History and Philosophy of Biology for the continuous teacher training about Biological Evolution.
	González-Galli, 2017	Use of History and Sociology for the teaching of Biological Evolution (Darwin and the naturalists of La Plata).
	Nelson et al., 2019	Nature of science to enhance the understanding of Biological Evolution.
Phylogenetic trees	Julius & Schoenfuss, 2006	Phylogenetic reconstruction as a widely applicable teaching tool in Biology.
	Gregory, 2008	Review of the understanding and use of phylogenetic trees in the classroom.
	Sandvik, 2008	Tree thinking as an acquired skill that requires intensive training.
	Meisel, 2010	Teaching of tree-thinking for the understanding of phylogenetic relationships and other concepts of Evolutionary Biology.
	Rosa & Tricarico, 2016	Evolutionary trees to explain plant biodiversity.
	Kong et al., 2017	Review of the use of phylogenetic trees in research, considerations and educational applications.
Argumentation	Quintana-Pinto, 2018	Toulmin's argumentation model to study biological inheritance.
Gamification	Molina-Martínez, 2015	"A journey through geological time" and "the labyrinth of evolution" to work on alternative ideas about directional and progressive evolution as well as geological time.
	Aivelo & Uitto, 2016	An augmented reality game about the race of parasites for the teaching of Evolutionary Biology.
	Vicente-Romero, 2016	"Survival & Evolution" game to work on alternative ideas in Evolutionary Biology.
	Ramírez-Olaya, 2017	Simulation game to understand the concept of biological evolution.
	García-Penas, 2019	Argumentative debate with elements of role play focused on the conflict between evolutionary theories and intelligent design.

Continued on following page

Table 3. Continued

Teaching Approach	Reference	Description
Project-based learning	McComas, 1991	Resources for teaching Biological Evolution through case studies and inquiry activities.
	Vondrasek et al., 2004	Laboratory course to learn Evolutionary Biology in an applied way.
	Toth et al., 2009	Virtual laboratories to explain concepts of molecular evolution through inquiry.
	Cooper et al., 2019	Inquiry through laboratory practices in Microbiology to understand evolutionary concepts.
	Cook, 2009	Project-based learning for the teaching of Evolutionary Biology in High School.
	Desantis, 2009	Science inquiry-based lessons for teaching Biological Evolution.
	Stewart, 2010	Teaching of Evolutionary Biology through real cases (Modeling for Understanding in Science Education-Project MUSE).
	Ratcliff et al., 2014	Teaching of the evolution of multicellularity through laboratory practices.
	Smith et al., 2016	An inquiry-based curriculum through the Avida-ED evolution digital platform.
	Yoho & Vanmali, 2016	Citizen science to learn Evolutionary Biology and discuss its relationship with climate change.
	Ahuja, 2021	Local flowers to learn Biological Evolution by projects.
Simulations	Soderberg & Price, 2003	Computer simulation (EVOLVE) to learn population genetics and evolution.
	Miglino et al., 2004	Use of artificial life software to teach Evolutionary Biology.
	Alonso et al., 2005	Gene pool simulation to address evolutionary mechanisms with the Hardy-Weinberg law.
	Mengascini & Menegaz, 2005	Simulation of population changes of *Biston betularia* (industrial melanism).
	Bixler, 2007	Science fiction to teach evolution.
	Latham & Scully, 2008	*Critters!* as a realistic simulation for learning Evolutionary Biology.
	Kerfeld & Scott, 2011	Use of BLAST to understand concepts such as divergence, gene duplication or orthologous genes.
	Brewer & Gardner, 2013	Activity to simulate the Hardy-Weinberg equilibrium through an active learning exercise in real time using response devices in the classroom (clicker).
	Ramírez-Olaya, 2014	Simulation game to deal with the concept of natural selection.
	Appleton & Mackie, 2019	Software review of evolutionary simulations.
	McGuire et al., 2022	Interactive applications for teaching Ecology and Biological Evolution.
Case study	Gendron, 2000 Medina-Talero, 2017	The evolutionary history of Caminalcules to address the concepts of speciation, natural selection, common ancestor and phylogenetic tree.
	Nadelson et al., 2009	Analysis of the similarities and differences in organisms using the forelimb of vertebrates and comparative anatomy of ancient and living birds as references.
	Pobiner, 2012 Pobiner et al., 2018	Examples of the human species to learn Evolutionary Biology.

Continued on following page

Table 3. Continued

Teaching Approach	Reference	Description
Analogies	Burks & Boles, 2007	Use of chocolate bars as "model organism" analogies to understand key concepts in evolution.
	Marcelos & Nagem, 2012	The analogy of the tree in the teaching of Evolutionary Biology.
	Campos & Sá-Pinto, 2013	Set of activities to explore key evolutionary concepts using analogies and familiar, real-life context.
	Sanabria-Totaitive & Arango-Martínez, 2021	Use of the analogy "evolution of video games" in the teaching of Evolutionary Biology.
Narrative and storybooks	Gómez-Moreno, 2020	Contextualized conceptual comics for the study of Biological Evolution.
	Kelemen et al., 2014 Emmons et al., 2017 Brown et al., 2020	Storybook intervention to teach about natural selection and reduce teleological misunderstandings
Cross-curricular	García de la Torre et al., 1995 Morón-Monge et al., 2012	"The evolution of living beings" and "Plate tectonics" to address Biology and Geology.
	Downie, 2004 Nesse & Natterson-Horowitz, 2019	Relationship between Evolutionary Biology and Medicine.
	Arias & Sánchez, 2016	Interdisciplinarity for curricular teacher training.
	Burmeister & Smith, 2016 De Fraga, 2018 Pierce & de Man, 2019 Cooper et al., 2019	Relationship between Evolutionary Biology and Microbiology.
	Sá-Pinto et al., 2021	Exploration of Malthus' principle and intraspecific diversity in heritable characters using mathematical models and linking the historical key evolutionary concepts.
Flipped classroom	Jung et al., 2017	Flipped classroom as a strategy for teaching Evolutionary Biology at university levels.

APPENDIX 2

Links to Download Each DNA Sequence Indicated in SHAPE MATTERS!!

COII gene:

- LMIG: https://www.ncbi.nlm.nih.gov/nuccore/X80245.1
- HSAP: https://www.ncbi.nlm.nih.gov/nuccore/AF181091.1
- MMUS: https://www.ncbi.nlm.nih.gov/nuccore/KY018919.1
- DMEL: https://www.ncbi.nlm.nih.gov/nuccore/GQ229520.2
- ATHA: https://www.ncbi.nlm.nih.gov/nuccore/Y08501.2

18S gene:

- LMIG: https://www.ncbi.nlm.nih.gov/nuccore/KM853191.1
- HSAP: https://www.ncbi.nlm.nih.gov/nuccore/NR_146146.1
- MMUS: https://www.ncbi.nlm.nih.gov/nuccore/NR_003278.3
- DMEL: https://www.ncbi.nlm.nih.gov/nuccore/NR_133559.1
- ATHA: https://www.ncbi.nlm.nih.gov/nuccore/NR_141642.1

Entry format required by the alignment software (example for COII):

```
>HSAP_COII
ATGGCACATGCAGCGCA...
>LMIG_COII
ATGGCAACATGATCAAA...
>MMUS_COII
ATGGCCTACCCATTCCA...
>ATHA_COII
ATGATTGTTCTAAAATG...
>DMEL_COII
ATGTCTACATGAGCTAA...
```

Format of the distance matrix for phylogenetic reconstruction:

```
    5
Lmig      0.000 2.370 5.810 2.760 5.680
Hsap      2.370 0.000 3.440 0.390 3.310
Dmel      5.810 3.440 0.000 3.050 0.130
Mmus      2.760 0.390 3.050 0.000 2.920
Atha      5.680 3.310 0.130 2.920 0.000
```

Section 2

Connections Between Science and Mathematics on Educational Practices

Chapter 5
Views From Singapore Students Towards the Use of a Mathematics-Based STEM Project to Demonstrate Their Creativity Using Scientific Principles

Nazir Amir
Academy of Singapore Teachers, Singapore

ABSTRACT

This chapter reports the views from 39 students, averaging 14 years of age, in a secondary school in Singapore towards the use of a mathematics-based STEM project in Design and Technology (D&T) lessons to demonstrate their creativity using scientific principles. An instrument in the form of a questionnaire was designed to gather the students' views. This instrument infused elements from Guilford's and Torrance's frameworks for creativity (fluency, flexibility, originality, and elaboration) and indicators of product creativity (novelty, resolution, and style) through Besemer's creative product analysis model. Findings from this classroom-based research highlight that students found the project useful in promoting their interest, confidence, and creativity. This study contributes a way of strengthening the links between D&T, mathematics, and science in a feasible manner in school practices.

DOI: 10.4018/978-1-6684-5765-8.ch005

1. INTRODUCTION

1.1 Conceptualization Towards the Development of a Mathematics-Based Teaching Aid

Many schools are gaining interest in maker-centred teaching and learning practices aimed at promoting student interest in STEM (*Science, Technology, Engineering & Mathematics*). While there have been articles highlighting interdisciplinary practices between Design & Technology (D&T), science and mathematics through design-and-make projects, very few articles have described how students can be guided to demonstrate their creative use of scientific principles through design-and-make projects that exhibit specific mathematical concepts. Guiding students to design and make such projects can help them deepen their understanding of mathematical concepts in a hands-on-minds-on-hearts-on manner, as well as providing them with an avenue to demonstrate their creativity by coming up with designs that utilize a variety of scientific principles.

In this regard, the author, who teaches science and D&T in a secondary school in Singapore, worked with a group of three students to develop a simple hand-held kit that demonstrates the mathematical concept of Pythagoras Theorem. The inspiration for this kit came about during a visit to the Singapore Science Centre where the author and his three students (who are in secondary two - equivalent to Grade 8 students, averaging 14 years of age) viewed an exhibit demonstrating Pythagoras Theorem as shown in Figure 1.

Figure 1. Pythagoras Theorem exhibit
Source: Author

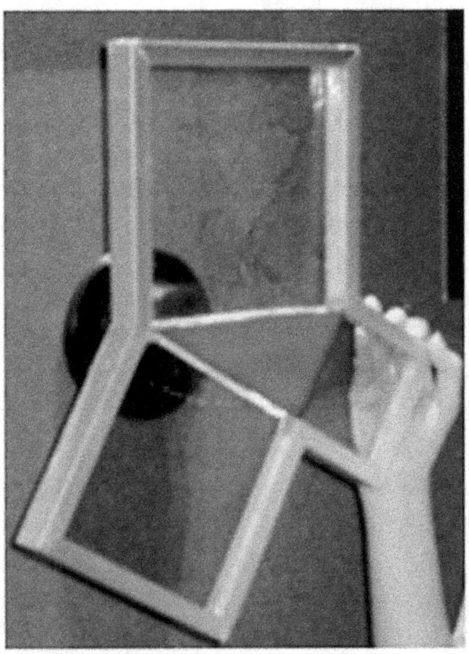

The students commented that while they appreciate being able to see a visual proof of the theorem through the exhibit, they felt that it would have been better if their mathematics teacher had a portable version of the exhibit as a teaching aid to introduce the concept (at the time of teaching) in the classroom. The visual scaffold could have helped them in grasping the concept. This led to the author and his three students to come up with an idea to conceptualise a smaller and portable version of the exhibit that can be used as a teaching aid for mathematics teachers to teach students the concept at the secondary two level. It seemed apt that the author and his three students reached out to mathematics teachers in the school to gather their views. Mathematics teachers welcomed this idea, which then led to a pilot study of the three students coming up with a prototype in the D&T workshop.

In the process of designing and making a hand-held prototype, the students communicated regularly with their mathematics teacher to gather ergonomics data, such as where the prototype would be used in a classroom, as well as anthropometry data, such as the average palm size of teachers who would be using the prototype. The prototype, shown in Figure 2, makes use of a 3cm-4cm-5cm Pythagorean triple combination (which formed the right-angled triangle) and tiny Styrofoam beads (found in a 'bean bag' doll) to demonstrate the concept. Apart from designing skills, students learnt plastic processing skills and took about three hours to develop the prototype.

Figure 2. The hand-held prototype developed by the three students
Source: Author

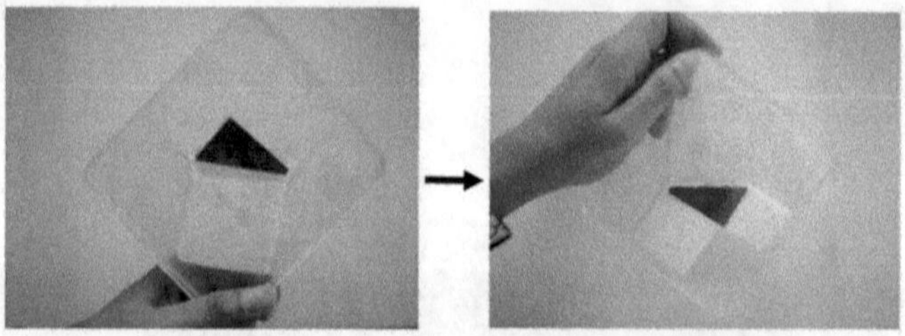

The three students presented the prototype to their mathematics teacher who was happy and satisfied with the design. This prototype, referred to as the Pythagoras Demo Kit, gained attention from the mathematics community in the school. Mathematics teachers thanked the students for developing a prototype that is valued in teaching and learning of mathematics.

1.2 Fostering Students' Creativity and Inventiveness Through the Use of Scientific Principles

As the author observed mathematics teachers using the prototype in the classroom, he felt that more could be done to add value to the prototype's functionality. He noticed how mathematics teachers had to flip the prototype along the hypotenuse (of the right-angled triangle) and partially rotate the kit in a clockwise and anti-clockwise manner in order to get all the beads to flow from the biggest square compartment to the other two smaller square compartments to completely fill them up, and vice versa. He felt that there could be more variations of the design which could function on a variety of scientific principles. This led the author to develop the Pythagoras Demo Kit as a D&T project. The author wanted to explore how he could guide more students to come up with creative designs of the initial prototype through a variety of scientific principles. This led him to carry out a classroom research study with a class of 39 secondary two students (14 girls and 25 boys). Several factors were taken into account to develop this project as one to be used as a lower secondary project in D&T lessons. These factors include the feasibility of making the prototype through use of available tools and materials in the D&T workshop in school; the time taken to complete the project (ten hours over ten weeks, averaging one hour per week in regular D&T lessons); and the scope that the project offers for students to come up with creative versions of the kit. These factors are in alignment with requirements of the lower secondary D&T syllabus in that the choice of lower D&T projects must offer opportunities for students to pick up sufficient design knowledge and skills, as well as exposing students to fabrication skills in working with resistant materials - namely wood, metal and plastic.

The author guided his students to design and make the kit over two phases. In the first phase, students made a similar version of the initial prototype. They learnt plastic processing which includes cutting, drilling, sanding, and polishing amongst other material processing skills. This first phase spanned five hours over five weeks (one hour of lessons per week). In the second phase, students were challenged to come up with design variations of this prototype. This was carried out over another five weeks (also one hour per week). In this phase, students brainstormed several ways for mathematics teachers to use their prototypes as teaching aids. This included incorporating creative use of mechanical principles such as pivots, wheel-and-axle, levers and springs in their designs, as well as improving the rate of flow of beads between the three compartments of the initial prototype. It became apparent to the students that they could improve the initial prototype's functionality through a variety of scientific principles. All these were in addition to the students exploring the configuration for shape, forms, and functionality of their prototypes with respect to the end users (teachers and students) and the environment (classroom). Students sketched their ideas in their design sheets and came up with mock-up models in the process of testing these ideas. They also discussed the choice of materials and construction details to make their prototypes.

1.3 Research Questions

In the context of this classroom action research which involves the use of the Pythagoras Demo Kit as a project that is carried out in regular secondary two D&T lessons with 39 students, the author wanted to find out the students' views in the following areas:

1. Learning mathematics, science and D&T through this project;
2. If going through the design-and-make process has helped them articulate their creativity through the use of scientific principles in this project;
3. The use of this project to promote interest, confidence, and creativity by linking D&T to mathematics and science

These three areas formed the research questions for this classroom-based research.

2. LITERATURE REVIEW

2.1 Linking Up Subjects

Many articles on subject interaction mention that learning becomes more meaningful for students when teachers design lessons to allow students to see connections between subjects (Fogarty, 1991; Lee, 2007) and seeing the relevance of the interaction between the subjects in their daily lives (Caine & Caine, 1991). D&T is a subject that could enhance students' understanding of mathematics and scientific concepts by guiding them to be more aware of how mathematics and science can contribute to their design of prototypes (Allison, 1986; Amir, 2020; Barlex and Pitt, 2000; Norman, 1993; Zubrowski, 2002). This could, in turn, create a positive impact on their attitudes towards the study of mathematics and science (Amir, 2020; Norton, 2007). Doppelt et al. (2008) highlighted that learning science through design-based approaches has the "*potential to increase students' desire to learn, enhance students' success in science class, and increase students' interest in science topics*".

The extent to which mathematics and science have been linked to D&T has been extensively reviewed by Sharkawy et al. (2009). In their paper, the authors revealed three claims to support the need for interaction amongst subjects: the first is that each of these three subjects has similar nature of disciplines such that linking the subjects may get students to view science and technology from an interactionist view - where the learning is enhanced in both science and D&T (Gardner, 1994); the second is creating positive impact in students' learning; and the third is in enhancing students' engagement in learning. Sage and Steeg (1993) further mentioned that benefits of linking up subjects in purposeful ways include increased motivation, enhanced learning and reduced assessment load for students, while benefits for teachers include improved understanding of each subject's aims and approaches and contribution to teacher development. They also discussed strategies to link up mathematics and science to D&T and raised some techniques in overcoming timetabling and collaboration problems that may arise between the departments. Davidson et al. (1998) recommends that an effective way of linking mathematics and science to D&T is by identifying a range of topics within mathematics and science which are frequently used in D&T. Ainley et al. (2006) showed that mathematics (and science) can be linked up to D&T through the development of a purposeful activity. Amongst many of such projects, Hansen (2009) highlighted this by showing how students learnt physics, specifically in the topics of electricity and electronics, in the context of making a simple badge project.

2.2 Maker Activities to Introduce Specific Mathematical Concepts

Many articles share how teachers have adopted creative teaching and learning approaches to engage students' understanding of a mathematical concept. For example, teachers have guided their students to deepen their understanding of the concept of Pythagoras Theorem. These include the use of a board game (Eriksen et al.,1995) and a 'Geoboard' as shown by Naraine (1993). While these are good examples to help students deepen their learning, the question arises as to how teachers can introduce the content in enjoyable ways at the time of teaching. Some articles that address this question feature creative approaches by teachers to introduce specific mathematical concepts through simple maker activities. This includes the article by Urich and Sasse (2011) that share how teachers teach the concept of surface areas of spheres through orange peels. The idea is similar to the following example. Teachers start the lesson by guiding students to peel the skin off an orange and cut them into small pieces. Students are then guided to place the peels over four circles, printed on a piece of paper (from 3D to 2D). Each circle has the same diameter as the orange. Students would eventually realise that all the peel would be able to fit into the four circles. Teachers then guide students to see a relationship between their prior knowledge of the area of a circle (πr^2) to the activity that they are doing, which could lead to an 'aha' moment as students make sense that the total area of the four circles (in 2D) is equivalent to the surface area of the sphere as $4\pi r^2$ (in 3D). From the student-end, learning through such a maker activity through the process of discovery using familiar materials may help them make sense of the equation, and perhaps even a more enjoyable learning experience as compared to just memorizing the equation. Such examples can inspire teachers to present mathematics concepts through other maker activities, including design-and-make projects.

While there are numerous articles in the literature that share approaches and benefits of teaching mathematics through design-and-make projects, such as those in Furner and Kumar (2007), James et al. (2000), Norton (2007), and Sharkawy et al (2009), what seems lacking are articles sharing how D&T teachers can guide students to design and develop prototypes that would exhibit specific mathematical concepts which are within the mathematics syllabus that their students are studying. Many projects presented in the literature seem to focus on students' application of skills in mathematics, such as measurements, during the design-and-make process. This is considerably different from guiding students to deepen their understanding of a specific mathematical concept through making a prototype that exhibits that particular concept. If we look at the woodpecker mechanism developed by students highlighted in Davidson et al.'s (1998) article, one could appreciate that teachers teaching science and D&T could use the students' prototype to introduce concepts related to mechanics in an enjoyable way to many other students. This could in turn enhance many students' understanding of such concepts in both science and D&T lessons. This is not so much the case with mathematics as this prototype does not explicitly highlight a mathematical concept.

It could thus be useful to explore the development of simple prototypes that exhibit specific mathematical concepts (that students are learning in mathematics lessons) that could be used as lower secondary projects in D&T. This could be a feasible way in guiding students to deepen their understanding of mathematics concepts that are being taught to them in mathematics lessons. Care must be taken that this needs to be done without compromising the objectives and teaching of content in the lower secondary D&T syllabus, and yet one that can be carried out in a feasible manner using available tools and machinery in the regular D&T workshop.

2.3 Fostering Creativity Through Design-and-Make Projects

Creativity in the areas of D&T has been discussed by Spendlove and Hopper (2005) where they emphasized the need for ways to nurture and encourage creativity through design-and-make activities. Students can be encouraged to come up with ideas that contribute to the functionality and contexts of their designs based on Guilford's (1959) and Torrance's (1979) framework for creativity (*fluency, flexibility, originality* and *elaboration*) and indicators of product creativity (*novelty, resolution*, and *style*) through Besemer's (1998) Creative Product Analysis Model (CPAM). During the brainstorming process, students can be encouraged to come up with many ideas (*fluency*), different ideas (*flexibility*), think of new ideas (*originality*), and describe these ideas in detail (*elaboration*). These four traits can be showcased through both the students' design sheets and prototypes. Fortus et al. (2004; 2005), Furner and Kumar (2007) and Sidawi (2009) are amongst a few researchers that highlight how design-based contexts can provide ways for students to learn science and solve problems through knowledge drawn from mathematics and science, which is typical in design-and-make projects. Examples of such projects are windmills presented in Zubrowski (2002) and a woodpecker mechanism as part of a money collecting box presented in Davidson et al (1998). These projects allow students to demonstrate creative use of scientific principles through design-and-make activities. The diagram of a student's woodpecker mechanism shown in Davidson et al.'s (1998) article describes how a student has made creative use of a pivot point in his woodpecker coin box such that the woodpecker 'pecks' each time a coin is dropped into the coin slot. Teachers teaching science and D&T could use such prototypes as teaching aids to introduce concepts related to mechanics in an enjoyable way. Doing this could excite their students in learning such concepts. The student inventors who came up with such prototypes could also be credited for their creative use of scientific principles in their prototypes. This could be done by showcasing their prototypes and highlighting their creativity to staff and students in school, the students' parents and members of the larger community (Amir, 2014; 2020; 2021a; 2021b). Doing this could boost the self-worth of the students in a STEM area.

Part of the requirements of the D&T syllabus in Singapore is for students to gain proficiency in skills on working with resistant materials (wood, metal and plastic). The author has shown how this can be achieved by getting students to design and fabricate (a) wooden toys that demonstrate the concept of centripetal (or centrifugal) force (Amir & Subramaniam, 2006) and (b) safe metallic gadgets that demonstrate physical changes through making candy floss kits (Amir & Subramaniam, 2009). Both these projects were designed based on the Relevant, Appealing and Personal (RAP) pedagogical guideline (Amir, 2022). These RAP-infused projects evoke emotional responses from students as the design of these projects take into account appealing factors which include elements of (i) novelty in demonstrating the science concepts, (ii) multi-sensory stimulus, and (iii) age-appropriateness. Apart from design knowledge and skills, students who went through both these projects picked up basic technical skills in wood (marking out, cutting drilling, sanding) and metal processing (marking out, cutting, bending), and reflected their learning through individual design sheets that formed their design journals. The students demonstrated creative use of scientific principles by coming up with variations of the initial prototypes that was made in the first phase. Results from these two studies reveal that students developed positive attitudes towards learning science and D&T through such projects, and desire to do more D&T projects of such nature (Amir, 2021a). Such projects illustrate a two-phase approach in teaching and fostering creativity utilizing scientific principles. In the first phase, students learn science concepts through making a prototype shown to them, and then guided to come up with various designs of the prototype through a variety of scientific principles in the second phase. One example is the pencil holder project (Amir & Subramaniam, 2012). In this project, students were guided to demonstrate mechanical principles in an interesting manner through their various designs of pencil holders carried out in D&T lessons right from the start of the project. Students were able to brainstorm creative use of scientific principles as they go through the design-and-make process. A discussion by science and D&T teachers at the end of the project highlight that the prototypes were valued by both these communities of teachers. These prototypes were eventually used as teaching aids (Amir, 2020). The cited projects are examples of linking science to D&T in a feasible manner in school practices. These classroom studies were in response to calls for better interaction between the two subjects over the last few decades (Allison, 1986; Barlex & Pitt, 2000; Gardner, 1994; Norman, 1993; Norman, 1998) and adds on to existing literature of strengthening the interaction between both subjects such as those in Doppelt et al. (2008), Hansen (2009), and Norton (2007).

3. METHODOLOGY

3.1 Development of a Questionnaire to Gather Students' Views

To address the research questions, the author developed a questionnaire comprising four categories. The questionnaire was validated by teachers and two members from academia in university. The questionnaire has 20 statements, which are placed on a six-point Likert Scale (Strongly Disagree = 1 and Strongly Agree = 6) and grouped in four categories. Elements from Guilford's (1959) and Torrance's (1979) framework for creativity (*fluency*, *flexibility*, *originality* and *elaboration*) and indicators of product creativity (*novelty*, *resolution*, and *style*) through Besemer's (1998) CPAM are infused in the second and third categories.

The first category in the questionnaire aims to gather students' views towards the learning mathematics, science and D&T through the project. The work of Fortus et al. (2004), Fortus et al. (2005), and Sidawi (2009) have shown that students could gain knowledge in science through their design-based projects. In this, the author placed the statement *'I learnt Pythagoras Theorem through this D&T project'* to find out if students felt that they learnt this mathematics concept through this project. Apart from gathering their views on gaining the concept, the author wanted to find out if students enjoyed learning the concept through this project. This led the author to place the statement *'I enjoyed learning Pythagoras Theorem through this D&T project'* has been placed in this category. Statements such as *'I enjoyed learning skills to design this prototype'* and *'I enjoyed learning skills to make this prototype'* attempt to gather students' views towards learning experiences in gaining design and make skills through this project. The two statements are essentially the same but with the words 'design' and 'make' placed in italic so that students can see a difference in them and respond accordingly. The author anticipates that students may deepen science concepts in the process of coming up with creative variations of their prototypes. This led the author to place the statement *'This D&T project has helped me increase my understanding of several science concepts'* has been placed in this category.

The second category in the questionnaire aims to gather students' views towards the use of design sheets in helping them articulate their creative use of scientific principles in the design-and-make process in the context of this project. Statements relating to the creativity traits of *fluency*, *flexibility* and *elaboration* are evident in this category. The statement *'I am able to come up with at least one idea that works on a scientific principle'* attempts to find out if students felt that they were able to come up with at least one design idea that incorporates the use of scientific principles, indicating flexibility. The statements *'I am able to show a description on how to make my chosen idea'* and *'I am able to explain how my chosen idea works using scientific principles'* attempts to find out if students felt that they were able to elaborate the use of scientific principles in their designs through their design sheets.

The third category in the questionnaire aims to gather students' views towards how the process of making their prototypes has helped them articulate their creative use of scientific principles in the design-and-make process in the context of this project. The statement *'I find that I get new ideas that work on scientific principles when I am exposed to science concepts during stages of making'* attempts to find out if students felt that they were able to come up with new ideas through the use of scientific principles (*fluency*, *flexibility* and *originality* since 'new' falls under *originality* sub-trait) when they are engaged in this project. Students could come up with ideas when they are exposed to mechanical toys (which was shown to them during the brainstorming and ideation process). The statements *'I find that I can incorporate more than one scientific principles during the process of making'* and *'I find that I can continuously improve my design using scientific principles during stages of making'* attempts to find out if students felt that they were able to improve their designs through the use of science principles when they were engaged in the making aspects. The words 'more' and 'continually improve' can be categorised under the *'fluency'* and *'flexibility'* sub-trait since these words suggest coming up with 'more' designs and 'different' designs. The statement *'My prototype works by combining what I have learnt in science to the design-and-make skills in D&T'* attempts to find out if students felt that they were able to see a link between the two subjects through this project. The statement *'My prototype will get students curious to find out how scientific principles makes it work'* attempts to find out if students felt that their prototypes would be of value for their peers to want to know more about how scientific principles contributes to the functionality of their designs ('value' falls under *originality* sub-trait).

The fourth category in the questionnaire aims to gather students' views towards the use of this project to promote interest, confidence and creativity by linking D&T to mathematics and science. The statement *'I am interested to do more D&T projects that can demonstrate mathematics concepts'* aims to find out if students are interested to do more similar projects that exhibit specific mathematics concepts. The statements *'I am more confident of learning Pythagoras Theorem after doing this D&T project'* and *'I find that D&T is a useful subject that can help me understand several mathematics and science concepts'* attempts to find out if students felt that D&T is a subject that can boost their confidence and get them to understand mathematics and science concepts. The statement *'I can demonstrate creative use of scientific principles through this mathematics-based D&T project'* attempts to find out if students felt that the project offered sufficient room for them to articulate their creativity utilizing scientific principles. The statement *'I am further interested to show how creative I can be by doing similar mathematics-based D&T projects in the future'* attempts to find out if students felt that they would want to further take up such mathematics-based D&T projects as a means to demonstrate their creativity.

There is also a blank space on the second page of the questionnaire for students to write down their comments from their experiences whilst doing the project.

3.2 Administration of Questionnaire

The questionnaire was distributed to the students at the end of the project during a D&T lesson. Students sat down with their design sheets and prototypes, as they filled up the questionnaire individually. They were told to give their honest views towards each of the statements in the questionnaire. They were also encouraged to write down their comments from their experiences whilst doing the project. While the questionnaire was anonymous, students were given the option to write their names next to their comments. Students took about 10 - 15 minutes to complete the questionnaire.

3.3 Analysing Students' Responses

The mean and standard deviation of students' response to each statement was calculated. In reporting the results, the term 'high score' is used for statements having mean scores above '5.50' and having standard deviations below '1' as this suggests the consistency in the number of students who opted for the same option. Students' written comments in the second page were analyzed qualitatively. Based on the work of conducting research through case studies by Stake (1995) and Yin (1989), such sentences can be broken down into phrases and interpreted to sieve out the key message. As such, several sentences in the comments given by students were broken down and key points of the comments interpreted. These comments were also used to support quantitative data that is obtained through the students' responses.

4. RESULTS AND DISCUSSION

4.1 Presentation of Results and Analysis

Table 1 presents the mean scores and standard deviations of the responses received from the 39 students. Based on the 39 responses, the questionnaire had a Cronbach alpha reliability of 0.95, which is above the recommended 0.70 (Cronbach, 1970). Results from the table indicate that, as a whole, students exhibited positive responses towards the project. This is evidenced by an overall high score of above '5.50' and standard deviation below '1' for each of the four categories.

Table 1. Students' views at the end of the Pythagoras Demo Kit project

	Categories	Mean	SD
	Learning maths, science and D&T through this project		
1.	I learnt Pythagoras Theorem through this D&T project.	5.84	0.44
2.	I enjoyed learning Pythagoras Theorem through this D&T project.	5.82	0.51
3	I enjoyed learning skills to *design* this prototype.	5.71	0.65
4.	I enjoyed learning skills to *make* this prototype.	5.81	0.40
5.	This D&T project has helped me increase my understanding of several science concepts.	5.86	0.42
	Average	**5.81**	**0.48**
	Use of design sheets		
6.	I am able to come up with a good design context for my project.	5.39	0.89
7.	I am able to come up with at least three ideas before choosing my final design.	5.58	0.76
8.	I am able to come up at least one idea that works on a scientific principle.	5.65	0.75
9.	I am able to show a description on how to make my chosen idea.	5.61	0.68
10.	I am able to explain how my chosen idea works using scientific principles.	5.55	0.76
	Average	**5.56**	**0.77**
	Making prototype through scientific principles		
11.	I find that I get new ideas that work on scientific principles when I am exposed to science concepts during stages of making.	5.84	0.44
12.	I find that I can incorporate more than one scientific principle during the process of making.	5.76	0.49
13.	I find that I can continuously improve my design using scientific principles during stages of making.	5.58	0.76
14.	My prototype works by combining what I have learnt in science to the design-and-make skills in D&T	5.76	0.49
15.	My prototype will get students curious to find out how scientific principles make it work.	5.61	0.72
	Average	**5.71**	**0.58**
	Promoting interest, confidence and creativity by linking D&T to mathematics and science through this project		
16.	I am interested to do more D&T projects that can demonstrate mathematics concepts.	5.66	0.63
17.	I am more confident of learning Pythagoras Theorem after doing this D&T project.	5.50	0.80
18.	I find that D&T is a useful subject that can help me understand several mathematics and science concepts.	5.66	0.67
19.	I can demonstrate creative use of scientific principles through this mathematics-based D&T project.	5.58	0.79
20.	I am further interested to show how creative I can be by doing similar mathematics-based D&T projects in the future.	5.50	0.95
	Average	**5.58**	**0.77**

4.2 Discussion in Response to Research Questions

Response to First Research Question Based on Analysis of Students' Responses in First Category

In response to the first research question, responses from students suggest that, as a whole, they are positive towards learning mathematics, science and D&T through this project. This is evidenced by an overall high score of 5.81 and standard deviation of 0.48 for the first category. Their responses to the first statement suggest that they are able to learn Pythagoras Theorem through this D&T project. This is evidenced by a high score of 5.84 and low standard deviation of 0.44. These high scores could be attributed to the students being able to visualize the concept of Pythagoras Theorem through the kit. Several written comments by the students further support the high agreement of students' response to this statement. These comments are as follows:

"I like the Pythagoras toy because it can teach math in an interesting way instead of using the white board and teach." - Male 13 years old

"I like this project because it make me and my classmates understand the mathematics concept." - Female 13 years old

"I like this project because I can learn math in a fun and easy way." - Male 13 years old

In all three comments we can link the words 'teach', 'understand' and 'learn' to 'teaching and learning'. The comment from the first student suggests that the project can be a useful teaching resource that could present the concept of Pythagoras Theorem in an interesting way, as the student mentioned '*it can teach math in an interesting way*'. The same comment suggests that this project can be a useful tool for students to understand Pythagoras Theorem as the student mentioned '*make me and my classmates understand the mathematics concept*'. In the third comment, we can see the student mentioned the word 'easy' which suggests that the prototype seems useful to help them scaffold the understanding of the concept.

The students' responses also suggest that they found that they enjoyed learning Pythagoras Theorem through this project. This is evidenced by a high score of 5.82 and standard deviation of 0.51 through the second statement. This high score could be attributed to the novelty of the pedagogical approach that involves making a kit whilst learning the concept, as well as articulating their creativity through knowledge from science. Several comments by the students further support the high agreement of students' response to this statement. These comments are as follows:

"I like this because maths, D&T, Science are combined together." - Female 14 years old

"I like this project because it is better learning this way than learning on the whiteboard and it's fun " - Female 13 years old

"I like this project because its fun and exciting." - Female 13 years old

In all three comments, we can link the word 'like' to 'enjoyment'. The first comment suggests that the student enjoys this project because it links mathematics and science to D&T. The second comment suggests that the student likes the project because she prefers learning through making the kit rather than through the whiteboard as she mentioned '*better learning this way than learning on the whiteboard*'. The third comment suggests the student liked the project because of appeal factors related to 'fun' and 'excitement'.

Responses from students also suggest that they enjoy learning D&T skills in designing and making through this project. This is evidenced by a high score of 5.71 and standard deviation of 0.65 through the third statement, and a high score of 5.81 and low standard deviation of 0.40 through the fourth statement. A possible reason for the slightly higher score towards the making aspects as compared to the design aspects could be due to students having more 'fun' making rather than designing their prototypes. Several comments by the students further support the high agreement of students' response to this statement. These comments are as follows:

"I like this project because doing project is fun and can learn science and math the different way" - Female 13 years old

"I like the maths toy because it is fun doing it and can understand more" - Male 13 Years old

"I like this project because its fun to make it and you can easily understand the concepts." - Female 13 years old

In the comments above, we can link the word 'doing' to 'making'. These comments above suggest that students enjoyed (*like*) the project. The first comment suggests that the student enjoyed doing the project because of the appealing nature of the project and that it offers a different way to learn mathematics and science ('*learn science and math the different way*'). The words 'understand more' and 'easily understand the concepts' in the second and third comments suggest that students were able to understand the concept of Pythagoras Theorem and science concepts through designing and making the kit.

The students' responses also suggest that they were also able to increase their understanding of several science concepts. This is evidenced by a high score of 5.86 and low standard deviation of 0.42 through the fifth statement. Several comments by the students further support this. These comments are:

"I like this project because its fun to make it and you can easily understand the concepts." - Female 13 years old

"I like this project because it helps me to learn math and science fun and I will try harder to make my toy creative." - Male 13 years old

"I like the science project because it helps me learn maths and science concepts" - Male 13 years old

In the comments, we can link the words 'easily understand', 'learn math and science' to 'understanding'. All three comments indicate students' understanding of science concepts. This could be attributed to students being able to see scientific principles 'in action' through designing and making the kit. These could include the use of mechanical principles and improving the rate of flow of beads to add value to the initial prototype.

Response to Second Research Question Based on Analysis of Students' Responses in Second and Third Categories

In response to the second research question, responses from students suggest that as a whole, they are positive towards the design-and-make process in helping them articulate their creativity through the use of scientific principles in this project. This is evidenced by an overall high score of 5.56 and standard deviation of 0.77 for the second category, and an overall high score of 5.71 and standard deviation of 0.58 for the third category.

Responses from students to statement 6 suggest that while they did not find it difficult to come up with a design context for their projects, it was also not too easy in coming up with design contexts. This is evidenced by a high score of 5.39 and standard deviation of 0.89. This is the only statement that had a score of under '5.50'. The design context in this project required students to come up with design brief and specifications of the project. Students could be limited in coming up with a good design context as the end-users for their prototypes are somewhat pre-defined (mathematics teachers to use the prototype in the classroom). This may limit students in being able to generate ideas for their prototypes to be used across different contexts. For example, if the end-user was not pre-defined, students could come up with prototypes that would suit various users and in different settings – such as using them as a mini exhibits in the mathematics corner in school, or sold as a commercially available product to be used at home, or even as a gift for the mathematics faculty in a university, or another school. Nevertheless, the mean score is still considerably high at 5.39 which indicates a positive response. Responses from students also suggest that they were able to come up with at least three ideas before choosing their final designs (*flexibility* trait). This is evidenced by a high score of 5.58 and standard deviation of 0.76 through statement 7. This could be attributed to the brainstorming sessions in the course of coming up with variations of their designs. The prototypes produced by the students suggest that they had taken some suggestions from the author during brainstorming sessions to come up with several of their own design ideas. For example, in one of the brainstorming sessions, the author suggested two ways to improve the functionality of the initial prototype. First was improving the rate of flow of beads between the three compartments of the initial prototype. One way this to do this is to reduce the surface friction between the Styrofoam beads (that is in the initial prototype) and the plastic surface in contact with the beads. A second suggestion is to improve the way the initial prototype is being rotated. The partial clockwise and anti-clockwise rotation that is required (so that the beads will flow from the largest compartment to the two smaller compartments) may tire off a mathematics teacher. These ideas were adapted by the students and incorporated into their designs. Responses from students also suggest that they were able to come up with at least one idea that utilizes scientific principles (*flexibility* trait). This is evidenced by a high score of 5.65 and standard deviation of 0.75 through statement 8. Again, this could be attributed to brainstorming sessions and the use of 'idea triggers' such as mechanical toys in the course of coming up with variations of the initial prototype. Responses from students also suggest that they were able to show a description as to how to make their chosen ideas in their design sheets (*elaboration* trait). This is evidenced by a high score of 5.61 and standard deviation of 0.68 through statement 9. This could be attributed to the monitoring process of the students' journals (which is a compilation of their design sheets). During the process of designing and making, students who did not met up to achieving certain standards in their design pages were given feedback to improve their work. The author did this through writing on sticky notes and attaching them on the design sheets as well as providing verbal feedback to students. Responses from students also suggest that they were able to explain how their chosen idea works using scientific principles (*elaboration* trait). This is evidenced by a high score of 5.55 and standard deviation of 0.76 through statement 10.

Responses from students suggest that they get new ideas that utilize scientific principles when they were exposed to science concepts during stages of making, and that they were able to incorporate more than one science concept in their designs during the process of making. This is evidenced by a high score of 5.84 and low standard deviation of 0.44 through statement 11, and a high score of 5.76 and low standard deviation of 0.49 through statement 12. There is evidence of these through the students' prototypes. Several students' designs featured the use of fine sand in place of the Styrofoam beads to increase the rate of flow between the compartments of the kit, as well as the use of mechanical principles. Responses from students also suggest that they were able to continuously improve their designs using science during stages of making ('continually improve' can be categorised under the *fluency* and *flexibility traits*). This is evidenced by a high score of 5.58 and standard deviation of 0.76 through statement 13. This can be attributed to encouraging students to 'mix and match' materials. Ideas were triggered when they 'mixed-and-matched' materials that were found in the D&T workshop, as well as other simple materials found at home or elsewhere. For example, one student came up with the design that initially only made her kit swivel along the hypotenuse. It was not towards the later part of the making stages that the student realized that her design would still require the user to make partial clockwise and anti-clockwise rotations of the kit in order to make the fine sand flow from the biggest square compartment to the two smaller compartments to completely fill them up, and vice versa. Her solution to this problem was to add another pivot and springs at the base of her design as shown in Figure 3. This multiple use of pivot points and springs allowed users to see the visual proof of the theorem through mechanical principles.

Figure 3. Using multiple pivot points and springs to aid the movement of fine sand flowing between the three compartments
Source: Author

Responses from students also suggest that their prototypes work by combining what they have learnt in science to the design-and-make skills in D&T. This is evidenced by a high score of 5.76 and low standard deviation of 0.49 through statement 14. Responses from students also suggest that their prototypes will get students curious to find out how science makes them work (*originality* trait). This is evidenced by a high score of 5.61 and standard deviation of 0.72 through statement 15. This is also evidenced by the 'element of surprise' sub-traits of *originality* that are evident in the students' designs that would evoke curiosity in their designs.

Response to Third Research Question Based on Analysis of Students' Responses in Fourth Category

In response to the third research question, responses from students suggest that as a whole, they find that the project promotes interest, confidence and creativity by linking D&T to mathematics and science. This is evidenced by a high score of 5.58 and standard deviation of 0.77 for this category. Responses from students suggest that they are interested to do more D&T projects that can demonstrate mathematics concepts. This is evidenced by a high score of 5.66 and standard deviation of 0.63 through statement 16. This can be attributed to the nature of the project being a simple yet interesting mathematics-based kit that offers a way for students to visualize the concept of Pythagoras Theorem, and one that can be made in an average D&T workshop in a feasible manner. Responses from the students suggest that they are more confident of learning Pythagoras Theorem after doing this D&T project, and that they find D&T a useful subject that can help them understand several mathematics and science concepts. This is evidenced by a high score of 5.50 and a standard deviation of 0.80 through statement 17, and a high score of 5.66 and a standard deviation of 0.67 through statement 18. Several comments by the students further support the high agreement of students' response to this statement. These comments are as follows:

"I like this project because it <u>helps me to learn</u> math and science fun and I will try harder to make my toy creative." - Male 13 years old

"I like the science project because it <u>helps me learn</u> maths and science concepts" - Male 13 years old

"I like this project because it is simple in Sec 2 <u>when we are learning</u> about Pythagoras Theorem it is <u>easy to understand</u>." - Female 13 years old

"I like the math toy because it <u>give me a confident to learn math</u>" - Male 13 years old

The first two comments suggest that the project has helped them learn Pythagoras Theorem. This can be attributed to how students may need help in visualizing algebraic concepts that may be abstract. Sudarshan and Khin (2008) suggested that educators look into how students can see the relevance of mathematics in their real life and make a *'conscious effort in crafting activities and tasks which will help the students appreciate and enjoy what they learn in school (p 17)'*. In addition, Butcher (2005) mentioned the need for minimal number of algebraic equations to be used in introducing mathematical concepts in physics. He highlighted that students may develop a fear of studying physics if the teacher places emphasis on just the mathematical equations. Naraine (1993) also highlighted the importance of teachers to get students to convince themselves on the validity of particular mathematical concepts through a hands-on approach. Presmeg (1986) and Trumbo (2006) have shown that in a number of instances, mathematics teachers tend to emphasise the teaching of algebraic concepts through verbal modes at the expense of visual and spatial modes in their classroom practices. The third comment suggests that the project could be useful in helping students understand the concept of Pythagoras Theorem by making it *'easy to understand'*. This, in turn, can help them boost their confidence in mathematics as evidenced by the fourth comment – *'give me a confident to learn math'*.

Responses from the students also suggest that they were able to demonstrate creative use of scientific principles through this mathematics-based D&T project. This is evidenced by a high score of 5.58 and standard deviation of 0.79 through statement 19. The variations of designs produced by the students indicates the potential for this project being useful to link D&T to mathematics and science. A comment from a student shows further evidence of this:

"I like this project because I can show we are creative*." - Male 13 years old*

The comment made by the student suggests that he enjoys the project as it had offered him and avenue to demonstrate his creativity. Students' responses to statement 20 suggest that they are further interested to show how creative they can be by doing similar mathematics-based D&T projects in the future. This is evidenced by a high score of 5.50 and a standard deviation of 0.95 for this statement.

5. CONCLUSION

It is important to gather students' views towards their experiences in learning across all subjects. Such information can be useful for teachers to affirm, refine and enhance their practices to better motivate, engage and foster creativity and inventiveness through a variety of pedagogical approaches. To date, little has been shared on students' views towards demonstrating their creative use of scientific principles through designing and developing prototypes in D&T that exhibit specific mathematical concepts which are within the mathematics syllabus that the students are studying. The positive views of students from this classroom-based research highlight that it is possible for them to be guided through the design-and-make process to demonstrate their creativity by coming up with design variations of a mathematics-based kit through a variety of scientific principles. As part of professional development in an area of STEM teaching, the positive views from students suggest that D&T, mathematics and science teachers can work together to identify toys, exhibits and other manipulatives that show visual proofs of specific mathematics concepts. D&T can be an anchor subject for the development of such projects in school practices. This is in view of introducing students to design and make skills in D&T (within the D&T syllabus) through the project, as well as linking the prototype in the project to the specific concept that mathematics teachers are teaching students (within the mathematics syllabus). Apart from exploring and identifying manipulatives that exhibit specific mathematics concepts, mathematics teachers can serve as the end users of this real-world design project, providing ergonomics and anthropometric data to the student inventors. Mathematics teachers can also contribute to the project by collaborating with D&T teachers to guide students to come up with various designs of the kit based on mathematical elements. For example, mathematics teachers could guide students to make use of other combinations of Pythagorean triples, instead of just sticking to the 3cm-4cm-5cm right-angled triangle combination. This can promote students' understanding of Pythagoras theorem in a deeper manner. Science teachers can use the creative prototypes developed by the students as teaching aids to highlight creative use of scientific principles 'in action'. Science notes and worksheets can be developed based on the students' creative designs, highlighting the scientific principles in the context of the D&T project. Seeing diagrams of their prototypes in science notes and worksheets can help students identify and relate with a familiar context. Doing this, along with other necessary scaffolds, can allow students to develop confidence and deepen their knowledge and understanding of science concepts through paper-and-pen modes of learning (Amir, 2021). When using students' prototypes as teaching aids, D&T, science and mathematics teachers can give credit to students who came up with the creative and useful designs. Students' design sheets and their prototypes can also be shown to parents of the students to highlight their children's inventiveness. Doing this can boost the self-worth of the students in an area of STEM (Amir, 2021a). These design sheets and creative prototypes can also serve as an inspiration for other students to develop even further creative variations.

This study contributes to strengthening the links between D&T, mathematics and science in a feasible manner in school practices.

6. LIMITATIONS

The findings from this classroom-based research are restricted to one class of students and must thus be viewed in this context.

REFERENCES

Ainley, J., Pratt, D., & Hansen, A. (2006). Connecting engagement and focus in pedagogic task design. *British Educational Research Journal*, *32*(1), 23–38. doi:10.1080/01411920500401971

Allison, J. C. (1986). CDT - craft, design and technology. *Physics Education*, *21*(6), 337–339. doi:10.1088/0031-9120/21/6/303

Amir, N. (2014). Showcasing the creative talents in science of the academically less-inclined students through a values-driven toy story-telling project. In L. C. Lennex & K. F. Nettleton (Eds.), Cases on Instructional Technology in Gifted and Talented Education (pp. 141-179). IGI Global Publishing. doi:10.4018/978-1-4666-6489-0.ch007

Amir, N. (2020). Strengthening the science-D&T interaction through simple maker-centred projects in a Singapore classroom. *Physics Education*, *55*(4), 045019. doi:10.1088/1361-6552/ab9212

Amir, N. (2021a). Fostering joy and creativity amongst students in a Singapore classroom through fun design-and-make STEM projects. In D. Ktoridou, E. Doukanari & N. Eteokleous (Eds.), Fostering Meaningful Learning Experiences through Student Engagement (pp. 51-73). IGI Global Publishing. doi:10.4018/978-1-7998-4658-1.ch003

Amir, N. (2021b). Views from parents of at-risk students in Singapore towards the use of fun design-and-make STEM projects to foster creativity and joy of learning. In C. Gaines & K. Hutson (Eds.), Promoting Positive Learning Experiences in Middle School Education (pp. 106-131). IGI Global Publishing. doi:10.4018/978-1-7998-7057-9.ch006

Amir, N. (2022). Fostering Student Motivation and Engagement Through the Relevant, Appealing, and Personal (RAP) Pedagogical Guideline. In Transformative Teaching Around the World: Stories of Cultural Impact, Technology Integration, and Innovative Pedagogy (pp. 87-91). Routledge.

Amir, N., & Subramaniam, R. (2006). Making physics toys fosters creativity. *Physics Education*, *41*(1), 18–20. doi:10.1088/0031-9120/41/1/F05

Amir, N., & Subramaniam, R. (2009). Making a low cost candy floss kit gets students excited about learning physics. *Physics Education*, *44*(4), 420–428. doi:10.1088/0031-9120/44/4/013

Amir, N., & Subramaniam, R. (2012). Fostering inquiry in science among kinaesthetic learners through Design & Technology. In L. C. Lennex & K. F. Nettleton (Eds.), Cases on Inquiry Through Instructional Technology in Math and Science: Systemic Approaches (pp. 221-257). IGI Global Publishing. doi:10.4018/978-1-4666-0068-3.ch009

Barlex, D., & Pitt, J. (2000). *Interaction: The relationship Between Science and Design and Technology in the Secondary School Curriculum*. Engineering Council.

Besemer, S. P. (1998). Creative product analysis matrix: Testing the model structure and a comparison among products – three novel chairs. *Creativity Research Journal*, *11*(4), 333–346. doi:10.120715326934crj1104_7

Butcher, F. (2005). Two personal reviews of the European Physics Education Conference in Bonn. *Physics Education*, *40*(5), 397–402.

Caine, R. N., & Caine, G. (1991). *Making connections: teaching and the human brain*. Association for Supervision and Curriculum Development.

Cronbach, L. J. (1970). *Essentials of psychological testing* (3rd ed.). Harper and Row.

Davidson, M., Evens, H., & McCormick, R. (1998). Bridging the gap: the use of concepts from science and mathematics in design and technology at KS 3. *International Design and Technology Education Research and Curriculum Development (IDATER), Loughborough University of Technology*, 48-53. Retrieved July 1, 2022, from https://dspace.lboro.ac.uk/dspace-jspui/handle/2134/1419

Doppelt, Y., Mehalik, M., Schunn, C. D., Silk, E., & Krysinski, D. (2008). Engagement and achievements: A case study of design-based learning in a science context. *Journal of Technology Education*, *19*(2), 22–39.

Eriksen, D., Stasiuk, J., & Frank, M. (1995). Bringing Pythagoras to life. *Mathematics Teacher*, *88*(9), 744–747. doi:10.5951/MT.88.9.0744

Fogarty, R. (1991). Ten ways to integrate curriculum. *Educational Leadership*, *49*(2), 61–65.

Fortus, D., Krajcikb, J., Dershimerb, R. C., Marxc, R. W., & Mamlok-Naamand, R. (2004). Design-based science and student learning. *Journal of Research in Science Teaching*, *41*(10), 1081–1110. doi:10.1002/tea.20040

Fortus, D., Krajcikb, J., Dershimerb, R. C., Marxc, R. W., & Mamlok-Naamand, R. (2005). Design-based science and real-world problem-solving. *International Journal of Science Education*, *27*(7), 855–879. doi:10.1080/09500690500038165

Furner, J. M., & Kumar, D. D. (2007). The mathematics and science integration argument: A stand for teacher education. *Eurasia Journal of Mathematics, Science and Technology Education*, *3*(3), 185–189. doi:10.12973/ejmste/75397

Gardner, P. (1994). Representations of the relationship between science and technology in the curriculum. *Studies in Science Education*, *24*(1), 29–47. doi:10.1080/03057269408560037

Guilford, J. P. (1959). Traits of Creativity. In H. H. Anderson (Ed.), *Creativity and Its Cultivation* (pp. 142–161). Harper & Row.

Hansen, P. J. K. (2009). Analysing cases in technology and design education: How could designing and making technological products be a vehicle for enhancing understanding of natural science principles? *Design and Technology Education: An International Journal*, *14*(2), 45–52.

Lee, M. (2007). Spark up the American revolution with math, science, and more: An example of an integrative curriculum unit. *Social Studies*, *98*(4), 159–164. doi:10.3200/TSSS.98.4.159-164

Naraine, B., Dotseth, K. A., Kim, G., & Moody, M. (1993). If Pythagoras had a geoboard. *Mathematics Teacher*, *86*(2), 137–139. doi:10.5951/MT.86.2.0137

Norman, E. (1993). Science for design. *Physics Education*, *28*(5), 301–306. doi:10.1088/0031-9120/28/5/010

Norman, E. (1998). The Nature of Technology for Design. *International Journal of Technology and Design Education*, *8*(1), 67–87. doi:10.1023/A:1008827820764

Norton, S. J. (2007). The use of design practice to teach mathematics and science. *International Journal of Technology and Design Education*, *18*(1), 19–44. doi:10.100710798-006-9019-8

Presmeg, N. C. (1986). Visualisation and mathematical giftedness. *Educational Studies in Mathematics*, *17*(3), 297–311. doi:10.1007/BF00305075

Sage, J., & Steeg, T. (1993). Linking the learning of mathematics, science and technology within key stage 4 of the national curriculum. *International Conference on Design and Technology Educational Research and Curriculum Development, Loughborough University of Technology*, 58 – 64. Retrieved July 1, 2022, from: https://dspace.lboro.ac.uk/dspace-jspui/bitstream/2134/1579/3/sage93.pdf

Sharkawy, A., Barlex, D., Welch, M., McDuff, J., & Craig, N. (2009). Adapting a Curriculum Unit to Facilitate Interaction Between Technology, Mathematics and Science in the Elementary Classroom: Identifying Relevant Criteria. *Design and Technology Education: An International Journal*, *14*(1), 7–20.

Sidawi, M. (2009). Teaching science through designing technology. *International Journal of Technology and Design Education, 19*(3), 269–287. doi:10.100710798-007-9045-1

Spendlove, D., & Hopper, M. (2005). D&T and ICT. In *Creativity in the Primary Curriculum* (pp. 75–90). Routledge.

Stake, R. E. (1995). *The art of case study research.* Sage Publications.

Sudarshan, A., & Khin, M. A. (2008). Teacher Practices in Mathematics Classrooms with At-risk Students. *Proceedings of the Meeting of the Australian Association for Research in Education.*

Torrance, E. P. (1979). *The search for satori and creativity.* Creative Education Foundation and Creative Synergetic Associates.

Trumbo, J. (2006). Making science visible: Visual literacy in science communication. In L. Pauwels (Ed.), *Visual culture of science: Re-thinking representational practices in knowledge building and science communication* (pp. 266–283). Dartmouth College Press, University Press of New England.

Urich, J. A., & Sasse, E. A. (2011). An Ap"peel"ing Activity. *Mathematics Teacher, 105*(3), 189–193. doi:10.5951/mathteacher.105.3.0189

Yin, R. K. (1989). *Case study research: Design and methods.* Sage.

Zubrowski, B. (2002). Integrating science into Design Technology projects: Using a standard model in the design process. *Journal of Technology Education, 13*(2), 48–67. doi:10.21061/jte.v13i2.a.4

ADDITIONAL READING

Amir, N. (2015). Using corrugated boards to introduce the concept of 'turning effect of forces' to the less-academically inclined students. *The School Science Review, 97*(358), 22–25.

Amir, N. (2018). Visualising 'work done' with a simple flying wheel toy. *Physics Education, 53*(4), 043003. doi:10.1088/1361-6552/aabb09

Amir, N., & Abdullah, N. (2021). 'Dancing dolls in resonance': A simple design-and-make STEAM toy project to promote student interest and engagement in physics. *Physics Education, 56*(2), 025022. doi:10.1088/1361-6552/abcf0f

Amir, N., & Subramaniam, R. (2007). Making a fun Cartesian diver: A simple project to engage kinaesthetic learners. *Physics Education, 42*(5), 478–480. doi:10.1088/0031-9120/42/5/004

Amir, N., & Subramaniam, R. (2014). The Structural Challenge: A simple design-based science activity to foster creativity among kinaesthetic learners. *The School Science Review, 95*(353), 73–78.

KEY TERMS AND DEFINITIONS

Design and Technology (D&T): A subject offered in secondary schools in Singapore where students learn mainly through design-and-make projects guided by a design process.

Design Journal: Compilation of students' design sheets.

Design Sheets: A document showing the thought processes of students as they come up with design ideas.

Fostering Creativity in Science: Process of guiding students to come up with original and valuable ideas that draws knowledge from science in the course of solving problems.

Prototype: A physical product developed by students that aims to address a design need.

STEM: Science, technology, engineering, and mathematics.

Variations of Design: A variety of designs that are different from a standard design shown to students.

Chapter 6

Design and Implementation of Practical Workshops for Teaching STEM Content:

Analysis of Cognitive and Emotional Variables

Milagros Mateos Núñez
University of Extremadura, Spain

Guadalupe Martínez Borreguero
University of Extremadura, Spain

ABSTRACT

The main objective of this work has been to assess the potential cognitive and emotional benefits of several practical experiences integrating STEM areas. A sample of 13 students in the 6th grade of primary school was accessed. Several questionnaires were designed, depending on the contents to be dealt with, and were applied at different times in order to ascertain the evolution of the variables under study. The data from the post-tests applied after the workshops confirm that the practical and integrated sessions carried out with the students improve the level of knowledge, even finding statistically significant differences (Sig.< 0.05) with respect to the initial values. The emotional results show that following an active and integrated methodology has benefited the promotion of positive emotions in the participants.

INTRODUCTION

The rapid development of science and technology in the current century has created great competition in the global economy. Many global challenges such as climate change, overpopulation, natural resource management, agricultural production, health or biodiversity require international engagement supported by further development of science and technology (Thomas & Watters, 2015). However, to solve these

DOI: 10.4018/978-1-6684-5765-8.ch006

continuous challenges, it is essential to have personnel with the necessary scientific-technological knowledge.

Considering the previous context, it could be accepted that experiences at an early age are determinant in later learning, so that focusing on integrated practical education during the early school years can be key to mitigate vocational declines in these areas (Bagiati et al. 2010) and promote scientific literacy in the youngest (Neira, 2021). For this reason, the main objective of this research has been to assess the possible benefits at the cognitive and emotional level of various practical experiences that integrate STEM areas in the 6th grade of primary education.

BACKGROUND

Currently, there is a deficit of entrepreneurial and creative individuals who possess critical thinking and communication skills, who can work cooperatively, find solutions to everyday problems, and use technology well in everyday situations (Tanik-Onal & Saylan- Kirmizigul, 2021). More recent studies confirm that as students move from elementary school to high school, many choose not to pursue science (King et al., 2015; Vazquez & Manassero, 2011). The causes that may lead to this academic decline are linked to students' affective domain (Mellado et al., 2014; Osborne et al., 2003). Experts in the field point out that there is an emotional decline throughout schooling (Murphy & Beggs, 2003; Vázquez & Manassero, 2011; Mateos-Núñez et al., 2019a). Generally, elementary school students tend to have positive emotions and attitudes towards mathematics and science learning (Mateos-Núñez et al., 2019a; Martínez-Borreguero et al., 2020), but these attitudes decrease with age, especially during secondary education (Beauchamp & Parkinson, 2008; Marbá & Márquez, 2010; Vázquez & Manassero, 2008). Specifically, previous studies show worrying results about the attitudes of secondary school students towards scientific-technological subjects (Mellado et al., 2014; Pérez & de Pro, 2013), especially towards the subject of Physics and Chemistry (Dávila et al., 2016), as most students describe the content of these classes as boring, difficult and distant from their own reality (Cleaves 2005; Lyons, 2006; Osborne et al., 2003). Consequently, the role played by the teacher is considered decisive in the face of this motivational problem, which places the teacher as one of the main actors in promoting vocations (Mellado et al., 2014) and fostering an adequate scientific-technological literacy.

The low levels of scientific literacy and the lack of interest on the part of citizens could be linked to the lack of contextualization in the classroom of scientific-technological topics with everyday life, through examples or nearby phenomena (Busquets et al., 2016). Today's students need to develop skills for an eminently scientific-technological society and be able to perform competitively in the new labor market. Therefore, effectively answering the questions that children ask about their environment is one of the main objectives of science education (Tanik-Onal and Saylan- Kirmizigul, 2021). Teachers have a great responsibility to keep children's interest and curiosity alive, answer their questions patiently, and instill in them an inclination for science. For science teaching to be effective, teachers need to ensure that children keep asking questions and are able to find answers. Therefore, the way we have seen and understood school education years ago cannot be the same today, and must urgently adapt to these new conditions (Fernández, 2006). In this sense, authors indicate that one solution could be to apply active methodologies that promote active learning (Freeman et al., 2014). The ideal would be to promote learning situations in which students are able to contrast and analyze different realities and this allows them to learn for life (Pozo and Gómez-Crespo, 1998). Consequently, education experts stress the importance of promoting education that is not only conceptual but also knowledge-integrative.

Numerous studies have shown that the integration of mathematics and science has a positive impact on students' attitudes and interest, motivation to learn and achievement (Honey et al., 2014; Mateos-Núñez et al., 2020). Activities that integrate different scientific areas are a teaching medium highly valued by teachers and researchers because they allow an adequate acquisition of knowledge, skills and competencies for the development of scientific literacy and, therefore, critical thinking. In addition, they are activities that encourage collective work, have the ability to adapt in real time to the conditions of the participants and, above all, combine practice and theory (action and reflection) in processes aimed at building very enriching scientific experiences. In this sense, STEM (Science, Technology, Engineering and Mathematics) projects emerge as a proposal to support scientific literacy in society. They are based on the integration of scientific disciplines (Bybee 2010) and have begun to be developed in formal education in many countries.

The STEM concept began to take shape in the 1990s by the National Science Foundation in order to modernize the school system within an increasingly industrial and technological and less agrarian society (NSB, 2004). However, according to Friedman (2005) it was not until the creation of a degree in STEM education at Virginia Tech in 2005 that the movement began to succeed, and it was an event that signified the gradual international expansion of STEM until it became widespread (Sanders, 2009). The term STEM then moved into the field of education as the essential tool to meet the growing demand for scientific-technological expertise (Chiu and Duit, 2011).

As STEM education becomes a greater focus for an increasing number of schools and teachers, the need to better define what is meant by the term STEM education becomes more apparent. As discussed above, the term STEM only alludes to the terms science, technology, engineering and mathematics. However, STEM education can be considered a meta-discipline based on the standards that reside at the school level, where all teachers, especially science, technology, engineering, and mathematics (STEM) teachers, take an integrated approach to teaching and learning, where discipline-specific content is not divided, but rather addressed and treated as a combined and dynamic subject (Kennedy and Odell, 2014). In this sense STEM education is recognized as a new way of jointly teaching Science, Mathematics and Technology with two distinct characteristics (Sanders, 2009): promoting a teaching-learning of Mathematics, Science and Technology in an integrated manner, rather than as separate knowledge areas and including Engineering approach, solving complex real-world problems through the design and construction of a working technological object.

STEM education is oriented to the development of STEM competence, understanding this as the ability to identify and apply both the key knowledge and the ways of doing, thinking, speaking and feeling of science, engineering and mathematics, in a more or less integrated way, in order to understand, decide and/or act in front of complex problems and to build creative and innovative solutions, taking advantage of personal synergies and available technologies (Couso, 2017). There are many benefits related to the use of STEM education (Stohlmann et al., 2012). Authors such as Furner & Kuman (2007) indicate that the use of an interdisciplinary or integrated curriculum provides opportunities for more relevant, less fragmented, and more stimulating experiences for students. Other authors point out that integrated STEM education is student-centered, enhancing higher-order thinking skills, problem-solving abilities, and concept retention (Fllis & Fouts, 2001; King & Wiseman, 2001). STEM education also involves students being innovators, inventors, self-reliant, logical thinkers, and technologically literate (Morrison, 2006).

The STEM approach must combine content from each of the disciplines mentioned above. It is not enough that the hypothetical possibility of connection exists; it is necessary to work the four disciplines explicitly (Ferrada et al., 2018). For this reason, educational institutions have begun to restructure the curriculum to include contextualized and content-integrating learning situations. Furthermore, it is evident that, in the case of scientific-technological-mathematical education, the ideal would be to propitiate learning situations in which students are able to contrast and analyze diverse realities and allow them lifelong learning (Pozo & Gómez-Crespo, 1998). Along these lines, any adaptation to the curriculum must consider that students who are literate under a STEM approach must be technically well prepared for the world of work in order for them to become creators, innovators and problem solvers (Bybee, 2013). Therefore, each area included in STEM education has to incorporate a particular set of skills so that, when managed together, the much demanded scientific literacy is achieved.

The treatment of STEM skills in the classroom inserts the student in a dimensional learning of science, allowing him/her to acquire a broad scientific-technological knowledge and to approach real situations without problems (Mateos-Núñez et al., 2019b). On the other hand, the need to put these skills into practice from early ages derives in the achievement of greater self-confidence and security when facing them in higher grades. Therefore, it is becoming increasingly necessary to train citizens in science and technology issues due to the numerous advances that arise every day in today's society as specified in different European reports (European Commission, 2010; OECD, 2018). The most adequate preparation does not imply a mere scientific knowledge from a cognitive point of view, but emotional and competence variables must also be taken into account, so that the student has a positive attitude towards scientific subjects and knows how to put into practice the knowledge worked in the classroom. STEM education is not just a "convenient integration" of the four disciplines, but encompasses real-world problem-based learning linked to the STEM disciplines. This educational approach implies that the STEM disciplines cannot and should not be taught in isolation, so the right approach in the classroom will require cohesive, active, and exciting teaching and learning approaches (English et al., 2017). Consequently, a change in curriculum approach must begin with well-trained and willing teachers (Borrego and Henderson, 2014). Pedagogical and interdisciplinary capacity building is needed to enable teachers to effectively plan learning activities that involve the integration of STEM disciplines in a given context (Sanders, 2009).

METHODOLOGY

The study carried out has a mixed character (quantitative and qualitative) and follows an exploratory design because it investigates a problem that is not clearly defined and the purpose is to better understand the usefulness of the STEM paradigm in the primary stage. For data collection, a pre-test and a post-test were applied. The variables under study were the cognitive domain and the affective component of the students.

Sample of Research

The sample participating in the research was selected on the basis of convenience sampling, due to the ease of access to it. Specifically, a sample of 13 students in the 6th grade of primary school between the ages of 11-12 years old was accessed. Of the 13 participants, seven were girls and six boys.

Measuring Instrument

Two questionnaires were designed for each STEM practice, one applied as a pre-test, before the corresponding workshop, and another, called post-test, applied several weeks later.

The purpose of the pre-test questionnaires was to check whether the students remembered the contents of the questions, since if difficulties were observed when answering these questions, it was understood that they had forgotten the contents and misconceptions had arisen. The pre-tests had a similar structure, but were composed of questions related to the scientific and mathematical contents worked on in each workshop. In short, they were intended to analyze the initial level of knowledge shown by the students.

The second questionnaire or post-test was structured in two sections in all cases. The first section included questions very similar to those of the pre-test. The idea was to find out whether, over time, the learning acquired in the workshops was maintained. In addition, the post-test contained a section to evaluate the emotions expressed by the students during the workshops. The emotions selected were curiosity, amusement, confidence, satisfaction, disgust, boredom, worry and anger. The students had to indicate whether or not they had felt each emotion during the didactic interventions received in each case. They also had to determine at which moment or moments they had felt the selected emotions. Finally, to assess the students' attitudes, the post-test included 10 statements on methodological, learning and competency aspects related to what was dealt with in the classroom. In short, the purpose of this questionnaire was to analyze, on the one hand, the level of knowledge shown by the students some time after the interventions, and on the other hand, the attitudes and emotions shown by the students towards the learning of scientific-mathematical contents and the development of technological skills.

The validity and reliability of the questionnaires was determined by a group of experts. Expert judgment is a validation method useful to verify the reliability of a research according to authors (Robles-Garrote and Rojas, 2015). It is defined as an informed opinion of people with trajectory in the subject, who are recognized by others as qualified experts in it, and who can give information, evidence, judgments and valuations (Escobar-Pérez and Cuervo-Martínez, 2008). Considering the above, the questionnaire was reviewed by expert teachers in didactics of experimental sciences with the aim of checking whether the questions included in the questionnaires were adapted to the level of the participating subjects or whether their wording was appropriate for the achievement of the objectives. The data provided by the group of experts indicate that the questionnaire is favorable for the research.

Design and Development of Interventions

The interventions designed were carried out during the 2022-2023 academic year, in order to work on topics of the curricular subjects Nature Sciences and Mathematics. However, the workshops were intended not only to consolidate the learning of scientific-mathematical content, but also to develop manipulative skills in engineering and technology through design and construction. A total of 5 STEM projects were carried out on the contents under study. Both the process and the product are evaluated, the former being considered essential.

The first intervention (STEM Workshop 1) was integrated into the didactic unit "Biodiversity and the environment" and was based on the design and construction of garbage containers in order to raise awareness about the importance of recycling different materials. This workshop deepened the learning of Nature Science contents related to the effects of people on the environment and its conservation. We also worked on contents included in the didactic unit "Area of plane figures" of Mathematics. Figure 1 shows the containers made by some students in the STEM Workshop 1.

Figure 1. Containers made in STEM Workshop 1
Source: Authors

The second intervention (STEM Workshop 2) was carried out during the didactic unit "The locomotor apparatus" and consisted in the realization of a robotic hand. During the realization of this project, much emphasis was placed on the usefulness of this object in real life, thus showing the usefulness of engineering in society. We also worked on contents included in the didactic unit "Operations with decimal numbers" of the Mathematics subject.

Figure 2 shows images of the elaboration and use of the robotic hand developed in the STEM 2 Workshop.

Figure 2. Realization of the robotic hands of STEM Workshop 2
Source: Authors

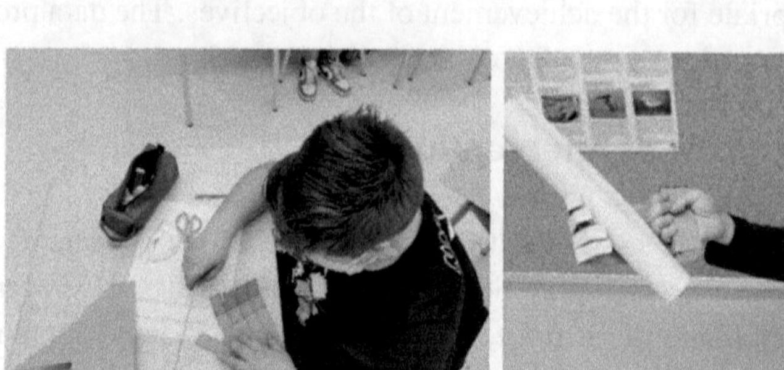

The third intervention (STEM Workshop 3) was carried out during the didactic unit "Matter and its changes" and was aimed at making a volcano to visualize the chemical changes of matter. Questions on the impact of volcanoes on geography and on people's lives were also developed. In the area of mathematics, this project worked on content related to quantities.

Figure 3 shows images of the volcano made by some students during the STEM Workshop 3.

Figure 3. Volcano made in STEM Workshop 3
Source: Authors

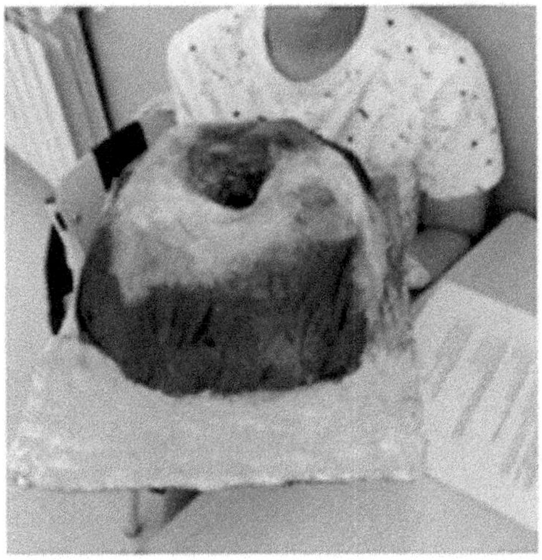

The fourth intervention designed (STEM Workshop 4) was carried out during the explanations of the didactic unit "Forces and their effects" and was based on the construction of a car that worked due to the action-reaction principle. In this project we also worked on contents included in the didactic unit "Area of plane figures" of Mathematics.

Figure 4 shows the car made by a student during the STEM Workshop 4.

Figure 4. Car made in STEM Workshop 4
Source: Authors

The fifth and last intervention (STEM Workshop 5) was framed within the Natural Sciences topic "Simple and compound machines". It also dealt with mathematical contents framed within the topic "Angles and Circles". The students built a simple catapult.

Figure 5 shows the catapult made by a student during the STEM Workshop 5.

Figure 5. Catapult made in STEM Workshop 5
Source: Authors

Finally, it should be remembered that for a project to be considered STEM, it must integrate content and/or skills from each of the disciplines that make up the STEM paradigm. For this reason, and taking into account that the curricular subjects in the primary stage that are directly linked to STEM education are Mathematics and Natural Sciences, the issues and skills that were worked on within the framework of Technology and Engineering in all the projects developed are detailed below.

The development of the Technology discipline is based on the set of knowledge and techniques that, applied in a logical and orderly manner, allow students to seek appropriate ways of working or to modify errors to meet needs, i.e., a combined process of thought and action with the aim of creating useful solutions. This dimension also appeals to analogical and visual thinking, particularly useful when adapting resources from an existing artifact in the production of a new one (Cajas, 2001). Specifically, work is done with the creation of different technological objects that involves the use of the appropriate tools and procedures for their construction. In addition, skills such as innovation, self-improvement and discipline, among others, are fostered.

The essence of the Engineering discipline is design (Ocaña et al. 2017). Its role in STEM Education is to incorporate design activities into the curriculum and provide a meaningful context for applying and making sense of Mathematics, Science and Technology learning (Carr and Strobel, 2011). An important pedagogical feature that contributes to the success of STEM programs is the opportunity for students to generate creative ideas and act on them. This is the basis of the Engineering competency, encouraging creativity and originality when creating the model, choosing materials appropriately and using them responsibly during the activity and in their daily lives. Likewise, this discipline is also enhanced by establishing short teacher-student discussions that highlight the scientific and engineering principles associated with the activity.

RESULTS

This section shows the results obtained after carrying out the different interventions with 6th grade students. First, the results regarding the level of knowledge are presented, distinguishing between pre-test and post-test. These data were analyzed descriptively and inferentially, using the IBM SPSS Statistics 23 software.

Level of Knowledge Variable

First, the cognitive results obtained in the pre-test are presented, distinguishing by workshop.

Table 1 shows the descriptive statistics obtained after analyzing the cognitive data of the pre-test.

Table 1. Descriptive statistics of the pre-test

Pre-test	Mean	Std. Error Mean	Std. Deviation
STEM Workshop 1	4.819	0.407	1.470
STEM Workshop 2	4.346	0.363	1.312
STEM Workshop 3	4.197	0.443	1.600
STEM Workshop 4	3.661	0.466	1.682
STEM Workshop 5	4.23	0.360	1.298

As shown in Table 1, the students had problems remembering most of the contents to be covered in the different didactic units. In no case was the minimum average grade, five points out of ten, exceeded. This leads us to think that the students had forgotten over time concepts dealt with previously, perhaps because the didactic intervention in previous courses had not led to significant learning or lasting memories that would make possible new learnings in the future.

It is also observed that the results of the pre-tests related to blocks of biology content (STEM Workshop 1 and STEM Workshop 2) are slightly higher, especially those related to biodiversity and the environment (Workshop 1), although it is true that the data are far from being considered desirable. In contrast, it is observed that the results of the pre-tests based on physics or chemistry contents (STEM Workshop 3 and STEM Workshop 4) are the lowest, especially those referring to forces and motion (STEM Workshop 3), which may be due to a greater rejection towards learning contents of these areas, and therefore, a greater difficulty in learning them (Dávila-Acedo, 2017).

Subsequently, Table 2 presents the descriptive statistics obtained after analyzing the cognitive data of the post-test.

Table 2. Descriptive statistics of the post-test

Pre-test	Mean	Std. Error Mean	Std. Deviation
STEM Workshop 1	7.307	0.343	1.238
STEM Workshop 2	5.753	0.533	1.924
STEM Workshop 3	6.980	0.617	2.225
STEM Workshop 4	6.088	0.496	1.789
STEM Workshop 5	6.980	0.493	1.779

Table 2 shows that, after the didactic interventions, the students have a much better mastery of the content covered in the different workshops. A notable increase in the scores of all students is observed, especially in the post-test referring to the topic of biodiversity and environment (STEM Workshop 1), but also in the case of STEM Workshop 3 in which chemical changes of matter were worked on and in the one referring to the topic of simple and compound machines (STEM Workshop 5).

Figure 6 shows a comparative graph of the results.

Figure 6. Comparison of cognitive results (Pre-test vs. Post-test)
Source: Authors

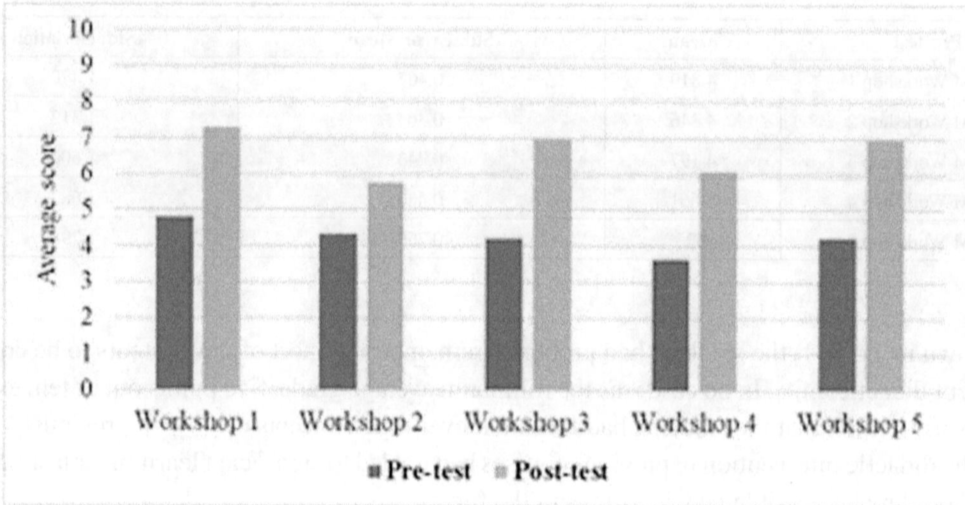

As shown in Figure 6, the average scores of the students improve significantly after the completion of all the STEM workshops: a difference of more than 2 points is observed in most of the cases analyzed. However, it seems that it is the contents related to physics and chemistry that were best remembered by the students after the interventions. The inferential analysis performed revealed the existence of statistically significant differences (Sig.$<$ 0.05) between the pre-test versus post-test scores, in favor of the latter.

The above data show the importance of using active hands-on methodologies that include manipulative workshops to achieve meaningful and long-term learning of the STEM content worked on in the classroom. Likewise, these results complement previous research showing that manipulation is a way to enhance meaningful learning (Martínez-Borreguero et al., 2019; Wade-Jaimes et al., 2018) because it favors the mental constructions of the most abstract contents of subjects considered complex such as physics or chemistry (Borrachero et al., 2014).

Affective Variable

The results obtained in the different emotional variables indicate that the primary school students expressed mainly positive emotions. One hundred percent of the students reported feeling fun during all the interventions. For example, some students indicated that they felt this emotion during STEM Workshop 3 when they created the volcano lava, or in STEM Workshop 4 when they put the car working. The emotion curiosity was also expressed by 100% of the participants, indicating in most cases that they felt this emotion during the scientific explanations related to the artifact, or at the beginning of each project of the didactic unit, due to the uncertainty of not knowing what we were going to do with the material requested by the teacher. In the case of the emotion of confidence, there were some students who did not feel this emotion during some of the workshops, because they believed, initially, that they were not going to make the model correctly. This result coincides with those obtained for the emotion worry, according to the qualitative analysis carried out. In this case, almost 50% of the students indicated that they were worried during the elaboration of the models. They believed that they would not make it correctly or that they would not finish it on time. On the contrary, the analysis of the data confirmed that the students did not manifest at any time the emotions disgust, anger and boredom.

Finally, Table 3 presents the results after analyzing the students' attitudes towards the workshops conducted. The percentage of students who answered *Yes* to the question is detailed.

Table 3. Analysis of student attitudes (percentage of Yes)

Question	Percentage of Students
1. Did you like the workshop/class you attended?	100%
2. Have you learned the contents explained?	92.3%
3. Would you like to do more of these activities in science and math classes?	100%
4. Do you think it's easier to learn science content by doing hands-on activities?	100%
5. Would you have learned the contents without doing the practical workshop?	15.38%
6. Do you think you will remember the contents you have learned more easily thanks to the practical workshop?	100%
7. Have you learned any content you didn't know about?	100%
8. Did you need help with any activity?	46.15%
9. Did you find it difficult to perform the activity?	38.46%
10. Could you make a model of your own related to the contents worked on?	92.3%

The results shown in Table 3 confirm that the participants showed interest in science when it was explained by combining theory and practice. In addition, the students indicated that it was easier for them to learn the contents in this way. On the other hand, the questionnaire also provided information related to perceptions of self-efficacy (Questions 8, 9, and 10). The results found suggest that the use of STEM workshops, based on cooperative and authentic learning, fosters a work environment in which student effort, improvement, and mastery take precedence and helps them not only to feel confident but also to become competent in STEM (Zollman, 2012).

DISCUSSION AND CONCLUSION

The findings found in this research highlight the importance of developing during the early school years a good level of abstraction in student's STEM reasoning. The ability to solve complex problems is fundamental to science, engineering and STEM education in general, therefore, it is concluded that in order to foster or maintain positive STEM dispositions in the elementary school levels, and later in the secondary stage, it will be of utmost importance to include in the educational process diverse methods that involve students and promote the construction of their own learning. In addition, good learning in these areas can serve as an incentive to choose to study STEM careers in higher education.

Variability in the use of didactic resources during class development ensures teaching effectiveness (Bulger et al. 2002; Corona-Cruz, 2008). However, students show greater interest if an experiential component is included in the learning process and if an active learning environment is promoted (Mateos-Núñez et al., 2019a). Incorporating in the school a practical teaching of scientific phenomena that occur daily is determinant in learning so that the student correlates the actions of their daily life with the theories that are learned and in this way, they acquire even more sense the scientific concepts evaluated in their training process. Therefore, it is essential to continue promoting ways of teaching science and technology that allow citizens to develop a comprehensive literacy, including that which allows students to build their own knowledge and apply it in their daily lives (Bencze, 2010).

On the other hand, it is concluded that the integration of STEM areas through hands-on and interactive workshops can increase students' enthusiasm for learning, since this didactic resource assumes a supporting role in the expository teaching of the contents as pointed out by recent studies (Taboada & Santiago, 2021; Yeboah et al., 2019). Active methodologies such as STEM integration contribute to increase students' motivation to learn, reduce boredom in the classroom and systematize the teaching-learning process. However, we agree with authors that awakening this interest in students through the use of active teaching methodologies and strategies can only be achieved if teachers learn and implement them (Li et al., 2022). Taking into account that the attitudes and emotions of primary and secondary students improve when expository sessions are complemented with experimental activities (Mateos-Núñez et al., 2020), it will be necessary to improve STEM teacher training to promote not only the interest and promotion of scientific vocations in their future students but also to improve their own attitudes towards educational practice related to STEM areas, because if teachers exhibit a positive assessment of their teaching practice, the same students will have a favorable assessment towards it (Hué, 2012).

LIMITATIONS AND FUTURE RESEARCH DIRECTIONS

In terms of limitations, we believe that it is necessary to continue contrasting the results with larger samples and in different contexts, in order to reaffirm the potential of STEM education in the classroom. Also, the inclusion of qualitative methods of data collection and analysis could provide new data on the role of emotions in the teaching-learning process.

REFERENCES

Bagiati, A., Yoon, S. Y., Evangelou, D., & Ngambeki, I. (2010). Engineering curricula in early education: Describing the landscape of open resources. *Early Childhood Research & Practice*, *12*(2), 1–15.

Beauchamp, G., & Parkinson, J. (2008). Pupils' attitudes towards school science as they transfer from an ICT- rich primary school to a secondary school with fewer ICT resources: Does ICT matter? *Education and Information Technologies*, *13*(2), 103–118. doi:10.100710639-007-9053-5

Bencze, J. L. (2010). Promoting student-led science and technology projects in elementary teacher education: Entry into core pedagogical practices through technological design. *International Journal of Technology and Design Education*, *20*(1), 43–62. doi:10.100710798-008-9063-7

Borrachero, A. B., Dávila-Acedo, M. A., Brígido, M., Gómez, R., & Mellado, L. (2014). Las emociones que experimentaban los futuros profesores de secundaria en el aprendizaje de las ciencias [The emotions that future secondary school teachers experienced in learning science]. *Revista INFAD de Psicología. International Journal of Developmental and Educational Psychology*, *4*(1), 271–280. doi:10.17060/ijodaep.2014.n1.v4.612

Borrego, M., & Henderson, C. (2014). Increasing the use of evidence-based teaching in STEM higher education: A comparison of eight change strategies. *Journal of Engineering Education*, *103*(2), 220–252. doi:10.1002/jee.20040

Bulger, S. M., Mohr, D. J., & Walls, R. T. (2002). Stack the deck in favor of your students by using the four aces of effective teaching. *The Journal of Effective Teaching*, *5*(2).

Busquets, T., Silva, M. y Larrosa, P. (2016). Reflexiones sobre el aprendizaje de las ciencias naturales: Nuevas aproximaciones y desafíos [Reflections on learning from the natural sciences: New approaches and challenges]. *Estudios pedagógicos, 42*, 117-135. doi:10.4067/S0718-07052016000300010

Bybee, R. W. (2013). *The Case for Education Challenges and Opportunities*. National Science Teachers Association.

Cajas, F. (2001). Alfabetización científica y tecnológica: la transposición didáctica del conocimiento tecnológico [Scientific and technological literacy: the didactic transposition of technological knowledge]. *Enseñanza de las ciencias: Revista de investigación y experiencias didácticas, 19*(2), 243-254.

Carr, R. L., & Strobel, J. (2011). Integrating Engineering Design Challenges into Secondary STEM Education. National Center for Engineering and Technology Education.

Chiu, M., & Duit, R. (2011). Globalization: Science education from an international perspective. *Journal of Research in Science Teaching*, *48*(6), 553–566. doi:10.1002/tea.20427

Cleaves, A. (2005). The formation of science choices in secondary school. *International Journal of Science Education*, *27*(4), 471–486. doi:10.1080/0950069042000323746

Corona-Cruz, A. (2008). ¿Qué hace al buen maestro?: La visión del estudiante de ciencias físico-matemáticas [What makes the good teacher?: The vision of the physical-mathematical science student]. *Latin-American Journal of Physics Education*, *2*(2), 148–151.

Couso, D. (2017). Per a què estem a STEM? Un intent de definir l'alfabetització STEM per a tothom i amb valors [What are we for in STEM? An attempt to define STEM literacy for all and with values]. *Ciències*, *34*(34), 22–30. doi:10.5565/rev/ciencies.403

Dávila, M. A., Cañada, F., Sánchez-Martín, J., & Mellado, V. (2016). Las emociones en el aprendizaje de física y química en educación secundaria. Causas relacionadas con el estudiante [Emotions in the learning of physics and chemistry in secondary education. Student-related causes]. *Educación en la Química*, *27*(3), 217–225. doi:10.1016/j.eq.2016.04.001

Dávila Acedo, M. A. (2017). Las emociones y sus causas en el aprendizaje de Física y Química, en el alumnado de Educación Secundaria [Emotions and their causes in the learning of Physics and Chemistry, in the students of Secondary Education]. *Revista Eureka sobre Enseñanza y Divulgación de las Ciencias, 14*(3), 570–586.

English, L. D., King, D., & Smeed, J. (2017). Advancing integrated STEM learning through engineering design: Sixth-grade students' design and construction of earthquake resistant buildings. *The Journal of Educational Research, 110*(3), 255–271. doi:10.1080/00220671.2016.1264053

Escobar-Pérez, J., & Cuervo-Martínez, A. (2008). Validez de contenido y juicio de expertos: Una aproximación a su utilización [Content validity and expert judgment: an approximation to its use]. *Avances en Medición, 6*, 27–36.

European Comission. (2010). *Science and Technology. Special Eurobarometer340.* European Commission.

Fernández, N. (2006). Estrategias de enseñanza para favorecer el aprendizaje significativo [Teaching strategies to promote meaningful learning]. *Revista Cognición, 1*(6), 12–28.

Ferrada, C., Díaz-Levicoy, D., & Salgado-Orellana, N. (2018). Análisis de actividades STEM en libros de texto chilenos y españoles de ciencias [Analysis of STEM activities in Chilean and Spanish science textbooks]. *Revista de Pedagogía, 39*(105), 111–130.

Fllis, A., & Fouts, J. (2001). Interdisciplinary curriculum: The research base: The decision to approach music curriculum from an interdisciplinary perspective should include a consideration of all the possible benefits and drawbacks. *Music Educators Journal, 87*(22), 22–26, 68. doi:10.2307/3399704

Freeman, S., Eddy, S. L., McDonough, M., Smith, M. K., Okoroafor, N., Jordt, H., & Wenderoth, M. P. (2014). Active learning increases student performance in science, engineering, and mathematics. *Proceedings of the National Academy of Sciences of the United States of America, 111*(23), 8410–8415. doi:10.1073/pnas.1319030111 PMID:24821756

Friedman, T. L. (2005). *The world is flat. A brief history of the twenty-first century*. Farrar, Straus and Giroux.

Furner, J. M., & Kumar, D. D. (2007). The mathematics and science integration argument: A stand for teacher education. *Eurasia Journal of Mathematics, Science and Technology Education, 3*(3), 185–189. doi:10.12973/ejmste/75397

Honey, M., Pearson, G., & Schweingruber, A. (2014). *STEM integration in K-12 education: status, prospects, and an agenda for research*. National Academies Press.

Hué, C. (2012). Bienestar docente y pensamiento emocional. *Revista Fuentes, 12*, 47–68.

Kennedy, T. J., & Odell, M. R. L. (2014). Engaging students in STEM education. *Science Education International, 25*(3), 246–258.

King, D., Ritchie, S., Sandhu, M., & Henderson, S. (2015). Emotionally intense science activities. *International Journal of Science Education, 37*(12), 1886–1914. doi:10.1080/09500693.2015.1055850

King, K., & Wiseman, D. (2001). Comparing science efficacy beliefs of elementary education majors in integrated and non-integrated teacher education coursework. *Journal of Science Teacher Education*, *12*(2), 143–153. doi:10.1023/A:1016681823643

Li, J., Luo, H., Zhao, L., Zhu, M., Ma, L., & Liao, X. (2022). Promoting STEAM education in primary school through cooperative teaching: A design-based research study. *Sustainability*, *14*(16), 10333. doi:10.3390u141610333

Lyons, T. (2006). Different countries, same science classes: Students' experiences of school science in their own words. *International Journal of Science Education*, *28*(6), 591–613. doi:10.1080/09500690500339621

Marbá, A., & Márquez, C. (2010). ¿Qué opinan los estudiantes de las clases de ciencias? Un estudio transversal de sexto de Primaria a cuarto de ESO [What do students think of science classes? A cross-sectional study from sixth grade of Primary to fourth year of ESO]. *Enseñanza de las Ciencias*, *28*(1), 19–30. doi:10.5565/rev/ensciencias.3618

Martínez-Borreguero, G., Mateos-Núñez, M., & Naranjo-Correa, F. L. (2019). Implementation and Didactic Validation of STEM Experiences in Primary Education: Analysis of the Cognitive and Affective Dimension. In Theorizing STEM Education in the 21st Century. IntechOpen.

Martínez-Borreguero, G., Mateos-Núñez, M., & Naranjo-Correa, F. L. (2020). Emotions of elementary school students: comparative analysis between different school subjects. In E. J. Byker & A. Horton (Eds.), *Elementary Education. Global Perspectives Challenges and Issues Of The 21st Century*. Nova Science Publishers, Inc.

Mateos-Núñez, M., Martínez-Borreguero, G., & Naranjo-Correa, F. L. (2019a). Comparación de las emociones, actitudes y niveles de autoeficacia ante áreas STEM entre diferentes etapas educativas [Comparison of emotions, attitudes and levels of self-efficacy in STEM areas between different educational stages]. *European Journal of Education and Psychology*, *13*(1), 251–267. doi:10.30552/ejep.v13i1.292

Mateos-Núñez, M., Martínez-Borreguero, G., & Naranjo-Correa, F. L. (2019b). Análisis cognitivo y emocional hacia competencias científico-tecnológicas en el alumnado de 6° de Educación Primaria [Cognitive and emotional analysis towards scientific-technological competences in the students of 6th year of Primary Education]. In A. B. Corsini (Ed.), *Conectando pilares para la educación* (pp. 113–124). Global Knowledge Academics.

Mateos-Núñez, M., Martínez-Borreguero, G., & Naranjo-Correa, F. L. (2020). Learning science in primary education with STEM workshops: Analysis of teaching effectiveness from a cognitive and emotional perspective. *Sustainability*, *12*(8), 3095. doi:10.3390u12083095

Mellado, V., Borrachero, A. B., Brígido, M., Melo, L. V., Dávila, M. A., Cañada, F., Conde, M. C., Costillo, E., Cubero, J., Esteban, R., Martínez, G., Ruiz, C., Sánchez, J., Garritz, A., Mellado, L., Vázquez, B., Jiménez, R., & Bermejo, M. L. (2014). Las Emociones en la Enseñanza de las Ciencias [Emotions in Science Education]. *Enseñanza de las Ciencias*, *32*(3), 11–36. doi:10.5565/rev/ensciencias.1478

Morrison, J. (2006). *TIES STEM education monograph series, Attributes of STEM education*. TIES.

Murphy, C., & Beggs, J. (2003). Children's perceptions of school science. *The School Science Review*, *84*(308), 109–116.

National Science Board. (2004). *Science and Engineering Indicators 2004 (NSB 04-01)*. National Science Foundation.

Neira, J. C. R. (2021). La experimentación en ciencias naturales como estrategia de alfabetización científica [Experimentation in natural sciences as a strategy for scientific literacy]. *UCMaule*, (60), 102–116. doi:10.29035/ucmaule.60.102

Ocaña, G., Romero, I.M. & Gil, F. (2017). Educación STEM para integrar conocimientos científicos en la asignatura "tecnología industrial" de bachillerato [STEM education to integrate scientific knowledge in the subject "industrial technology" of baccalaureate]. *Enseñanza de las ciencias: revista de investigación y experiencias didácticas*, (Extra), 5327-5336.

OECD. (2018). The Future of Education and Skills. *Education.*

Osborne, J., Simon, S., & Collins, S. (2003). Attitudes towards science: A review of the literature and its implications. *International Journal of Science Education*, *25*(9), 1049–1079. doi:10.1080/0950069032000032199

Pérez, A., & de Pro, A. (2013). Estudio demoscópico de lo que sienten y piensan los niños y adolescentes sobre la enseñanza formal de las ciencias [A demographic study of what children and adolescents feel and think about formal science education]. In V. Mellado, L. J. Blanco, A. B. Borrachero, & J. A. Cárdenas (Eds.), *Las Emociones en la Enseñanza y el Aprendizaje de las Ciencias y las Matemáticas* (pp. 495–520). DEPROFE.

Pozo, J. I., & Gómez- Crespo, M.A. (1998) *Aprender y enseñar ciencia. Del conocimiento cotidiano al conocimiento científico* [Learn and teach science. From everyday knowledge to scientific knowledge]. Ediciones Morata.

Robles-Garrote, P., & Rojas, M. C. (2015). La validación por juicio de expertos: Dos investigaciones cualitativas en Lingüística aplicada [Validation by expert judgment: two qualitative investigations in applied linguistics]. *Revista Nebrija de Lingüística Aplicada a La Enseñanza de Las Lenguas*, *18*, 1–16.

Sanders, M. (2009). STEM, STEM Education, STEMAnia. *Technology Teacher*, *68*(4), 20–27.

Stohlmann, M., Moore, T. J., & Roehrig, G. H. (2012). Considerations for teaching integrated STEM education. *Journal of Pre-College Engineering Education Research*, *2*(1), 28–34. Advance online publication. doi:10.5703/1288284314653

Taboada, N. J., & Santiago, V. A. O. (2021). Evaluación de talleristas científicos: Propuesta desde la educación y la comunicación científica [Evaluation of scientific workshops: Proposal from education and scientific communication]. *Emerging Trends in Education*, *4*(7). Advance online publication. doi:10.19136/etie.a4n7.4458

Tanik-Onal, N., & Saylan-Kirmizigul, A. (2021). A Makey-Makey based STEM activity for children. *Science Activities*, *58*(4), 166–182. doi:10.1080/00368121.2021.2011086

Thomas, B., & Watters, J. J. (2015). Perspectives on Australian, Indian and Malaysian approaches to STEM education. *International Journal of Educational Development*, *45*, 42–53. doi:10.1016/j.ijedudev.2015.08.002

Vázquez, A., & Manassero, M. A. (2008). El declive de las actitudes hacia la ciencia de los estudiantes: Un indicador inquietante para la educación científica [The decline of students' attitudes towards science: a disturbing indicator for science education]. *Revista Eureka*, *5*(3), 274–292. doi:10.25267/Rev_Eureka_ensen_divulg_cienc.2008.v5.i3.03

Vázquez, A., & Manassero, M. A. (2011). The decline of girls' and boys' attitudes toward science in compulsory education. *Ciência & Educação (Bauru)*, *17*(2), 249–268. doi:10.1590/S1516-73132011000200001

Wade-Jaimes, K., Demir, K., & Qureshi, A. (2018). Modeling strategies enhanced by metacognitive tools in high school physics to support student conceptual trajectories and understanding of electricity. *Science Education*, *102*(4), 711–743. doi:10.1002ce.21444

Yeboah, R., Abonyi, U. K., & Luguterah, A. W. (2019). Making primary school science education more practical through appropriate interactive instructional resources: A case study of Ghana. *Cogent Education*, *6*(1), 1–14. doi:10.1080/2331186X.2019.1611033

Zollman, A. (2012). Learning for STEM Literacy: STEM Literacy for Learning. *School Science and Mathematics*, *112*(1), 12–19. doi:10.1111/j.1949-8594.2012.00101.x

ADDITIONAL READING

Çiftçi, A., Topçu, M. S., & Foulk, J. A. (2022). Pre-service early childhood teachers' views on STEM education and their STEM teaching practices. *Research in Science & Technological Education*, *40*(2), 207–233. doi:10.1080/02635143.2020.1784125

Li, Y., Wang, K., Xiao, Y., & Froyd, J. E. (2020). Research and trends in STEM education: A systematic review of journal publications. *International Journal of STEM Education*, *7*(1), 1–16. doi:10.1186/2196-7822-1-1

Özdinç, F., Kaya, G., Mumcu, F., & Yıldız, B. (2022). Integration of computational thinking into STEM activities: An example of an interdisciplinary unplugged programming activity. *Science Activities*, *59*(3), 1–9. doi:10.1080/00368121.2022.2071817

Toma, R. B., & Greca, I. M. (2018). The effect of integrative STEM instruction on elementary students' attitudes toward science. *Eurasia Journal of Mathematics, Science and Technology Education*, *14*(4), 1383–1395. doi:10.29333/ejmste/83676

Vlasopoulou, M., Kalogiannakis, M., & Sifaki, E. (2021). Investigating Teachers' Attitudes and Behavioral Intentions for the Impending Integration of STEM Education in Primary Schools. In S. Papadakis & M. Kalogiannakis (Eds.), *Handbook of Research on Using Educational Robotics to Facilitate Student Learning* (pp. 235–256). IGI Global. doi:10.4018/978-1-7998-6717-3.ch009

Yıldırım, B. (2021). Preschool STEM activities: Preschool teachers' preparation and views. *Early Childhood Education Journal*, *49*(2), 149–162. doi:10.100710643-020-01056-2

KEY TERMS AND DEFINITIONS

Emotions: Reaction to the information received from our environment, the intensity of which depends on the subjective evaluations we make and where previous knowledge and beliefs have a great influence.

Level of Knowledge: Learning acquired after the formal explanation of a series of contents.

Primary Education: Stage of the Spanish educational system comprising six academic years to be taken ordinarily between six and twelve years of age.

Self-Efficacy: Confidence in one's own ability to achieve the objectives set.

STEM Education: Educational paradigm aimed at teaching science, mathematics, technology, and engineering together.

Teacher Training: Set of courses or academic events with official validity, which aim to enable the practice and/or update those who perform teaching functions, in the theories, procedures and techniques for teaching.

Teaching Methodologies: A set of strategies, procedures and actions organized by the teacher to facilitate student learning.

Workshops: Science and technology dissemination media that seeks to leave behind the one-way transmission of information, in order to build experiences and knowledge relevant to the participants.

Chapter 7
Different Teaching Methods: A Step for Student Motivation

Inês Teixeira
University of Minho, Portugal

Violeta Meneses Carvalho
University of Minho, Portugal

Nelson Rodrigues
University of Minho, Portugal

Ana Cristina Ferreira
https://orcid.org/0000-0002-4775-5713
University of Minho, Portugal

Senhorinha Teixeira
https://orcid.org/0000-0002-7464-3944
University of Minho, Portugal

Cristina S. Rodrigues
University of Minho, Portugal

ABSTRACT

Teaching is made by providing students with theoretical lessons and giving them the basic concepts of a subject and potentiating further development. Although this is an important step, student learning differs, and they have difficulty in transposing their theoretical knowledge into real-world problems. To help them make a bridge between the two realities, practical activities are essential. However, due to class restrictions, such as time and number of students, the activities may fall far from the expected results. In this matter, the University of Minho has been introducing different methods of teaching, interacting, and captivating students through learning-by-doing approaches. In light of these experiences, the objective of this work is to provide an overview of how with simple activities the correlation between science and mathematics can be demonstrated to high school students and also to undergraduate students. These types of activities open their horizons and promote their interest and willingness to continue studying and learning actively.

DOI: 10.4018/978-1-6684-5765-8.ch007

INTRODUCTION

Over the years, traditional teaching based on the expositive method of content has prevailed. However, with technological advances the digital tools available increased, and the concept of teaching has been changed, with the themes of "active learning" (Finbråten et al. 2022; López-Fernández et al. 2019; Nasr El-Din et al. 2022; Ratan et al. 2022), "problem-based learning" (Jonassen, 2010), "game-based learning" (Glendon and Ulrich 2005; Grimley et al. 2011; Hartikainen et al. 2019) "project-based learning"(Alrahlah 2016; Andersen and Rösiö 2021; Hussain et al. 2019; Ricaurte and Viloria 2020), "team-based learning" (Cunha et al. 2018; Miller et al. 2017; Sharma et al. 2017) and "blended learning" (Al-Kahtani et al. 2022; Janes et al. 2022; Müller and Wulf 2022; Truss and Anderson 2023) being increasingly addressed. The use of these methodologies has also been increasing since the confinement due to the COVID-19 pandemic (Bestiantono et al. 2020; Pawlicka et al. 2022). During this novel reality, the teaching method was inevitably modified (Blythe and Thompson 2022; El-Soussi 2022; Klusmann et al. 2022; Nurse-Clarke and Joseph 2022). On the bright side, professors had to recreate and improve the teaching-learning processes in order to maintain the student's engagement and motivation and some of them currently apply digital technologies combined with traditional ones (Hasnine et al 2022; Lytras et al. 2022; McKay et al. 2022; Succar et al. 2022).

Some authors have investigated different approaches to teaching, and interesting outputs have been revealed. Capone, (2022) presented a case study with undergraduate engineering students in which a "blended" teaching approach was implemented through both active learning and online learning through activities in the platform. In this case, students showed more enthusiasm and engagement associated with the application of realistic and practical problems, as well as the use of digital tools to discuss the concepts in a group. Beneroso and Robinson, (2022) investigated how online tools can support the design abilities of students through project-based learning, and its impact on students' performance. The results showed that the combination of online and face-to-face teaching provided similar results to those under fully in-person conditions. It is important to notice this works well when online activities also include social connections and prompt feedback. Lino-Neto et al., (2022) also compared the virtual and online components but in a team-based learning approach. In general, students' feedback was positive regardless of the situation, fully in-person or fully online. This proves that if the online component is well-defined and planned, the student's outputs and engagement can be similar.

Hammad et al., (2022), tested a novel approach for mathematics teaching to freshmen university students. The teacher's teaching and attitude, methodology, group work, and gamification were the elements with the most impact on self-efficacy, motivation, and success. Hossein-Mohand et al., (2021) focused on secondary mathematics' teachers and the use of different methodologies for teaching. The aim of the study consisted in identifying the teaching practices, ratios, and mathematics teacher training indicators that could influence the choice of different practices. A positive relation between the Flipped Learning model, Project-Based Learning, and Gamification with the evaluated items was found. However, Project-Based Learning showed a negative association with two items of the teaching practice indicator. Furthermore, it was found that the participation in projects connected to digital technologies for teaching had a significant influence on the choice of Flipped Learning model and active methodologies. Martín-Garin et al., (2021) presented a descriptive study of how problem-based learning (PBL) or research-based learning (RBL), and environmental tools can be used to promote the education for sustainable development in University courses such as Civil engineering. Despite not analyzing quantitatively the impact of these practices, students became more active and participated in research projects which resulted in a collaboration in research and conference papers.

These practices have also been explored at the engineering school of the University of Minho for innovating and improving teaching methods and also for activities with secondary students and children. The i9Masks summer course was a great example of a learning-by-doing approach to face new demands that arrived with COVID-19 whose results can be found elsewhere (Carvalho et al. 2021). Examples of the last case are the "Uminho with open doors", "Summer on Campus" and "European Research Night".

The Department of Production and Systems – DPS has been involved in these activities to show how engineering can be put into practice in different scientific areas, namely, medicine, mathematics and chemistry, through empirical educational activities. Higher education is not always considered by the students, or it is not their first choice. Showing them prospects about a future career and what high degree courses include along with the institution facilities helps them to take a more reasoned decision along with the advantages to motivate them. Additionally, the DPS has been involved in short-period research projects with undergraduate students. These activities motivate them to work together with researchers and teachers to find new ideas and viewpoints about current topics. It is also important to understand their efficacy and effects on the student's motivation. In light of this, the present chapter aims to provide different knowledge transfer methodologies stressing the connection of science and mathematics in engineering problems. To this end, a synopsis of different activities carried out in 2022 with the involvement of both high school and undergraduate students is provided.

BACKGROUND

Active learning is an approach that actively seeks to engage students through problem-solving, case studies, simulations, brainstorming, and other methods (Bruff 2009; Harmin and Toth 2006; Howell 2021). As pointed out by Felder and Brent (2016), active learning is "a teaching approach that encompasses anything students might be called on to do in class besides watching and listening to an instructor and taking notes".

Problem-based learning (PBL) as a curricular innovation emerged in medical education being learning anchored in an authentic problem. As an active learning method, problem-based learning assumes that students will master content while solving a meaningful problem (Fernandes 2021; Liu and Pásztor 2022; Lonergan et al. 2022). Jonassen (2010) considered problem-based learning a "paradigm shift" since it is an instructional solution designed to improve learning by requiring students to learn content while solving problems.

Research studies have shown that active learning outperforms traditional lecturing in promoting almost every learning outcome examined (Brent and Felder 2016; Freeman et al. 2014; Prince 2004; Theobald et al. 2020; Wieman 2014). Active learning also resulted in higher long-term retention, skill development, and students' satisfaction (Brent and Felder 2016; Freeman et al. 2014; Prince 2004; Theobald et al. 2020; Wieman 2014). Felder and Brent, (2016) pointed out two reasons behind active learning success: 1) it reduces the cognitive load on working memory, making retention and storage of new information more likely, and 2) learning requires attentiveness and it is difficult or impossible for students to pay attention to anything for very long while they are passive.

The use of a large variety of teaching methodologies has also been increasing since the confinement due to the COVID-19 pandemic (Bestiantono et al. 2020; Pawlicka et al. 2022). During this novel reality, the teaching method was inevitably modified (Blythe and Thompson 2022; El-Soussi 2022; Klusmann et al. 2022; Nurse-Clarke and Joseph 2022). On the bright side, professors had to recreate and improve the teaching-learning processes in order to maintain the student's engagement and motivation and some of them currently apply digital technologies combined with traditional ones (Hasnine et al. 2022; Lytras et al. 2022; McKay et al. 2022; Succar et al. 2022).

Fink (2003) argued that teachers should be responsible for designing learning experiences and supporting learning on multiple levels. Therefore, a good course has five principles: 1) Challenge students to significant kinds of learning; 2) Use active forms of learning; 3) Have teachers who care—about the subject, their students, and about teaching and learning; 4) Have teachers who interact well with students. 5) Have a good system of feedback, assessment, and grading.

The taxonomy of significant learning proposed by Fink (2003) identifies six kinds of learning that can contribute to this requirement in figure 1.

Figure 1. Taxonomy of significant learning
Source: Fink (2003, p.35)

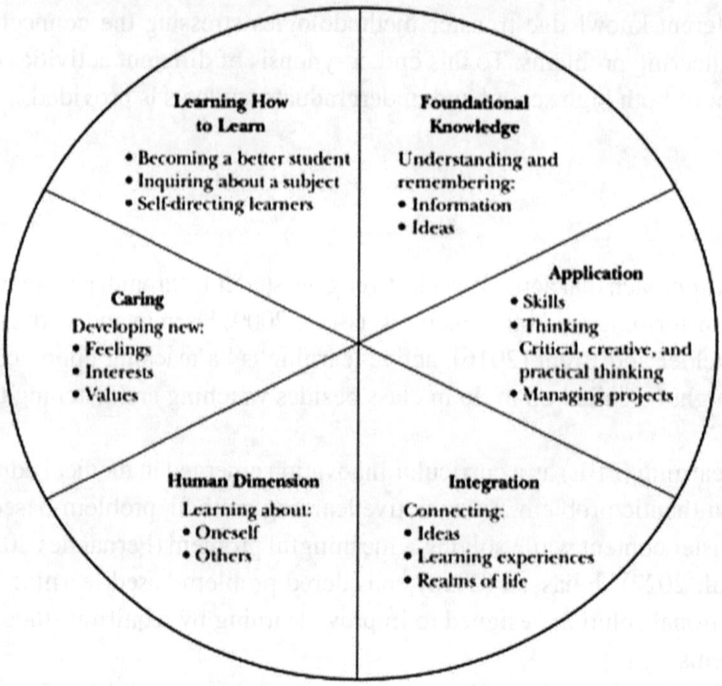

METHODS AND MOTIVATION

The methods used were focused on the integration of students from their first academic years into a research environment. The method can be called engagement of students, i.e., opening opportunities to discover the world of research, and motivating them with activities that meet the student's area of interest, innovation, and proactivity. Living in a world where the curriculum weights academic and professional life, providing them with opportunities for research integration that not only enriches it in that aspect but also shapes and matures them in a disciplinary aspect. This recognition comes from participation certificates, presence in events and conferences and even some get to be part of publications giving their contribution to their elaboration. Fink (2003) stated that for learning to occur, there must be some kind of change in the learner: "No change, no learning". Therefore, he stated that the key to quality in the educational program is creating significant learning experiences, "that learning experience resulted in something truly significant in terms of the student's lives".

The experiences presented in this chapter focus mainly on the application to a real example and on learning "how to do it", bearing in mind the different domains of the Fink model (2003) and with the support of a multidisciplinary team composed of master's students, PhD students, Professors and Researchers of different areas of research. By promoting student participation and interaction with a multidisciplinary team, students are invited to connect old and new concepts and reflect on the relevance of information, thus developing their critical thinking.

ACTIVITIES CARRIED OUT

The activities described in this work occurred during the 2021/2022 academic year and follow the research activities performed in previous events already reported going from the I9Masks Summer Course (Carvalho et al. 2021) and Summer grants (Rodrigues et al. 2022). Both projects integrated undergraduate students in research projects, with the first activity comprising classes as a course. The students involved in these previous events were encouraged to present their results to a younger generation through two activities organized by the University of Minho to high school students (ODU - University with Open Doors and Summer on Campus). The main objectives of these two activities were to motivate students to enroll at the University and provide them with initial contact of those students with some research fields.

In these events, the authors' team intended that participants in the different proposed experiences could explore and understand the connection of mathematics and Science, Technology, Engineering, and Mathematics (STEM), with science and technology and daily life problems such as forest fires, and thermal comfort. A summary of these activities is presented in Table 1.

Table 1. Summary of the activities carried out at the University of Minho.

Activity	ODU	UMinho Prize	Summer on Campus	Summer Grants
Team	4 teachers from subject areas of Operational Research, Numerical and Statistical Methods and Mechanical Engineering 2 undergraduate students from summer grant 2021 1 undergraduate student from UMinho Prize 2021 3 PhD students 2 BI grants 2 researchers from ALGORITMI R&D Center	6 teachers from subject areas of Operational Research, Numerical and Statistical Methods and Mechanical Engineering 2 Master student 3 PhD students 2 BI grants 2 researchers from ALGORITMI R&D Center 1 researcher from CMEMS R&D Center	3 teachers from the subject areas of Numerical and Statistical Methods and Mechanical Engineering 2 Ph.D students 2 BI grants 5 undergraduate students from UMinho Prize 2022	3 teachers from the subject areas of Numerical and Statistical Methods and Mechanical Engineering 2 Ph.D students 2 BI grants
Students	Best high school students from Braga	5 Undergraduate students	12 High school students	3 Undergraduate and 2 graduate students
Duration	1 day	3 months	1 week	1 month
Objectives	The development of new interests, feelings, and values with engineering The ability to recognize the connection between science and mathematics knowledge with different engineering applications	Understand the research process Interact with others regarding the team success Be more interested in doing research Becaming a better student	Understand the research process Identify the interaction between science and mathematics with research on enginnering The ability to connect one body of knowledge with other ideas and bodies of knowledge	Understand the research process Learning how to communicate research outputs

The first activity was organized in April 2022 and high school students were invited to visit the Azurém Campus of the University of Minho and learn about the main research projects taking place at each school.

In the second activity, a group of high school students was allocated by their choices to a Department, during the last week of June 2022 and they were integrated with the research activities.

On the other hand, the form of incentive is the BII scholarship (scholarship for the initiation of research) and BI scholarship (research scholarship), The scholarships are directed towards undergraduate students without previous research experience, providing them with a small grant to help with day-to-day expenses.

The activity team also participated in two new opportunities involving undergraduates as well as master students (UMinho prize and Summer with science 2022) throughout the year. The first activity is a context launched by the University of Minho for undergraduate students to get involved with a project in a research group. This activity took place at the end of the academic year between May and August 2022. Summer with science 2022 is a special program launched for the 2nd year by the Foundation for Science and Technology (FCT) to finance some research grants with a duration of one month during September 2022, just before the next academic year. The timeline of the activities is presented in Figure 2.

Figure 2. Timeline of the activities carried out at the DPS.
Source: Authors

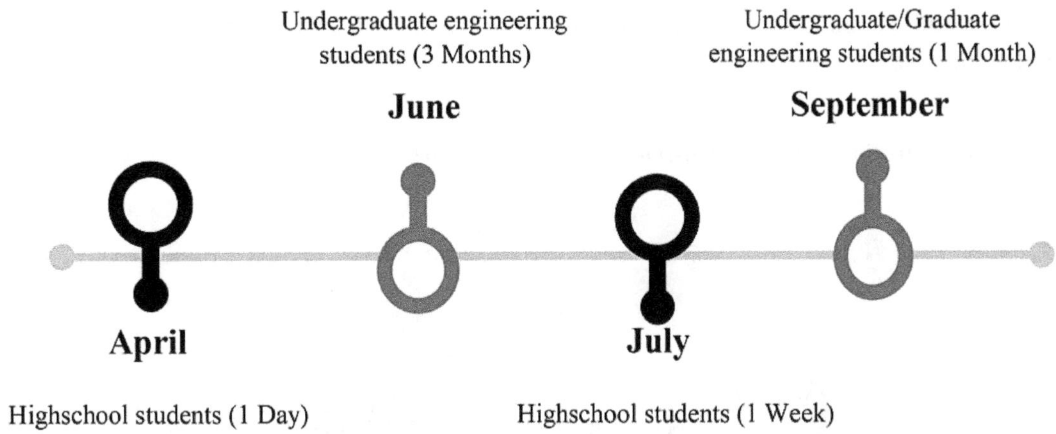

ODU – Open Doors at University

The ODU - Open Doors at University (translation of the original in Portuguese *UPA, Universidade de Portas Abertas*) took place over three days during the Easter holidays. The target participants of this activity were high school students. The activity consisted of the presentation of the University of Minho in a comprehensive perspective, combining the presentation of its educational offer, through an exhibition, where all Schools and Institutes of UMinho were represented, with the visit to the facilities such as classrooms and laboratories, and the possibility of getting to know the research being conducted and the equipment used, through the participation in guided tours, workshops, lectures, among other dynamics.

Thus, this event became an excellent opportunity to get to know the best that is done and produced at the University of Minho, as well as to allow contact with the academic environment and experience.

On the other hand, it was also an opportunity to get to know the infrastructures and services that UMinho makes available to the student community, for the continuation of the studies, and for the development of transversal tools such as teamwork, a sense of responsibility and knowledge transfer, so important for the integration of future professionals in the labor market.

The face-to-face and online activities were free of charge, with limited vacancies, and available upon registration (individual or group). The activities were aimed at 9th-grade and high school students, teachers, vocational advisors, and tutors.

The adhesion to the proposed activities was considerable, having been necessary to limit the groups to enter the rooms and laboratories in groups of 15 to 20 students, having about 4 group entries, in a total of 100 students in the activities.

In this sense, some activities were proposed by the authors's team from the Department of Production and Systems and they were included in the School of Engineering activities program. The three activities proposed were related to three research projects going on: a Ph.D. project on the development and optimization of a preclinical organ-on-a-chip platform for nanomedicine; another Ph.D. project on a technology-based approach to reduce energy poverty according to the region and characteristics of existing buildings and a financed research project entitled "an Optimization Framework to reduce Forest Fire".

The activity dedicated to the organ-on-a-chip devices' work consisted of the presentation of the device, a brief explanation of the functioning and the importance of dynamic systems in drug efficacy studies, applicability, and the relevant role of these devices in the advancement of science in the fight of countless diseases, namely cancer (Figure 3).

Figure 3. Demonstration of how organ-on-a-chip devices work
Source: Authors

In the activity dedicated to thermal comfort, it was presented briefly what thermal comfort is, how can take some ideas home to improve the performance of residences, and raise awareness of the students for energy poverty, a current sustainability problem. The students were also invited to answer a questionnaire through their mobile phones and indicate their age, gender, clothing, and how they evaluated the thermal comfort of the room where the activity took place with the purpose of interacting and raising awareness about the importance of buildings and work thermal comfort and how it influences the performance, health, and well-being of everyone (Figure 4).

Figure 4. Equipment and materials used in the thermal comfort explanation
Source: Authors

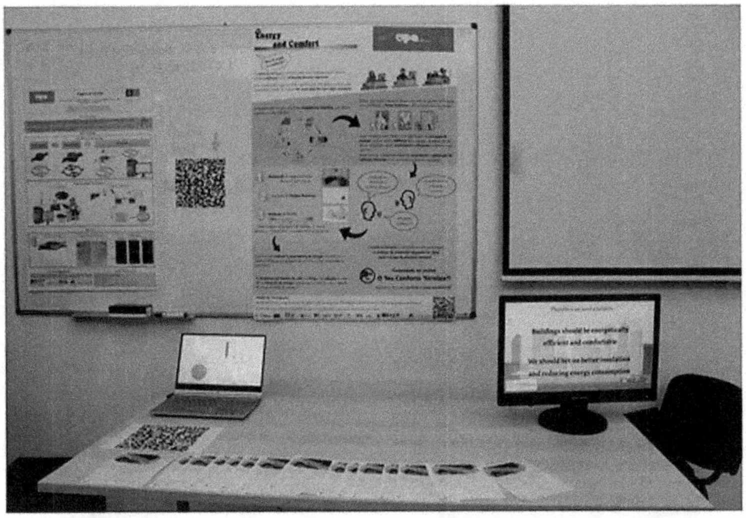

Finally, in the activity about forest fire propagation prediction, the students were sensitized to a problem that occurs in Portugal every summer and introduced the Numerical Simulation as a tool for modeling prevention and fire propagation prediction (Figure 5). The students were also invited to answer a quiz about what they have learned and a general culture game on the theme of forest fires in the entrance hall, while waiting to enter the room they signed up to attend.

Figure 5. Fire forest propagation poster
Source: Authors

UMinho Prize

At the end of each academic year, the University of Minho opens a very challenging context among its undergraduate students. The UMinho Prize for Scientific Research Initiation was created in 2020 at the University of Minho with the aim of providing an opportunity to bring its undergraduate students closer to real contexts of scientific research and insertion in research teams that work in the creation of new and relevant knowledge. This can help to foster student interest in research, and it can be accomplished under the supervision of experienced and recognized researchers in the scientific area of each project. Therefore, the students selected for the Prize have the possibility to get involved in research activities that will demonstrate the connection between different subjects, such as social sciences, mathematics, physics, or chemistry. The Prize is sponsored by the Dean board, and involves the different research centers of the university. Each center collects and disseminates a set of projects to be carried out at its facilities with students. After an application period, the students which apply for the Prize are evaluated and selected by their academic merit and *curriculum vitae*.

The experience presented here concerns projects of ALGORITMI research centre, more specifically the five research projects that were developed by authors' team from May to August 2022 with five undergraduate students from the School of Engineering.

One of the Prize research projects was part of an I&D research project from ALGORITMI center, "O3F: An Optimization Framework to reduce Forest Fire", PCIF/GRF/0141/2019, financed by national funds through the Portuguese *Fundação para a Ciência e a Tecnologia* (FCT). It was entitled "Mapping of fire propagation rate and intensity", and its main objective was to implement in Python language the main equations related to fire propagation. Besides this main objective, it was intended that the student would consolidate multidisciplinary knowledge of Computational Fluid Dynamics (CFD) simulation, heat transfer, and programming languages.

The second project entitled "Influence of the presence of the human body in thermal environments using numerical software" was developed along with a Ph.D. project from the Doctoral Program on Industrial and Systems Engineering entitled "A technology-based approach to reduce energy poverty according to the region and characteristics of existing buildings". The main idea behind the project was to discover the thermal comfort topic, and the main concepts associated and to use a software to quantify the interaction between the human body and the environment.

Following a master thesis project finished in the previous academic year, a third research project was proposed "Simplified model of flow in a stenosis". The student worked with a team of 4 Ph.D. researchers and a Ph.D. student. Their work was part of an I&D research project "The next generation of in vitro models for wall deformability and blood flow studies in aneurysms", EXPL/EME-EME/0732/2021, financed by FCT. The work consisted of a literature review on stenosis modeling, and the adaptation of an existing program to model a venturi geometry to simulate the stenosis flow in Matlab.

Another student was involved in the fourth project, a Ph.D. project running in the ALGORITMI research center: "Development and optimization of innovative nanoparticles and a preclinical organ-on-a-chip platform for nanomedicine", UI/BD/151028/2021, and the main objective was to optimize the OoC geometry and understanding the influence of the various parameters involved in this type of fluid flow. The Ansys software was used for fluid flow simulation and optimization.

The final research project proposed to the students involved a social association of former students from Industrial and Commercial School from Braga, Portugal. The project was developed in collaboration with a master's student from the Master of Industrial and Management Engineering from the School of Engineering, at the University of Minho. It was entitled "A collaborative study of Industrial Schools and its impact in the Braga region", and it proposed the study the Industrial School influence considering the experiences of its former student's, a database development and the application of statistical techniques through the statistical software SPSS to treat results.

All the undergraduate students taking part in these projects were integrated within each research team, working in a research laboratory, and participating in extra activities taking place during the time of the research period. In the beginning of the different Prize projects, the authors' team proposed introductory activities to better introduce scientific researchers and understand how theoretical subjects from the classes could help to develop new ideas and real products. During the Prize period, the students were also incentivized and taught how to write a scientific paper.

At the end of the Prize period, taking the student's final report and the main activities developed, the ALGORITMI research centre selected two students to present their results before a jury. In 2022, fifteen UMinho students were selected for the 3rd edition of the UMinho Prize for Initiation in Scientific Research. From the different areas of knowledge of the University of Minho, the jury selected eight students to win a monetary value of €500 and a certificate. One of these students, was one of the students supervised by the authors.

Summer on Campus

The "Summer on Campus" activity takes place every year, during a week in July, in Braga and Guimarães. This activity aims to be a learning space for young high school students and foster teamwork, bonds of friendship and mutual trust, through the various educational, recreational, and cultural activities carried out in a University environment. Students also have the opportunity to get to know the cities of Braga and Guimarães, socialize with colleagues from different geographic regions, and learn while having fun trying out the scientific, cultural, and sports activities proposed.

The program "Summer in Campus" covers different scientific areas, where non-formal education activities are promoted in the areas of architecture, mathematics, chemistry, archaeology, sociology, law, education, engineering, and letters, among others.

Within the School of Engineering, the Department of Production and Systems has proposed a week of activities showing ways how engineering can be learned and put into practice, through the topic "Industrial Engineering and Management A step forward towards sustainability". Twelve students that choose this topic were invited to meet University students from some degree courses of the Department, teachers, and researchers, and they were incentivized to participle in different activities from different research groups from the department.

The authors' team proposed three activities for the Department of Production and Systems: 1) "Strategies to reduce forest fires", 2) "Organ-on-a-chips" and 3) "Energy, and thermal comfort in buildings", all to be presented by undergraduate students already working with the authors' team. These undergraduate students were very engaged in these activities with high school students and this promoted exchange of experiences.

In the first activity, it was demonstrated how fire trajectory can be predicted using numerical simulation, considering site conditions and meteorological phenomena. The activity also included didactic moments in which students were invited to follow a path that went back and forward according to the correct and wrong answers given by them.

In the second activity, the concept of an organ-on-a-chip was explained to students, and a demonstration was carried out. Students had the opportunity to understand how the mechanobiology and physiology of the human body can be mimicked at a small scale through microfluidics, cell culture, medicine, and engineering knowledge. Furthermore, they learned important scientific concepts and realized the importance of combining different expertise.

To prepare them for the third activity (which took place over two days), the students were introduced to the topic on the first day, when the needs and requirements to ensure thermal comfort in buildings were presented. Two extra topics were explained namely how to use SPSS software and Initiation to scientific research. In this last topic, students were incentivized to do some research on the theme: Planning and building a home that is sustainable in terms of energy and comfort. For this purpose, the students had a short lecture regarding the effects of materials, orientations to the environment, and building strategies on the energy demands of buildings. On the next day, the students were divided into groups and provided with recyclable materials to build a model of a house, aiming to have the best insulation, design, and fulfilling the established requirements.

On the following day and applying the acquired learning and the research done, the students created hand-house models per group for two hours. Then, they had to present, expose, and defend their ideas.

The house models they created had to go through test periods to determine which of them had the best insulation capacity, considering the previously established requirements and design (Figure 6).

Figure 6. Explanation of thermal comfort, a) the group photo after the presentation and explanation of the student's ideas, b) a group oral presentation with their home model and c) a test of the model developed by the research team
Source: Authors

The test to determine the most sustainable house model was done using a model made by a group of researchers. The maquette was created to study the importance of the covering in a building to help heat it. In this sense, this was built using acrylic plastic (to be translucent and allow the visualization of the test at the level of the walls and top of the maquette), a wooden base for easy drilling (to put the lamps inside), two heating lamps and 1 Arduino + display with connection to the electric network. The coating used to seal the maquette was a standard thermal insulation sponge capable of being removable to analyse its presence and absence in testing periods (Figure 7).

Figure 7. Maquette model built and used to realize the tests
Source: Authors

Summer Grants

Following the previous year's initiative of FCT, one-month research incentive grants (Summer with Science 2022) have been launched during September 2022 (before classes start), to provide the grantees with a brief research experience, but also to provide a team spirit, a working environment of research, search, and creation of academic knowledge.

In this experience, five scholarship themes (Applied Statistics: a characterization study of a population; Simplified model of flow in a stenosis; Thermo-economic analysis of an interior space and optimization of integrated technological solutions; Optimizing the geometry of an Organ-on-a-Chip device; Use of commercial software to study the phenomena associated with the spread of forest fires) were launched through the ALGORITMI research center as BII grants. Undergraduate and master students applied for the scholarship and they were selected according to academic merit, CV, and motivation letter, by a public contest.

These five grants were attributed to five university students and besides having contact with research teams, composed of professors, Ph.D., and undergraduate students, students had the opportunity to participate in national events of dissemination and sharing of knowledge, collaborating on posters, conference presentations, and scientific papers, giving them a strong research background and also gaining essential lines in the curriculum that can be an asset if they intend to pursue an academic career.

RESULTS AND ACHIEVEMENTS

High School Students' Achievements

Different practices and teaching methods were used during the academic year from April to September 2022, with undergraduate and graduate students from the engineering school of the University of Minho and receiving high school students.

In the two activities promoted by the University of Minho for receiving high school students, one of the main achievements was the contact between the research Ph.D. and master students from the research team with the high school students. The use of questionaries through mobile phones and the quiz on forest fires were just two different methods implemented by students to students.

During the Summer on Campus, the high school students had contact with software from numerical, simulation, and statistic areas and they realize that mathematics, physics, and chemistry can be applied in different scientific areas such as medicine, energy, and comfort in buildings, combustion fires. They were introduced to Matlab, Ansys, and SPSS applications.

The activity Planning and building a home that is sustainable in terms of energy and comfort was very well received by the students who got involved in a group task of planning, designing, and building by hand a house. They were also introduced to informatic tools that can bring ideas to a new research project.

The Energy and Comfort in Buildings activity was a follow-up of the previous activity and therefore aimed at challenging the high school students to build an energy-efficient house in teams using materials provided, partly, with recycled materials. The students were firstly given a contextualization of the need for buildings to be sustainable, namely, for the reduction of the Ecological Footprint, as well as to reduce the costs associated with air conditioning needs. Also, some parameters and requirements were provided, namely minimum dimensions and number of openings (such as windows) minimum.

Undergraduate Students' Activities

The UMinho Prize contest and the Summer grants were very interesting activities for the undergraduate students involved. This active learning environment created by the student insertion in a research team was very good to be able to tackle a real engineering problem. Some achievements can be highlighted here:

- A phyton computer program to solve the differential equations describing the boundary layer growth in a venture;
- A phyton computer code to solve the main equations modeling the fire propagation;
- A CFD code in Ansys software to simulate and optimize the fluid flow in an organ-on-a-chip device;
- A questionnaire was implemented in Google forms to access the literacy on Information Technologies (IT) of an elderly population of a Former Students association and the use of SPSS software to treat the results;
- A database of the Former Students Association implemented in an excel file;

Furthermore, undergraduate students had the opportunity to learn how to create a scientific curriculum approved by Portugal's scientific community, how to use scientific databases such as Mendeley to search for research papers in an optimized way, and how to write a systematic review.

Dissemination

At the end of September, the students who applied for the award were also invited to participate in the 9th edition of the National Meeting of Engineering and Industrial Management (ENEGI), at the *Instituto Superior de Engenharia do Porto* to present the database created for the Former Students Association members (Martins et al. 2022).

ENEGI is an annual event that brings together students, teachers, researchers, and professionals from all over the country in the area of Industrial Engineering and Management, where everyone is invited to present in public the work they have done in the area and to divulge the results they have obtained. At the end of the event, the students had their participation certificate, which contributes to the academic recognition and merit in taking part in these events.

In addition, the analysis of the data gathered with the questionnaire was presented at the 11th EAI International Conference on Wireless Mobile Communication and Healthcare (Martins et al. 2022).

European Researchers' Night

In the European Researchers' Night (translation of the original in Portuguese NEI, *Noite Europeia dos Investigadores*) that took place at the end of September, the Research students, candidates for the UMinho prize, and teachers were present at the event that took place in Braga, at the Altice Fórum.

NEI had more than 50 exhibitors, set up by more than 300 researchers, mostly young Ph.D., available to explain their research to the public. From the work developed, three exhibitors were set up: "Organ-on-the-chip", "Ways to prevent Fireforests" and "Energy Poverty and Comfort in Buildings". The event offered 1 roll-up to publicize each exhibition and some information and awareness leaflets for the public to take home.

In this sense, the first exhibitor, "Organ-on-the-chip", explained the need to create the equipment, how it works and how the device can help the prevention and diagnosis of diseases, such as cancer (Figure 8).

Figure 8. Organ-on-a-chip expositor a) and its flyer b)
Source: Authors

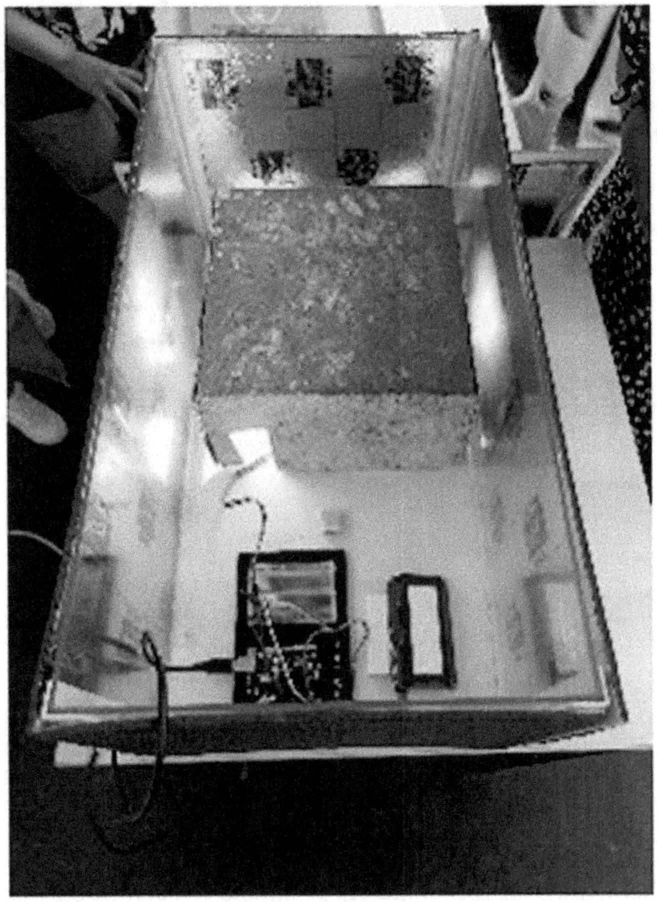

In the second display, "Ways to prevent Fireforest", a video demonstration of the numerical process that allows predicting the movement of a fire and how it can spread according to the climate, vegetation, and specific conditions of a space/forest under analysis was shown (Figure 9).

Figure 9. Forest fire prevention expositor a) and its flyer b)
Source: Authors

The third exhibiton, "Energy Poverty and Comfort in Building", was also presented using a video demonstration, as well as the model described in figure 7, to explain the need to adopt measures to reduce energy poverty in our houses.

For further dissemination, it was also created a children's activity book of content Energy Poverty in Buildings, a frame to promote the theme and the event, as well as a Poster explaining the work done numerically in the context of thermal comfort (Figure 10).

Figure 10. Energy Poverty and Comfort in Building expositor a) and its flyer b).
Source: Authors

Following the taxonomy proposed by Fink, the authors' team has sought in the experiences developed with secondary school students to contribute to their learning process with meaning, namely: 1) to help students understand and recognize the field of engineering knowledge, 2) contribute to the application of critical, creative and practical thinking, 3) contribute to the integration of different mathematics and STEM concepts with practical and useful engineering examples, 4) contribute to providing students with a new vision about their future academic and professional, 5) contribute to students' interest in learning something new and interesting, and 6) contribute to students' awareness of the scientific method.

In activities with undergraduate students, the team seeks to 1) help students validate basic knowledge acquired as well as ideas and perspectives in engineering knowledge, 2) contribute to the development of their ability to investigate and communicate, 3) contribute for the integration of the different concepts or disciplines of engineering courses, 4) to contribute to providing a new understanding of themselves, as well as a better perception of others and how to work in a team, 5) to contribute to the students' interest in research and communicate their work and 6) help students to become aware of the research process, and the importance of continuing to learn.

DISCUSSION AND FINAL REMARKS

The academic education is living in a time of profound cognitive and social transformations, associated with a fast technological development and an exponential increase in ever-changing information. This transformation in the field of knowledge has repercussions on the requests and demands placed on students, who will become the professionals of tomorrow. These professionals will be asked to consider having international mobility experiences and how to think critically, look for information, carry out continuous training, take decisions, work as a team and propose creative and innovative solutions (González-Gómez and Jeong, 2022).

The reported activities in the present work explore several of these aspects. The UMinho Prize and Summer Grants activities involved several graduate and undergraduate students working with a multidisciplinary team in order to develop research skills. The activity aims for the assessment of intellectual learning, as well as the ability to investigate and apply knowledge. The retention of engineering students to develop research is a very important aspect so that higher education institutions can continue to develop both fundamental and applied research. Thus, the most important aspect of UMinho Prize or Summer Grants activities was to improve students' engagement. Students demonstrated increased levels of engagement to their projects due their technology-based character, together with the possibility in receiving an award or to be recompensed by their research work contribution. In this sense, they developed analytical knowledge gained by investigating problem solutions. This positive outcome is aligned with the conclusions of Goyal et al. (2022), which results show that the undergraduate students' performance explores the creative and critical thinking, collaboration and communication in outcome-based educational environments. According to Maia and Dos Santos (2022) the cognitive dimension is related to the learner's intentionality and self-direction since higher education is undergoing a growing movement to shift the traditional paradigm of knowledge-based teaching to competency-based education, using active learning models.

Another aspect is related with the need to integrate knowledge from different scientific fields, so the students can build a bridge between the conceptual and application. Such activities can promote the subjects learning (e.g., physics, heat transfer, mathematics, statistics or fluid mechanics) by applying technological tools (e.g., computation simulation tools or prototypes) that otherwise these so-called "digital students" would not be challenged to study. Entrepreneurial education motivates change, innovation and integrates knowledge, which together provide students with the development of a mindset that encourages the creation of new solutions to a problem or its optimization. There are some successful examples which main objectives include building an extensive and flexible knowledge base, leading to students to learn and develop effective problem-solving skills, especially in engineering (Morais et al., 2021). Weber (2019) developed several experiences from the implementation of active learning in sophomore-level of Mechanical Engineering by using a web-based programming software. The approach consisted in the incorporation of PBL into the thermodynamics courses, so undergraduate students relate and engage with the thermodynamics subject, on a motivated way. The students applied the knowledge from the lecture videos to "real" examples, leading to use analysis and evaluation, taking into account that the problem has not only one definitely-correct answer. The PBL approach, the flipped classes reinforced the engagement of students in problem solving, leveraging their comprehension through the use of the material built outside of class. It is also important is this type of projects to attain the student's perspective.

With the UMinho Prize and Summer Grants activities, it is concluded that the interdisciplinary nature of projects and the different roles have when supporting students does not require a specific planning of content so students can propose creative solutions. This is corroborated by the study of Sukackė et al. (2022) and Moreno et al. (2017). Moreno et al. (2017) evaluated the impact of learning experiences from annual industrial sponsored projects developed in academia by mechanical engineering students. When asked about their satisfaction with the learning program, students stated that the experience allow them to have a good balance between the technical analysis and the experimentation. They also mentioned that the easier communication with the team management helped with the project planning despite the difficulties in sticking with the schedule. This aspect arises a very interesting issue in developing such activities, which is the time frame that should be adopted for this type of activities. In further activities this aspect should be addressed.

Regarding ODU and Summer on Campus, although the objective of these two initiatives is also related to the need to attract students' interest in the areas of knowledge in Engineering, there must be a set of different concerns in the implementation of these activities. Thus, the approach lays on guided tours, workshops, lectures, among other dynamics. It is also important to consider the diversity of students who nowadays enrol in higher education. Valuing previous knowledge and experiences proves to be an important factor in promoting more meaningful learning, by facilitating the establishment of connections between the "new" and the "acquired" and, consequently, the internalization of knowledge. Also, these activities provide to high school students the opportunity to explore their academic interests and spark new ones, allow them to "taste" the contact with the university life and meet other students and share experiences. Also, the students are able to take lectures and seminars from college professors, researches and alumni on emergent and important topics. Yet, further effort must be done to increase the interaction level of high school students regarding the activities. Thus, the objective is to promote an exciting learning experience through the concept of "learn by seeing", "learn by doing" and "learn by feeling" (Othman and Mohammad 2022).

In conclusion, teaching methods have changed at a rapid pace and several methodologies have been implemented to improve the student's learning process, namely active learning methods. The present chapter shows different STEM activities and projects developed by a team of teachers and researchers of the Production and Systems Department of the University of Minho. The goal of these activities and projects is that students of different graduate levels can understand the importance of mathematics in engineering problems as well as the importance of research in the development of improved solutions to real-life problems. The activities "Summer on Campus" and "Open Doors University" were important for high school students because they had the opportunity to clarify some doubts regarding the educational offers at the University of Minho and the professional perspectives, from the experience and knowledge of previous graduates, with both Ph.D. students and professors in a friendly environment. Furthermore, students had an overview of the research teams and ongoing projects in different scientific areas and realized the interdisciplinarity involved and the importance of learning different subjects during their education.

On the other hand, the UMinho prize and Summer with Science are initiatives that aim to promote the interest of undergraduate students in different courses in research. The results were beyond expectations given the short duration of the projects. Students showed commitment and enjoyment for the work carried out and the outcomes were rewarding for the research team.

ACKNOWLEDGMENT

This research was supported by the Portuguese Foundation for Science and Technology (FCT), within the Project Scope: UIDB/00319/2020 (ALGORITMI) and R&D Units Project Scope UIDP/04077/2020 (METRICS). Violeta Carvalho acknowledges the Ph.D. scholarship UI/BD/151028/2021 attributed by FCT.

REFERENCES

Al-Kahtani, N., Abdullah, A., Arun, V. S., Tunny, S., Hend, A., & Duaa, A. (2022). Sustaining Blended and Online Learning during the Normal and New Normal Conditions in a Saudi Higher Education Institution: Health Science Students' Perspectives. *Heliyon*, *8*(10), e10898. doi:10.1016/j.heliyon.2022.e10898 PMID:36247156

Alrahlah, A. (2016). How Effective the Problem-Based Learning (PBL) in Dental Education. A Critical Review. *The Saudi Dental Journal*, *28*(4), 155–161. doi:10.1016/j.sdentj.2016.08.003 PMID:27872545

Andersen, A. L., & Carin, R. (2021). Continuing Engineering Education (CEE) in Changeable and Reconfigurable Manufacturing Using Problem-Based Learning (PBL). *Procedia CIRP*, *104*, 1035–1040. doi:10.1016/j.procir.2021.11.174

Beneroso, D., & Robinson, J. (2022). Online Project-Based Learning in Engineering Design: Supporting the Acquisition of Design Skills. *Education for Chemical Engineers, 38*, 38–47. . doi:10.1016/j.ece.2021.09.002

Bestiantono, Agustina, & Cheng. (2020). Online Learning amid the COVID-19 Pandemic: Students' Perspectives Muhammad. *Journal of Pedagogical Sociology and Psychology*, *2*(1), 45–51. doi:10.46627ilet.v1i3.46

Blythe, A. J. C., & Thompson, T. (2022). Virtual Teaching in the COVID Era: Providing Surgical Core Trainee Teaching via Online Webinars and Videoconferencing. *The Surgeon*, *20*(6), e405–e409. doi:10.1016/j.surge.2022.03.007 PMID:35431109

Brent, R., & Felder, R.M. (2016). *Teaching and Learning in STEM: A Practical Guide*. Academic Press.

Bruff, D. (2009). *Teaching with Classroom Response Systems: Creating Active Learning Environments* (Vol. 33). Jossey-Bass.

Capone, R. (2022). Blended Learning and Student-Centered Active Learning Environment: A Case Study with STEM Undergraduate Students. *Canadian Journal of Science, Mathematics and Technology Education*, *22*(1), 210–236. doi:10.100742330-022-00195-5

Carvalho, V., Rodrigues, C., Minas, G., Lima, R., Teixeira, J. C., & Teixeira, S. (2021). I9Masks: From a Multidisciplinary Summer Project to a Non-Accredited Short Course. In *Engineering Education* (Vol. 9). American Society of Mechanical Engineers.

El-Soussi, A. (2022). The Shift from Face-to-Face to Online Teaching Due to COVID-19: Its Impact on Higher Education Faculty's Professional Identity. *International Journal of Educational Research Open*, *3*(January), 100139. doi:10.1016/j.ijedro.2022.100139 PMID:35233543

Fernandes, H. V. J. (2021). From Student to Tutor: A Journey in Problem-Based Learning. *Currents in Pharmacy Teaching & Learning*, *13*(12), 1706–1709. doi:10.1016/j.cptl.2021.09.037 PMID:34895682

Finbråten, H. S., Grønlien, H. K., Pettersen, K. S., Foss, C., & Guttersrud, Ø. (2022). Nursing Students' Experiences with Concept Cartoons as an Active Learning Strategy for Developing Conceptual Understanding in Anatomy and Physiology: A Mixed-Method Study. *Nurse Education in Practice*, *65*(November), 103493. doi:10.1016/j.nepr.2022.103493 PMID:36401995

Fink, D.L. (2003). *Creating Significant Learning Experiences: An Integrated Approach to Designing College Courses*. Jossey Bass.

Freeman, S., Eddy, S. L., McDonough, M., Smith, M. K., Okoroafor, N., Jordt, H., & Wenderoth, M. P. (2014). Active Learning Increases Student Performance in Science, Engineering, and Mathematics. *Proceedings of the National Academy of Sciences of the United States of America*, *111*(23), 8410–8415. doi:10.1073/pnas.1319030111 PMID:24821756

Glendon, K., & Ulrich, D. (2005). Using Games as a Teaching Strategy. *The Journal of Nursing Education*, *44*(7), 338–339. doi:10.3928/01484834-20050701-11 PMID:16094796

González-Gómez, D., & Jeong, J. S. (2022). Approaches and Methods of Science Teaching and Sustainable Development. *Sustainability (Switzerland)*, *14*(3), 1546. Advance online publication. doi:10.3390u14031546

Goyal, M., Gupta, C., & Gupta, V. (2022). A Meta-Analysis Approach to Measure the Impact of Project-Based Learning Outcome with Program Attainment on Student Learning Using Fuzzy Inference Systems. *Heliyon*, *8*(8), e10248. doi:10.1016/j.heliyon.2022.e10248 PMID:36042720

Grimley, M., Green, R., Nilsen, T., Thompson, D., & Tomes, T. (2011). Using Computer Games for Instruction: The Student Experience. *Active Learning in Higher Education*, *12*(1), 45–56. doi:10.1177/1469787410387733

Hammad, S., Graham, T., Dimitriadis, C., & Taylor, A. (2022). Effects of a Successful Mathematics Classroom Framework on Students' Mathematics Self-Efficacy, Motivation, and Achievement: A Case Study with Freshmen Students at a University Foundation Programme in Kuwait. *International Journal of Mathematical Education in Science and Technology*, *53*(6), 1502–1527. doi:10.1080/0020739X.2020.1831091

Harmin, M., & Toth, M. (2006). *Inspiring Active Learning: A Complete Handbook for Today's Teachers*. Association for Supervision and Curriculum Developmen.

Hartikainen, S., Rintala, H., Pylväs, L., & Nokelainen, P. (2019). The Concept of Active Learning and the Measurement of Learning Outcomes: A Review of Research in Engineering Higher Education. *Education Sciences*, *9*(4), 9–12. doi:10.3390/educsci9040276

Hasnine, M. N., Ueda, H., & Ahmed, M. M. H. (2022). Adaptation of AL-TST Active Learning Model in Hybrid Classroom: Findings from Teaching during COVID-19 Pandemic in Egypt. *Procedia Computer Science*, *207*, 3226–3233. doi:10.1016/j.procs.2022.09.380 PMID:36275371

Hossein-Mohand, H., Trujillo-Torres, J. M., Gómez-García, M., Hossein-Mohand, H., & Campos-Soto, A. (2021). Analysis of the Use and Integration of the Flipped Learning Model, Project-Based Learning, and Gamification Methodologies by Secondary School Mathematics Teachers. *Sustainability (Switzerland)*, *13*(5), 1–18. doi:10.3390u13052606

Howell, R. A. (2021). Engaging Students in Education for Sustainable Development: The Benefits of Active Learning, Reflective Practices and Flipped Classroom Pedagogies. *Journal of Cleaner Production, 325*(October), 129318. doi:10.1016/j.jclepro.2021.129318

Hussain, M., Sahudin, S., Samah, N. H. A., & Anuar, N. K. (2019). Students Perception of an Industry Based Approach Problem Based Learning (PBL) and Their Performance in Drug Delivery Courses. *Saudi Pharmaceutical Journal, 27*(2), 274–282. doi:10.1016/j.jsps.2018.11.009 PMID:30766440

Janes, G., Ekpenyong, M. S., Mbeah-Bankas, H., & Serrant, L. (2022). The Role of Blended Learning in Preparing Nursing and Midwifery Students for Initial Professional Registration: An International, Cross-Sectional, Sequential Mixed Methods Study. *Nurse Education in Practice, 103514*. Advance online publication. doi:10.1016/j.nepr.2022.103514

Jonassen, D. H. (2010). *Learning to Solve Problems: A Handbook for Designing Problem-Solving Learning Environments* (1st ed.). Routledge. doi:10.4324/9780203847527

Klusmann, B., Trippenzee, M., Fokkens-Bruinsma, M., Sanderman, R., & Schroevers, M. J. (2022). Providing Emergency Remote Teaching: What Are Teachers' Needs and What Could Have Helped Them to Deal with the Impact of the COVID-19 Pandemic? *Teaching and Teacher Education, 118*, 103815. doi:10.1016/j.tate.2022.103815 PMID:35874644

Lino-Neto, T., Ribeiro, E., Rocha, M., & Costa, M. J. (2022). Going virtual and going wide: Comparing Team-Based Learning in-class versus online and across disciplines. *Education and Information Technologies, 27*(2), 2311–2329. doi:10.100710639-021-10683-0 PMID:34421327

Liu, Y., and Pásztor, A. (2022). Effects of Problem-Based Learning Instructional Intervention on Critical Thinking in Higher Education: A Meta-Analysis. *Thinking Skills and Creativity, 45.* . doi:10.1016/j.tsc.2022.101069

Lonergan, R., Cumming, T.M., and O'Neill, S.C. (2022). Exploring the Efficacy of Problem-Based Learning in Diverse Secondary School Classrooms: Characteristics and Goals of Problem-Based Learning. *International Journal of Educational Research, 112.* doi:10.1016/j.ijer.2022.101945

López-Fernández, D., Ezquerro, J. M., Rodríguez, J., Porter, J., & Lapuerta, V. (2019). Motivational Impact of Active Learning Methods in Aerospace Engineering Students. *Acta Astronautica, 165*, 344–354. doi:10.1016/j.actaastro.2019.09.026

Lucas da Rocha Cunha, M., Amendola, F., Fernandez Samperiz, M. M., & Gomes da Costa Mohallem, A. (2018). Evaluation of Student Perception of the Team-Based Learning Method (APA-TBL): Instrument Construction and Validation. *Nurse Education in Practice, 33*(April), 141–147. doi:10.1016/j.nepr.2018.09.008 PMID:30296726

Lytras, M. D., Serban, A. C., Torres Ruiz, M. J., Ntanos, S., & Sarirete, A. (2022). Translating Knowledge into Innovation Capability: An Exploratory Study Investigating the Perceptions on Distance Learning in Higher Education during the COVID-19 Pandemic - the Case of Mexico. *Journal of Innovation and Knowledge, 7*(4), 100258. doi:10.1016/j.jik.2022.100258

Maia, D., & Dos Santos, S. C. (2022). Monitoring Students' Professional Competencies in PBL: A Proposal Founded on Constructive Alignment and Supported by AI Technologies. *Proceedings - Frontiers in Education Conference*. 10.1109/FIE56618.2022.9962459

Martín-Garin, A., Millán-García, J. M., Leon, I., Oregi, X., Estevez, J., & Marieta, C. (2021). Pedagogical Approaches for Sustainable Development in Higher Studies in Building. *INTED2021 Proceedings, 1*, 6777–83. 10.21125/inted.2021.1348

Martins, G., Carvalho, V., Afect, C., Teixeira, S. F. C. F., & Rodrigues, C. (2022). Getting Contact to Elderly Associates through the ICT: An Exploratory Study. *EAI MobiHealth 2022 - 11th EAI International Conference on Wireless Mobile Communication and Healthcare*.

Martins, G., Carvalho, V., Afect, C., Teixeira, S. F. C. F., & Rodrigues, C. (2022). *Um Estudo Colaborativo Das Escolas Industriais e Do Seu Impacto Na Região de Braga. 9º Encontro Nacional de Engenharia e Gestão Industrial.*

McKay, M.A., Pariseault, C.A., Whitehouse, C.R., Smith, T., and Gunberg Ross, J. (2022). The Experience of Baccalaureate Clinical Nursing Faculty Transitioning to Emergency Remote Clinical Teaching during the COVID-19 Pandemic: Lessons for the Future. *Nurse Education Today, 111*. doi:10.1016/j.nedt.2022.105309

Miller, D. M., Khalil, K., Iskaros, O., & Van Amburgh, J. A. (2017). Professional and Pre-Professional Pharmacy Students' Perceptions of Team Based Learning (TBL) at a Private Research-Intensive University. *Currents in Pharmacy Teaching & Learning, 9*(4), 666–670. doi:10.1016/j.cptl.2017.03.001 PMID:29233441

Morais, P., Ferreira, M. J., & Veloso, B. (2021). Improving Student Engagement with Project-Based Learning: A Case Study in Software Engineering. *Revista Iberoamericana de Tecnologias Del Aprendizaje, 16*(1), 21–28. doi:10.1109/RITA.2021.3052677

Moreno, V., Ferreira, M., DePasquale, J., Malicki, P., Schier, L., Morris, S., & Smith, H. (2017). Learning Experiences from an Industrially Sponsored Senior Design Capstone Program - A Student Perspective. *ASME International Mechanical Engineering Congress and Exposition Proceedings, 5*, 1–6. doi:10.1115/IMECE2017-70916

Müller, F. A., & Wulf, T. (2022). Blended Learning Environments and Learning Outcomes: The Mediating Role of Flow Experience. *International Journal of Management Education, 20*(3), 100694. doi:10.1016/j.ijme.2022.100694

Nasr El-Din, W.A., Atwa, H., Potu, B.K., Deifalla, A., and Fadel, R.A. (2022). Checklist-Based Active Learning in Anatomy Demonstration Sessions during the COVID-19 Pandemic: Perception of Medical Students. *Morphologie : Bulletin de l'Association Des Anatomistes*. doi:10.1016/j.morpho.2022.10.001

Nurse-Clarke, N., & Joseph, M. (2022). An Exploration of Technology Acceptance among Nursing Faculty Teaching Online for the First Time at the Onset of the COVID-19 Pandemic. *Journal of Professional Nursing, 41*, 8–18. . doi:10.1016/j.profnurs.2022.04.002

Othman, O., Iksan, Z. H., & Yasin, R. M. (2022). Creative Teaching STEM Module: High School Students' Perception. *Creative Teaching STEM Module: High School Students' Perception, 11*(4), 2127–2137.

Pawlicka, A., Tomaszewska, R., Krause, E., Jaroszewska-Choraś, D., Pawlicki, M., & Choraś, M. (2022). Has the pandemic made us more digitally literate? *Journal of Ambient Intelligence and Humanized Computing*, 1–11. doi:10.100712652-022-04371-1 PMID:35971559

Prince, M. (2004). Does Active Learning Work? A Review of the Research. *Journal of Engineering Education*, *93*(3), 223–231. doi:10.1002/j.2168-9830.2004.tb00809.x

Ratan, R., Ucha, C., Lei, Y., Lim, C., Triwibowo, W., Yelon, S., ... Chen, V. H. H. (2022). How do social presence and active learning in synchronous and asynchronous online classes relate to students' perceived course gains? *Computers & Education*, *191*, 104621. doi:10.1016/j.compedu.2022.104621

Ricaurte, M., & Viloria, A. (2020). Project-based learning as a strategy for multi-level training applied to undergraduate engineering students. *Education for Chemical Engineers*, *33*, 102–111. doi:10.1016/j.ece.2020.09.001

Rodrigues, N., Teixeira, I., Carvalho, V., Abreu, I., Gonçalves, I., Graçoeiro, D., ... Teixeira, S. (2022). Summer Grants 'Verão Com Ciência', from Foundation for Science and Technology, in Portugal: Experience and Achievements. *Proceedings of the ASME 2022 - International Mechanical Engineering Congress and Exposition*.

Sharma, A., Janke, K. K., Larson, A., & Peter, W. S. (2017). Understanding the early effects of team-based learning on student accountability and engagement using a three session TBL pilot. *Currents in Pharmacy Teaching & Learning*, *9*(5), 802–807. doi:10.1016/j.cptl.2017.05.024 PMID:29233307

Succar, T., Beaver, H. A., & Lee, A. G. (2022). Impact of COVID-19 pandemic on ophthalmology medical student teaching: Educational innovations, challenges, and future directions. *Survey of Ophthalmology*, *67*(1), 217–225. doi:10.1016/j.survophthal.2021.03.011 PMID:33838164

Sukackė, V., Guerra, A. O. P. D. C., Ellinger, D., Carlos, V., Petronienė, S., Gaižiūnienė, L., Blanch, S., Marbà-Tallada, A., & Brose, A. (2022). Towards active evidence-based learning in engineering education: A systematic literature review of PBL, PjBL, and CBL. *Sustainability*, *14*(21), 13955. doi:10.3390u142113955

Theobald, E. J., Hill, M. J., Tran, E., Agrawal, S., Arroyo, E. N., Behling, S., Chambwe, N., Cintrón, D. L., Cooper, J. D., Dunster, G., Grummer, J. A., Hennessey, K., Hsiao, J., Iranon, N., Jones, L. II, Jordt, H., Keller, M., Lacey, M. E., Littlefield, C. E., ... Freeman, S. (2020). Active learning narrows achievement gaps for underrepresented students in undergraduate science, technology, engineering, and math. *Proceedings of the National Academy of Sciences of the United States of America*, *117*(12), 6476–6483. doi:10.1073/pnas.1916903117 PMID:32152114

Truss, A., & Anderson, V. (2023). The navigational challenges of a blended learning approach to teaching in business and management. *International Journal of Management Education*, *21*(1), 100733. doi:10.1016/j.ijme.2022.100733

Weber, B. W. (2019, November). Climbing Bloom's Taxonomy With Jupyter Notebooks: Experiences in Mechanical Engineering. In *ASME International Mechanical Engineering Congress and Exposition* (*Vol. 59421*, p. V005T07A022). American Society of Mechanical Engineers. 10.1115/IMECE2019-10615

Wieman, C. E. (2014). Large-scale comparison of science teaching methods sends clear message. *Proceedings of the National Academy of Sciences of the United States of America, 111*(23), 8319–8320. doi:10.1073/pnas.1407304111 PMID:24853505

ADDITIONAL READING

Alves, A. C., & van Hattum-Janssen, N. (2022). *Training Engineering Students for Modern Technological Advancement*. IGI Global.

Alves, A. C., Van Hattum-Janssen, N., & Fernandes, S. (2021). Teacher collaboration in PBL: Setting the example for the students. *International Symposium on Project Approaches in Engineering Education, 11*, 6–14. 10.5281/zenodo.5093791

Barbosa, F. V., Lobarinhas, P. A. M., Teixeira, S. F., & Teixeira, J. C. F. (2022). Project-Based Learning in a Mechanical Engineering Course: A new proposal based on student's views. *International Journal of Mechanical Engineering Education, 50*(4), 767–804. doi:10.1177/03064190221078259

Chen, J., Kolmos, A., & Du, X. (2021). Forms of implementation and challenges of PBL in engineering education: A review of literature. *European Journal of Engineering Education, 46*(1), 90–115. doi:10.1080/03043797.2020.1718615

Cruz, S., Viseu, F., & Lencastre, J. A. (2022). Project-Based Learning Methodology as a Promoter of Learning Math Concepts: A Scoping Review. *Frontiers in Education, 7*(July), 1–11. doi:10.3389/feduc.2022.953390

Cukurbasi, B., & Kiyici, M. (2018). High School Students' Views on the PBL Activities Supported via Flipped Classroom and LEGO Practices. *Journal of Educational Technology & Society, 21*(2), 46–61. https://www.jstor.org/stable/26388378

Letzel-Alt, V., Pozas, M., Schwab, S., Schneider, C., Lindner, K.-T., Dias, P., & Cadime, I. (2022). Exploring inclusive education in times of COVID-19: An international comparison of German, Austrian and Portuguese teachers. *Frontiers in Education, 7*(October), 1–16. doi:10.3389/feduc.2022.969737

Mesquita, D., Chagas, R. L. C. P., Lima, R. M., & Chagas, J. M. (2018). Assessment models in two project based learning (PBL) approaches: An exploratory study | Modelos de avaliação em duas abordagens de aprendizagem baseada em projetos (PBL): Um estudo exploratório. *International Symposium on Project Approaches in Engineering Education, 8*, 612–619.

Navas, H. V. G., Hasegawa, H., Watanabe, D., Khantachawana, A., & Alves, A. C. (2020). Global PBL: Cross-cultural educational project for engineering students. *International Symposium on Project Approaches in Engineering Education, 10*, 319–326.

KEY TERMS AND DEFINITIONS

Active Learning: Active learning is a teaching approach that consists in students being actively involved in their own learning process. Instead of simply being presented with information, students in an active learning environment are encouraged to engage with the data and participate in activities that help them understand and apply what they are learning by analyzing, synthesizing, and evaluating.

Engineering Education: Engineering education is designed to provide the principles and practices of engineering to the future engineers, who use their skills and knowledge to design, build, and maintain the systems and technologies needed around the world.

Experimental Learning: This approach is also known as experiential learning or hands-on learning, and it is based on the idea that people learn best when they are actively engaged in the learning process. Students engage in activities that allow them to directly apply what they are learning, such as building a model, conducting an experiment, or creating a product.

Laboratory Learning: Laboratory learning, refers to the use of hands-on activities and experiments to teach scientific concepts and principles. They provide students with an opportunity to apply what they are learning in a practical setting and to develop important skills such as problem-solving and critical thinking.

Learning by Doing: Learning by doing is a teaching and learning approach that involves students actively participating in activities and experiences by applying new knowledge and skills.

Problem-Based Learning: Problem-based learning is an instructional approach that involves students learning through the process of solving real-world problems. Students are presented with a complex, open-ended problem or challenge, and they are asked to work together in small groups to develop a solution.

STEM: STEM education is designed to help students develop the skills and knowledge they need to succeed in the fields of science, technology, engineering, and mathematics, and to prepare them for careers in these fields.

Student Engagement: Student engagement refers to the level of involvement, interest, and enthusiasm that students show in their learning. When students are actively engaged in their education, their understanding and ability to retain new information is improved, and they are more likely to develop a positive attitude towards learning.

Chapter 8
Integrated Physics Learning Using an Interdisciplinary Inquiry Learning Space:
An Exploratory Study Using Computer Programming

João Robert Nogueira
Lusófona University, Portugal

Pedro Carmona Marques
https://orcid.org/0000-0003-4891-1754
Lusófona University, Portugal

Cristina Guerra
Lusófona University, Portugal

ABSTRACT

This work aims to understand how the use of problem-based learning (PBL), from an interdisciplinary point of view, can enhance the integration of concepts in different disciplines. In addition, the study strives to verify whether the use of computer programming can enhance this integration. In this sense, an interdisciplinary inquiry learning space (ILS) was implemented in the physics module of some lectures of undergraduate engineering degrees. For this study, the authors analyzed the learning evolution of two groups: the first with the application of ILS without a computer programming component and the second with the same ILS but with a computer programming component in MATLAB©. The ILS was organized around Newtonian mechanics, focused on the problem of planetary orbits and the possibility of orbiting around a black hole. From the observation conducted and the results of a survey, it is concluded that this type of approach can enhance the learning of physics, not only by increasing the motivation to learn, but also by the use of computer programming.

DOI: 10.4018/978-1-6684-5765-8.ch008

INTRODUCTION

The education of students in engineering degrees involves acquiring a robust background knowledge in the subjects of physics and mathematics, which allows students to deepen the specific disciplines of their field of study. The time available for teaching these basic subjects is, in general to short, forcing educators to describe many subjects with not much depth, leaving to the student the responsibility of autonomously complementing their training.

Experience has shown that students, in general, are unable to expand the subjects of physics and mathematics, resulting in a disintegrated knowledge of these subjects, from an interdisciplinary point of view (physics and mathematics), or in its connection with the specific subjects of their graduation (Lichtenberger et al., 2017).

Thus, there is a need to promote, in these fundamental thematic, interdisciplinary learning models that allow an integration of concepts that can be extended to their training. A widely used way of solving this type of difficulty is problem-based learning (PBL), in which students are guided to solve a problem, generally external to the subject, based on the knowledge acquired in this subject, as well as in other subjects in their training extent, whose application indicates a high level of satisfaction, either for the students or for the tutors (Sattarova et al., 2021). In order to support interdisciplinary learning from PBL, more research is needed (Stentoft, 2017).

Based on this premises, the authors proposed the reflection and the evaluation of the process of implementing a research (inquiry) project-based type in the modules of the course of physics of two engineering curricula. The study is exploratory, in which the authors follow the scheme of an Inquiry learning space (ILS), proposed by Pedaste et al. (2015) and grounded in the foundations of gravitation and planetary orbits.

It should be highlighted that with this ILS, students can confront ideas from physics concepts, in particular those proposed by Newton and Einstein, regarding the possibility of orbiting around a black hole. The topic is not addressed in physics modules and is not within the mathematical reach, by the students, of an analytical resolution. The choice of the subject is also related to the misconceptions that students have on the subject, the result of a disjointed basic education, combined with great speculation provoked by non-scientific popularization articles.

The authors have also introduced to this study a computer programming component in MATLAB©, providing students with basic concepts of computer programming language and allowing them to build a program to simulate the orbit of a celestial body around another one with a much higher mass. Throughout the ILS project, the concepts covered in the module of physics are developed in association with the mathematical concepts of vector and derivative, using computer programming as an integrating element.

The study was conducted in two groups from different engineering focuses – electricity and industrial management, each with various levels of learning in computer programming. The main research question that guided this study was:

- Does this type of approach can enhance the learning of Physics, not only by increasing the motivation to learn, but also by the use of computer programming?

And the objectives of this empirical study are:

1. To observe the satisfaction and motivation of students, associated with the use of this type of strategies.
2. To evaluate the benefits for the learning of physics that this approach brings, in particular, the importance that computer programming has in this process.
3. To observe the level of integration of concepts, in the sense of meaningful learning (Ausubel, 1963), that students have associated with this type of strategies.
4. To observe the level of autonomy that students have at this stage of education.

BACKGROUND

Inquiry learning (IL) enables students to study a topic in a manner that replicates scientific inquiry, which is an appropriate procedure for teaching and learning knowledge. In fact, IL can provide peer feedback, scientific reasoning and conceptual understanding, aspects that a traditional lecture normally fail to achieve (Friesen & Scott, 2013). IL have also the capability to involve students in achieving better learning results in particular those regarding online solutions, such as the case of an ILS, as this process is normally anonymous and time consume less (de Jong, 2006; Rosa et al., 2016; Li et al., 2020).

To sum up, this type of education is believed to increase students' motivation to learn (Duarte and Nogueira, 2019) and able to promote an integrated learning, as suggested by Maknun (2021). Despite being more used in pre-university students (Major & Mulvihill, 2018; Ayyildiz &Tarhan, 2018; Siagian et al., 2019), this methodology has had an increasing adherence in university education according to Bijsmans & Schakel (2018), and Luo (2019).

Associated or not with this teaching typology, computer programming has played a key role in the teaching of physics and mathematics as studied by Yetilmezsoy & Mungan (2018), Malgieri & Onorato (2020), and diSessa (2018).

To end up, this tool has shown to be able to expand the curricular concepts of the subjects, to levels of knowledge, which are beyond the mathematical and physical possibilities, that students currently have, as well as has shown to endorse the computational thinking, necessary for the coherent construction of scientific rational (Nogueira et.al, 2019; diSessa, 2018). However, the difficulties associated with learning computer programming can comprise an obstacle to its implementation in the syllabus of the above-mentioned topics, removing time availability for learning specific contents of physics and/or mathematics (Chabay & Sherwood, 2008).

MATERIALS AND METHODS

Presentation of the ILS Project

The project called "Universal Gravitation Law" was built in the form of an ILS, developed on the GO-Lab platform (http://graasp.eu), following the typology proposed by Pedaste et al. (2015), being the work built by four main phases: (1) Orientation, (2) Conceptualization, (3) Research, and (4) Conclusion and Discussion. At each of these phases, different challenges were posed to students, such as answering questions and questionnaires, producing texts, drawing pictures, writing computer programs, among others.

In the (1) Orientation phase, students were led to search for information concerning orbital movements, stars, and black holes, as searching for a historical approach to these issues. Also, during this phase, students reviewed the basic concepts of mechanics, introducing the topic of computer programming as a support for solving problems in physics. The semi-implicit Euler method was presented in this stage, leading students to a reflection on the feasibility of this numerical method for solving the differential equations associated with dynamics. The mathematical concept of vector was developed, linking it with the concepts of velocity, acceleration, and force in physics. Also at this point, students were guided in the construction of an orbital simulation computer program, with which they performed simulations for the earth's orbital movement around the sun, using real data. The program was built on Newton's Law of Universal Gravitation section.

In the (2) Conceptualization phase, two questions were asked:

- Would it be possible for a planetary orbit to pass within the event horizon (black hole)?
- What makes us stop using Newton's theory and start using Einstein's?

Based on these two questions, students were asked to build hypotheses, and to develop, based on the models they used previously, a plan that can put the hypotheses they made to the test – (3) Research phase.

Conclusions were recorded in the next section and discussion was held with educators and peers – phase 4. Throughout the ILS, students must answer open and closed questions that were registered on the platform, as well as writing short opinions about the documents they read during the ILS.

Description

The application of the ILS was conducted in two groups of engineering curriculums at a Portuguese private university, one in Industrial Engineering and Management (IEM) and the other in Electrical Engineering (EE). The students worked on the ILS project during the second semester of the first year (freshmen) in the module of Physics. The syllabus of the two groups is similar, both belonging to the same department of the university. In this module, it is intended to reinforce the learning that they bring from minor education in classic mechanics, ending up with an introduction to mechanical waves. The two groups differ in the number of working hours, as for the case of EE students having one more hour per week (a total of five and a half hours). Groups were divided into theoretical, theoretical-practical and laboratory. Thus, while the IEM group has a weekly hour in the laboratory, the EE group has two weekly hours. The laboratory lessons, in both degrees, were divided into two parts, a physical laboratory, and a virtual laboratory, in which students develop simulations in MATLAB©. In the IEM degree, simulation lessons consist of solving problems using programs provided by the lecturer. In the EE degree, making use of the extra weekly hour, students were guided in the construction of these programs to be able to use them in solving problems. Note that in this first year both degrees have an introductory module in computer programming. The IEM group has sixteen students, seven females and nine males, while the EE group consists of only eleven males. To proceed with the application of the ILS, educators organized the groups into small groups, i.e., seven groups in IEM and five groups in EE. The ILS project was developed by the students during the last four weeks of the semester. The first week was conducted by the educators of the module, in which they explained the scope of the work to the students, made a presentation about the GRAASP platform, and showed how they would have to develop the work. The following sessions were students' autonomous work, always accompanied by educators to clarify issues. The project ended with a face-to-face session to present and discuss the work developed.

The material collected for analysis consists of the questionnaires that the students answered, the short texts they produced and the programs they write during the ILS, the discussion of the research work, and an online anonymous satisfaction questionnaire after submission of work, in which the last two questions were open. The content analysis of the last two questions was based on Bardin (2013). From data analysis emerge indicators that are presented in the results section. The analysis is qualitative, seeking to triangulate the diverse sources of information.

Participants

To distinguish the groups, the authors used a numbering from 1 to 5 in the EE groups (EE-1, EE-2, EE-3, EE-4, and EE-5) and from 1 to 7 in the other group (IEM-1, IEM-2, IEM-3, IEM-4, IEM-5, IEM-6, and IEM-7) as presented in Table 1.

Table 1. Composition of groups of the ILS.

Groups	Group	Number of Students per Gender	
		Male	Female
Electrical Engineering (EE)	EE-1	2	0
	EE-2	2	0
	EE-3	2	0
	EE-4	3	0
	EE-5	2	0
Industrial Engineering and Management (IEM)	IEM-1	2	0
	IEM-2	0	2
	IEM-3	2	0
	IEM-4	2	2
	IEM-5	2	0
	IEM-6	0	2
	IEM-7	1	1

Regarding the action of groups in the ILS, on the one hand, the authors present the evolution of the groups throughout each task of the ILS, and on the other hand, highlight the evolution of each group, to analyze differences between them.

Results Observed During the ILS

This set includes the results of the questionnaires answered in the ILS, the texts and computer programs produced, as well as the analysis and discussion of their work. It should be noted that not all groups finished the ILS, as well as the two major groups differ in the production of computer programs, since the IEM groups did not have the same basic training in MATLAB©.

The results are presented sequentially, in the same order as they appear throughout the ILS.

At first, students were asked to provide a text on the differences in thought between Newton and Einstein, only two groups (EE-5 and IEM-1) did not respond, and the rest showed work with curiosity in the constitution of our planetary system. The answers were poorly constructed and denoting an elevated level of common sense, apart from IEM-5 group, from which the following excerpt is presented:

Both Newton and Einstein sought to describe the gravitational interaction in order to understand the reason the attraction of masses between celestial bodies occurred and also sought to explain the orbital movements that were observed in the sky...

In the questionnaire that follows this point, it was asked which of the two previous scientists can explain the existence of a black hole. Apart from two groups (EE-5 and IEM-2), all respond correctly.

In the second text, which asks for a reflection on what distinguishes a black hole from a normal star, students show an evolution in the construction of ideas, denoting a lower level of common sense. It should be noted that the evolution is more significant in EE groups, and in this case all groups responded. For example, group EE-1 answered with the following statement:

With this we conclude that there are four major differences that distinguish a black hole from an ordinary star, its age, density, gravitational force and luminosity.

Regarding its age, an ordinary star is just the beginning of the cycle that can lead to the formation of ablack hole, so the black hole is much older than an ordinary star. From a density point of view, a black hole has a much higher density than an ordinary star. The gravitational force of the black hole is also much higher, so even light cannot escape the event horizon. Finally, a black hole does not emit any type of light due to its gravitational force, but on the contrary, an ordinary star has the emission of light as a characteristic.

The next two questions seek to see if students have mastered the basic concepts of kinematics. The two questions were:

1. If the position of a car is described by the function $x(t) = 2t^2 + 2t + 4$, indicate which of the following sentences is correct:
    a. The velocity of the car is given by the function $v(t) = 2t + 2$;
    b. The velocity of the car is constant;
    c. The acceleration of the car is constant.
2. Indicate if the following statements are true or false:
    a. An object can have an accelerated motion with constant velocity;
    b. If the position of a body varies with a quadratic time function, its acceleration can be constant.

All groups responded correctly.

Next, the groups were asked to run two MATLAB© programs that generate graphs of lines and parabolas. Apart from one group (IEM-2), all produced the figures and downloads them onto the platform.

Then, from a given algorithm, the groups were asked to produce in MATLAB© a program that determines the roots of a quadratic equation. In this case, the results are quite different in each group. While in EE group everyone produced the program, in the other case, only one group did it (IEM-1). Note that of the programs presented, two show some lack of logical structure (EE-4 and EE-5).

Also, at this stage of computer programming development, the were asked to present a reflection on the use of the computer programming. Only three groups from EE did it (EE-2, EE-3, and EE-4) and only one from IEM did it (IEM-7). The answers were very descriptive of the programming processes, except for one group (EE-2), whose answer shows maturity:

To make these programs we had to start by thinking about the formulas used that will be the basis of the programs. After solving these issues, we can then develop the programs further. Finally, we can evaluate the programs to see the errors and resolve them with the help of comments. Thus, reaching the delivery stage, we can conclude that the programs helped to better understand mathematical formulas and have a graphic image.

In the next challenge where students were expected to use computer programming to solve physics problems, the authors started by asking students, using a given program, to determine the instants of intersections of two cars that move with different position, velocity, and acceleration. Only three groups failed to solve the problem (EE-4, IEM-2, and IEM-4).

The next task consisted of answering three questions based on the analytical solution of the problem posed above. Seven groups answered all the questions correctly (EE-1, EE-3, IEM-1, IEM-3, IEM-5, IEM-6, and IEM-7), one answered correctly only to the last question (EE-2), and the remaining four did not respond correctly to any (EE-4, EE-5, IEM-2, and IEM-4).

After that, students were asked to answer a question that intended to see if the students understood the definitions of velocity and acceleration in one-dimensional motion. The question was:

1. In the situation where the red car (with constant velocity of 90 km/h) is 150 m ahead of the blue one when the latter starts (with constant acceleration), select the true statements:

    a. The blue car never overtakes the red one;
    b. The blue car will eventually overtake the red one;
    c. The blue car only passes the red after a given value of its acceleration.

In this case, four groups answered wrongly (EE-3, EE-4, EE-5, and IEM-5).

At the end of this section, approaching the two-dimensional kinematics, and using a simulation program for the oblique launch of a projectile, three more questions were asked whose answers were different between the various groups, highlighting the most consistent from the groups EE-1, EE-2, EE-3, EE-4, IEM-3, IEM-5, and IEM-7.

Then, for introducing the Euler method, the students downloaded a computer program, given by the authors, in which they could compare graphically the analytical solution against the numerical solution, using different time-based steps for the numerical one. Only five groups correctly answered one of the three questions posed in a questionnaire related to the method (EE-1, EE-2, EE-4, EE-5, and IEM-1). The remaining groups did not have any correct answers.

After this questionnaire, the students were proposed to produce a program in MATLAB© to determine the position and velocity of a falling body with significant air resistance. Only three groups uploaded the program (EE-2, EE-3, and EE-4). The authors present below the program developed by the EE-2 group:

Computer Code

```
clc
clear all
disp('This determines the graph of the height of a falling body as a function
of time')
disp('It is assumed that the body is dropped and falls with air resistance')
disp('You will have to enter the initial height value as well as the time
step')
h=input('Enter a positive value for the initial height: ');
if length(h)==0 || h<=0
  error('You have to enter a positive value')
end
g=9.8;
tmax=sqrt(2*h/g);
dt=input('Enter a positive value for the time step: ');
if length(dt)==0 || dt<=0
  error('You have to enter a positive value')
elseif dt>tmax/2
  error('You have to enter a smaller value')
end
b=input('Enter a positive value for the drag coefficient between 0 and 1: ');
if length(b)==0 || b<0 || b>=1
  error('You must enter a valid value')
end
m=10;
t(1)=0;
y(1)=h;
v(1)=0;
a(1)=-g;
i=1;
while y(i)>0
  t(i+1)=t(i)+dt;
  v(i+1)=v(i)+a(i)*dt;
  y(i+1)=y(i)+v(i+1)*dt;
  a(i+1)=-g+((b/m)*v(i+1)^2);
  i=i+1;
end
t_max=max(t);
```

```
v_min=min(v);
subplot(3,1,1)
plot(t,y)
legend('Position')
axis([0 t_max 0 h])
subplot(3,1,2)
plot(t,v)
legend('Velocity')
axis([0 t_max v_min 0])
subplot(3,1,3)
plot(t,a)
legend('Acceleration')
```

Using the MATLAB© program produced by the students, or the one given by the authors, for simulating the fall with significant air resistance, the students had to answer three more questions, to which only three groups answered all questions correctly (EE-1, EE-2, and EE-3), and of the rest, only one answered a question correctly (IEM-1).

Regarding the concept of energy, it was proposed that students produce a program to simulate kinetic energy, potential energy, and variation in mechanical energy, as well as the work performed by forces. As for the previous situation, the majority of the groups did not develop a computer program, being only done by groups EE-1, EE-2, EE-3, and EE4. From the program done by the students or the one given by the authors for the intended situation, four questions were asked related to the conservation of energy in the fall of a body with significant air resistance. In the answers given, only one group correctly answered the four questions (EE-1), two groups correctly answered three (EE-2 and IEM-6), two groups correctly answered two (EE-3 and IEM-3), and three groups correctly answered one question (EE-5, IEM-1, and IEM-7).

With the given program, in the situation where a body (with a mass of 10 kg) is dropped from a height of fifty meters, with significant air resistance, students can graphically observe the variations in energy and the work of forces as shown in Figure 1.

Figure 1. Graphs obtained for the variations of energy and for the work of forces, from the program given by the educators.
Source: The authors

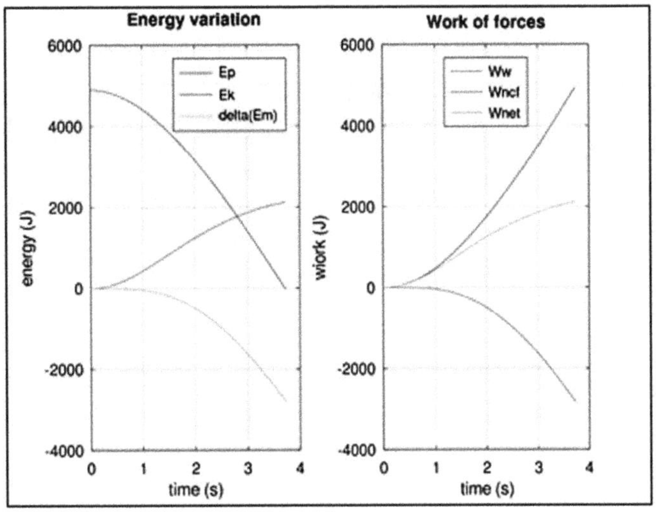

When entering orbital motion, the authors tried to see if the students had acquired the form of writing a vector. Students were asked to write the gravitational force vector and the corresponding acceleration vector on a Cartesian basis. Only two groups correctly write the vectors (EE-4 and EE-5), and six groups did not even respond (IEM-1, IEM-2, IEM-3, IEM-4, IEM-5, and IEM-6).

The task that followed consisted of producing a program in MATLAB© to simulate the Earth's orbit around the Sun using Newton's Law of Universal Gravitation. Only the groups from the Electrical Engineering group done the computer program, and all of which had several limitations.

Using a program given by the educators and tabulated values of the Earth's position and velocity, the students were asked about the Earth's orbital motion, having to answer four questions. The EE-1 group consistently answered three questions, the EE-3 group consistently answered two questions, and the EE-2 group consistently answered one question

The program given by the teachers, allows you to graphically observe the orbit of a planet around any star, particularly, it allows you to simulate an orbit around a black hole, with the representation of the event horizon, as shown in Figure 2.

Figure 2. Graphic representation of a planet's orbit around a black hole, made using the program provided by the lecturer. In this figure, the black hole is represented in the center by a small circle, the surrounding circumference represents the event horizon, defined by the Schwarzschild radius, and the ellipse represents the planet's orbit.
Source: The authors

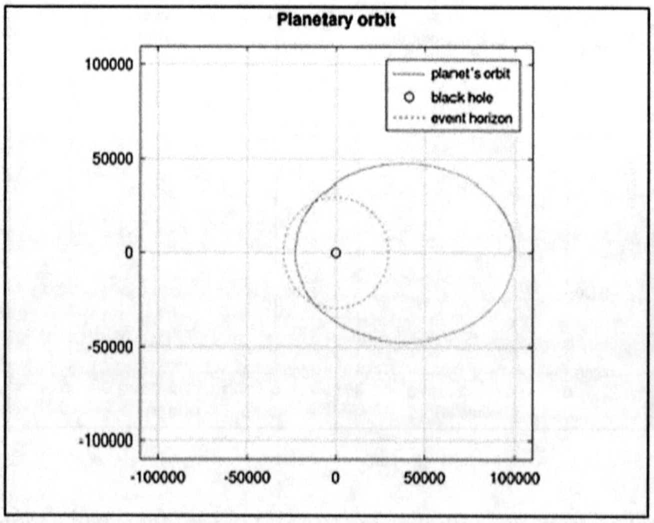

From the point of view of the performance of each group in the answers to the various questions posed throughout the ILS, the authors can see that it is irregular in each of the groups, being weaker in the Industrial Engineering and Management group, as is clear in the percentage of correct answers given by each group, regarding the twenty-six questions asked, and shown in Table 2. Note that the only three groups with a satisfactory performance in the answers were EE-1, EE-2, and EE-3, which had percentages higher than 65%.

Regarding the hypotheses formulated by the students, the authors found that four groups did not present hypotheses (EE-5, IEM-2, IEM-5, and IEM-6), and those presented by the rest were not very consistent, thus failing to logically formulate a hypothesis.

The last two texts requested in this ILS are in the final phase of the work, in the first, each group was asked to make a research plan based on the orbital simulation program given by the teachers, and in the second, they are asked to make the discussion and analysis of research results. These texts were made by only two of the groups (EE-1 and EE-2). Regarding the research plan, the texts presented are reasonable. As for the analysis of the results, only the EE-2 group presents robust conclusions with consistent conclusions regarding the presented hypotheses. The execution of the investigation plan of both groups is good.

The final presentation did not add information relevant to the work that each group had done throughout the ILS.

Table 2. Performance of each group in the answers to the twenty-six questions of the ILS.

Group	Correct Answers (%)
EE-1	86.5
EE-2	69.2
EE-3	67.3
EE-4	28.8
EE-5	25.0
IEM-1	48.1
IEM-2	15.4
IEM-3	53.8
IEM-4	21.2
IEM-5	42.3
IEM-6	51.9
IEM-7	50.0

Source: The authors

Anonymous Survey Results

In this survey, seventeen closed questions were asked to evaluate on a Likert scale from one to four, and two open questions at the end. In the Likert scale used, four corresponds to totally agree, three to agree, two to not agree and one to totally disagree. The questions are presented in Table 3.

Table 3. List of questions of the anonymous questionnaire

Question	
Q1	Did you find the theme proposed in this ILS interesting?
Q2	In general, were you satisfied with ILS?
Q3	Was ILS relevant to your studies?
Q4	Has ILS increased your motivation in learning physics?
Q5	Would you like to use other ILS in other disciplines?
Q6	Would you advise your colleagues to use ILS?
Q7	Has ILS made it possible to improve communication with your colleagues?
Q8	Has ILS improved your understanding of the theoretical concepts of physics that have been addressed in practice?
Q9	Has ILS contributed to your learning?
Q10	Was ILS an effective learning experience?
Q11	Has the knowledge acquired been valuable for your learning?
Q12	Do you consider that the realization of ILS has helped you to understand the concept of force better?
Q13	Do you consider that the realization of ILS has helped you to understand the concept of trajectory better?
Q14	Do you think that the realization of ILS has helped you to better understand the link between acceleration and velocity?
Q15	Do you think that the realization of ILS has helped you to better understand the notion of vector?
Q16	Do you consider that ILS has enabled you to better understand the notion of derivative?
Q17	Do you think computer programming can help you understand physics?
Q18	Indicate positive aspects of the use of ILS
Q19	Indicate negative aspects of the use of ILS

Source: The authors

Of the twenty-seven participating students, twenty-five responded to the survey, two of which are null. The universe of responses is therefore twenty-three, corresponding to 85% of the students participating in the activity. The percentages of answers to each of the questions are related to the twenty-three answers, except for question 12, in which one of the students did not answer, so it is related to twenty-two answers. Figure 3 presents the total four satisfaction levels in the first seventeen questions of the ILS.

Figure 3. Satisfaction levels in the first seventeen questions of the ILS.
Source: The authors

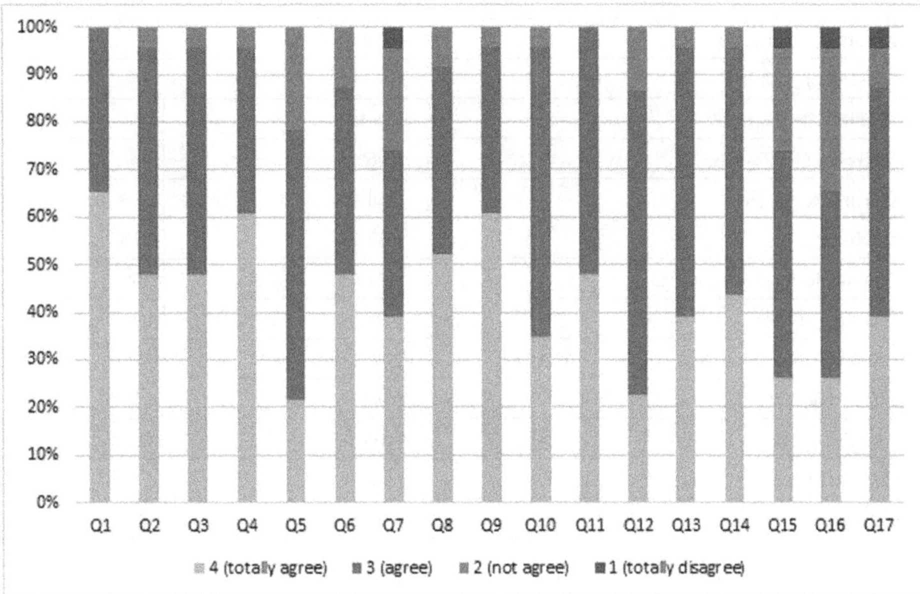

In Table 4, the authors present a summary of the students' agreement expressed in the first seventeen questions, adding, in the third column, the answers totally agree and agree, and in the fourth column, the answers do not agree and do not agree completely.

Table 4. Answers to the first seventeen questions of the anonymous survey.

Question	Subject	Agree	Disagree
Q1	Importance of Theme	100%	0%
Q2	Satisfaction with ILS	96%	4%
Q3	Relevance to studies	96%	4%
Q4	Motivation for Physics	96%	4%
Q5	Use in other subjects	78%	22%
Q6	Use by others	87%	13%
Q7	Communication with colleagues	74%	26%
Q8	Concepts of Physics	91%	9%
Q9	Contribution to learning	96%	4%
Q10	Effectiveness of ILS	96%	4%
Q11	Knowledge value	100%	0%
Q12	Concept of Force	87%	13%
Q13	Concept of trajectory	96%	4%
Q14	Velocity and Acceleration Concepts	96%	4%
Q15	Concept of Vector	74%	26%
Q16	Concept of Derivative	65%	35%
Q17	Computer programming Utility	87%	13%

Source: The authors

In the last two questions, students were asked to indicate the aspects they considered positive in using the ILS (first) and the aspects they considered negative in using the ILS (second). Not all students responded, with nineteen responding to the first and twelve to the second.

- In the first open-answer question, eighteenth of the questionnaire, all answers show satisfaction with the work, highlighting aspects such as: use of simulators; learning about black holes and relativity; practical and interactive nature of the ILS; development of work and discussion with colleagues; expansion of knowledge of Physics. An example of the first open-answer question follows:

It forces the user to research by himself and in this way study and select information from different sources, which leads to a more in-depth study than usual, in addition to that the proposed programs are important for the understanding of the subject.

- In the second open-answer question, nineteenth of the questionnaire, two of the answers say there are no negative aspects, and the rest highlight aspects such as: bugs in the software; use of computer programming; brief time to conduct the project; little interaction with teachers; overly complex concepts. An example of the second open-answer question follows:

There should be more work and support with the teacher, because although it is a research work, there were numerous doubts and programming errors that required greater communication with the teacher.

SOLUTIONS AND RECOMMENDATIONS

It is important in this section to start by denoting that the time devoted to this activity seems to have been insufficient, which may have limited the completion of the work by most groups. Note that this point not only appears in the students' answers to the last question of the anonymous questionnaire but was also expressed by the students' answers throughout the activity.

Regarding the questions that the authors proposed to observe, one can start by verifying that the satisfaction with which the students' learned physics is high, as expressed by the answers to the anonymous questionnaire. This aspect is quite evident in questions 1, 2, 3, 4 and 11 of the above-mentioned questionnaire (see Table 3), where students mostly consider the topic interesting, being satisfied with the completion of the ILS, claiming to have been important for their studies, motivated them to learn physics, and considered that the acquired knowledge was valuable for their learning. This satisfaction was also expressed by the texts produced in the initial phase of orientation, on Newton and Einstein, and on black holes, where despite denoting some common sense, students showed research and analysis work, as well as interest in the physics issue, highlighting the importance of this PBL. This result is in line with a recent study (Sattarova et al., 2021), in which a considerable number of students and tutors reveal an elevated level of satisfaction in the use of PBL strategies such as the one proposed by this ILS.

With regard to the enhancement of learning in physics, despite the answers to the anonymous questionnaire expressing that it took place (questions 8, 9, 12, 13 and 14), it is not completely clear in the performance of the groups throughout the ILS. In fact, in the answers to the questions posed throughout the project (see Table 2), the distinct groups do not show much consistency, especially in IEM group, in which the percentage of correct answers is no more than 53.8%. However, we can attribute this deficient performance to the greater difficulty that this group has regarding computer programming, which may have prevented these students from correctly answering the questions whose answer depended on the use of programs. Regarding groups EE-4 and EE-5, whose performance is also weak (28.8% and 25%), it should be noted that during the sessions the elements of these groups were not always present. Based on this information, the authors consider that this ILS has enhanced the learning of physics, making it more significant. However, the authors believe that the result could be improved with more interaction between students and educators.

As for the use of computer programming in the learning process of physics, the authors found that 87% of students believe that computer programming helps in understanding physics (question 17 of the anonymous questionnaire – see Table 3). In addition, regarding the performance of the groups in the ILS, the authors found that the group with the best performance is the one with the greatest training in computer programming, not only because they are able to produce the programs, but mainly because they show greater ease in using the simulation programs, given by educators, for problem solving. Note that, despite an introduction to computer programming in the orientation phase of this ILS, students in the IEM group were unable to program the simplest examples. This could be related to the time it took to accomplish it. Also in this case, the use of computer programming is shown to enhance the learning of physics, as previously verified (Nogueira et al., 2019).

Regarding the interdisciplinary between Physics and Mathematics, despite the majority of students claiming that the completion of the ILS was important for understanding the concepts of vector and derivative (questions 15 and 16 of the anonymous questionnaire – see Table 3), the authors were unable to observe throughout the ILS evidence of an increase in the understanding of these two concepts. This may be related to the lack of time to mature the concepts associated with the ILS extension. A significant part of the students could not reach the part related to the planetary orbit, in which the concept of gravitational force in space promotes the notion of vector and also the derivative (associated with the Euler method used). It should be noted that interdisciplinary is not guaranteed with the use of a PBL (Stentoft, 2017), requiring that the tasks that ensure interdisciplinary are conducted. The authors believe that this ILS could be more effective in this domain if students had more time to complete the work.

An issue that the authors also tried to observe is related to the level of autonomy that students acquire with this type of activities. Despite the lack of interaction with the educators, the capacity for autonomous research, as well as the ability to solve complex problems, some of them using computer simulation, may be indicators of some expected autonomy in participating students.

FUTURE RESEARCH DIRECTIONS

As a future work, this ILS will be used again, increasing the extent given for its accomplishment, and the period for discussion among students and educators, as present results seem to indicate that these aspects could lead to further meaningful ones. Being this study explorative, and mostly qualitative, the authors suggest the inclusion of a pre-test and post-test to statically achieve the validation of this ILS, in particular focusing on the advantages of this approach in student learning, insofar as the results are not conclusive. The authors also believe that the scope of this project can be larger, emphasizing the positive aspects found so far in the education of physics.

CONCLUSION

As a synopsis of what was stated during solution and recommendations section, the authors believe that the use of an ILS, assisted by computer programming, enhanced the interest of the students in learning physics, answering this way to the main research question that guided this study.

It should be noted that the four objectives of this exploratory study were completed. First, authors observed the motivation to learn physics with this type of approach, as stated in solution and recommendations section - paragraph two – and the enhancement of learning physics – paragraph three, despite some difficulties. Secondly, the benefits for the learning of physics that this approach brings, in particular, the importance that computer programming has in this process was achieved as stated in paragraph three. The observation of the level of integration of concepts, that students have associated with this type of strategies, showed difficulties in being perceived, as the authors acknowledge, in paragraph four. Finally, the authors observed the level of autonomy that students have at this level of education, giving positive indications as referred in paragraph five.

ACKNOWLEDGMENT

The authors are grateful to all the students that have participated in this study.

This research received no specific grant from any funding agency in the public, commercial, or not-for-profit sectors.

REFERENCES

Ausubel, D. P. (1963). *The psychology of meaningful verbal learning*. Grune & Stratton.

Ayyildiz, Y., & Tarhan, L. (2018). Problem-based learning in teaching chemistry: Enthalpy changes in systems. *Research in Science & Technological Education*, *36*(1), 35–54. doi:10.1080/02635143.2017.1366898

Bardin, L. (2013). *L'analyse de contenu* [Content analysis]. PUF. doi:10.3917/puf.bard.2013.01

Bijsmans, P., & Schakel, A. H. (2018). The impact of attendance on first-year study success in problem-based learning. *Higher Education*, *76*(5), 865–881. doi:10.100710734-018-0243-4

Chabay, R., & Sherwood, B. (2008). Computational physics in the introductory calculus-based course. *American Journal of Physics*, *76*(4), 307–313. doi:10.1119/1.2835054

de Jong, T. (2006). Technological advances in inquiry learning. *Science*, *312*(5773), 532–533. doi:10.1126cience.1127750 PMID:16645080

diSessa, A. A. (2018). Computational Literacy and "The Big Picture" Concerning Computers in Mathematics Education. *Mathematical Thinking and Learning*, *20*(1), 3–31. doi:10.1080/10986065.2018.1403544

Duarte, J., & Nogueira, J. R. (2019). From the Hidden Protest of Students towards Problem-Based Learning. *Education Sciences*, *9*(3), 215. doi:10.3390/educsci9030215

Friesen, S., & Scott, D. (2013). *Inquiry-based learning: A review of the research literature*. Paper prepared for the Alberta Ministry of Education.

Li, H., Xiong, Y., Hunter, C. V., Guo, X., & Tywoniw, R. (2020). Does peer assessment promote student learning? A meta-analysis. *Assessment & Evaluation in Higher Education*, *45*(2), 193–211. doi:10.1080/02602938.2019.1620679

Lichtenberger, A., Wagner, C., Hofer, S. I., Stern, E., & Vaterlaus, A. (2017). Validation and structural analysis of the kinematics concept test. *Physical Review. Physics Education Research*, *13*(1), 010115. doi:10.1103/PhysRevPhysEducRes.13.010115

Luo, Y. (2019). The influence of problem-based learning on learning effectiveness in students of varying learning abilities within physical education. *Innovations in Education and Teaching International*, *56*(1), 3–13. doi:10.1080/14703297.2017.1389288

Major, T., & Mulvihill, T. M. (2018). Problem-Based Learning Pedagogies in Teacher Education: The Case of Botswana. *The Interdisciplinary Journal of Problem-Based Learning*, *12*(1). Advance online publication. doi:10.7771/1541-5015.1543

Maknun, J. (2021). Physics Problem Solving Skills of Civil Engineering Students by Implementing Problem-Based Learning. *Review of International Geographical Education*, *11*(3), 594–603.

Malgieri, M., & Onorato, P. (2020). Teaching the heat transfer law using a stochastic toy model. *European Journal of Physics*, *41*, 015103.

Nogueira, J. R., Alves, R., & Marques, P. C. (2019). Computational Programming as a Tool in the Teaching of Electromagnetism in Engineering Courses: Improving the Notion of Field. *Education Sciences*, *9*(1), 64. doi:10.3390/educsci9010064

Pedaste, M., Mäeots, M., Siiman, L. A., de Jong, T., van Riesen, S. A. N., Kamp, E. T., Manoli, C. C., Zacharia, Z. C., & Tsourlidaki, E. (2015). Phases of inquiry-based learning: Definitions and the inquiry cycle. *Educational Research Review*, *14*, 47–61. doi:10.1016/j.edurev.2015.02.003

Rosa, S. S., Coutinho, C. P., & Flores, M. A. (2016). Online peer assessment: Method and digital technologies. *Procedia: Social and Behavioral Sciences*, *228*, 418–423.

Sattarova, U., Groot, W., & Arsenijevic, J. (2021). Student and Tutor Satisfaction with Problem-Based Learning in Azerbaijan. *Education in Science*, *11*, 288.

Siagian, M. V., Saragih, S., & Sinaga, B. (2019). Development of Learning Materials Oriented on Problem-Based Learning Model to Improve Students' Mathematical Problem Solving Ability and Metacognition Ability. *International Electronic Journal of Mathematics Education*, *14*(2), 331–340.

Stentoft, D. (2017). From saying to doing interdisciplinary learning: Is problem-based learning the answer? *Active Learning in Higher Education*, *18*(1), 51–61.

Tsourlidaki, E. (2015). Phases of inquiry-based learning: Definitions and the inquiry cycle. *Educational Research Review*, *14*, 47–61.

Yetilmezsoy, K., & Mungan, C. E. (2018). MATLAB time-based simulations of projectile motion, pendulum oscillation, and water discharge. *European Journal of Physics*, *39*, 065803.

ADDITIONAL READING

Bryan, L. A., Moore, T. J., Johnson, C. C., & Roehrig, G. H. (2016). Integrated STEM education. In C. C. Johnson, E. E. Peters-Burton, & T. J. Moore (Eds.), *STEM road map: A framework for integrated STEM education* (pp. 23–38). Routledge.

Caballero, M. D. (2015). Computation across the Curriculum: What Skills Are Needed? In *2015 Physics Education Research Conference Proceedings*. (pp 79–82) American Association of Physics Teachers. 10.1119/perc.2015.pr.015

Cavadas, B., Rézio, S., Nogueira, J. R., & Branco, N. (2022). A Framework and a Research Design Proposal to Identify Preservice Teachers' Integration Performance of Science and Mathematics. *Canadian Journal of Science. Mathematics and Technology Education, 22*(1), 101–129. doi:10.100742330-022-00198-2

Halloun, I., & Hestenes, D. (1985). The Initial Knowledge State of College Physics Students. *American Journal of Physics, 53*(11), 1043–1055. doi:10.1119/1.14030

Hung, W., Dolmans, D. H. J. M., & van Merriënboer, J. J. G. (2019). A review to identify key perspectives in PBL meta-analyses and reviews: Trends, gaps and future research directions. *Advances in Health Sciences Education: Theory and Practice, 24*(5), 943–957. doi:10.100710459-019-09945-x PMID:31768786

Valadares, J. (2011). A teoria da aprendizagem significativa como teoria construtivista [Meaningful learning theory as a constructivist theory]. *Aprendizagem significativa em Revista/Meaningful Learning Review, 1*, 36-57.

KEY TERMS AND DEFINITIONS

Computer Programming: Is the process of performing a particular computation usually by designing and building an executable computer program.

Engineering Education: Is the activity of teaching knowledge and principles to the professional practice of engineering. It includes an initial education (bachelor's and/or master's degree), and any advanced education and specializations that follow.

Inquiry Learning Space (ILS): Are personalized learning resources for students, including a lab, apps, and any other type of multimedia material. ILSs follow an inquiry cycle. Inquiry cycles can differ, but the basic Go-Lab cycle consists of the phases of Orientation, Conceptualization, Investigation, Conclusion and Discussion.

Interdisciplinary: Involves the combination of two or more academic disciplines into one activity.

MATLAB©: Is a programming and numeric computing platform used by millions of engineers and scientists to analyze data, develop algorithms, and create models.

Motivation to Learn: Has been described as a student's 'energy and drive to learn, work effectively and achieve to their potential, in addition to the behaviors associated with this energy and drive.

Problem-Based Learning (PBL): Is a student-centered approach in which students learn about a subject by working in groups to solve an open-ended problem.

Chapter 9
Science and Mathematics Teacher Collaboration in Higher Education:
A Pedagogical Experience Using Inquiry Learning Spaces

Bento Cavadas
https://orcid.org/0000-0001-6021-6581
Lusófona University, Portugal & Polytechnic Institute of Santarém, Portugal

Neusa Branco
https://orcid.org/0000-0003-4695-0098
Polytechnic Institute of Santarém, Portugal & CIAC, Algarve University, Portugal

ABSTRACT

Collaboration between teachers is essential for professional development. This work presents two collaborative teacher cycles of inquiry, with the aim of characterizing the dynamics between four higher education teachers involved in a collaborative experience related to the creation, implementation, and evaluation of interdisciplinary inquiry learning spaces (ILS). Written communications between the participants, field notes, the ILS produced, ILS assessment rubrics, and meetings summary were used as data sources. Initially, teachers met to organize the interdisciplinary work, and two subteams were formed, each one with one math teacher and one science teacher. Collaborative work is evident in each subteam and between both teams on dialogue, decision making, action, and evaluation components. The interdisciplinary collaboration had a strong focus on mathematical and science content knowledge as well as pedagogical content knowledge of the ILS. Furthermore, it promoted the reflection on teachers' own practice and their effects on student learning.

DOI: 10.4018/978-1-6684-5765-8.ch009

INTRODUCTION

Increasing the quality of collaboration that occurs in instructional teams is a promising approach to educational improvement (Ronfeldt et al., 2015). In fact, Woodland et al. (2013) adds that collaboration seems to be an essential requisite for achieving school improvement. Therefore, shifting from teachers working alone to professional ethics that emphasizes collaboration is needed (Hattie, 2015). In higher education contexts, collaboration between faculty is relevant to foster innovative educational practices. The importance of the existence of dynamics in higher education that foster the interdisciplinary dialogue and collaboration between higher education teachers is highlighted by researchers as Vieira (2009). Ronfeldt et al. (2015) suggest that it is critical to understand the extension to which teachers collaborate about certain issues and how that collaboration is useful to support their practice. Furthermore, Levine and Marcus (2010) stated that research about teacher collaboration should seek to link specific kinds of collaborative activities to observable changes in classroom practice.

The previous framework shows that interdisciplinary collaboration between higher education teachers for pedagogical purposes is a research field that deserves further attention, which this research aims to give a contribute to. As suggested by Slater (2004), studies that provide rich descriptions of participants' collaborative experiences may lead to new understandings of the process. The main goal of this work is to characterize the dynamics that occurred between higher education teachers from different scientific fields due to their engagement with an interdisciplinary pedagogical experience of collaboration. This experience involved the planning, creation, implementation, and evaluation of interdisciplinary inquiry learning spaces (ILS). The research question that guided this work is:

- Which practices of dialogue, decision-making, action taking, and evaluation occurred during a pedagogical experience of collaboration between higher education science teachers and mathematics teachers regarding the use of inquiry learning spaces (ILS)?

THEORETICAL FRAMEWORK

Teacher Collaboration

Teacher collaboration, as defined by Woodland et al. (2013), assumes that there are "teachers working together, and engaging in reflective dialogue, with the common goal of improving practice and increasing students learning" (p. 443). For the authors, in teacher collaboration teachers are working together in continuous cycles of inquiry. These cycles integrate four inter-related components: dialogue, decision-making, action, and evaluation (Woodland et al., 2013). These elements emerge when teachers collaborate within teams with the following characteristics:

- Dialogue is related with collective dialogue within the teacher teams about student learning, the effects of instruction on student achievement, and how to provide support and challenge to each student.
- Decision making is related with allocation, security, boundary, evaluation, and instructional decisions. The most relevant elements in decision making are the teacher teams' decisions about their

individual and collective instructional practices and their effects on student learning, in order to improve their teacher practice.
- Action focus is the actions that teachers take due to their team decisions, which should be directly related with the improvement of practice and entail innovation.
- Evaluation occurs when the teacher teams systematically collect and analyze both quantitative and qualitative information, instead of episodically and superficial information about their teaching performance and students' achievement.

The research highlighted relevant factors that can increase teacher collaboration: formal and informal time and access to a physical space to meet (Shakenova, 2017); joint work and interdependence (Slater, 2004); parity, which involves shared power and equality amongst participants (Slater, 2004); the opportunity to observe other colleagues' lessons (Shakenova, 2017), and the voluntary basis of collaboration, due to the perceived value and understanding that working together is beneficial for the participants involved (Slater, 2004). The collaboration could be related with different aspects. Slater (2004) presented five categories for the content of collaboration: pedagogical (e. g., curriculum planning, assessment of students), professional development (e.g., coaching and peer observation), building and sustaining relationships, governance (e.g., school organization) and events or projects.

In the framework presented by Slater (2004) the trigger for collaboration could be choice, goals, needs, and relationships amongst the participants. Martin and Bradbeer (2016) reinforced that creating and maintaining positive relationships between teachers is a critical element of positive school global change. Slater (2004) also noted that a successful collaboration depends on the personal interaction of the participants. Martin and Bradbeer (2016) suggested that effective collaboration occurs when the team members have the capability of engaging in conversations that might be avoided and try to do synergetic sensemaking that is engaging in a "conversation that needs to occur when there is a point of difference or a point of "not understanding" between colleagues" (p. 50). Paralleling these findings, in the literature review conducted by Shakenova (2017) it was noticed that some conditions related with interpersonal dynamics can also support collaboration, such as dialogue and active participation, trust, openness and respect, positive attitudes towards teaching, shared beliefs and understandings, and teachers' initiatives.

One mode of collaboration could be realized through professional learning communities. Woodland (2016) characterized professional learning communities as networked and collaborative communities, in which teachers look to ensure students learn through systematic collective inquiry into practice to improve instructional quality, students' achievement and curricular outcomes. Woodward and Mazur (2015) stated that professional learning communities could impact the school culture positively and enhance teacher satisfaction as well as student achievement. Woodland (2016), supported by other researchers' work, remarked that professional learning communities are one of the most powerful strategies for achieving substantive instructional improvement and critical student learning outcomes. Professional learning communities offer teachers high-quality collaboration opportunities when they are based on structured cycles of inquiry into students' work (Woodland & Mazur, 2015).

Ronfeldt et al. (2015) also highlighted teachers demonstrated better achievement when working in schools with a quality collaborative environment. This collaborative environment, when it is focused on instructional decision making and characterized by teachers that prepare instruction together, observe one another, and do co-teaching, may improve students' outcomes (Ronfeldt et al., 2015). Therefore, school leaders and teachers should find ways to encourage collaboration (Ronfeldt et al., 2015) and be aware that the school culture is one of the main factors that affect teacher collaboration (Shakenova,

2017). Furthermore, the literature identified skills that can benefit collaboration, such as communication skills, emotional intelligence, and conflict management competencies (Martin & Bradbeer, 2016; Slater, 2004), decision making and problem-solving skills and teambuilding (Slater, 2004). A highly developed self-awareness sense, characterized by openness to learning, reflection, self-questioning, and the capacity to adapt to changing environments, are also teacher features that can promote collaboration (Martin & Bradbeer, 2016).

Some studies also pointed out the importance of teachers' collaboration for professional development. Sharing of ideas, experiences, and resources, within a context of reflective practice, and help between teachers (Shakenova, 2017; Woodland et al., 2013) could improve the teachers' professional development. Levine and Marcus (2010) suggested that teacher learning is enhanced by more frequent and transparent access to other teachers' practices. Slater (2004) also agreed that opening classroom doors to collaborative work may lead teachers to greater satisfaction and increased professional development. However, Hattie (2015) suggested that the teachers' dialogue should be focused on the impact of teaching upon student learning, and not been dominated by themes such as curriculum, students' behaviour, or assessment.

Shakenova (2017) stated that collaboration involves shared values through teachers' learning which influences teaching practice and students' achievements. Therefore, one focus of the studies about teacher collaboration is how it contributes to students' achievement (e.g., Goddard et al., 2007; Ronfeldt et al., 2015). In fact, research suggests that there is an association between teacher collaboration and students' academic success (e.g., Goddard et al., 2007; Ronfeldt et al., 2015; Woodard et al., 2013). For example, Ronfeldt et al. (2015) found a positive relationship between teachers and schools that engage in quality collaboration and students' better performance in reading and math. One aspect that explains this relation is teachers' regular meetings with their peers to discuss ways to increase achievement and to examine students' work (Ronfeldt et al., 2015). The improvement in students learning results from changes in classroom practice and their positive effects on the quality of teachers' skills and knowledge, due to teacher collaboration (Woodland et al., 2013). Ronfeldt et al. (2015) also suggested that collaboration across many instructional domains is more successful at improving students' achievement than collaboration in only one single domain. However, Levine and Marcus (2010) have a different opinion and suggest that having a specific focus on collaboration is more useful than just organising teachers into teams.

Concerning teacher collaboration in higher education, the work by Chaovanapricha and Chaturongakul (2020) has shown that the collaboration of teachers of different fields, for example, between English for Specific Purposes and subject teachers, has the benefit of offering instant feedback from the subject teacher. The authors mentioned that this interdisciplinary teacher collaboration also results in increased confidence and reduced anxiety when a teacher is approaching the other teacher's subject.

Despite the important role that teacher collaboration can play, time and work overload were some of the main factors restricting teacher collaboration found in the review study done by Shakenova (2017). Chaovanapricha and Chaturongakul (2020) specifically pointed out some constraints, such as difficulties in the conciliation of teachers' schedules and a more time-consuming planning of the interdisciplinary lessons. Micropolitics can be understood as the use of formal or informal power exercised by individuals or groups inside schools to achieve their goals are also mentioned as obstacles to teacher collaboration (Shakenova, 2017).

Collaboration Between Science and Mathematics Teachers

Looking specifically for collaboration by science and mathematics teachers, Wong and Dillon (2020) defined collaboration as "any form of working together by teachers from mathematics and science disciplinary backgrounds and departments which focuses on teaching and learning" (p. 398). These authors identified four types of collaboration between science and mathematics teachers: joint projects or lessons, visiting expert, informal conversations and collaboration in the curriculum, and scheme of work.

The need for interaction and teacher collaboration could provide new alternatives to traditional mathematics and science courses, and therefore those teachers should seek collaboration (Bobonová et al., 2019). In fact, growing consensus among STEM education reforms indicates that teaching of mathematics should be done in context (McHugh et al., 2018). Engaging teachers in this way may lead them to benefit from a view of how science provides mathematics with interesting problems to investigate, and reciprocally: mathematics provides science with powerful tools to use in analysing science problems (Rutherford & Ahlgren, 1990).

In fact, teacher collaboration and support, such as technical support or an adequate class schedule to participate in collaboration experiences, are some of the key aspects that have an impact on the integration of science and mathematics education (Ferri, 2019; Ríordáin et al., 2016) and are imperative when attempting a new educational innovation related to integration of content from another subject (Weinberg & McMeeking, 2017). Furthermore, the proximity of classrooms to teachers from other disciplines as well as a shared office space with teachers from different subjects could encourage collaboration (Weinberg & McMeeking, 2017). So, a common departmental room used by science and mathematics teachers could be a promotor of collaboration (Wong & Dillon, 2020). However, it is necessary to clarify that are various teacher collaboration models that could emerge in different circumstances of interdisciplinary collaboration, as suggested by study by the Hobbs et al. (2019) about STEM projects. Furthermore, collaboration seems to be easier for teachers who have previously shared professional experiences (Nelson & Slavit, 2007).

The importance of collaboration between different educational actors in teams for the development and implementation of integrated units or activities is also highlighted in the literature (e.g., Broggy et al., 2017; Lederman & Lederman, 2013; Ríordáin et al., 2016; Navy & Kaya, 2020). Teaching in an integrative way and planning interdisciplinary units seems to be more effective than traditional approaches and has the advantage of developing teaching efficacy (King & Wiseman, 2001). Frykholm and Glasson (2005) also noted that the collaborative work between prospective science and mathematics teachers in their teacher preparation programs resulted in the creation of interdisciplinary units connecting mathematics and science concepts through curriculum projects and activities. Moreover, this collaboration extended to their internships. Aguirre-Muñoz et al. (2021) study mentioned that teacher collaboration on joint design of integrated lessons in cross-disciplinary teams could support transformative learning. The Association for Science Education [ASE] (2016b) presented fruitful examples of collaboration between science and mathematics teachers involving drawing charts and graphs, and teaching equations and graphs. The study by McHugh et al. (2018) also supported that a promising strategy for STEM teachers to promote higher order thinking skills, such as the graphical interpretation of data, is collaborative work in authentic contexts.

The benefits of science and mathematics teacher collaboration could be reflected on professional development through the creation of a collegiality of different educational actors who support and stimulate each other (Bell & Denniston, 2002). Pyke and Lynch (2005) showed that when mathematics and science teachers have a collaborative approach to the preparation for the National Board for Professional Teaching Standards, the respondents usually have higher scores and higher passing rates. They also valued the motivational and instrumental support which emerged from the collaborative preparation. The interdisciplinary approach promotes teacher collaboration, mitigating the sense of isolation (St. Clair & Hough, 1992; Frykholm & Glasson, 2005) and reducing subject fragmentation (Frykholm & Glasson, 2005). Furthermore, a collaborative approach in the context of methods courses in teacher education focused on modelling, reflection and extensive discussions can lead to changes in preservice teachers' understanding of constructivism (Jones et al., 2005).

Yet, collaborative partnerships between science and mathematics teachers can be challenging to establish and maintain (Weinberg & McMeeking, 2017). Frequently they start as a response to an external stimulus and not simply because a science and a mathematics teacher decided to work together (Wong & Dillon, 2020). Consequently, when the support for collaboration is removed or reduced, collaboration seems to collapse rapidly, according to Wong e Dillon (2020).

Successful collaboration requires similar instructional philosophies and commitment from participating teachers (Frykholm & Glasson, 2005). As stated by Ríordaín et al. (2016), the technical nature of science and mathematics and the intricate links within the subject areas can create problems for teachers who have a limited understanding of the respective subjects, consequently affecting the teaching and learning of an integrated unit of learning. These difficulties could result, for example, from differences in the way that some terms are used in mathematics and science (ASE, 2016a) or from concerns related with the loss of the rigor of subject-specific content in common projects (Wong & Dillon, 2020). Wong and Dillon (2020) also reported teacher' concerns with the loss of focus on the specific content of each subject when the overlap or joint work is prioritized. Paralleling previous findings, Tytler et al. (2019), in a study related with STEM teaching in interdisciplinary settings, found barriers such as the challenge for teachers of generating productive and coherent mathematics learning in interdisciplinary settings. Sometimes it is hard for the teacher to define a common inquiry question in collaborative experiences (Nelson & Slavit, 2007). So, making connections across science and mathematics could be a challenging endeavor because finding those connection points is time consuming and sometimes frustrating (Wong & Dillon, 2020). Other studies identified that time is a significant constraint for science and mathematics teacher collaboration (Frykholm & Glasson, 2005; Weinberg & McMeeking, 2017), and a huge constraint for teacher planning, as reported by the science teachers in Davis (2003) study.

Hargreaves and Macmillan (1995) suggest that the school organization in several departments promotes balkanization which compromises collaboration between teachers of different departments. Therefore, a good relation between the science and mathematics departments could enable collaboration, possibly promoted by the heads of the departments (Wong & Dillon, 2020). Another barrier is a lack of administrative support in the implementation of interdisciplinary STEM programs (Mayes, 2019).

RESEARCH METHODOLOGY

Research Design

The present work is based on an educational experience implemented in an initial teacher education context. The work crossed two academic years and followed a qualitative research design, considering the goal and research question defined, producing knowledge about the dynamics of collaboration between science and mathematics teachers. Concretely, this work presents the analyzes of two collaborative teacher cycles of inquiry, structured in the dialogue, decision making, action and evaluation components proposed by Woodland et al. (2013). The two teachers team cycles of inquiry organized the specific main dynamics that emerged from the science and mathematics teachers' collaboration (Figure 1).

Figure 1. Two cycles of team inquiry resulting from the science and mathematics teachers' collaboration
Source: Authors (reproduction of the teacher team cycle of inquiry visuals authorized by Rebecca H. Woodland).

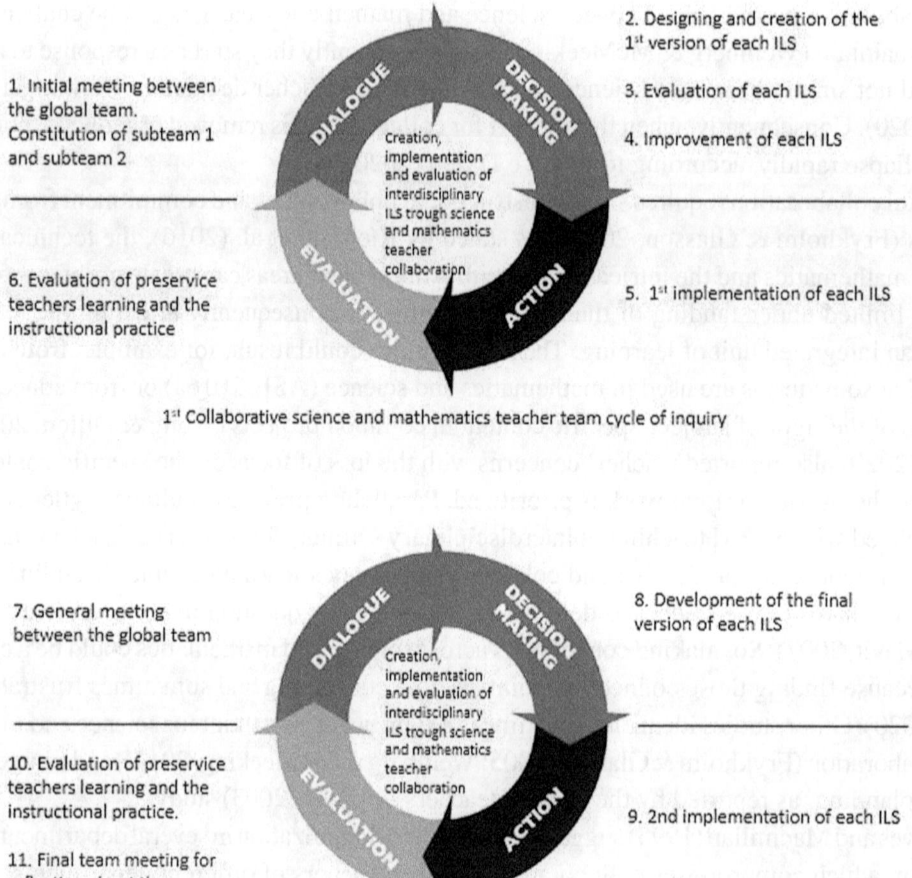

The processes within the two cycles of inquiry are numbered (from 1 to 11) to facilitate the analysis of each cycle and the transposition from one cycle to the other one. The shared goals of the teacher team cycles of inquiry were the creation, implementation, and evaluation of interdisciplinary ILS through science and mathematics teacher collaboration.

Participants and Context

The participants in this work were four teacher educators of three different institutions of higher education, which constituted the team. Two teachers are from the fields of mathematics and mathematics education (MT1 and MT2), the third from the areas of biology and geology and science education (ST1) and the fourth from the field of chemistry and physics and science education (ST2). The participants acted as a professional learning community, and were organized in two interdisciplinary subteams: subteam 1 (ST1 + MT1) and subteam 2 (ST2 + MT2). Each subteam had to design, create, implement, and evaluate one interdisciplinary ILS for initial primary teacher education. This collaboration emerged due to a project financed by the research center of three of the teachers of the global team.

Inquiry Learning Spaces

The ILS are online spaces which include digital laboratories, apps and other multimedia resources useful for students' learning. There is much evidence concerning students' learning with the use of digital environments (e.g., de Jong et al., 2013; Gillet et al., 2013). The ILS also have the advantage of delivering an online environment to promote that integration. The ILS follows an inquiry approach organized in the phases of orientation, conceptualization, investigation, conclusion, and discussion (Pedaste et al., 2015). The Go-Lab and Next-Lab projects have been promoting ILS creation in different areas, and we hypothesize it could be a useful tool to integrate science and mathematics. Thus, we consider it to be an important context to use in initial teacher training, providing preservice teachers (PSTs) with the experience of using it and discussing its relevance for working with their future students, particularly with an interdisciplinary focus.

Literature mentions that an integrated approach could be based on the identification of common learning goals and concepts across multiple subjects to address big ideas, concepts, or themes (Beane, 1992; Berlin & White,1995; Broggy et al., 2017). With that aim, the teachers of the team identified connections in science and mathematics curricula, as proposed by Ríordáin et al. (2016), and created two ILS. The ILS named "STEM Bees" was based on the theme of bees, focusing on connections between evolution and the mathematics of honeycombs. The other one integrates physics and mathematics and was named "Jumping ball". Each ILS was implemented by a subteam with a class of preservice teachers.

Data Collection

Several types of data were collected for this work. Email communication between teachers, meetings summary, teachers' course materials (e.g., lesson plans), the ILS created by the teachers, and evaluation instruments of the work done by PST in the ILS were used to characterize the process of collaboration between the teachers. A semi-structured final interview was conducted by one of the science teachers (ST1) to the other participants to collect additional data about the process of collaboration. To reduce bias, in data analysis only the ideas presented in the interview of the other participants (ST2, MT1 and MT2) were considered. The interview questions were based on the teacher collaboration assessment survey presented by Woodland et al. (2013), which provide content validity to the group interview questions.

Data Analysis

The analysis of the dynamics which emerged from the collaborative process was based on the four interrelated components of the teacher collaboration cycle of inquiry presented by Woodland et al. (2013): dialogue, decision-making, action, and evaluation. Teachers' conceptions about collaboration were also analyzed.

To characterize the dynamics of an interdisciplinary pedagogical experience of collaboration between higher education science and mathematics teachers regarding the use of ILS, the Teacher Collaboration Assessment Rubric (TCAR) presented by Woodland (2016) was used. The TCAR rubric presents several descriptors which guided the organization of data and the analysis of the science and mathematics teachers' dynamics of collaboration. According to Woodland (2016), "ongoing embedded evidenced-based assessment of teaming using a conceptually grounded tool such as the TCAR can help foster conditions in which educators are able to create, develop, and participate in engaging, efficient, and effective professional inquiry with their colleagues" (p. 9). Woodland (2016) also suggested that the TCAR rubric could be used for developmental evaluation, for example, to pilot innovations prior to scaling up.

FINDINGS

Dialogue

The team, constituted by the four teacher educators, initially meet to organize the interdisciplinary teams (subteam 1 and subteam 2) and structure the work (Figure 2). Afterwards, the dialogue dimension of the collaborative teacher team cycle of inquiry was evident, within subteam 1 and subteam 2 and within the global team. During the creation, implementation, and evaluation of each ILS the teacher educators discussed the scientific rigor of the science and mathematics content, the best use of the Graasp platform resources to achieve the PSTs' learning goals, and the didactical sequence of the ILS, with a focus on integration of science and mathematics and PSTs learning. This process was concluded with a final meeting were the four teachers discussed and reflected on the global collaboration process.

Figure 2. Dialogue interactions between the teachers in the two cycles of inquiry
Source: Authors

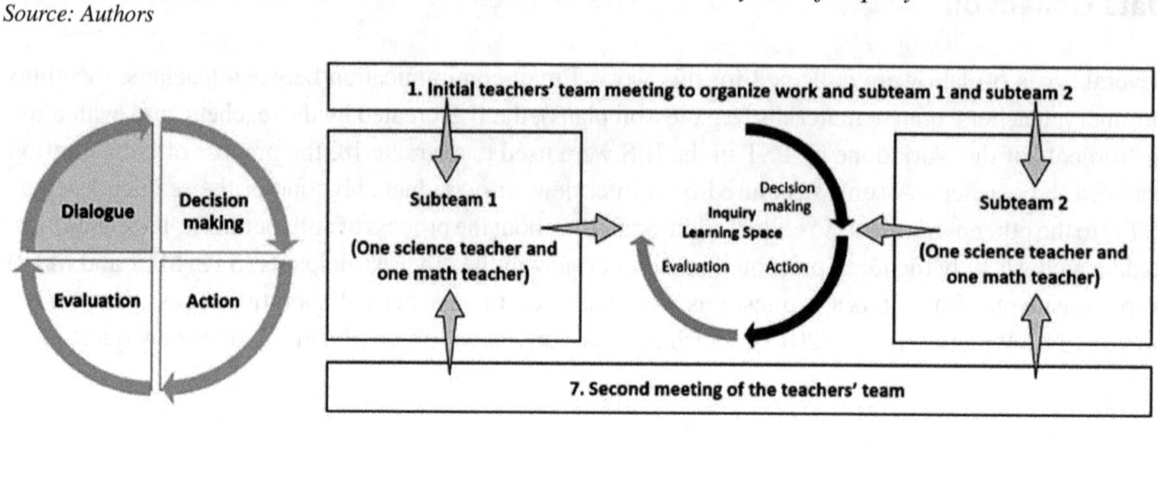

As the four teachers already knew each other, the initial meeting of the first cycle of inquiry was used by ST1 to summarize the project goals, constitute the two interdisciplinary teams and to organize the teamwork, due to its role as principal investigator. The meeting followed a written agenda prepared by ST1.

The four teachers met regularly in-person or online for a briefing about the work done and future work, namely, to clarify technical and pedagogical issues about the ILS creation in the decision-making process, to share pedagogical strategies to implement it and to clarify doubts about the evaluation process of the ILS content. To better keep the schedule of creation and implementation for both ILS, the regularity of team meetings was flexible. That flexibility was pointed out by Martin and Bradbeer (2016) as favorable to collaboration.

All meetings had a defined purpose, but they were not oriented by a protocol used to facilitate or structure team dialogue. The team dialogue wasn't recorded, but minutes of the meetings, with decisions and intended actions, were taken by one of the tecahers (ST1).

After the initial meeting from each cycle of inquiry, a major part of work and dialogue was done within each team, with the support of the other team, when necessary. During the ILS creation and implementation, the team dialogue consistently addressed questions of practice, ILS instructional and student learning, with a focus on science and mathematics integration, which is reflected in the two final ILS produced.

The email correspondence and feedback about the ILS showed some discussions due to different points of view that occurred within subteam 1 and subteam 2. These discussions, some of them quite lengthy, were about how to structure the ILS with an integration of science and mathematics focus and a coherent didactical sequence. However, with a continuous dialogue the different perspectives led to convergent points of view which result in the final structure of the ILS in the second cycle of inquiry. The final version of these ILS was published it in the Go-Lab platform (see Cavadas & Branco, 2020; and Rézio et al., 2020) and implemented with preservice teachers in co-teaching, as descripted below.

It was noted that all team members participated in group dialogue equally. Maybe this was because they already knew each other, and the small size of the team and the subteams enabled full participation of all members in an atmosphere of trust, openness, and respect between all members.

Decision Making

The major practices of the decision-making process in the first and second cycles of team inquiry are summarized in figure 3.

Figure 3. Interactions between teachers in the decision-making process in the two cycles of inquiry
Source: Authors

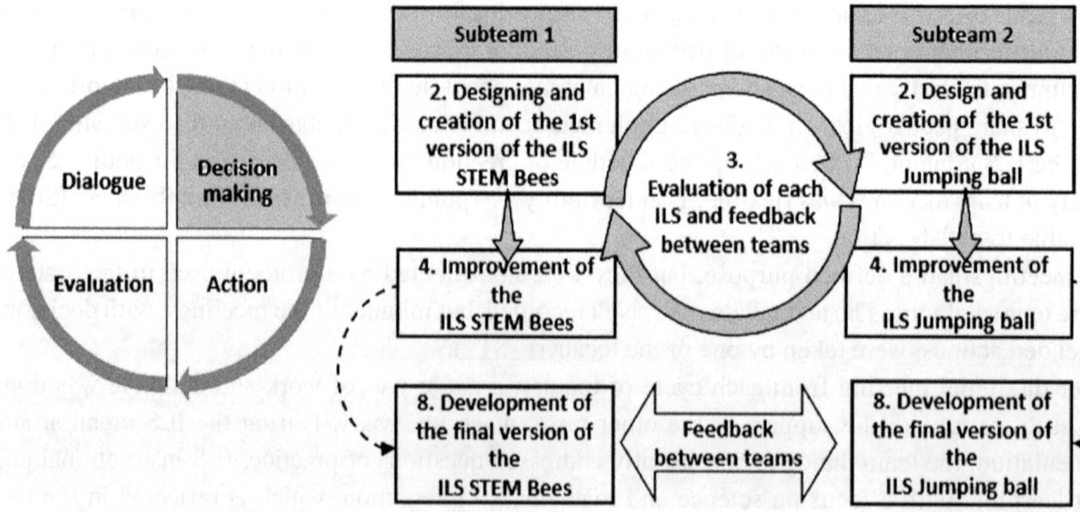

During the initial creation and improvement of the ILS in the first inquiry cycle and the final creation of the ILS in the second inquiry cycle, the members of the two teams regularly identified and determined specific actions to improve the instructional quality of the ILS and student learning, with a focus on interdisciplinarity between science and mathematics. For instance, in the interview, MT1 highlighted some aspects of this decision making process focused on decisions about the instructional practice, mainly concerning the use of manipulatives to PSTs to explore the problems posed in the ILS STEM Bees:

One of the focuses of discussion was about the tasks, the introduction of the tasks, the theoretical basis that supports the tasks - this was our first approach. We deepened several aspects of the content that we wanted to integrate in those tasks and shared various resources (books, articles, ...), which we discussed. We also discussed the use of manipulatives: Which manipulatives should be used? Who would create them? Us or the PSTs? How would they be used? In what tasks? We also discussed the organization of the ILS, the testing and the various stages we wanted PSTs to go through. (MT1, Interview)

In fact, Ríordáin et al. (2016) had already pointed out that teacher perspective and knowledge have an important impact on the implementation of integration and are core components of the successful implementation at classroom level and were key focal points in the collaboration process.

Another important decision-making action was the evaluation and feedback between subteams regarding the scientific content and the instructional design of each ILS, summarized in the process represented in figure 2. Initially each subteam planned and drafted an interdisciplinary ILS, which was reviewed by the other subteam using an ILS evaluation rubric. Some decisions were made by each subteam to structure the first version of ILS and then, within the team, the four teacher educators discussed decisions related with the main instructional practices. After that, each subteam had another decision making moment. They analyzed the feedback and made changes to the scientific and pedagogical content of the ILS to improve it for more impact on the PSTs learning and understanding of mathematics and science integration. This process was clearly emphasized by ST1 as having the advantage of improving the instructional approach of the ILS created by sub-team 1:

In a first stage, after the dialogue we had with you [Subteam 1] and your feedback, we improved the ILS, its tasks, and the way these tasks were introduced. This led us back to our phase of dialogue in our team [Subteam 2], to think again about the task and how we could reformulate and restructure it so that the PSTs who would solve had an active role, exploring the concepts and building their own knowledge. So, during the construction of the ILS, we went through these various phases of dialogue with different teachers of both teams, and this was quite important in this collaboration. (ST1, interview)

Since the ILS were addressed to PSTs, both subteams reflected on ways to provide them with an enriching experience, not only to improve their science and mathematics knowledge, but also to deliver a significant pedagogical experience, as clearly stated by MT1:

We were always thinking about the experience we are going to provide them [preservice teachers] with, so that the experience itself can also promote learning for their education as teachers. We tried to make sure that the experience we were going to offer was also significant for them to think about how they can propose that kind of work in their future practice. (MT1, Interview)

The preservice teachers' awareness of the importance of science and mathematics integration was another focus of attention in the decision making process of the work presented:

What to teach and how to teach is one of the things we must reflect upon. About how to teach, it is important to teach preservice teachers in an integrated approach, because the future teachers in general have a very compartmentalised idea of each subject - one content is physics, another content is mathematics, still another content is biology - and they must realise that it is possible to teach in a way that combines these subjects. (ST2, Interview)

The basic idea that emerged for these preservice teachers was that mathematics is not only theoretical, but it is present in all other sciences through mathematical modelling. (MT2, Interview)

In fact, one focus of the decision making process was how to integrate science and mathematics effectively to have a greater impact on preservice teachers' understanding of both areas and the power of their integration. These decisions involved a curricular analysis of both subjects and taking options, as sated by MT2: "Since the ILS was going to be applied in a class of the Algebra curricular unit, where functions are studied, it made sense to reconcile the modelling of physical phenomena with learning the notion of function" (MT2, Interview).

As expressed in the previous statement, the interdisciplinary content knowledge of the higher education teachers which collaborated in this interdisciplinary pedagogical experience was mobilized frequently to achieve better integration in the ILS. The next statement is another example: "I think that it was also important in this ILS to try to integrate Mathematics in Physics and Physics in Mathematics" (ST2, Interview). Although integration was reflected in the decision making process, there were also opportunities to reflect about the specific content of mathematics and the specific content of science in the creation of the ILS. For example, this idea was reflected in ST2's point of view about the physics that the preservice teachers should learn to teach their students adequately:

Our concern is that (…) without clear ideas of physics, the PSTs will not be able to explain the physics-related content well (...) if they do not know what a movement is, what energy is and what it is not (…) they cannot satisfy the natural curiosity of younger students if they themselves have not learned these concepts (for me, the fundamental concepts of Physics) and understood them well. (ST2, Interview)

Adaptations of the tasks to online teaching due to the COVID-19 pandemics were other aspects that emerged in the decision making process:

The initial idea was that PSTs would collect data from a sensor, but because it was not possible to do this, we changed the strategy and created a simulator. I created a bouncing simulator for different types of ball, taking the characteristics of the ball, of the bounce, etc. (…) we had to do this, it [teaching] was totally online. (ST2, Interview)

From the initial creation to the improvement of each ILS in the first cycle of team inquiry it was noted that each subteam regularly made decisions about how to improve the instructional approach of the ILS. Sometimes the decisions were related to the discontinuity of an initial approach. These features of the decision process also characterize the transition from the improved to the final version of each ILS. For example, in the ILS STEM Bees, initially the problem posed to PSTs was "Which is the best shape for the alveoli in beehives?". After reflection, the final problem was defined as "Why is the shape of the alveoli in beehives efficient?", which better relates evolution and mathematics.

Most team decisions were informed by some level of team dialogue, as exemplified by the statement of MT1:

In this collaborative process, there is joint work, there is a process and, at the end of it, there will be a product of that collaboration. In relation to this part of the process, there is a set of meanings that need to be discussed (...), it is necessary, within this collaborative process, to agree on a set of ideas and conceptions, to produce a joint product. (MT1, Interview)

The collaborative work was also expressed in an equitable division of work and responsibilities during the ILS creation, implementation, and evaluation, which was quite important. The sharing of tasks between the participants made them feel they were working in an effective team and that no member had a passive role. The participants also mentioned that the small size of the teams facilitate the equitable sharing of decisions and tasks.

The decisions were quite transparent due to the small size of the subteams and of the global team. Therefore, each teacher knew exactly what decisions were made, and where, how, and why they were made, because all teachers had an active part on this decision making process. Consensus was the specific process used to achieve decisions. Consensus was achieved by full group dialogue in each subteam.

Concerning the decisions about the collection and analysis of information, the team decided to collect evidence of the PSTs' work and analyze the PSTs' problem-solving skills, from the beginning of this interdisciplinary pedagogical experience. In fact, the team created a specific rubric to evaluate the PSTs' problem-solving performance through their productions in the ILS.

Action

The ILS were implemented using the strategy of co-teaching, with active collaboration from both the science and mathematics teacher of each subteam. The support between subteams occurred when the subteam that implemented the ILS first (subteam 1) made several suggestions to subteam 2 about the operationalization of the ILS implementation and the discussion moment of the PSTs' work in the different ILS tasks. The feedback received by the other subteam after implementation was also used to improve the ILS implementation in the second team cycle of inquiry (Figure 4).

Figure 4. Interactions between teachers in the action component in the two cycles of inquiry
Source: Authors

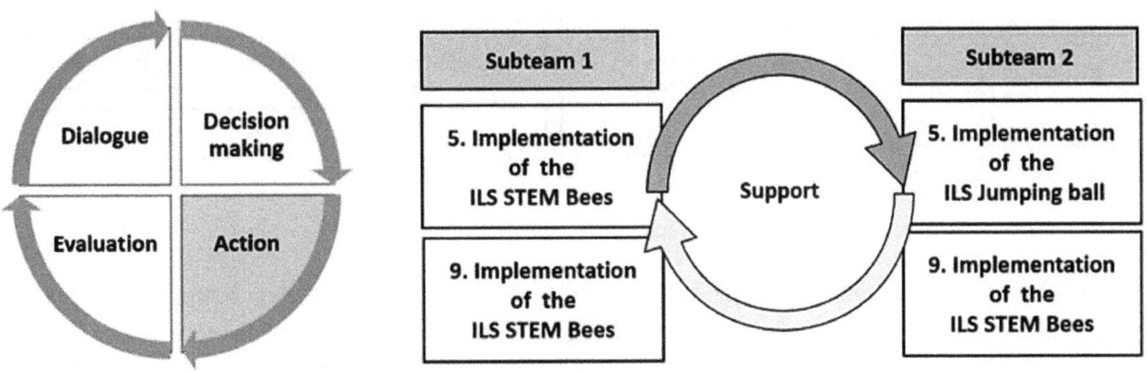

Each subteam implemented the ILS collaboratively in online environments, due to the COVID-19 pandemic, as shown in the teachers' course materials, such as lesson plans. New strategies were defined to implement the ILS exclusively in online environments, and the balance was positive:

From the perception I have, it does not seem that it would have been possible to do better about the implementation of the ILS remotely. We did a follow-up of the PSTs' work, despite the distance. We had several synchronous moments that were complemented with the autonomous work of the PSTs in asynchronous moments. This structure seems to have been well organized because those moments were not too far apart in time. These moments were done in a regular way and were well articulated in relation to what we had planned to do face-to-face. (MT1, Interview).

Therefore, subteam member actions were coordinated and interdependent. Each subteam member knew the specific individual actions that should be taken because of the lesson planning process in which the ILS was to be implemented. The ILS was used to present the tasks to the PSTs, for them to introduce their work, for asynchronous teachers' feedback to the PSTs' work, and to support the group presentation and collective discussion in the synchronous or face-to-face class. All those practices were done in co-teaching: both teachers in online synchronous classes or face-to-face classes to present the tasks to the PSTs and give them feedback. In the discussion moment of the PSTs' work, both science and mathematics teachers contributed to manage the discussion, comment on PSTs' problem-solving strategies and promote PSTs' reflection on those strategies and on the connections between mathematics and science that became evident.

Figure 5 presents the co-teaching process used by one subteam during the implementation of the ILS in both team cycles of inquiry.

Figure 5. Co-teaching process in the action component of both team cycles of inquiry
Source: Authors

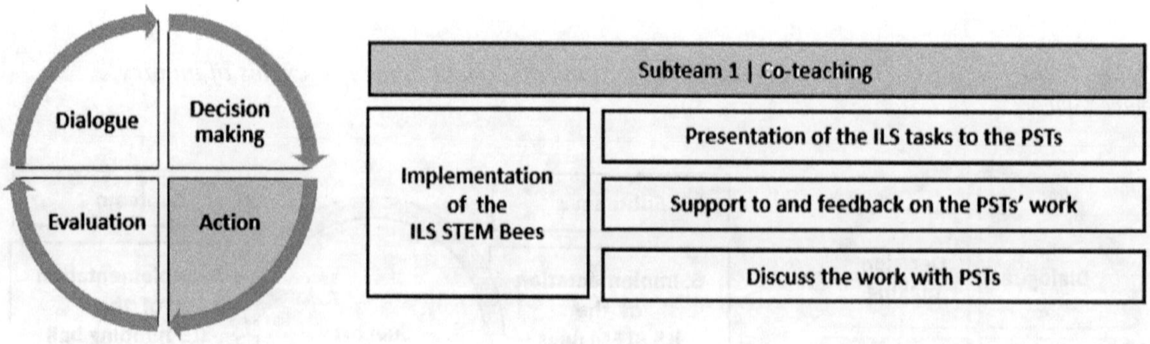

In the interview it was noted that the members of both subteams believe their actions could directly improve instructional practice, which suggests a high leverage of intended actions. For example, MT1 argues that the discussion moment that was thought during the instructional design of the ILS was valuable for the PSTs' integration of science and mathematics in the resolution of the problems posed:

In addition, we also had a moment of discussion between the groups, which seemed very important to me and which I think also worked quite well precisely to allow us to strengthen the link between science and mathematics and to clarify the concepts explored ... such as the concept of evolution, but also some ideas of mathematics ... at that moment of the discussion, it was possible to address and clarify them. The discussion emerged not only because the groups shared different solution strategies to the problems posed, but also because afterwards we asked them some questions to make them think more ... I also think that the moment of discussion we facilitated among the PSTs of each group was also very important. After the two moments of discussion, the preservice teachers drew up a conclusion in which many elements emerged that came precisely from these discussions and from some of the work they did after reviewing their resolution strategy. (MT1, Interview)

Along the same lines, ST2 also argues that this discussion moment during implementation was very important for the PSTs' reasoning:

I also think it is very important that the groups discuss how they came up with the strategies to solve the problem, why they chose that strategy and not another, why they followed that plan, why they raised that question, what led them to do that, etc. (ST2, Interview)

Team members' actions are specific and observable due to the co-teaching strategy adopted, as shown by this statement:

This is extremely important, especially if the two teachers are in agreement, if there is a dialogue and it is not the case of one teacher reciting one thing and the other reciting another, but rather if there is an explanation and a collaborative interaction with the PSTs ... we kept talking and interacting and communicating with each other and we kept asking each other: "but how do you do this here?" and this part can also be formative for future teachers to understand this interaction. (ST2, Interview)

In each team's cycle of inquiry, each teacher acted to improve individual instructional practice and group performance because of team decision making, highlighting the inter-related nature of components of teacher team's cycles of inquiry.

The email communications between the teachers show that both subteams' members had continuous and accessible information about and documentation of the instructional practices over time, since they planned, implemented, and evaluated collaboratively the work done by the preservice teachers in the ILS. Any change was documented in the ILS structure and was a result of reflections about their own practice and the PSTs' learning.

Evaluation

The evaluation dimension was expressed in the analysis of the data about the PSTs' learning, through the evaluation instruments of the work done by the PSTs. It was also based on the reflection about the educational approach used to inform the decision making process, expressed in the email communications between teachers, in the teachers' courses materials and in the interview (Figure 6).

Figure 6. Interactions between teachers in the evaluation component in the two cycles of inquiry
Source: Authors

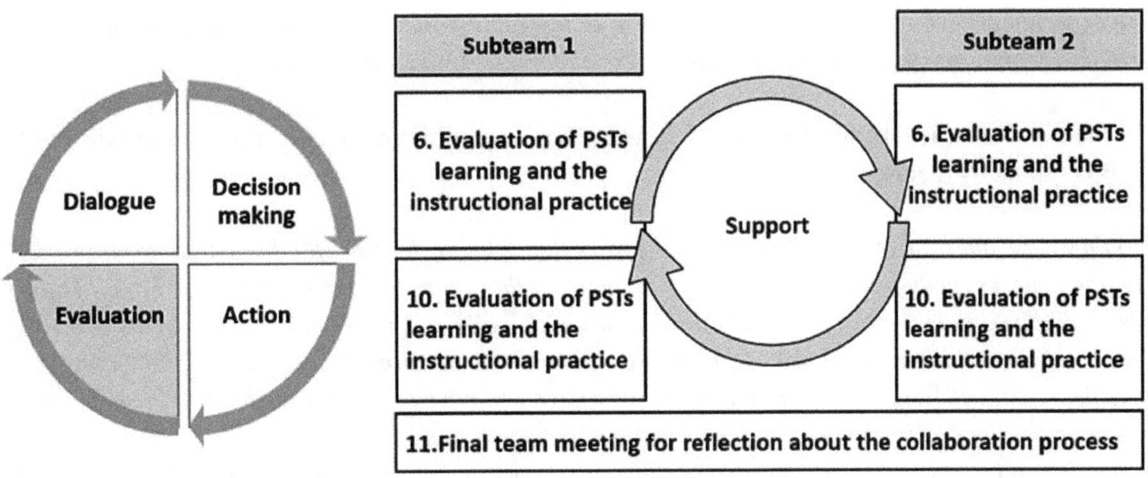

The team members collected data about the PSTs' work. In fact, one of the major advantages on an ILS is that students' can record or upload all their productions into the ILS for each task proposed. Furthermore, subteam 1 had the initiative to record the online synchronous moment of presentation and discussion of the results of each PST group on the "STEM Bees" ILS, to support the evaluation process.

In the evaluation element of the first and the second cycles of team inquiry, each team analyzed the PSTs' work in each ILS, considering the expected learning goals. The performance of the PSTs was analyzed and discussed within the teams, with the support of the other team to clarify doubts, based on the evidence of their outputs in the various tasks of each interdisciplinary ILS. Regarding this process, MT2 mentioned that the initial content knowledge of the PSTs was taken into consideration in the creation of the didactic sequence of the ILS Jumping ball: "the PSTs might not have prerequisites in terms of the minimum scientific knowledge necessary to understand what we were going to write" (MT2, Interview). This difficulty led the teachers of subteam 2 to reflect about new ideas to overcome these potential difficulties of the PSTs, thinking of new ways to approach some concepts. The contribution of subteam 1 to those ideas in the evaluation and feedback process of the ILS Jumping ball was very valuable, as highlighted by ST2: "The ILS was highly enriched with your feedback [Subteam 1], because there were, in fact, aspects that had got stuck in our discussion. These aspects, with an external view from two other teachers who are also were thinking about the same thing, enriched our work" (ST2, Interview).

The interdisciplinary feedback within subteams and within the global team was valued as an opportunity for improving the instructional design of each ILS, as illustrated by MT1's statement: "This analysis of the scientific content from a person who is not from the specific area also helps us to think about it and reflect on it more deeply" (MT1, Interview). Also, the collaboration between mathematics and science teachers within subteams and within the global team was relevant to analyze the PSTs' productions and reflect about teaching performance.

The discussion moment with the PSTs in co-teaching was very important to give feedback about their work, as emphasized by MT1:

Concerning assessment, we had the opportunity to make formative assessment in those moments of discussion. We give feedback to the PSTs, and they completed some elements of their answers, they made new experiments, we gave them suggestions also to test other situations to consolidate or check if they were really integrating all the variables that were necessary. (MT1, Interview)

One aspect that led the teachers to reflect was when to give more direct instructions or when to let the PSTs think by themselves:

Should we ask questions or give clues about the work that is being done or should we leave it more open? And then check whether or not we will have to confront them with the decisions taken? I don't know if this should be an aspect that we should discuss more, at which moments we intervene in the work they are developing or what the nature of this monitoring and intervention should be. So, I think that this is an aspect that, with a new phase of implementation, we will have to discuss further. (MT1, Interview)

In relation to the PSTs' work evaluation and the question you were raising just now, whether or not to correct the students' mistakes immediately, I think that I may feel this at certain times. Perhaps we need more moments of letting them make mistakes but being able to redirect a little more and remove, avoid some common-sense ideas that they have. If we let these ideas persist without any redirection, I think it may not work and this worries me. (ST2, Interview)

A rubric was used to evaluate the problem solving skills of the preservice teachers, also created in collaboration between the science and mathematics teachers. The rubric was applied to their productions in the ILS. This rubric made it possible to give the PSTs not only quantitative but also qualitative feedback to their work.

The evaluation component in each team's cycle of inquiry was used as trigger that led to the analysis of the quality of the classroom-based instructional practice, as shown by this extract: "I think the fact that we have analyzed the productions in such detail and reflected on this, on the level of integration we managed to achieve, will undoubtedly make us improve our practice" (MT1, Interview).

Some teachers' reflections about the instructional practice focused on the structure of the ILS, as exemplified bellow:

The division of the PSTs' work in two problems, problem 1 and problem 2, was beneficial because the first moment of discussion after the PSTs have solved problem 1 resulted in improvements which were then evident in the second moment of the PST' work. (MT1, Interview)

Although this instructional practice was beneficial to the sharing of ideas and the pedagogical development of the teachers involved, it was quite time-consuming. In fact, the relevant amount of time that the teacher educators had to dedicate to this interdisciplinary pedagogical experience was mentioned as a possible constraint to it development.

Nevertheless, some benefits of collaboration were also reported, such as the exchange of ideas, as highlighted by MT2: "Collaboration, for me, is working together, so it is a constant exchange of perspectives, of ideas, of sharing individual research". On their own, ST2 focused collaboration on the opportunities it allows for interdisciplinarity: "in collaboration, in fact, you can find ways of making associations between different subjects. So, different teachers and different subjects can find ways of teaching the other subject (…) In other words, we can understand better the learning that the other teacher is trying to implement.". Another idea expressed was the feedback from the other colleagues as a benefit of collaboration, as exemplified below:

collaboration has, for me, a huge advantage, which is that we can count on the colleagues we are collaborating with to give feedback about what we are doing (…) this feedback, for me, was one of the most valuable things in our project. Because it is extremely enriching, and it is something that we can't do alone. (ST2, Interview)

Co-teaching was the teaching strategy selected by the team members to implement the ILS. This strategy allowed each sub-team member to share reflections about the moments when the other member was teaching in different moments of the class. Because of the COVID-19 pandemic, some classes were mediated by technology:

There is an aspect that was very important, the fact that both of us are present in class and in the discussion and we have already experienced this on several occasions. And the fact that we are both in the class discussing and interacting with the PSTs is an aspect that is quite significant and that is not always easy to achieve to complete and enhance this integration. I think this process is also quite significant and we have to continue to improve it and it obviously requires practice so that it can be a harmonious collaboration within the classroom. (MT1, Interview)

For ST2, the co-teaching instructional strategy was also important for PSTs to experience how teachers of different subjects can work together: "I think co-teaching is extremely important for the PSTs, because they observe two teachers talking about the same thing, one in the role of a science teacher and the other of a mathematics teacher" (ST2, Interview).

Furthermore, academic research was done about how the instructional practice affects PSTs' learning. Therefore, the team members reflected about this issue and substantiated it with evidence from the PSTs' work.

SOLUTIONS AND RECOMMENDATIONS

As noticed by Ríordáin et al. (2016), teachers themselves are core to successful integration, and a community of practice is a major contributing factor to achieving science and mathematics integration. Although it is highlighted in the literature that the creation of professional learning communities is a difficult process, the results of the collaboration process presented in this work suggest that the four teachers were able to create them because they acted in accordance with features of these communities presented by Woodland (2016). In science and mathematics teacher collaboration, mutual support and encouragement were evident, as expressed by Bell e Denniston, 2002. As pointed out by the results of participants' conceptions about collaboration, and the other results of this work, the science and mathematics teacher educators involved in this collaborative approach created a collaborative community through systematic collective inquiry into practice to improve instructional quality using ILS. The engagement in two teacher teams' cycles of inquiry helps to strengthen the interdisciplinary approach and the PSTs' learning.

The interdisciplinary teams acted as instructional teams, which can be defined as teams or groups of colleagues that work together on instruction (Roenfeldt et al., 2015). Therefore, team dialogue was focused on developing instructional strategies and reviewing assessments, which are domains of collaboration also found to be common to other studies (e.g., Roenfedlt et al., 2015). The continuous dialogue between the team and within each subteam allowed the exchange of ideas, the reflection about the integration of science and mathematics, and scientific and pedagogical feedback from the other teachers. The teachers met when necessary, and this flexibility was pointed out by Martin and Bradbeer (2016) as favorable to collaboration.

Furthermore, having a common goal is instrumental to the formation of collaborative relationships (Slater, 2004), and this was noticed in this collaborative experiment. The goal of the dialogue was the same presented in the work done by Frykholm and Glasson (2005): to create an opportunity for preservice teachers to work together, share ideas, collaborate, and participate in conversations about connecting science and mathematics. As stated by Ríordáin et al. (2016), teacher collaboration and planning integrated connections between science and mathematics contribute to improve the teaching and learning experience of the students.

In the two cycles of inquiry, disagreements sometimes emerged between the science and mathematics teachers. This is quite natural because, as highlighted in the literature (e.g., Wong & Dillon, 2020;

Tytler et al., 2019), making connections across science and mathematics can be a challenging endeavor as finding those connection points is time consuming and sometimes frustrating. This also happened in the process of creating the interdisciplinary ILS, which involved quite long discussions. This diversity of opinions and roles is common in collaborative dynamics, but the appreciation of other perspectives provides the framework for a shared vision (Slater, 2004). These discussions resulted in a final consensus due to an atmosphere of trust, openness, and respect, which are features that can support teacher collaboration (Shakerova, 2017; Slater, 2004). The participants in this collaborative experiment felt comfortable sharing their scientific knowledge of science or mathematics and their perspectives about the pedagogical experiences that should guide the ILS, and this atmosphere of absence of judgement was beneficial to the dialogue process and the increasing sense of trust. Creating positive relationships is beneficial to collaboration (Slater, 2004), and a strong relationship emerged between the teachers due to their participation in this experiment and their willingness to collaborate with each other. Furthermore, the participants' perception is that this process led to more robust products that are more appropriate to the learning objectives for students and for promoting the integration of mathematics and science.

In the process of decision making, the negotiation and agreement between teams during the collaborative work was essential to its success. The feedback of subteam1 about the work done by subteam 2 and vice-versa was crucial to the improvement of each ILS. The process of creating and improving the ILS with purpose of integrating mathematics and science involved different decision making moments between members of each sub-team and between members of the entire team. The dominant areas of the team decision making were related to instructional decisions, choices about what to teach and how to teach, and assessment decisions, focused on how to assess the PSTs' performance, which are areas also highlighted in the work carried out by Woodland et al. (2013). Furthermore, the decisions about the integration of science and mathematics in each ILS focused on the connections of science and mathematics, a process characterized by Frykholm and Glasson (2005) as connections between science and mathematics that teachers see as intuitive and relevant but situated in authentic contexts. Slater (2004) also highlighted that collaboration seems to work better when experiential content and a real need are at the core of collaboration and learning how to collaborate. As stated by Martin and Bradbeer (2016), "there needs to be a reason to collaborate" (p. 51). In the case of the ILS STEM Bees, the connections were about evolution, bees and geometry, and in the case of the ILS Jumping ball, the connections were about movement, energy and equations. These rich contexts are essential to promote deep content learning and recognition of the close relationship between science and mathematics (Frykholm & Glasson, 2005), as intended in the present educational experience, in which the PSTs' problem solving skills were promoted and evaluated. Moreover, during the decision making component of the two cycles of team inquiry, the teachers focused on fostering the PSTs' engagement with the problems posed in the interdisciplinary ILS to develop the content knowledge and pedagogical content knowledge necessary for their future practice. By offering the PSTs the experience of reflecting on the links between mathematics and science during teacher education programs, we hope they can guide better their own students through rich and authentic integrated experiments with science and mathematics, which are benefits of the collaboration process also noticed by Frykholm and Glasson (2005).

In this collaborative experiment, the science and mathematics teacher educators had to push forward their skills of creative collaboration, which involves bringing together two or more seemingly unrelated ideas, as proposed by Hattie (2015). Nevertheless, the science and mathematics teachers experienced some difficulties in articulating the knowledge and pedagogical strategies between mathematics and natural or physical-chemical sciences in some tasks in a common ILS. The COVID-19 pandemic also brought some constraints to its implementation, namely regarding tasks that implied the use of specific equipment.

Since a successful collaboration depends on the personal interaction of the participants (Slater, 2004), the fact that the teams were small in this collaborative experiment was useful because the work proceeded smoothly and was relatively easy to achieve consensus after a few moments of discussion. Furthermore, since the teams where small there was also an equitable distribution of the work and responsibilities, and this was an aspect valued by the teachers who participated in this interdisciplinary pedagogical experiment. The size of the teams must be taken into consideration in these collaboration processes. Maybe small and focused interdisciplinary teams facilitate the decision making process in teacher collaboration and are, therefore, more effective than larger teams.

One concern of the four teachers regarding the decision making process was to offer a good experience of integration of science and mathematics to the future teachers. This is very important because some researchers in this area suggest that, for a successful integration of subject areas in their practices, PSTs must experience integrated methods in their teacher education programs (Koirala & Bowman, 2003; McHugh et al. 2018). With that aim, the teachers had to assume some risks, such as connect some topics of science and mathematics and not others, or using this didactical sequence and not another; however, the literature (e.g., Slater, 2004) mentions that risk taking is part of the collaborative process and is even related to the building of trust in a collaboration relationship.

The results show that teachers engaged in actions related to the improvement of the produced ILS and in the discussion about their implementation in initial teacher education contexts. Subsequently, each team implemented the ILS with their group of PSTs, in the co-teaching modality. This aims to promote PSTs' mathematics and science knowledge and also provide them with more experience of the integration approach during their teacher education program. In fact, Riórdáin et al. (2016) had already noticed that the success of the implementation of integrated lessons is associated with teachers' opportunity to acquire knowledge of both the content of science and mathematics. Frykholm and Glasson (2005) also pointed out that the work done by PSTs in integrated science and mathematics contexts is valuable for them, as they have explored ways of mathematizing scientific phenomena, as well as ways of situating mathematics in relevant contexts.

The results also show that the action process is closely related with the interpersonal dynamics that supports collaboration in general, and co-teaching in particular, such as dialogue and active participation, trust, openness and respect, positive attitudes towards teaching, shared beliefs and understandings, teachers' initiatives, and clear specific individual and common actions during the lessons. A positive feeling of support and the opportunity to share ideas and develop them with other teachers were mentioned as beneficial to science and mathematics integration in the study by Ríordáin et al. (2016).

One desired goal of the action process was that the PSTs experience collaboration, and to that end an integrated context of science and mathematics is favorable, as stated by Frykholm and Glasson (2005).

One aspect that concerned the teachers was feedback, namely when and what type of feedback should be given to the preservice teachers' during their work. Should it be more direct, with clues for them to advance in their work? Or should feedback be given that allows them to reflect on their own on the best strategies to achieve the solutions to the problems posed? In this process, one aspect that could be improved is valuing error. The teacher educators analyzed the collected information about the PSTs' work to reflect about their practice, specifically if they should let the PSTs make some mistakes, so that they realized they had made them, thus giving them the opportunity to reformulate their approach, or give them a clue so that they could take the appropriate path, if necessary. Furthermore, there are learning goals that the teachers want PSTs to achieve, and if they are always producing poor quality products, and constantly making mistakes or being stuck due to a mistake, this will not be productive. Moreover, the work in the ILS is time consuming, and the curricular units where it was implemented had time constraints. So, this balance between giving to PSTs autonomy to make and correct their mistakes and the intervention of the teacher with small clues, is something that this research team think deserves more reflection.

Time consumption was an aspect that emerged often in the teacher educators' evaluation of this interdisciplinary collaborative experiment. Collaboration is not an easy and quick process and demands a lot from participants in terms of time and energy (Slater, 2004). While collaboration across a range of instructional domains is likely beneficial, time and resources for collaboration are limited, so teachers and schools often have to prioritize some instructional topics over others, as stated by Ronfeldt et al. (2015). Paralleling the parameters of Ríordáin et al. (2016), the teachers participated in this collaborative experience on their own time and out-of-school hours. As in the study by Davis (2003), the teachers considered that interdisciplinary planning requires a lot of time. Nevertheless, it was the teachers' willingness to participate in this experiment and give up their own time that ensured its implementation, as was the case with the teachers that participated in the study of Ríordáin et al. (2016).

Another aspect which emerged during the evaluation of the instructional practice was related with the transposition of the integration of science and mathematics and the collaboration process experienced by the PSTs to their one practice. This is an issue that also deserves further research and clarification because, as Frykholm and Glasson (2005) stated, difficulties of content and pedagogical knowledge emerge when PSTs are asked to connect and contextualize science and mathematics.

FUTURE RESEARCH DIRECTIONS

One recommendation for future educational experiences or studies about collaboration is to collect and analyze the PSTs' perceptions about the collaboration process of their teachers. Also, it will be relevant to follow up PSTs careers to identify if and how they implement mathematics and science integration practices in their teaching practice using ILS. The collaboration between science and mathematics teachers in higher education, while unusual, is beneficial for the integration of science and mathematics and for teachers' professional development, so future research could identify the specific contributions of participating in team teacher cycles of inquiry.

CONCLUSION

Integration is a difficult task, but this educational experience showed that collaboration between science and mathematics teachers is key to successful integration. This work shows the dynamics of collaboration which emerged between higher education science and mathematics teachers in the different moments of planning, creating and implementing interdisciplinary inquiry learning spaces. The dialogue, decision making, action and evaluation components of the team cycle of inquiry used to foster collaboration showed, globally, some constraints and many benefits of this interdisciplinary collaboration. These dynamics, despite some difficulties, contributed to deepen the way of presenting content and to improve the pedagogical approach of each one of the ILS, creating an environment of reflection on the practice of the teachers involved, promoting their professional development as well as innovative pedagogical experiences for future teachers.

Although this is a time-consuming process and the workload is considerable, this collaboration created synergies and brought benefits highlighted in the literature, such as diminished sense of isolation (St. Clair & Hough, 1992; Frykholm & Glasson, 2005), better relationships between the teachers (Martin & Bradbeer, 2016), and greater pedagogical development of the teachers involved (Bell & Denniston, 2002). In this collaboration process, the science and mathematics teachers engaged in dialogue, planned and solved problems together. This interdisciplinary pedagogical collaboration experiment provided opportunities for the science and mathematics teachers to learn from and with each other and share their ideas in a friendly and trusting environment. The continuous decision making process engaged the teachers in constructive pedagogical conversations about the connections of science and mathematics and the best pedagogical approach to impact preservice teachers' achievement. This collaboration experiment created a supportive environment for reflection about the instructional practice which contributed to deepen the way the science and mathematics content was presented to the PSTs and improved the pedagogical approach of each interdisciplinary ILS. The teacher educators worked together, and engaged in reflective dialogue, with the common goal of improving their practice and increasing the PSTs' learning, as characterized by Woodland et al. (2013) as an effective collaboration process. The teachers who participated in this educational experience considered the collaboration in instructional teams quite helpful, a result also reported in Roenfeldt et al. (2015) work.

The work shows dynamics of collaboration between science and mathematics teachers in higher education, and its benefits for successfully developing and implementing pedagogical proposals of integration of science and mathematics as well as for the teachers' professional development. However, in order to have a greater impact on the teachers who participated, we suggest that this collaboration should be based on clear connections between the science and mathematics content, and be framed by a common pedagogical approach, such as the creation, implementation and evaluation of inquiry learning spaces.

ACKNOWLEDGMENT

This work was funded by the Seed Project – Innovation in science and mathematics teaching through interdisciplinary Inquiry Learning Spaces (UIBD-04114-2020) of CeiED—Interdisciplinary Research Centre for Education and Development of Lusófona University, Portugal, through the Seed Projects grants.

REFERENCES

Aguirre-Muñoz, Z., Yeter, I. H., Garro, E. S. L., & Koca, F. (2021). Building teachers' capacity to integrate science and math content: Implications for professional development and learning. *Journal of Science Teacher Education*, *32*(1), 62–84. doi:10.1080/1046560X.2020.1789814

Association for Science Education. (2016a). *The language of mathematics in science. A guide for teachers of 11-16 science*. Association for Science Education.

Association for Science Education. (2016b). *The language of mathematics in science. Teaching approaches*. Association for Science Education.

Beane, J. (1992). Creating an integrative curriculum: Making the connections. *NASSP Bulletin*, *76*(547), 46–54. doi:10.1177/019263659207654706

Bell, C., & Denniston, K. J. (Eds.). (2002). *Journeys of transformation II: The impact of the Maryland Collaboration for Teacher Preparation on science and mathematics instruction. Successful strategies and results of a reform-based effort to improve science and mathematics instruction and teacher preparation.* Towson University, Center for Science and Mathematics Education. https://eric.ed.gov/?id=ED467606

Berlin, D. F., & White, A. L. (1995). Connecting school science and mathematics. In P. A. House & A. F. Coxford (Eds.), *Connecting mathematics across the curriculum. 1995 National Council of Teachers of Mathematics Yearbook* (pp. 22–33). National Council of Teachers of Mathematics.

Bobonová, I., Ceretková, S., Tirpáková, A., & Markechová, D. (2019). Inclusion of interdisciplinary approach in the mathematics education of biology trainee teachers in Slovakia. In B. Doig, J. Williams, D. Swanson, R. B. Ferri, & P. Drake (Eds.), *Interdisciplinary Mathematics Education. The state of the art and beyond* (pp. 263–279). Springer Open. doi:10.1007/978-3-030-11066-6_17

Broggy, J., O'Reilly, J., & Erduran, S. (2017). Interdisciplinarity and science education. In K. S. Taber & B. Akpan (Eds.), *Science Education* (pp. 81–90). Sense Publishers. doi:10.1007/978-94-6300-749-8_6

Cavadas, B., & Branco, N. (2020). *Abelhas STEM (STEM Bees)*. GO-LAB. https://www.golabz.eu/ils/abelhas-stem-rp

Chaovanapricha, K., & Chaturongakul, P. (2020). Interdisciplinary teacher collaboration in English for specific purposes subjects in a Thai University. *English Language Teaching*, *13*(5), 139–148. doi:10.5539/elt.v13n5p139

Davis, K. S. (2003). "Change is hard": What science teachers are telling us about reform and teacher learning of innovative practices. *Science Education*, *87*(1), 3–30. doi:10.1002ce.10037

Ferri, R. B. (2019). Teacher education and teacher development. In B. Doig, J. Williams, D. Swanson, R. B. Ferri, & P. Drake (Eds.), *Interdisciplinary Mathematics Education. The state of the art and beyond* (pp. 259–261). Springer Open. doi:10.1007/978-3-030-11066-6_16

Frykholm, J., & Glasson, G. (2005). Connecting science and mathematics instruction: Pedagogical context knowledge for teachers. *School Science and Mathematics*, *105*(3), 127–141. doi:10.1111/j.1949-8594.2005.tb18047.x

Goddard, Y. L., Goddard, R. D., & Tschannen-Moran, M. (2007). A theoretical and empirical investigation of teacher collaboration for school improvement and student achievement in public elementary schools. *Teachers College Record*, *109*(4), 877–896. doi:10.1177/016146810710900401

Hargreaves, A., & Macmillan, R. (1995). The Balkanization of Secondary School Teaching. In L. Siskin & J. W. Little (Eds.), *The Subjects in Question* (pp. 141–171). Teachers College Press.

Hattie, J. (2015). *What Works Best in Education: The Politics of Collaborative Expertise*. Pearson.

Hobbs, L., Doig, B., & Plant, B. (2019). The successful students STEM project: A medium scale case study. In B. Doig, J. Williams, D. Swanson, R. B. Ferri, & P. Drake (Eds.), *Interdisciplinary Mathematics Education. The state of the art and beyond* (pp. 209–227). Springer Open. doi:10.1007/978-3-030-11066-6_13

Jones, I., Lake, V. E., & Dagli, U. (2005). Integration of science and mathematics methods and preservice teachers' understanding of constructivism. *Journal of Early Childhood Teacher Education, 25*(2), 165–172. doi:10.1080/1090102050250209

King, K. P., & Wiseman, D. L. (2001). Comparing science efficacy beliefs of elementary education majors and non-integrated teacher education coursework. *Journal of Science Teacher Education, 12*(2), 143–153. doi:10.1023/A:1016681823643

Koirala, H. P., & Bowman, J. K. (2003). Preparing middle level preservice teachers to integrate mathematics and science: Problems and possibilities. *School Science and Mathematics, 103*(3), 145–154. doi:10.1111/j.1949-8594.2003.tb18231.x

Lederman, N. G., & Lederman, J. S. (2013). Is it STEM or "S & M" that we truly love? *Journal of Science Teacher Education, 24*(8), 1237–1240. doi:10.100710972-013-9370-z

Levine, T. H., & Marcus, A. S. (2010). How the structure and focus of teachers' collaborative activities facilitate and constrain teacher learning. *Teaching and Teacher Education, 26*(3), 389–398. doi:10.1016/j.tate.2009.03.001

Martin, S., & Bradbeer, C. (2016). Creating collaborative effectiveness. One school's approach. *Teachers as Communities of Learning professionals, 2*, 48-52. doi:10.18296/set.0046

Mayes, R. (2019). Quantitative reasoning and its role in interdisciplinarity. In B. Doig, J. Williams, D. Swanson, R. B. Ferri, & P. Drake (Eds.), *Interdisciplinary Mathematics Education. The state of the art and beyond* (pp. 113–133). Springer Open. doi:10.1007/978-3-030-11066-6_8

McHugh, L., Kelly, A. M., & Burghardt, M. D. (2018). Professional development for a middle school mathematics-infused science curriculum. *Journal of Science Teacher Education, 29*(8), 804–828. doi:10.1080/1046560X.2018.1514825

Pyke, C. L., & Lynch, S. (2005). Mathematics and science teachers' preparation for national board of teaching standards certification. *School Science and Mathematics, 105*(5), 25–35. doi:10.1111/j.1949-8594.2005.tb18033.x

Rézio, S., Cavadas, B., Nogueira, J. R., & Branco, N. (2020). *Bola Saltitona*. GO-LAB. https://www.golabz.eu/ils/bola-saltitona

Ríordáin, M. N., Johnston, J., & Walshe, G. (2016). Making mathematics and science integration happen: Key aspects of practice. *International Journal of Mathematical Education in Science and Technology, 47*(2), 233–255. doi:10.1080/0020739X.2015.1078001

Ronfeldt, M., Farmer, S., McQueen, K., & Grissom, J. (2015). Teacher collaboration in instructional teams and student achievement. *American Educational Research Journal, 52*(3), 475–514. doi:10.3102/0002831215585562

Shakenova, L. (2017). The theoretical framework of teacher collaboration. *Khazar Journal of Humanities and Social Sciences, 20*(2), 34-48. http://jhss-khazar.org/wp-content/uploads/2016/11/2.Lyaila.20.02.2017.2.pdf

Slater, L. (2004). Collaboration: A framework for school improvement. *International Electronic Journal for Leadership in Learning, 8*(5), 1-19. https://files.eric.ed.gov/fulltext/EJ984546.pdf

St. Clair, B., & Hough, D. L. (1992). *Interdisciplinary teaching: A review of the literature.* Southwest Missouri State University Department of Curriculum and Instruction. https://files.eric.ed.gov/fulltext/ED373056.pdf

Tytler, R., Williams, G., Hobbs, L., & Anderson, J. (2019). Challenges and opportunities for a STEM interdisciplinary agenda. In B. Doig, J. Williams, D. Swanson, R. B. Ferri, & P. Drake (Eds.), *Interdisciplinary Mathematics Education. The state of the art and beyond* (pp. 51–81). Springer Open. doi:10.1007/978-3-030-11066-6_5

Vieira, F. (2009). Em contracorrente: O valor da indagação da pedagogia na universidade. *Educação Sociedade & Culturas, 28,* 107–126. https://www.fpce.up.pt/ciie/revistaesc/ESC28/28_flavia.pdf

Wong, V., & Dillon, J. (2020). Crossing the boundaries: Collaborations between mathematics and science departments in English secondary (high) schools. *Research in Science & Technological Education, 38*(4), 396–416. doi:10.1080/02635143.2019.1636024

Woodland, R., Lee, M. K., & Randall, J. (2013). A validation study of the teacher collaboration assessment survey. *Educational Research and Evaluation, 19*(5), 442–460. doi:10.1080/13803611.2013.795118

Woodland, R., & Mazur, R. (2015). Beyond hammers versus hugs: Leveraging educator evaluation and professional learning communities into job-embedded professional development. *NASSP Bulletin, 99*(1), 5–25. doi:10.1177/0192636515571934

Woodland, R. H. (2016). Evaluating PK–12 professional learning communities: An improvement science perspective. *The American Journal of Evaluation, 37*(4), 505–521. doi:10.1177/1098214016634203

ADDITIONAL READING

An, S. A. (2017). Preservice teachers' knowledge of interdisciplinary pedagogy: The case of elementary mathematics-science integrated lessons. *ZDM Mathematics Education, 49*(2), 237–248. doi:10.100711858-016-0821-9

Berlin, D. F., & White, A. L. (2010). Preservice mathematics and science teachers in an integrated teacher preparation program for grades 7-12: A 3-year study of attitudes and perceptions related to integration. *International Journal of Science and Mathematics Education, 8*(1), 97–115. doi:10.100710763-009-9164-0

Cavadas, B., Rézio, S., Nogueira, J. R., & Branco, N. (2022). A framework and a research design proposal to identify preservice teachers' integration performance of science and mathematics. *Can. J. Sci. Math. Techn. Educ., 22*(1), 101–129. doi:10.100742330-022-00198-2

Davidson, D. M., Miller, K. W., & Metheny, D. L. (1995). What does integration of Science and Mathematics really mean? *School Science and Mathematics, 95*(5), 226–230. doi:10.1111/j.1949-8594.1995.tb15771.x

Furner, J. M., & Kumar, D. D. (2007). The mathematics and science integration argument: A stand for teacher education. *Eurasia Journal of Mathematics, Science and Technology Education, 3*(3), 185–189. doi:10.12973/ejmste/75397

GO-LAB. (n.d). https://www.golabz.eu/

Honey, M., Pearson, G., & Schweingruber, A. (2014). *STEM integration in K-12. Status, Prospects, and an agenda for research.* National Academies Press.

Huntley, M. A. (1998). Design and implementation of a framework for defining integrated mathematics and science education. *School Science and Mathematics, 98*(6), 320–327. doi:10.1111/j.1949-8594.1998.tb17427.x

Mason, T. C. (1996). Integrated curricula: Potential and problems. *Journal of Teacher Education, 47*(4), 263–270. doi:10.1177/0022487196474004

Samson, G. (2014). From writing to doing: The challenges of implementing integration (and interdisciplinarity) in the teaching of mathematics, sciences, and technology. *Canadian Journal of Science. Mathematics and Technology Education, 14*(4), 346–358. doi:10.1080/14926156.2014.964883

Weinberg, A., & McMeeking, L. (2017). Toward meaningful interdisciplinary education: High school teachers' views of mathematics and science integration. *School Science and Mathematics, 117*(5), 204–213. doi:10.1111sm.12224

KEY TERMS AND DEFINITIONS

Action Taking: Element of the cycle of inquiry in which teachers take actions directly related to the improvement of practice which entail a degree of innovation (according to Woodland et al., 2013).

Cycle of Inquiry: Process of high-quality teacher collaboration, in which teachers work closely with colleagues during the workday to examine student learning data and solve problems of instructional practice through a continuous cycle of dialogue, decision making, action taking, and evaluation (according to Woodland et al., 2013).

Decision Making: Element of the cycle of inquiry related with allocation, security, boundary, evaluation, and instructional decisions. The most relevant element in decision making is the teacher teams' decisions about their individual and collective instructional practices and their effects on student learning (according to Woodland et al., 2013).

Dialogue: Element of the cycle of inquiry in which teachers engage in collective dialogue about student learning, the effects of instruction on student achievement, and how to provide an appropriate level of challenge and support to each student (according to Woodland et al., 2013).

Evaluation: Element of the cycle of inquiry in which teachers collect and analyze data about student learning and their teaching performance. Teachers in high-functioning teams will systematically collect and analyze both quantitative information (such as scores on formative and summative assessments of student learning) and qualitative information (such as notes taken during a classroom observation of a colleague and student written work) (according to Woodland et al., 2013).

Inquiry Learning Space (ILS): Personalized learning resources for students, including a lab, apps, and any other type of multimedia material. ILS follow an inquiry cycle. Inquiry cycles can differ, but the basic GO-LAB cycle consists of the phases Orientation, Conceptualization, Investigation, Conclusion, and Discussion. The aim of an ILS is to provide students with an opportunity to conduct scientific experiments, with guidance through the inquiry process and support at each step (GO-LAB, n.d.).

Integration of Science and Mathematics: Working in the context of complex phenomena or situations on tasks that require students to use knowledge and skills from science and mathematics (adapted from Honey, 2014).

Teacher Collaboration: Teachers working together, and engaging in reflective dialogue, with the common goal of improving practice and increasing students' learning (according to Woodland et al., 2013).

Chapter 10
Malthus' Principle in the Teaching of Evolution as an Integrative Context of Science and Mathematics in Elementary Education

Nelson Mestrinho
Polytechnic Institute of Santarém, Portugal

Joana Ribeiro
Polytechnic Institute of Porto, Portugal

Alexandre Pinto
Polytechnic Institute of Porto, Portugal

Inês Sarmento
Polytechnic Institute of Porto, Portugal

Xana Sá-Pinto
University of Aveiro, Portugal

ABSTRACT

The Malthus principle, which determines that the natural populations grow at a faster rate than the resources of which they depend, was a crucial concept for Darwin to discover natural selection. A previous study shows that an educational activity between biology and mathematics that explores the Malthus principle, implemented with elementary school students, effectively promoted their understanding of natural selection. In the chapter, the authors analyse the impacts of this activity in students' abilities to solve mathematical problems involving geometric progressions. For that, they applied a pre- and post-test approach in two target groups. The answers were ranked and coded by levels of performance. The analysis consisted in comparing order statistics (boxplots) and by the application of Wilcoxon signed rank tests. The results show that the educational activity impacted significatively the student ability to solve mathematical problems involving geometric progressions.

DOI: 10.4018/978-1-6684-5765-8.ch010

INTRODUCTION

Inter and transdisciplinarity is a feature of scientific research and has been fundamental for several breakthrough discoveries in science (MacKinnon et al., 2013). The discovery of evolution by natural selection is one excellent example of how thinking across disciplinary border is important for boosting creativity and innovation in science. Darwin's scientific career was deeply marked by transdisciplinarity with his research and publications ranging from geology to biology and psychology (MacKinnon et al., 2013).

Darwin himself describes how knowledge from other disciplines influenced his ideas about evolution (Barlow, 1958). In particular, he explains how the work of the economist Malthus (1798) was fundamental for his discovery of the process of natural selection (Barlow, 1958, p.120):

In October 1838, that is, fifteen months after I had begun my systematic enquiry, I happened to read for amusement 'Malthus on Population,' and being well prepared to appreciate the struggle for existence which everywhere goes on from long-continued observation of the habits of animals and plants, it at once struck me that under these circumstances favourable variations would tend to be preserved, and unfavourable ones to be destroyed. The result of this would be the formation of new species. Here then I had at last got a theory by which to work.

Malthus' principle is based on mathematic models that contrast populations' geometric growth with their needed resources arithmetic growth, thus offering an interesting opportunity for interdisciplinary work in the classroom, involving biology and mathematics, even in elementary school. Previous studies showed that interdisciplinary problem-based learning (PBL) activities that explore Malthus principle through simple mathematical modelling and graphic representations allowed elementary school students to learn about natural selection and improve their understanding of evolution (Sá-Pinto et al., 2021).

Mirroring the scientific discoveries processes, several researchers and institutions have been supporting the importance of interdisciplinary/transdisciplinary education (MacKinnon et al., 2013; ME, 2018 a and b, Takeuchi, 2020). This led to the emergence of educational approaches such as STEM (Takeuchi, 2020; Tytler et al., 2021) and more recently STEAM (Perignat & Katz-Buonincontro, 2019; Belbase et al., 2022) that argue for educational activities that integrate different areas of knowledge, such as science, technology, engineering, arts and mathematics, with an emphasis on real contexts and problems, linked to students' experiences. Besides promoting contextualized learning of the distinct disciplinary fields, these approaches aim to foster the development of students' creativity and problem-solving skills and to engage diverse students including those coming from underrepresented minorities (Perignat & Katz-Buonincontro, 2019).

The integration of mathematics and science in teaching promotes students' conceptual understanding of these two subjects, since science provide students with concrete examples that offer meaning to abstract mathematical ideas, while mathematics can enable students to deepen understanding of science concepts and relations by means of quantification (Ríordáin et al., 2016). Some mathematics and science educators contend that this integration promotes students' motivation, enabling a more positive attitude toward meaningful learning those subjects (Bossé et al., 2010), therefore facilitating students' engagement, and the development of critical thinking and problem-solving skills (Ríordáin et al, 2016).

In elementary school, children learn about the concept of arithmetic growth when they count, for example 2 by 2 or 5 by 5. In such counting growth occurs in arithmetic progression since the difference between each number and the previous is always constant. Arithmetic growth is also present when they deal with families of multiples of integer numbers in multiplication. Mathematically speaking, geometric growth, or growth in geometric progression, is linked to the concept of power as representation of iterated multiplication of a fixed factor, and, for that reason, is not foreseen in the curriculum for elementary education. However, the exploration of growth sequencies, another topic in elementary school curriculum, allows to accommodate both types of progressions, the widely common arithmetic and the geometric, since this one is formed by repeated multiplication.

The exploration of regularities in numerical growth sequences, as foreseen in elementary school curriculum, contributes to the development of algebraic thinking of students, opening horizons for later learnings and allowing students to progress from recursive reasoning to reasoning involving functional relationships (Ponte et al, 2009). Numerical and geometric sequencies have been used in problem-solving situations to promote the development of algebraic reasoning in elementary education (Bautista-Perez et al., 2021). Since geometric growth is such an important concept in biology and evolution, it would be interesting to get some insight on how interdisciplinary PBL activities can help students to develop understanding of geometric progression and its mobilization in problem-solving.

In the present chapter we will contribute to this aim by answering the following research questions:

- How elementary school students address mathematical problems involving geometric progressions?
- What are the contributions of an interdisciplinary PBL activity that explore Malthus principles to the development of students' understanding of geometrical progressions in a mathematical problem-solving context?

BACKGROUND

The Malthus' Principle

Thomas Malthus (1766-1834) was an English demographer and economist who lived and worked in a period of increasing optimism (Allaby & Gjertsen, 2002). The social, political, and technological changes resulting from the industrial revolution had significant implications both in increasing food production and in progressive improvement in sanitary conditions and controlling diseases and epidemics (Spielvogel, 2015). With this, conditions were created for an accelerated population growth, by combining a decrease in mortality and an increase in birth rates (Spielvogel, 2015). On the other hand, the social movements that set-in motion the French and the American Revolutions, focused on valuing the individuals, fostered the belief in the ability of societies to solve the problems of humanity, including those that emerged from the new production paradigm (Allaby & Gjertsen, 2002).

In his 1798 *An Essay on the Principle of Population*, Malthus reflects on the sustainability of this trend of population growth, concluding that human populations increase geometrically while food supplies only increase arithmetically, meaning that the increase of food supplies will eventually be unable to match increases in this populations (Malthus, 1798). This idea, known as the Malthus' principle, allowed him to conclude that, in the long run, natural population growth would be restricted by food shortage. Individuals would have to compete for the limited amount of food available and this would lead to war and disease (Malthus, 1798). It was this idea of competition for the available resources that inspired Darwin and Wallace to propose, in 1858, the process of natural selection (Darwin & Wallace, 1858).

Geometric Progressions

The concept of geometric progression is not foreseen in the mathematics curriculum for elementary education. However, it is easy to frame this concept in the curricular standards for the teaching of mathematics in the early grades of basic education in Portugal. The guiding official document prepared and approved by the Portuguese Ministry of Education (Ministério da Educação), referred to as "Essential Learnings" (Ministério da Educação, 2021), presents the development of the ability to establish mathematical connections, both internal and external, as one of the general objectives for the teaching of mathematics. While internal connections refer to the establishment of relationships between different subjects in mathematics, external connections refer to the relationships between mathematics and other subjects or with situations in different contexts of reality. This enables "mathematical knowledge to be used to understand, model and act in various areas or disciplines" (Ministério da Educação, 2021, p. 3).

Another of the general objectives of mathematics teaching is the development of the ability to solve problems of various types and in various contexts (Ministério da Educação, 2021). Problem solving is a central activity in mathematics that provides opportunities for students to acquire and apply mathematical knowledge. Interpreting real-world situations mathematically, constructing appropriate mathematical models, using multiple representations, and using them to solve problems enables the usefulness and power of mathematics to be recognised in predicting and intervening in those situations, as well as recognising the importance of mathematics and its relevance in the construction of knowledge in various other domains (Ministério da Educação, 2021).

Regarding mathematical content in elementary education, the idea of geometric progression can be associated with topics such as Algebra and Numbers and Operations. In Algebra, one of the featured topics is number sequences, in particular growing sequences. It's expected that students became able to identify and describe regularities in sequences and continue a growing sequence respecting a rule of formation or identified regularities (Ministério da Educação, 2021). At this level of education, it is fundamental to develop an understanding of the different meanings of the arithmetic operations, in our specific case of the operation of multiplication, using it as a mathematical model in specific situations and in problem solving.

METHODOLOGY

The Didactic Sequence

The educational activity was composed by three sessions, each lasting approximately 150 minutes. The three sessions took place in 3 consecutives weeks, one session per week. The activity was implemented by a researcher and an experienced teacher. The interdisciplinary project-based learning activity was designed to simultaneously address learning goals from mathematical education and biology (see Table 1). The full scope of this didactic sequence and the principles that guided its design to promote students' learning on evolution are described in detail in Sá-Pinto et al. (2021).

Table 1. Biology and mathematics learning goals for each of the three sessions and its contextualization in Portuguese curriculum (Ministério da Educação, 2018a and b; 2021)

Sessions	Mathematics Learning Goals	Curricular Contextualization for Portugal	Biology Learning Goals	Curricular Contextualization for Portugal
Session 1	1) Length measurements and estimation 2) Spatial Scales 3) Mathematical problem-solving skills	"Estimate the measure of a length, and explain the reasons for your estimate." "Interpret and model situations related with length" "Solve problems involving lengths, critically comparing different resolution strategies."	1) Scientific instrument manipulation skills 2)Designing solutions for problems 3) Exploring spatial scales	"Manipulate, imagine, create or transform simple technical objects" "Mobilize cultural, scientific and technological knowledge to understand reality and to solve everyday situations and problems"
Session 2	1) Algebraic operations 2) Geometric progressions 3) Graphic representation of data 4) Mathematical problem-solving skills 5) Mathematical modelling skills	"Apply mathematical ideas to solve problems from different contexts." "Interpret real world situations mathematically, build appropriate mathematical models, and recognize the usefulness and power Mathematics in prediction and intervention in these situations." "Interpret and model situations with operations and solve associated problems, comparing critically different resolution strategies." "Identify and describe regularities in sequences of growth, explaining your ideas."	1) Intraspecific diversity 2) Heritable traits 3) Using mathematical thinking and modelling 4) Population growth	"Properly communicate their ideas, through the use of different languages ([...] mathematics, [...]) supporting them and arguing against the ideas of others." "Recognize that living beings reproduce and that their descendants have characteristics similar to their parents, but also differ in some of them." "Knowing how to ask questions, raise hypotheses, do inferences, check results and know how to communicate them, recognizing how knowledge is constructed." "Mobilize cultural, scientific and technological knowledge to understand reality and to solve everyday situations and problems"
Session 3	1) Algebraic operations 2) Geometric progressions 3) Graphic representation of data 4) Mathematical problem-solving skills 5) Mathematical modelling skills	"Apply mathematical ideas to solve problems from different contexts." "Interpret real world situations mathematically, build appropriate mathematical models, and recognize the usefulness and power Mathematics in prediction and intervention in these situations." "Interpret and model situations with operations and solve associated problems, comparing critically different resolution strategies." "Identify and describe regularities in sequences of growth, explaining your ideas."	1) Environmental carrying capacity 3) Natural selection 4) Using mathematical thinking and modelling.	"Properly communicate their ideas, through the use of different languages ([...] mathematics, [...]) supporting them and arguing against the ideas of others." "Understand that living beings depend on each other, namely through food relationships…" "Mobilize cultural, scientific and technological knowledge to understand reality and to solve everyday situations and problems"

Source: Authors

In the case of the proposed educational activity, the situations presented growth in geometric progression as the iterated multiplication of equal factors, a prelude to the concept of power of a natural exponent, allowing at the same time to promote the ability to use multiple mathematical representations as tools to support reasoning and communication (Ministério da Educação, 2021). Other mathematical content emerging from the proposed activity, such as length measurement, estimations, scales, and graphical representations of data, are equally well framed in the curriculum, since these topics are commonplace in pupils' mathematical work throughout this schooling cycle.

In the first session, the students were introduced to the two-spotted spider mite - *Tetranychus urticae* - an agricultural pest species. This mite species was used at this activity as some of its biological features makes it a nice model to study and observe evolution and the consequences of geometrical populational growth. These features include: i) its short life cycle that allows to follow evolution in short-time periods; ii) its high fecundity which may result in populations' fast exponential growth and quickly depletion of resources; iii) the existence of two varieties within the species who differ in the food resources they can feed from (the citrus leaves and the bean leaves variety; Wrensch & Young, 1975; Magalhães et al., 2009). Students were asked to observe spider mites using distinct instruments (naked eye, hand magnifier, and mobile phone magnifier), without being informed of the magnifications used. They were then asked how much a mite of this species measures. In a class discussion, students were asked to collaboratively propose strategies to estimate the size of the spider mites, to discuss the strengths and problems of each strategy. After this discussion, the class choose one of the proposed approaches and used it estimated the size of at least two spider mites and the average individual's size.

During the second session, a short movie introduced students to the research group MITE2: Multidisciplinary Investigation Targeting Ecology and Evolution from the Centre for Ecology, Evolution and Environmental Changes based at the University of Lisbon (https://ce3c.ciencias.ulisboa.pt/sub-team/mite2). The two main goals of this video were: *i)* to allow students to learn about the nature of science, namely about the researchers' diversity (to deconstruct gender stereotypes for example) and about the collaborative nature of scientific research (NRC, 2012); *ii)* to provide a real context for the problem posed to students. Students were told that, in 45 days, we would go to other classes to explore the mites with other students, but the MITE2 team had sent us, on that day, 8 spider mites: two individuals of the citrus leaves variety and six of the bean leaves variety. We asked students' help to estimate the number of mites of each variety we would have in 45 days. In a class discussion, students identified the information about the species biology needed to solve the problem. Among other parameters that were proposed (but later discarded) during the discussion, in both classes, the students requested information about the number of descents left by males and by females, the life expectancy, the number of males and females in each population, and the age at which individuals start to reproduce (generation time). To simplify the problem resolution, we told students to consider a sex ratio of 1:1, a generation time equal to a life expectancy of 15 days, and that each female lays approximately 100 eggs, all of which hatch and survive until reproduction. The teacher of the class divided students into groups of 4 to 6 members and asked them to collaboratively estimate the number of individuals of the citrus leaves variety that we would have in 45 days. After reaching a consensual solution, each group presented to the other groups their strategy and the obtained results. After all groups presented their resolution of the problem, in a class discussion, the students decided on the best strategy. Applying this strategy, each group was then asked to estimate the number of spider mites of the bean leaves variety that would exist in 45 days and to graphically represent the number of mites of each variety for each 15-days period. For biology, the learning goals for this session were related with species' intraspecific diversity, traits' heritability (in this case related with the ability to feed on different types of leaves), species reproduction, and use mathematical modelling to estimate population growth (see Table 1). At the same time, such activity offers a strong relationship with mathematics by means of measurement and problem-solving skills. For students at this grade level, the concept of exponential growth can emerge from the study of geometric progressions, giving an additional meaning to the operation of multiplication and contributing to the development of their number and operation sense.

In the third session, students were told that we could only provide 100 citrus tree leaves and 10 bean leaves per week to all the mites and were asked if the estimations that they had made in the previous session were still hold under these circumstances. The availability of each of these two food resources was set so that students could predict a change in the relative frequencies of the two varieties, with the rarest one becoming the most frequent after 45 days. This would allow students to discuss and learn about evolution by natural selection. In a class discussion, students decided about the species biology information they needed to answer the question and were told that each leaf (independently of the plant type or mite variety) could feed up to 100 spider mites per week. The students were again assigned to small groups and asked to collaboratively build on the strategies used in session 2, to estimate the number of mites from the citrus leaves variety expected under this limited resource scenario, 45 days later. Again, each group presented their strategy and results to the other groups, and, in a class discussion, the best strategy was chosen and used by all the groups to estimate the number of mites from the bean leaves variety expected in 45 days. At the end of the session, students were asked to graphically represent the results obtained for the two spider mite varieties, discuss it in their groups, and explain why the least frequent spider mite variety become the most frequent one. They were also asked to discuss the reasons for the differences observed between the results obtained in sessions 2 and 3. During this session, besides scaffolding the learning made in the previous session, students were also expected to learn the concepts of environmental carrying capacity and evolution by natural selection (see Table 1).

Participants

The present activity was developed in two fourth grade classes (students' age ranging from 9th and 10th years old) from the North of Portugal. In Portugal, this grade is the last of the four years of the first cycle of basic education, where most of the subjects are taught by a single teacher (Portuguese, Mathematics, Study of the Environment, Artistic education and Physical Education; Gabinetes da Secretária and de Estado Adjunta e da Educação e do Secretário de Estado da Educação, 2018). The two schools were chosen from the partner schools of the research team members thus consisting of a convenience sampling (Cohen et al., 2007). School A was a private school located in an urban area and the target class (from now one called SAT) had 23 students. School B was a public school located a rural area and the target class there (from now one called SBT) had 25 students. The research procedures followed were approved by both school boards and are in agreement with the Helsinki Declaration of 1975, as revised in 2008.

Implementation Procedures of the Educational Activity

Before implementing the didactic activity, students answered a pre-test in which they were asked to solve a mathematical problem. The proposed situation required the students to use a numerical sequence whose formation law translates a growth according to a geometric progression, resulting from iterated multiplication of a fixed factor. The resolution of this problem was not discussed with the students afterwards and the same test was implemented after the implementation of the didactical activity as a post-test. The pre-test and post-tests were also applied to another class from the private school (SAC; N= 22) that was not involved in the educational activity in the same dates it was applied to SAT. This procedure intended to seek validation of the evaluation instrument (Lahm, 2007) and ensure that any eventual improvements in the performance of the participating students were not due to its double exposure to the same problem.

The problem proposed in both pre and post-tests was as follow:

On a beautiful summer's afternoon, João decided to go for a walk in the street to enjoy the good weather. As he passed a street, he saw a poster which said that there was going to be a moonlight cinema session. He thought it would be interesting to bring together many colleagues to watch this cinema session, and he took his mobile phone and sent the following message to 3 friends: "There will be a moonlight cinema session tonight. Send this message to 3 friends so that we can fill the place". Knowing that the message takes 1 minute to be received and forwarded to colleagues, and that everyone has immediately sent the messages as requested, how many messages will be sent in the 5th minute?

The situation is well modelled by the product 3×3×3×3×3×3=729. John starts by sending the message to 3 friends, in the following minute, the first minute, the time between receiving the message and forwarding it, the message reaches 3×3=9 people; in the second minute the message reaches 3×3×3=27 people, and so forth. At the end of the 5th minute, 729 messages were sent. The product can be obtained from the interpretation of the situation as a product in the combinatory sense or as the result of the construction of a numerical sequence with this type of formation law. A different interpretation of the wording of the task could lead to a different solution, considering the counting of the first minute from the first sending of the message to the first three receivers. However, despite changing the solution of the problem (3×3×3×3×3=243), it doesn't modify the reasoning or the way the situation is translated mathematically.

The purpose of applying this task is to assess students' problem-solving skills - understanding the problem, defining, and implementing a strategy and formulating a plausible answer considering the presented context - with special focus on how students mathematically translate and represent the situation presented.

Data Analysis

The students' approaches, in particular the interpretations of the proposed problem and the definition and implementation of the strategies were classified, coded and ranked according to a focused holistic scale (Ponte et al., 1997). These scales reflect the students' global performance level and their problem-solving skills, considering the fundamental aspects involved in carrying out this type of mathematical task, such as understanding the problem, choosing and executing effective strategies, and the teacher builds them accounting for the aspects he intends to value (Ponte et al., 1997, Silva & Dalto, 2017). Despite the numerical encoding, it has a qualitative character (Ponte et al., 1997). Using this strategy a focused holistic scale was developed, with students' performance levels classified into six classes described in Table 2.

Table 2. Holistic scale with the description of the classes of students' performance levels and its code.

Rank Code	Description of the Level of Performance
31	Presents an appropriate and complete strategy for solving the problem, with evidence of having reached a correct answer or with small calculation errors.
21	Presents an appropriate strategy without, however, being able to complete it in order to get an answer OR completes it incorrectly.
12	Reveals some understanding of the problem but presents an incorrect multiplicative strategy, operating the values provided in the wording and responding in accordance with the value obtained, providing a direct answer.
11	Reveals some understanding of the problem but presents an incorrect strategy, operating the values provided in the wording with an unclear perception of their meaning.
10	Reveals some understanding of the problem but fails to formulate a strategy in order to provide an answer OR provides an answer that has no relation to the data or context.
00	No answer or production revealing poor or absent understanding of the problem.

Source: Authors

An appropriate strategy for solving this problem requires students to associate the solution to the result obtained from a numerical sequence that grows according to a geometric progression. Students at this level of education can translate this reasoning by iterated multiplication of the factor '3'. This multiplication can assume the form of a numerical expression or can be constructed step by step by successive multiplications by 3 or representing the situation through a tree diagram or a table where intermediate products and final answer are gathered.

To each class of students' performance level, it was attributed a code whose numerical value corresponds to a rank. To higher level of problem-solving skills corresponds a higher order value. Descriptive statistics, namely extremes and quartiles, allowed a global comparation of the outcome on both tests by the three groups. Wilcoxon Signed Rank tests were used to test for the existence of significant differences in students' performance levels between pre and post-test in the classes SAT and SBT and in the class SAC. This non-parametric statistical test is appropriate to compare two paired samples of qualitative variables, expressed in ordinal scale, establishing (or not) the significance of the difference between them (Kraska-Miller, 2014). The statistical analysis was done using the software *Jamovi*.

RESULTS AND RECOMMENDATIONS

Results

Looking at the pre-tests applied to the three classes under analysis, to which a total of 70 students responded, we can observe that more than 85% presented an answer to the problem situated in performance level 1 (table 3).

Table 3. Levels of performance and percentage of students' answers (%) attributed to each of these.

Levels of Performance	0	1			2	3
Sublevels	00	10	11	12	21	31
%	1.4	21.4	34.3	30.0	11.4	1.4

Source: Authors

Of these, 80% developed their reasoning based on the 5×3 multiplication, representing it in different ways, answering explicitly or not to the question posed. This corresponds to 48 answers, more than 68% of the students who answered the pre-test. Figures 1 and 2 show some examples of students' productions.

Figure 1. Examples of two students' productions.
Source: Authors

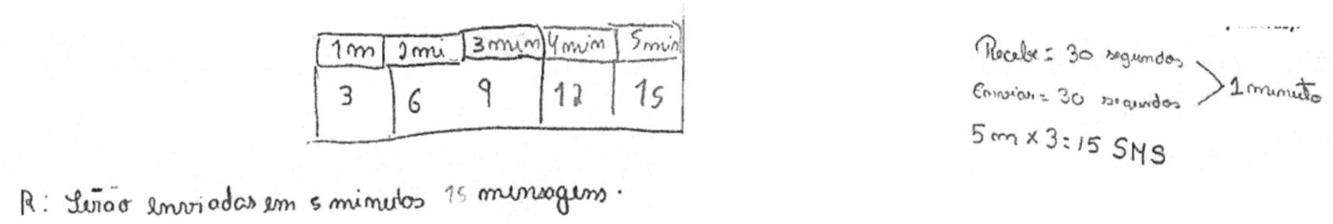

On Figure 1, the image on the left, shows how a student represents the first five terms (one for each minute) of the numerical sequence of the multiples of 3, starting with 3, and presents a final answer. On trying to represent the growth of the number of messages sent, it creates a model for an arithmetic, not geometric progression. The answer depicted on the right side of Figure 1, a student multiplies the number 3 and 5, showing the understanding of what each of the elements represents, including the product. Both were classified as rank 12.

Figure 2. Examples of three students' productions
Source: Authors

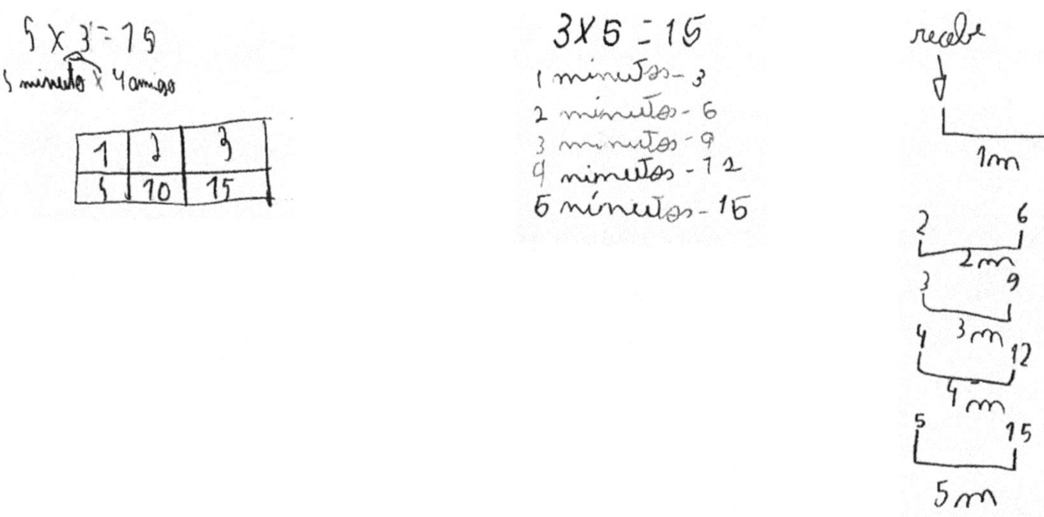

Depicted in the right side of Figure 2 are examples of two representations produced by two students, of the same kind of approach referred before. In these two cases, students don't present any explicit interpretation of the several elements to extract an answer to the question in hand. These two were ranked as 11. In the example in the left side of the Figure 2, the student reaches the product 15 as 5 times 3, but presents a table with the multiples of 5 and not 3, as if he was considering 3 iterations of the factor 5 and not the otherwise. This was ranked 10.

Other residual strategies used by students performing at level 1 involved variations of the "5×3" approach and additive approaches using different interpretations of the data provided in the wording of the problem. Some examples are provided in Figure 3.

Figure 3. Examples of two students' productions
Source: Authors

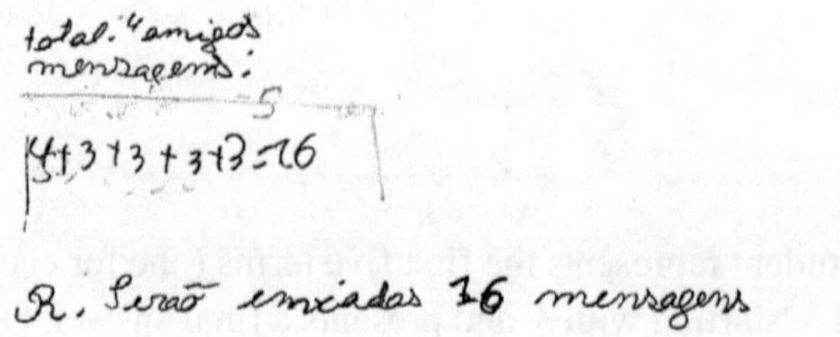

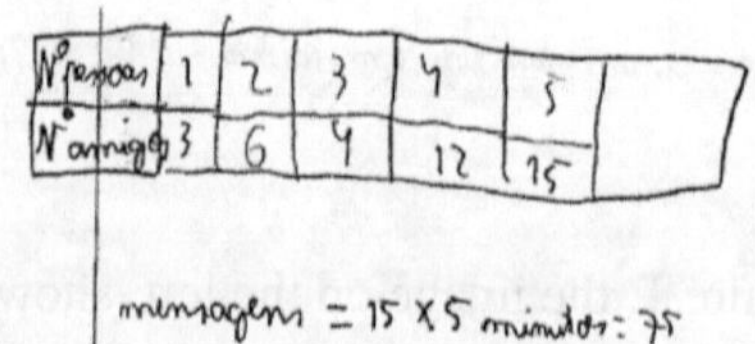

At the pre-test, more than 12% of students were able to approach the problem using a suitable strategy (ranks 21 and 31; Figure 4). One student from the SAC class achieved the maximum performance level 31 by exposing a full understanding of the type of growth involved in the context of this problem, not presenting the correct result just due to the lack of some calculations (Figure 4 on the right). Interestingly, it is possible to see, in the faded background, an attempt of this student to represent the situation through a tree diagram that was erased and replaced by the table representation of the construction of the first five terms of this geometric progression.

Figure 4. Examples of three students' productions
Source: Authors

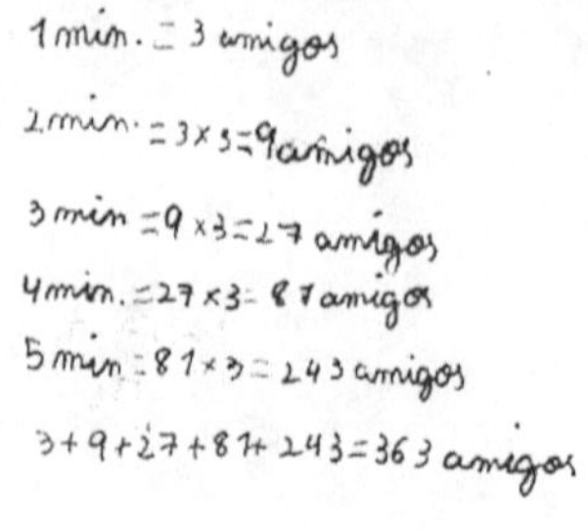

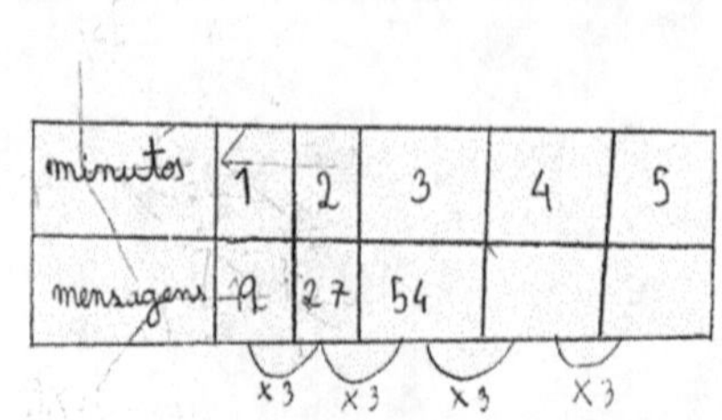

On the left-hand side of Figure 4 we can observe a students' attempt to represent the situation described in the problem by means of a tree diagram. Although the student was not able to complete his reasoning correctly, he/she properly understood the mathematical structure of the problem presented. In the centre of Figure 4, we can observe a construction, by recurrence, of the numerical sequence, that translates the conditions of the problem, starting from the initial message and not from the first three transmitter-receptors as it should. As expected, the process is iterated five times, meaning that the student understood the geometric nature of the growth of the number of messages. For some reason, the student incorrectly assumed that the final answer would result from adding the five terms of this sequence and proceeded accordingly.

Clearly, the approach to mathematical problems involving geometric progressions by elementary school students is marked by a set of misconceptions. In general, it was recognized that the growth problem presented in the problem was related to multiplication, but the choice was just to directly multiply the given numbers. Different representations and organization of the information provided students, in some cases, insight about the nature of geometric progression growth. However, at the pre-test, very few were able to turn that understanding in a viable solution.

In order to compare the results of the pre-tests with those of the post-tests, the authors obtained the descriptive statistics for these three groups of pairs of ranks, corresponding respectively to the SAT, SAC and SBT classes. Being these variables qualitative in ordinal scale, the extremes (maximum and minimum) were obtained, as well as the quartiles of the rank distributions, as can be seen in Table 4.

Table 3. Pre-tests and post-tests results of the SAT, SAC and SBT classes

DESCRIPTIVES	SAT_PRE	SAT_POST	SAC_PRE	SAC_POST	SBT_PRE	SBT_POST
N	23	23	22	22	25	25
Median	12	21	12.0	16.5	11	31
Minimum	0	12	10	10	10	21
Maximum	12	31	31	31	21	31
25th percentile	10.5	12.0	11.0	12.0	11.0	31.0
50th percentile	12.0	21.0	12.0	16.5	11.0	31.0
75th percentile	12.0	21.0	21.0	21.0	11.0	31.0

Source: Authors

Figure 5 presents the previous results in comparative box plots.

Figure 5. Comparative plot boxes of the pre-tests and post-tests results of the SAT, SAC and SBT classes
Source: Authors

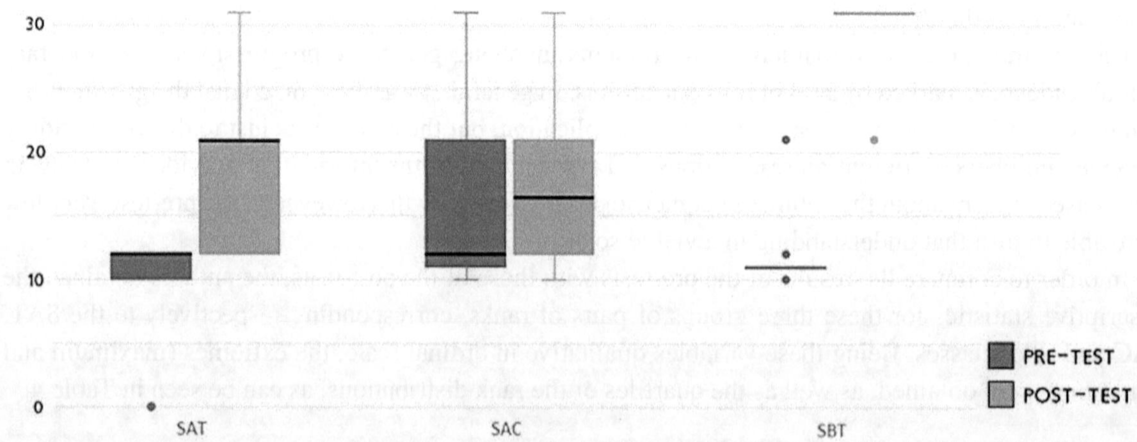

Figure 5 shows that at post-tests, students in the target groups (SAT and SBT) performed better at solving the mathematical problem, with more of them using adequate strategies and more clear representations to obtain a valid solution. The case of SBT is striking, since all the students performed at level 21 or 31 in the post-test. This can be seen in the visible increase in the median rank and the slight rise in the first quartile rank. The validity of the test is ensured by the substantially lower improvements observed for SAC from pre to post-test.

Now considering the paired data, Wilcoxon Signed Rank tests were applied in order to assess the significance of these differences in student performance. We took a significance level of α=0.01. From Jamovi output (Tables 4, 5 and 6) it's clear that the differences observed between pre and post-tests are significant in the case of SAT and SBT but not significant for SAC as we compare the respective p-values with the significance level. This suggests that the implemented interdisciplinary activity had a positive effect on the development of students' understanding of growth in geometric progression and its use in problem solving.

Table 4. Paired samples T-Test of SAT-PRE and SAT-POST.

Paired Samples T-Test				
			Statistic	P
SAT_PRE	SAT_POST	Wilcoxon W	0.00 [a]	<.001

Note. H_A $\mu_{\text{Measure 1 - Measure 2}} < 0$
[a] 3 pair(s) of values were tied
Source: Authors

Table 5. Paired samples T-Test of SAC-PRE and SAC-POST.

Paired Samples T-Test				
			Statistic	p
SAC_PRE	SAC_POST	Wilcoxon W	9.50 [a]	0.019

Note. H_A $\mu_{\text{Measure 1 - Measure 2}} < 0$
[a] 11 pair(s) of values were tied
Source: Authors

Table 6. Paired samples T-Test of SBT-PRE and SBC-POST

Paired Samples T-Test				
			Statistic	p
SBT_PRE	SBT_POST	Wilcoxon W	0.00	<.001

Source: Authors

It seems clear that the involvement of these students in this interdisciplinary educational activity that explores the Malthus' principle to foster evolution understanding also promoted the development of their understanding of the mathematical concept of geometric progression and geometric growth and their ability to mobilizing it in a problem-solving situation.

FUTURE RESEARCH DIRECTIONS

Despite our encouraging results, we should highlight the low number of students involved in this study. Future studies should engage a higher number of students. Attention should also be given to the integration of mathematics and science in teacher education of primary school teachers (Mestrinho & Cavadas, 2018) and studies of this nature could provide valuable insights into the learnings of these students in both areas.

Future studies could explore situations involving population growth and decline related with socio-scientific issues (SSI) such as natural resources exploitation (for example through fishing or hunting activities), invasive species and pathogenic agents' replication during infections or epidemics (such as the current COVID19 pandemics). By focusing on these SSI, the educational activities may allow students to also develop other skills such as socioscientific reasoning and argumentation, perspective taking and ethical reasoning among others, but additional research is needed to test ideas. Furthermore, research shows that interdisciplinary education practices are still scarce in formal education contexts and teachers struggle to plan and implement these types of practices in their classes (Turkka et al., 2017). Several factors contribute to this such as teachers' self-confidence, self-efficacy perceptions as well as the way school day is programmed in traditional school systems (Turkka et al., 2017; Russell & Zembylas, 2007). This suggests the importance of research on the methods and practices to empower teachers to plan and implement interdisciplinary education practices in their classrooms.

CONCLUSION

As seen here, interdisciplinary PBL activities based on science contexts provides a perfect set for the emergence of more advanced mathematical ideas, giving them meaning, enabling mathematical knowledge to be used to understand, model and act in various areas, and helping students to recognize the relevance of mathematics in the construction of knowledge. These results support the educational potential of inter and transdisciplinary activities that explore biological scenarios through mathematical modelling to promote mathematical learning.

REFERENCES

Allaby, M., & Gjertsen, D. (Eds.). (2002). *Makers of Science* (Vol. 2). Oxford University Press USA.

Barlow, N. (Ed.). (1958). *The autobiography of Charles Darwin 1809-1882. With the original omissions restored.* Collins.

Bautista-Perez, J. L., Bustamante-Rosario, M., & de Armas, T. (2021). Desarrollo de razonamiento algebraico elemental a través de patrones y secuencias numéricas y geométricas [Development of elementary algebraic reasoning through numerical and geometric patterns and sequences]. *Educación Matemática, 33*(1), 125–152. doi:10.24844/EM3301.05

Belbase, S., Mainali, B. R., Kasemsukpipat, W., Tairab, H., Gochoo, M., & Jarrah, A. (2022). At the dawn of science, technology, engineering, arts, and mathematics (STEAM) education: Prospects, priorities, processes, and problems. *International Journal of Mathematical Education in Science and Technology, 53*(11), 2919–2955. doi:10.1080/0020739X.2021.1922943

Bossé, M., Lee, T., Swinson, M., & Faulconer, J. (2010). The NCTM Process Standards and the Five Es of Science: Connecting Math with Science. *School Science and Mathematics, 110*(5), 262–276. doi:10.1111/j.1949-8594.2010.00033.x

Cohen, L., Manion, L., & Morrison, K. (2017). *Research methods in education* (6th ed.). Routledge.

Darwin, C., & Wallace, A. R. (1858). On the tendency of species to form varieties; and on the perpetuation of varieties and species by natural means of selection. *Journal of the Proceedings of the Linnean Society Zoology, 20*, 45–62.

Kraska-Miller, M. (2014). *Nonparametric statistics for social and behavioral sciences*. CRC Press.

Lahm, K. (2007). Validity, quantitative. In G. Ritzer (Ed.), *The blackwell encyclopedia of sociology* (pp. 5173–5175). Blackwell Publishing. doi:10.1002/9781405165518.wbeosv002

MacKinnon, P. J., Hine, D., & Barnard, R. T. (2013). Interdisciplinary science research and education. *Higher Education Research & Development, 32*(3), 407–419. doi:10.1080/07294360.2012.686482

Magalhães, S., Blanchet, E., Egas, M., & Olivieri, I. (2009). Are adaptation costs necessary to build up a local adaptation pattern? *BMC Evolutionary Biology, 9*(1), 182. doi:10.1186/1471-2148-9-182 PMID:19650899

Malthus, T. (1798). *An essay on the principle of population, as it affects the future improvement of society with remarks on the speculations of Mr. Godwin, M. Condorcet and other writers.* Eletronic Scholarly Publishing Project.

Mestrinho, N., & Cavadas, B. (2018). Innovation in teacher education: An integrative approach to teaching and learning science and mathematics. *Proceedings*, *2*(21), 1343.

Ministério da Educação (ME). (2018a). *Aprendizagens Essenciais. Terceiro ano: Educação Básica. Primeiro Ciclo* [Essential Learning. Third year: Basic Education. First Cycle]. Estudo do Meio. https://www.dge.mec.pt/sites/default/files/Curriculo/Aprendizagens_Essenciais/1_ciclo/3_estudo_do_meio.pdf

Ministério da Educação (ME). (2018b). *Aprendizagens Essenciais. Quarto ano: Educação Básica. Primeiro Ciclo* [Essential Learning. Fourth year: Basic Education. First Cycle]. Estudo do Meio. https://www.dge.mec.pt/sites/default/files/Curriculo/Aprendizagens_Essenciais/1_ciclo/3_estudo_do_meio.pdf

Ministério da Educação (ME). (2021). *Aprendizagens Essenciais. Quarto ano: Educação Básica. Primeiro Ciclo* [Essential Learning. Fourth year: Basic Education. First Cycle]. Matemática. https://www.dge.mec.pt/sites/default/files/Curriculo/Aprendizagens_Essenciais/1_ciclo/ae_mat_4.o_ano.pdf

Perignat, E., & Katz-Buonincontro, J. (2019). STEAM in practice and research: An integrative literature review. *Thinking Skills and Creativity*, *31*, 31–43. doi:10.1016/j.tsc.2018.10.002

Pessoa da Silva, K. A., & Dalto, J. O. (2017). Uma estratégia de Avaliação de Atividades de Modelagem Matemática [A strategy of Evaluating Mathematical Modeling Activities]. *Revista Electrónica de Investigación en Educación en Ciencias*, *12*(2), 1–14. doi:10.54343/reiec.v12i2.228

Ponte, J. P., Boavida, A. M., Graça, M., & Abrantes, P. (1997). *Didáctica da Matemática* [Mathematics didactics]. Editorial do Ministério da Educação.

Ponte, J. P., Branco, N., & Matos, A. (2009). *Álgebra no Ensino Básico* [Algebra in elementary school]. Direção Geral da Inovação e do Desenvolvimento Curricular.

Riordáin, M. N., Johnston, J., & Walshe, G. (2016). Making mathematics and science integration happen: Key aspects of practice. *International Journal of Mathematical Education in Science and Technology*, *47*(2), 233–255. doi:10.1080/0020739X.2015.1078001

Rocha, J., Pessoa, P., Gomes, J., Sá-Pinto, X., & Lopes, J. (2020). BiblioLab Project: Teachers, Parents and Students' Perspectives About the Usability and Usefulness of an Educational Distance Learning Platform. In A. Reis, J. Barroso, J. B. Lopes, T. Mikropoulos, & C. W. Fan (Eds.), *Technology and Innovation in Learning, Teaching and Education. TECH-EDU 2020. Communications in Computer and Information Science* (Vol. 1384, pp. 90–110). Springer.

Russell, J., & Zembylas, M. (2007). Arts integration in the curriculum: A review of research and implications for teaching and learning. In L. Bresler (Ed.), *International Handbook of Research in Arts Education* (pp. 287–312). Springer Netherlands. doi:10.1007/978-1-4020-3052-9_18

Sá-Pinto, X., Pinto, A., Ribeiro, J., Sarmento, I., Pessoa, P., Rodrigues, L., Vázquez-Bem, L., Mavrikaki, E., & Lopes, J. B. (2021). Following Darwin's footsteps: Evaluating the impact of an activity designed for elementary school students to link historically important evolution key concepts on their understanding of natural selection. *Ecology and Evolution, 11*(18), 12236–12250. doi:10.1002/ece3.7849 PMID:34594496

Silva, K. A. P., & Dalto, J. O. (2017). Uma estratégia de avaliação de atividades de modelagem matemática [A strategy for evaluating mathematical modeling activities]. *Revista Electrónica de Investigación en Educación en Ciencias, 12*(2), 1–17. doi:10.54343/reiec.v12i2.228

Spielvogel, J. (2015). *Western Civilization* (9th ed., Vol. 2). Cengage Learning.

Takeuchi, M. A., Sengupta, P., Shanahan, M.-C., Adams, J. D., & Hachem, M. (2020). Transdisciplinarity in STEM education: A critical review. *Studies in Science Education, 56*(2), 213–253. doi:10.1080/03057267.2020.1755802

Turkka, J., Haatainen, O., & Aksela, M. (2017). Integrating art into science education: A survey of science teachers' practices. *International Journal of Science Education, 39*(10), 1403–1419. doi:10.1080/09500693.2017.1333656

Tytler, R., Prain, V., & Hobbs, L. (2021). Rethinking Disciplinary Links in Interdisciplinary STEM Learning: A temporal model. *Research in Science Education, 51*(S1, Suppl 1), S269–S287. doi:10.100711165-019-09872-2

Wrensch, D. L., & Young, S. S. Y. (1975). Effects of quality of resource and fertilization status on some fitness traits in the two-spotted spider mite, *Tetranychus urticae* Koch. *Oecologia, 18*(4), 259–267. doi:10.1007/BF00345850 PMID:28308917

ADDITIONAL READING

Darwin, C. (1859). *On the origin of species by means of natural selection, or, The preservation of favoured races in the struggle for life*. J. Murray. doi:10.5962/bhl.title.82303

Doig, B., Williams, J., Swanson, D., Borromeo-Ferri, R., & Drake, P. (2019). *Interdisciplinary Mathematics Education – The State of the Art and Beyond*. Springer. doi:10.1007/978-3-030-11066-6

Hmelo-Silver, C. (2004). Problem-based learning: What and how do students learn? *Educational Psychology Review, 16*(3), 235–266. doi:10.1023/B:EDPR.0000034022.16470.f3

Sá-Pinto, X., Beniermann, A., Børsen, T., Georgiou, M., Jeffries, A., Pessoa, P., Sousa, B., & Zeidler, D. L. (Eds.). (2022). *Learning evolution through socioscientific issues*. UA Editora., doi:10.48528/4sjc-kj23

KEY TERMS AND DEFINITIONS

Geometric Progression: Sequence of non-zero numbers in which the ratio between two consecutive terms is constant. As a consequence, each term after the first can be obtained by multiplying the previous one by the referred ratio, called the common ratio.

Malthus' Principle: Human populations increase geometrically while food supplies only increase arithmetically, meaning that the increase of food supplies will eventually be unable to match increases in this populations (Malthus, 1798).

Section 3

Integration of Science and Mathematics in Teacher Education

Chapter 11
Integrating Science and Mathematics in Teacher Education Programs

Fulden Guler-Nalbantoglu
https://orcid.org/0000-0002-1233-6666
Ege University, Turkey

ABSTRACT

The main purpose of this chapter is to examine the science and mathematics integration models and summarize the relevant literature on the integration of science and mathematics with respect to teacher education. The obstacles reported in these studies were analyzed from the teachers' perspectives, and a model of an integrated teaching method course was offered by considering the drawbacks in the literature. The proposed course is pedagogical content knowledge-based and enriched with the engineering design process to integrate science and mathematics effectively. The main parts of the course are (1) understand, (2) engage, (3) collaborate, (4) apply, and (5) reflect. Recommendations for research and practice are provided at the end of the chapter.

INTRODUCTION

Integrating science and mathematics is not a brand-new idea; however, attempts to strengthen the connection between the two disciplines have been strongly continuing (Lee et al., 2013; McHugh et al., 2018; Offer & Vasquez-Mireles, 2009). National educational reforms emphasized integrating different content areas and relating different disciplines to each other (Ministry of National Education [MoNE], 2018a; National Research Council, 2012). In parallel with these efforts, curriculum integration has gained more popularity with the appearance of STEM education in recent years (Czerniak & Johnson, 2014). People have been confronting a growing number of global problems, such as pandemic, global warming, waste management, and energy crisis. These problems cannot be solved by utilizing the knowledge and practices of one discipline; solving these complex problems requires the embracement of different

DOI: 10.4018/978-1-6684-5765-8.ch011

disciplines. Therefore, interdisciplinary teaching that underlines creating connections between science and mathematics becomes more necessary.

Integrated curricula provide opportunities with a more relevant and authentic context for students (Kim & Bolger, 2017). One definition of curriculum integration is "engage students as active learners who make the most of the decisions about what they study" (Brown, 2016, p. 123). Another definition provided by Honey et al. (2014) underlines the importance of using various disciplines to solve complex situations (p.52). On the other hand, school curriculum is generally developed based on the isolation of different disciplines and fragmentation (Breiner et al., 2012). Furner and Kumar (2007) labeled this issue a "layer cake" approach. They asserted that teaching disciplines in isolation from one another are no more different than completing a puzzle without giving the whole picture (p. 186). Currently, a consented clear definition related to curriculum integration is unavailable.

There are different perspectives and debates on curriculum integration. Some epistemological concerns are expressed as "classroom activities must foster active engagement with the content and processes of the discipline, with students developing and testing ideas in ways consistent with the paradigms of the disciplines they study" (Schoenfeld, 2004, p. 238). This perspective questions whether the integration fails to provide specialized knowledge in one particular discipline. On the other hand, advocates of curriculum integration explain that if the subject area lines do not blur between different disciplines, it would be difficult for students to relate the concepts to their daily-life experiences (Beane, 1995; Hurd, 1991). The researchers in this line point out that curriculum integration transcended the boundaries of the disciplines and concentrated on the commonalities between them. Moreover, the integrated curriculum draws upon each discipline's knowledge, application and practices and provides different points of view (Lederman & Lederman, 2013). It could be inferred that an integrated curriculum enriches the learning environment and allow full student engagement as opposed to traditional approaches (Czerniak & Johnson, 2014; Gresnigt et al., 2014).

With respect to science and mathematics integration, science enables authentic context for mathematics, and through the use of mathematics, students can establish relationships (Bosse et al. 2010); therefore, the curriculum of these two disciplines are considered compatible with each other (Krajcik & Czerniak, 2014). Integrating these two closely-tied disciplines effectively increased students' motivation, engagement, problem-solving skills, and achievements (Berlin & White, 1994; Ríordáin et al., 2016). The purpose of this chapter is to provide an understanding of curriculum integration and the science and mathematics integration models and summarize the related literature. Moreover, a model for integrating science and mathematics in teacher education programs by considering the drawbacks in the literature is proposed at the end of the chapter.

BACKGROUND

The number of studies concerning the integration of science and mathematics in teacher education programs is limited in the last two decades. Some studies that might help make recommendations about how to integrate science and mathematics purposefully were summarized below.

One of the earliest studies focused on developing science and mathematics integrated method courses (Koirala & Bowman, 2003). The aim of the course was to demonstrate the similarities between some concepts and skills in science and mathematics. The chosen topics were pendulums, rockets, density, and environmental science with decimals, ratios, data analysis, and graphing. The participants majored in mathematics, social studies, and English, and there were no participants with a science major. The course structure included carrying out activities on predetermined topics, preparing reflection papers throughout the semester, field experience, teaching in a real classroom environment, and discussing with the course instructors. The findings of the study suggested that participants valued the integration of science and mathematics, but they were also faced with some difficulties and frustration while integrating science and mathematics. However, their understanding of integration was improved after completing the method course.

The other main study in this field was conducted by Frykholm and Glasson (2005). The purpose of the study was to introduce a new model for teacher education at the secondary level to connect science and mathematics and examine pre-service teachers' perspectives on the issue. They utilized pedagogical context knowledge as a framework. They underlined that "students and teachers construct meaningful knowledge of science and mathematics through interaction– then rich contexts become essential to promote deep thinking and learning and recognition of the symbiotic relationship between mathematics and science" (p.130). In order to reach the purposes of the study, the researchers created groups of five to seven pre-service teachers from different disciplines (science and mathematics) in integrated science and mathematics methods courses. The groups developed integrated units and implemented them in practicing schools. According to the study results, the participants felt inadequate in terms of content knowledge. For instance, pre-service mathematics teachers were not comfortable with their science content knowledge which was an expected result. On the other hand, collaborative work enhanced their perceptions with respect to the integration of science and mathematics, and the quality of collaboratively developed curricular units was increased at the end of the study.

Cady and Rearden (2007) investigated the pre-service teachers' beliefs in a content method course. The study participants completed their studies in content areas such as mathematics, biology, history, and psychology and took a method course to become elementary teachers. The method course focused on mathematics, science, social studies, and language arts and four different instructors from each discipline taught the course. Classroom assignments and Learning Context Questionnaire were administered to 68 pre-service teachers. The results of the study showed that pre-service teachers found the integration of science and mathematics difficult and contrived and had greater anxiety when teaching mathematics content compared to science.

Berlin and White (2010) developed courses for pre-service teachers to integrate science, mathematics, and technology. Six courses were offered, namely, Integrated Content I-II-III and Integrated Pedagogy I-II-III in the teacher education program. The focus of the courses showed variances. For instance, Integrated Content I course took science as the overarching discipline, and Integrated Content II course concentrated on mathematics discipline, and other disciplines were integrated into mathematics. On the other hand, the Pedagogy I course included the educational philosophy of each discipline, and Pedagogy II and III courses were designed for instructional strategies, classroom management, and assessment in integrated teaching supported by instructional technologies. Participants were asked to participate in field practice and complete an action research project at the end of the study. The results of the study suggested that participants' attitudes and perceptions did not significantly change during the study; they were aware of the importance of integrating science, mathematics, and technology disciplines both at the beginning and end of the study. On the other hand, participants' attitudes in terms of challenges encountered in integrating these disciplines were shifted and perceived as more difficult after participating in the study. Participants mentioned some barriers, such as the lessons being taught in a discipline-based approach in a real classroom environment or having difficulty in making appropriate relations across disciplines.

On the other hand, Sackes et al. (2012) studied pre-service early childhood teachers. The researchers offered a science and mathematics integrated method course and investigated the effect of participating in this course on pre-service teachers' efficacy beliefs. The method course was mainly developed to increase participants' PCK for integrating science and mathematics. There were 34 participants involved in the study, and they were assigned to these two groups randomly: Science First and Mathematics First. In the first four weeks of the study, participants in the Science First group took a science method course while the other group received a mathematics method course. The groups were changed in the second four weeks of the study. During the eight weeks, both groups also took extra instruction for integrating science and mathematics. The study results indicated no significant difference between the abovementioned groups. Taking science or mathematics content first did not make any difference. On the other hand, both groups' efficacy beliefs for integrated teaching were improved after attending the course. The researchers pointed out that having high efficacy beliefs in science and mathematics separately did not refer to having high efficacy beliefs in the integration of science and mathematics.

Furthermore, Berlin and White (2012) conducted a longitudinal study with pre-service STEM teachers to examine their attitudes and perceptions of integration. The participants took six integrated courses and completed their field experiences. A semantic differential developed by the researchers was applied to 229 participants. The results indicated that the participants appreciated the integration, and their attitudes and perceptions changed positively about the effectiveness of integration. Based on the findings of the study, the researchers made recommendations for teacher education programs, such as giving weight to content area courses, providing opportunities to apply practices of different disciplines, and encouraging collaboration among pre-service teachers.

A recent study carried out by An (2017) aimed to investigate the changes in elementary pre-service teachers' views regarding teaching strategies after attending interdisciplinary activities. The researchers used the "Interdisciplinary pedagogical content knowledge" framework, which was developed based on Shulman's (1987) and Frykholm and Glasson's (2005) studies. The researchers supported the mathematics teaching methods course with science-centered activities, such as making water membranes in space. Twenty-eight participants were involved in the study and were required to design mathematics lesson plans using science content knowledge. study's results demonstrated that pre-service teachers could use science-supported activities as an instructional approach for mathematics teaching. It could be inferred that participants' interdisciplinary PCK was improved, and they began to make meaningful connections across disciplines at the end of the study. The depth of activities developed in the intervention study was also increased.

Another study created a community of practice named Teaching and Learning Network, composing five science and mathematics teachers (Johnston et al., 2020). The purpose was to increase their capacity for integrating science and mathematics through the use of technology. The researchers considered the first-year results of the study and carried out the second year of the study in Ireland. For instance, integrated units were generally prepared by researchers in the first-year study while they were designed by the participating teachers in the second-year study. The second-year study concentrated on teamwork and collaboration to enhance participants' technological PCK. They pointed out that working with teachers of different subjects might help improve participants' knowledge and perspective about integrated teaching. The researchers were interested in whether the community of practice increased participants' attitudes, improved their knowledge, and changed their practice from single-subject focused instruction to integrated teaching instruction. The participants designed a unit on heat and energy transfer. The study results indicated that participants' content and pedagogical knowledge improved, and teachers from different disciplines created a common language for integrated teaching through collaboration and discussions. Moreover, their attitudes changed positively towards integrated teaching, which was then reflected in their practice.

Moreover, Aydin-Gunbatar et al. (2020) designed an elective course for pre-service chemistry teachers to integrate science, mathematics, engineering, and technology and aimed to develop their PCK for STEM. Chemistry was an overarching discipline; engineering and mathematics were integrated into chemistry in the context of this course. The researchers used a PCK framework and provided an environment where 13 pre-service teachers engaged in the STEM integration activities and planned integrated lesson plans in a research-based lesson. Based on the study's results, the intervention was effective for pre-service teachers in terms of connecting different disciplines effectively. The majority of the pre-service teachers developed their understanding of the subject matter knowledge in science, mathematics, and engineering and designed integrated lesson plans where each discipline had equal emphasis.

Cavadas et al. (2022) lately studied pre-service teachers. They aimed to develop a framework (IPoSM) to determine pre-service teachers' science and mathematics integration performance in each step of the problem-based learning activity. The problem-based activity called "STEM Bees" was prepared and implemented with 13 pre-service teachers to reach this purpose. A global research problem was presented to the participants in the STEM Bee activity, and a group of pre-service science and mathematics teachers worked with teacher educators. While solving the problem, the steps of Inquiry Learning Space were followed: orientation, conceptualization, investigation, conclusion, and discussion. According to the study's finding, integrating these two disciplines was more evident in some parts of the problem-based activity, especially in the conclusion and discussion phases. The framework (IPoSM) developed in the study effectively identified pre-service teachers' integration performance in every phase of problem-based activity.

Studies conducted with teachers also might be utilized for revising the teacher education programs in terms of the interaction of science and mathematics. For instance, Lee et al. (2013) designed a professional development program (Integration of Science, Mathematics and Reflective Teaching) with middle school mathematics and science teachers. The program focused on improving subject knowledge in science and mathematics and curriculum and inquiry. Teachers from different disciplines were required to work collaboratively throughout the study. According to the results, the participating teachers felt less confidence in teaching the content that they were not experts. Moreover, the collaboration between science and mathematics teachers worked well, and the researchers found it to be effective for successful integration.

One of the points suggested by Lederman and Lederman (2013) was that science teachers connect science and mathematics more easily because they have solid content knowledge in mathematics and the science curriculum allows them to implement mathematics concepts more comfortably. On the other hand, the mathematics teachers had difficulty integrating disciplines since they were unfamiliar with science concepts. It was concluded that working as a team of teachers with different specializations was one of the effective ways to overcome this barrier.

In summary, some studies designed integrated method courses in teacher education programs and examined their effects on pre-service teachers' subject matter knowledge and PCK. In contrast, some studies concentrate on participants' perceptions, attitudes, self-efficacy, and beliefs about integrated science and mathematics teaching. The results of studies displayed that having improved content knowledge in science and mathematics disciplines and PCK for integrated teaching are necessary. The teacher education programs should enable rich opportunities for pre-service teachers to equip them with the necessary knowledge. On the other hand, collaboration and community of practice were common in the studies. Practitioners and pre-service teachers from science and mathematics disciplines worked in groups and learned from each other in some studies. How these features should be infused into teacher education programs needs further research.

MAIN FOCUS OF THE CHAPTER

Science and Mathematics Curriculum Integration

Different definitions and perspectives are available in the literature regarding curriculum integration. Some researchers consider it as a continuum (Brown & Knowles, 2014; Fogarty, 1991), while others as pyramids (Manning & Bucher, 2012). Czerniak et al. (1999) pointed out that the lack of a clear definition makes it hard to conduct research and interpret the results in this area. Moreover, several notions have been used to address integration, such as interdisciplinary, multidisciplinary, and cross-curricular (Frykholm & Glasson, 2005). Springer (2013) remarked his thoughts on this issue as follows:

The central problem inherent in any attempt to analyze curriculum integration, i.e., no unified definition of curriculum integration, has taken hold. No single set pedagogy has been established. Instead, the concept remains a loose philosophy of student-centered education with several general philosophical consistencies but a myriad of actual, practical manifestations (p. 194).

A broad definition of curriculum integration that encompasses alternative views can be described as: "an integrated curriculum enables students to look toward multiple dimensions that reflect the realities of their experiences outside and inside the school" (Venville et al. 2008, p. 860). The researchers argue that narrow definitions might limit the possible powerful approaches regarding integration.

In parallel with the messy terms about curriculum integration, various models for curriculum integration with respect to science and mathematics are found in the literature, and there is no one widely used model (An, 2017; Treacy & O'Donoghue, 2014). Some models were developed specifically to integrate science and mathematics (Davison et al., 1995), whereas some models have started integrating engineering in parallel with the recent reforms (Next Generation Science Standards [NGSS], 2013).

One of the well-known models for science and mathematics integration was suggested by Lonning and DeFranco (1997). They explained that "curricular integration deals with the nature of the relationship between the concepts being drawn from the different disciplines in interdisciplinary curricula (p.212). This model is important since the researchers used the term "interdisciplinary" and many models were developed later by addressing Lonning and DeFranco's study. Moreover, they pointed out two important questions that should be asked when discussing the integration of disciplines: 1) what is being integrated and 2) to what extent each subject is being integrated (p. 212). The model underlines that the activities prepared for integration should be appropriate for students' age and grade level in each curriculum being integrated. Researchers proposed a continuum for science and mathematics integration based on this approach. One side of the curriculum integration involves "independent mathematics" and the opposite involves "independent science". For instance, conducting an activity regarding magnets is independent of mathematical concepts. Between these two extremes are "mathematics focus" and "science focus". For example, an activity about speed concept is appropriate for students in the 6th grade level according to the Turkish science curriculum (MoNE, 2018a). However, the mathematical concepts used in the speed activity are not in the same grade.; they are in 7th grade in the Turkish mathematics curriculum (MoNE, 2018b). According to the model, this type of integration refers to "science focus" integration because science concepts are given priority. There is balanced mathematics and science integration at the center of the continuum. Science and mathematics contents are at the same grade level and the instruction purposefully connects these two disciplines. Similarly, Huntley's model (1998) of integration was built based on the understanding of Lonning and DeFranco's model (1997) and named as "Mathematics/Science" continuum. There are five degrees of integration according to the model; mathematics and science are independent disciplines at the outsets of the continuum, and the degree of integration of science and mathematics is enhanced towards the model's center.

The other model specifically developed for science and mathematics integration was a balanced model (Kiray & Kaptan, 2012). The model's dimensions are content, skills, the teaching and learning process, affective characteristics, and measurement and evaluation. It is a more detailed model when compared to others. The content dimension includes seven sub-components; science and mathematics are separate disciplines at the outsets and total integration at the center of the model. Based on this model, one cannot distinguish whether the class is science or mathematics in total integration; two disciplines should be intertwined. In the right part of the model, two levels of integration are found: science-intensive, mathematics-connected integration (SIMCI) and science-centered, mathematics-assisted integration (SCMAI). In the former one, objectives of the two disciplines might be co-taught, if possible. Reaching the science objectives at the end of the lesson should be teacher's primary concern; reaching the mathematics objectives is optional. In the latter, the weight of mathematics is decreased, the mathematics concepts are used where appropriate and science lessons might be supported with mathematics-related daily life examples even if they are not covered in the mathematics curriculum. The left part of the model is developed with the same understanding.

With respect to integrating different disciplines, Bybee (2013) presented several perspectives on integrating science, engineering, mathematics, and technology. The first perspective advocates that integrating these four disciplines equals one discipline, generally science. The following one asserts that integrating STEM disciplines means using science and mathematics separately. The third perspective takes science as an overarching discipline that encompasses other STEM disciplines, while the fourth perspective includes all STEM disciplines independently and is considered silos. This perspective implies taking four independent courses for each STEM discipline, as in the case of many teacher education programs. The other perspective is called "coordination across disciplines". It refers to using mathematics concepts in science lessons or vice versa—for example, the use of drawing graphs in science lessons with little emphasis. The next one is more complex and needs infusing of two or more STEM disciplines equally. For instance, developing a course on science and mathematics in which both disciplines are used in balance. The overlapping across disciplines perspective integrates STEM disciplines by sequencing them and overlap between disciplines occurs. Students working on a design problem that overlaps through the disciplines could be an example. The most developed one is "STEM as a Transdisciplinary Course," which emphasizes real-world problems like global warming, access to clean water and energy crisis. Solving these problems involves harmonizing the knowledge and skills of all STEM disciplines. The analogy of the "quartet of musicians playing together" was used to explain this perspective (p.79). As seen above, Bybee (2013) presents various perspectives on integrating STEM disciplines and provides insight for future studies.

Kim and Cho (2015) proposed a model to integrate science and mathematics named the "Integrated Concept Understanding Model". The model consists of three main phases. Phase 1 offers "exploring the phenomenon," where teachers should create an environment for students to explore the scientific and mathematical phenomena by drawing upon daily-life experiences (p. 8). Phase 2 is labeled "understanding". The students are expected to connect their experiences from the first phase to scientific and mathematical concepts. Phase 3 is the "application" part, where students are asked to elaborate their understanding and apply the concepts from science and mathematics to new situations.

Challenges of Integration of Science and Mathematics from Teachers' Perspectives

Mathematics and science are considered naturally related disciplines (Pang & Gong, 2000; MacMath et al., 2009; Orton & Roper, 2000); however, there are some challenges in integrating science and mathematics from the perspective of teachers, such as, assessment, teaching practices and etc.

Frykholm and Glason (2005) asked the fundamental question regarding integrated teaching: "What kinds of knowledge structures are necessary for this kind of teaching" (p. 129). The answer to this question addresses two main barriers to integrated teaching: lack of sufficient subject matter knowledge and PCK (An, 2017; Stinson et al., 2009; Frykholm & Glasson, 2005). These two-knowledge bases gain importance when teaching two different disciplines in an integrated way. The common problem arises when teachers are trained to teach the knowledge and practices of one discipline. However, they are expected to have multidisciplinary content knowledge when exposed to an integrated curriculum (Moore et al., 2014). Therefore, they feel adequate to teach the content of other disciplines (Czerniak & Johnson, 2014; Furner & Kumar, 2007; Lederman & Lederman, 2013). Limited experiences for pre-service teachers also cause problems in improving subject matter knowledge from an interdisciplinary perspective (Brown & Bogiages, 2019). The situation is similar to the pedagogical content knowledge (PCK). PCK is defined as "the ways representing and formulating the subject that make it comprehensible to the learner" (Shulman, 1987, p. 9). Teachers are expected to have knowledge of curriculum, instructional strategies, learners, and assessment in their major (Magnusson et al. 1999). Since teachers are educated to teach their respective disciplines in traditional teacher education programs, their PCK for teaching integrated science and mathematics remains undeveloped (Frykholm & Glasson, 2005). For instance, teachers experience difficulty in teaching science while applying the methods of mathematics teaching or vice versa (Lee et al., 2013). The other problem regarding PCK is assessing students' learning in integrated science and mathematics lessons. Teachers feel unprepared to decide which part of the learning should be assessed in integrated lessons and apply formative assessment techniques (Dare et al., 2014). Moreover, teachers have some concerns about the structured curriculum they use, and they reported that flexibility in the curriculum is necessary to integrate different disciplines (Lehman et al., 2014). For example, solving the authentic problem of global warming requires integration of earth, life sciences, and mathematics; however, these topics are not in the same grade level in the science and mathematics curriculum (Lesseig et al., 2016).

Besides problems with knowledge bases, contextual factors also play an essential role. The planning and instruction in an integrated way require teamwork. The other challenges reported in the literature are the school structure, time setting for group work, and increased workload (Dare et al., 2014; Margot & Kettler, 2019; Lesseig et al., 2016). Other obstacles might be listed as preparing appropriate materials for integrated science and mathematics teaching, administrative problems, classroom management issues, and subject-based exams at the national and international levels (Berlin & White, 2010; Johnston et al., 2020; Ní Ríordáin et al., 2016; Park et al., 2017; West et al., 2006). Although teachers are willing to implement an integrated curriculum, these challenges might result in focusing on one discipline they feel is sufficient enough (Judson, 2013). Therefore, studies conducted with teachers and pre-service teachers have vital importance in overcoming these barriers.

RECOMMENDATIONS FOR TEACHER EDUCATION PROGRAMS

Czerniak and Johnson (2014) noted the importance of incorporating integrated approaches in teacher education programs. On the other hand, the ways of supporting pre-service teachers to make connections among disciplines still need further investigation (An, 2017). Pre-service teachers do not feel ready to implement an interdisciplinary curriculum (Gresnigt et al., 2014) because "if nobody teaches students how to integrate the different disciplines, students are prevented from discovering and creating links between relevant subject matter." (p. 73). In other words, pre-service teachers had limited knowledge and experiences in terms integrating science and mathematics (Radloff & Guzey, 2017; Fan & Yu, 2019). The literature demonstrated that when teachers and pre-service teachers are provided with the opportunity to gain experience, their understanding of integrating science and mathematics is progressed (Baxter et al., 2014).

Some studies in literature utilize technology to connect science and mathematics (Johnston et al., 2020), while others use problem-based learning (Cavadas et al., 2022) and engineering design process (Guzey et al., 2017; NGSS, 2013). The present proposal uses engineering as a connector of science and mathematics because infusing engineering provides a fruitful way to connect these two disciplines in teacher education programs (Aydin-Gunbatar et al., 2020; Czerniak & Johnson, 2014). Bryan et al. (2015) pointed out that there should be "anchor" disciplines while integrating different disciplines. Science and mathematics are considered anchor disciplines since the learning objectives are taken from these disciplines in STEM education. Then, the engineering design problem is solved by using the knowledge and skills from these disciplines in STEM lessons. The literature signifies that engineering is a core component of STEM education and effective way of learning science and mathematics concepts and the engineering design process helps to solve real-world problems (Guzey et al., 2014). The critical characteristics of engineering design process are listed as 1) highly iterative, 2) open to the idea that a problem may have many possible solutions, 3) a meaningful context for learning science and mathematics concepts, and 4) a stimulus to systems thinking, modeling and analysis (National Research Council, 2012, p.4). The engineering design process enables to use of knowledge and practices from science and mathematics disciplines, and it allows developing a holistic understanding of the subject to make explicit connections between disciplines. Therefore, the present study recommends revising science and method courses enriched with engineering discipline from the perspective of curriculum integration.

Furthermore, the literature highlighted the importance of developing subject matter knowledge and PCK while integrating science and mathematics, as given in the previous section. Undeveloped PCK is reported as one of the main barriers to integrating science and mathematics disciplines (Fan & Yu, 2019). In addition to having solid subject matter knowledge in science and mathematics, improved PCK is required (Shulman, 1987). The deficiencies in these two knowledge bases should be considered carefully in teacher education programs. In order to address this issue, Magnusson et al. (1999) PCK framework was used to organize the content of the proposed course in the present study, which is one of the widely used models in science education literature (Henze & Barendsen, 2019). This framework underlines that teachers should be knowledgeable about the objectives and the curriculum, learners' prerequisite knowledge, difficulties and misconceptions, variety of instructional strategies, and formative and summative assessment to have strong PCK. While talking about PCK for integrated STEM, Lau and Multani (2018) suggested that "teachers must have the pedagogical content knowledge, or PCK, to help them expand beyond science or mathematics to include defining and delimiting engineering problems, designing solutions, and optimizing designs" (p. 196). Therefore, the method course proposed in this chapter aims to use engineering as a context to effectively integrate science and mathematics and enhance pre-service teachers' PCK for STEM.

In Turkey, middle school science and mathematics pre-service teachers are educated in different departments and take method courses separately. From this point of view, these two method courses might be combined in teacher education programs, and pre-service teachers from science and mathematics education departments might enroll together. The content of the course might focus on making interdisciplinary connections between science and mathematics through applying the engineering design process, collaboration, solving real-world problems by using content and practices from different disciplines, developing curricular materials, and applying formative and summative assessment in an integrated lesson in parallel with Magnusson et al. (1999) PCK framework. In this way, the course provides a rich environment to enhance pre-service teachers' content knowledge in science, mathematics, and engineering and contribute to their development of PCK, which are the main obstacles reported in the literature for effective integration of science and mathematics. The essential parts of the integrated method course are planned to be as follows in Turkey:

Understand

The first step in the course should be to make pre-service teachers understand the basic features of STEM education, engineering and engineering design process, and subject matter knowledge in science (science practices, disciplinary core ideas, crosscutting concepts) and mathematics (i.e., model with mathematics, display numerical data in plots and so on) required to solve authentic problems through the use of engineering design process. Moreover, the pre-service teachers' subject matter knowledge about engineering and engineering design is superficial (Ring et al., 2017); therefore, special attention should be given to this issue in this part.

Engage

Participants should experience and participate in engineering-enriched science and mathematics integrated activities after grasping the key elements indicated in the first step. The real-world, open-ended problems might be chosen from local or global contexts to engage pre-service teachers fully. Half of the activities might concentrate more on science discipline, and the other half might feature mathematics content while integrating disciplines. Engineering should be used as a vehicle in every activity. The pre-service teachers are expected to appreciate the idea of integration.

Collaborate

Collaboration and working with partners give a significant contribution to teacher knowledge development (Loughran, 2004; Marble, 2007) and also overcome the barriers mentioned in the previous section while integrating science, mathematics, and engineering (Rennie et al., 2012). A group of pre-service teachers from science and mathematics fields should work collaboratively and plan common lesson. The collaborative environment enables rich opportunities to discuss, learn, encourage and motivate (Johnston et al., 2020). Pre-service teachers share their ideas and help each other; in other words, they "become a consultant to each other" as Frykholm and Glasson (2005, p. 139) refer. Additionally, using Content Representation (CoRe) as a lesson planning format with pre-service teachers is recommended (Adadan & Oner, 2014; Aydin-Gunbatar et al., 2020; Juhler, 2016) in parallel with Magnusson et al. (1999) PCK framework. The CoRe was initially developed by Loughran et al. (2004) to portray teachers' PCK; however, Aydin et al. (2013) revised it to be used with pre-service teachers. The big idea that is intended to be taught is placed in the rows of CoRe. The fundamental concepts of science and mathematics planned to teach should be chosen and written on these rows. On the other hand, there are nine columns that include such prompts: as a teacher, what should you know about this topic, what misconceptions and difficulties students might experience, how the students' learning is assessed, and the materials, as seen in Table 1. These prompts in CoRe are thought to be helpful in overcoming the barriers, such as teaching out-of-discipline subject matter knowledge and PCK for integrated teaching, including assessment and the materials needed for the integrated science and mathematics lesson.

Table 1. Content Representation (CoRe) (Aydin et al., 2013)

<table>
<tr><td>Topic/Content Area:
Grade Level:</td><td colspan="3">Curriculum objectives
to be addressed:</td></tr>
<tr><td>1. What concepts/big ideas do you intend students to learn?</td><td>Concept/Big Idea 1</td><td>Concept/Big Idea 2</td><td>Concept/Big Idea 3</td></tr>
<tr><td>2. What do you expect students to understand about this concept and be able to do as a result?</td><td colspan="3"></td></tr>
<tr><td>3. Why is it important for students to learn this concept? (Rationale)</td><td colspan="3"></td></tr>
<tr><td>4. As a teacher, what should you know about this topic?</td><td colspan="3"></td></tr>
<tr><td>5. What difficulties do students typically have about each concept/ idea?</td><td colspan="3"></td></tr>
<tr><td>6. What misconceptions do students typically have about each concept/ idea?</td><td colspan="3"></td></tr>
<tr><td>7. Which teaching strategy and what specific activities might be useful for helping students develop an understanding of the concept?</td><td colspan="3"></td></tr>
<tr><td>8. In what ways would you assess students' understanding about this concept?</td><td colspan="3">Formative Assessment:
Summative Evaluation:</td></tr>
<tr><td>9. What materials/equipment are needed to teach the lesson?</td><td colspan="3"></td></tr>
</table>

The group of pre-service teachers would discuss the subject matter knowledge required to teach the topic in science, mathematics, and engineering disciplines, develop teaching materials and prepare an assessment for the integrated lessons with the help of the course instructor. The CoRe should involve objectives from science, mathematics, and engineering disciplines, authentic problem and engineering design process to solve it, appropriate instructional strategy for integrated teaching, and deciding upon which parts of the integrated lesson are worth assessing and what are the ways of assessing them. Collaboration combined with CoRe is considered an effective way of fostering pre-service teachers' subject matter knowledge and PCK to integrate science and mathematics in a method course in the present study. Through collaborative CoRe, pre-service teachers might realize learners' misconceptions and difficulties in science, engineering, and mathematics and consider ways of identifying and eliminating them while planning their integrated lessons. Moreover, while filling out the prompt regarding instructional strategies, pre-service teachers might seek alternative strategies, representations, and activities to integrate science, mathematics, and engineering design process. They should explicitly infuse the engineering design process with science and mathematics content using CoRe because they have to match the big idea of the lesson and instructional strategies. For instance, one of the big ideas of the lesson might be related to "designing the product to prevent heat-loss in buildings by using engineering design process". In this way, pre-service teachers should plan effective strategies to employ the engineering design process, such as using open-ended and authentic problems and relevant science and mathematics content to solve the given problem. Moreover, pre-service teachers should pay attention to how they should assess science, engineering, and mathematics contents while designing the collaborative CoRe. In this way, they are expected to develop a holistic understanding of integrating science and mathematics with the help of engineering and augment their PCK for integrated teaching. Therefore, CoRe is recommended in this chapter to support pre-service teachers in planning a lesson in out-of-discipline content.

Apply

This step includes implementing at least one integrated lesson plan in a real classroom, if possible. If it is impossible, micro-teaching sessions should be organized in the context of an integrated method course. Since experience is another essential factor contributing to improving subject matter knowledge and PCK (Aydin et al., 2013; Juhler, 2016), it should be a part of the integrated method course. One pre-service teacher in the group should carry out the collaboratively designed CoRe while the other group members observe and take field notes to be used in the reflection step.

Reflect

The last step is reflecting on the learning process, which also shapes pre-service teachers' PCK (Lau & Multani, 2018; Nilsson, 2008). After collaboration and application steps, pre-service teachers should reflect upon what they have learned in terms of subject matter knowledge in science and mathematics and incorporate the engineering design process to integrate these two disciplines. Additionally, pre-service teachers should share their observations and reflections about teaching strategy, objectives, and big ideas of the lesson and assessment. Feedback should be provided to pre-service teachers by teacher educators at this step.

FUTURE RESEARCH DIRECTIONS

This study aims to provide an initial model for integrating science and mathematics teaching method courses. The proposed course is PCK-based and enriched with the engineering design process. The model has not been used practically in a teacher education program; therefore, future studies are recommended to implement it and report their findings. The literature calls for studies with replicable methods regarding science and mathematics integration (An, 2017). The model of integrated method lesson is prepared by considering the situation in the context of Turkey; however, researchers from different contexts might modify the proposal based on their needs and apply it.

Another point worth investigating is how pre-service teachers' subject matter knowledge and PCK develop after completing the integrated method course proposed in the present study. Because it contains the collaboration of pre-service teachers from different disciplines, teaching and reflection phases are significant sources of improved PCK for integrated teaching. To achieve this purpose, pre-service teachers might be asked to prepare CoRe at different times of the study (i.e., at the beginning, middle, and end of the study) to track their developments. Moreover, observation protocols, reflection papers, pre and post-interviews, and content knowledge tests in science and mathematics might be used as data collection tools to support the data coming from CoRe's.

Pre-service teachers and mentors (course instructors) work with each other in the proposed model. Future studies might increase the number of "Apply" sessions in real classroom environments, and experienced science and mathematics teachers might be invited to the collaboration groups with pre-service teachers. Future studies might concentrate on the effect of collaboration between experienced and pre-service teachers from science and mathematics disciplines on creating and implementing an integrated curriculum.

CONCLUSION

This chapter focused on the significance of an integrated curriculum for science and mathematics with the recent educational reforms. Some of the science and mathematics integration models were examined, and the current research on the integration of science and mathematics with respect to teacher education models was provided. The main obstacles reported in these studies were summarized, and a model of integrated teaching method course by considering the drawbacks in the literature was offered at the end of the chapter. The primary purpose of the proposed course is to improve pre-service teachers' subject matter knowledge in science and mathematics disciplines and PCK for teaching an integrated curriculum. The proposed course has five steps: understand, engage, collaborate, apply and reflect. The distinctive features of the proposed course are: (1) the use of CoRe as a lesson planning tool, (2) allowing both pre-service teachers from science and mathematics fields to work and learn together, and (3) require implementation in the real classroom context and (4) use the engineering design process as a bridge to connect science and mathematics effectively. It could be concluded that re-visiting the method courses in teacher education programs is critical when the relevant literature is examined, and the present study proposes a model of integrated method courses. Recommendations for research and practice are provided at the end of the chapter.

REFERENCES

Adadan, E., & Oner, D. (2014). Exploring the progression in preservice chemistry teachers' pedagogical content knowledge representations: The case of "behavior of gases". *Research in Science Education, 44*(6), 829–858. doi:10.100711165-014-9401-6

An, S. A. (2017). Preservice teachers' knowledge of interdisciplinary pedagogy: The case of elementary mathematics–science integrated lessons. *ZDM Mathematics Education, 49*(2), 237–248. doi:10.100711858-016-0821-9

Aydin, S., Demirdogen, B., Tarkin, A., Kutucu, S., Ekiz, B., Akin, F. N., Tuysuz, M., & Uzuntiryaki, E. (2013). Providing a set of research-based practices to support preservice teachers' long-term professional development as learners of science teaching. *Science Education, 97*(6), 903–935. doi:10.1002ce.21080

Aydin-Gunbatar, S., Ekiz-Kiran, B., & Oztay, E. S. (2020). Pre-service chemistry teachers' pedagogical content knowledge for integrated STEM development with LESMeR model. *Chemistry Education Research and Practice, 21*(4), 1063–1082. doi:10.1039/D0RP00074D

Baxter, J. A., Ruzicka, A., Beghetto, R. A., & Livelybrooks, D. (2014). Professional development strategically connecting mathematics and science: The impact on teachers' confidence and practice. *School Science and Mathematics, 114*(3), 102–113. doi:10.1111sm.12060

Beane, J. A. (1995). Curriculum integration and the disciplines of knowledge. *Phi Delta Kappan, 76*(8), 616–622.

Berlin, D. F., & White, A. L. (1994). The Berlin-White integration science and mathematics model. *School Science and Mathematics, 94*(1), 12–14. doi:10.1111/j.1949-8594.1994.tb12280.x

Berlin, D. F., & White, A. L. (2010). Preservice mathematics and science teachers in an integrated teacher preparation program for grades 7–12: A 3-year study of attitudes and perceptions related to integration. *International Journal of Science and Mathematics Education, 8*(1), 97–115. doi:10.100710763-009-9164-0

Berlin, D. F., & White, A. L. (2012). A longitudinal look at attitudes and perceptions related to the integration of mathematics, science, and technology education. *School Science and Mathematics, 112*(1), 20–30. doi:10.1111/j.1949-8594.2011.00111.x

Bosse, M., Lee, T., Swinson, M., & Faulconer, J. (2010). The NCTM process standards and the five Es of science: Connecting math and science. *School Science and Mathematics, 110*(5), 262–277. doi:10.1111/j.1949-8594.2010.00033.x

Breiner, J. M., Harkness, S. S., Johnson, C. C., & Koehler, C. M. (2012). What is STEM? A discussion about conceptions of STEM in education and partnerships. *School Science and Mathematics, 112*(1), 3–11. doi:10.1111/j.1949-8594.2011.00109.x

Brown, D., & Knowles, T. (2014). *What every middle school teacher should know* (3rd ed.). Heinemann.

Brown, D. F. (2016). Curriculum integration. In S. B. Mertens, M. Caskey, & N. Flowers (Eds.), *The Encyclopedia of Middle Grades Education* (2nd ed., pp. 122–125). Information Age Publishing.

Brown, R. E., & Bogiages, C. A. (2019). Professional development through STEM integration: How early career math and science teachers respond to experiencing integrated STEM tasks. *International Journal of Science and Mathematics Education, 17*(1), 111–128. doi:10.100710763-017-9863-x

Bryan, L. A., Moore, T. J., Johnson, C. C., & Roehrig, G. H. (2015). Integrated STEM education. In C. C. Johnson, E. E. Peters-Burton, & T. J. Moore (Eds.), *STEM road map: A framework for integrated STEM education* (pp. 23–37). Routledge. doi:10.4324/9781315753157-3

Bybee, R. (2013). *The case of STEM education: Challenges and opportunities*. NSTA Press.

Cady, J. A., & Rearden, K. (2007). Pre-service teachers' beliefs about knowledge, mathematics, and science. *School Science and Mathematics, 107*(6), 237–245. doi:10.1111/j.1949-8594.2007.tb18285.x

Cavadas, B., Rézio, S., Nogueira, J. R., & Branco, N. (2022). A framework and a research design proposal to identify preservice teachers' integration performance of science and mathematics. *Canadian Journal of Science. Mathematics and Technology Education, 22*(1), 101–129. doi:10.100742330-022-00198-2

Czerniak, C. M., & Johnson, C. C. (2014). Interdisciplinary science teaching. In S. K. Abell & N. G. Lederman (Eds.), *Handbook of Research on Science Education* (2nd ed., pp. 395–411). Routledge.

Dare, E. A., Ellis, J. A., & Roehrig, G. H. (2014). Driven by beliefs: Understanding challenges physical science teachers face when integrating engineering and physics. *Journal of Pre-College Engineering Education Research, 4*(2), 47–61. doi:10.7771/2157-9288.1098

Davison, D. M., Miller, K. W., & Metheny, D. L. (1995). What does integration of science and mathematics really mean? *School Science and Mathematics, 95*(5), 226–230. doi:10.1111/j.1949-8594.1995.tb15771.x

Fan, S. C., & Yu, K. C. (2019). Teaching engineering-focused STEM curriculum: PCK needed for teachers. In Y. S. Hsu & Y. F. Yeh (Eds.), *Asia-Pacific STEM Teaching Practices* (pp. 103–116). Springer. doi:10.1007/978-981-15-0768-7_7

Fogarty, R. (1991). Ten ways to integrate curriculum. *Educational Leadership, 49*(2), 61–65.

Frykholm, J., & Glasson, G. (2005). Connecting science and mathematics instruction: Pedagogical context knowledge for teachers. *School Science and Mathematics, 105*(3), 127–141. doi:10.1111/j.1949-8594.2005.tb18047.x

Furner, J. M., & Kumar, D. D. (2007). The mathematics and science integration argument: A stand for teacher education. *Eurasia Journal of Mathematics, Science and Technology Education, 3*(3), 185–189. doi:10.12973/ejmste/75397

Gresnigt, R., Taconis, R., van Keulen, H., Gravemeijer, K., & Baartman, L. (2014). Promoting science and technology in primary education: A review of integrated curricula. *Studies in Science Education, 50*(1), 47–84. doi:10.1080/03057267.2013.877694

Guzey, S. S., Harwell, M., Moreno, M., Peralta, Y., & Moore, T. J. (2017). The impact of design-based STEM integration curricula on student achievement in engineering, science, and mathematics. *Journal of Science Education and Technology, 26*(2), 207–222. doi:10.100710956-016-9673-x

Guzey, S. S., Tank, K., Wang, H. H., Roehrig, G., & Moore, T. (2014). A high-quality professional development for teachers of grades 3–6 for implementing engineering into classrooms. *School Science and Mathematics, 114*(3), 139–149. doi:10.1111sm.12061

Henze, I., & Barendsen, E. (2019). Unravelling student science teachers' pPCK development and the influence of personal factors using authentic data sources. In A. Hume, R. Cooper, & A. Borowski (Eds.), *Repositioning pedagogical content knowledge in teachers' knowledge for teaching science* (pp. 201–221). Springer. doi:10.1007/978-981-13-5898-2_9

Honey, M., Pearson, G., & Schweingruber, A. (2014). *STEM integration in K-12. Status, Prospects, and an agenda for research.* National Academies Press.

Huntley, M. A. (1998). Design and implementation of framework for defining integrated mathematics and science education. *School Science and Mathematics, 98*(1), 320–327. doi:10.1111/j.1949-8594.1998.tb17427.x

Hurd, P. D. (1991). Why we must transform science education. *Educational Leadership, 49*(2), 33–35.

Johnston, J., Walshe, G., & Ríordáin, M. N. (2020). Supporting key aspects of practice in making mathematics explicit in science lessons. *International Journal of Science and Mathematics Education, 18*(7), 1399–1417. doi:10.100710763-019-10016-1

Judson, E. (2013). Development of an Instrument to Assess and Deliberate on the Integration of Mathematics into Student-Centered Science Learning. *School Science and Mathematics, 113*(2), 56–68. doi:10.1111sm.12004

Juhler, M. V. (2016). The use of lesson study combined with content representation in the planning of physics lessons during field practice to develop pedagogical content knowledge. *Journal of Science Teacher Education, 27*(5), 533–553. doi:10.100710972-016-9473-4

Kim, D., & Bolger, M. (2017). Analysis of Korean elementary pre-service teachers' changing attitudes about integrated STEAM pedagogy through developing lesson plans. *International Journal of Science and Mathematics Education, 15*(4), 587–605. doi:10.100710763-015-9709-3

Kim, M. K., & Cho, M. K. (2015). Design and implementation of integrated instruction of mathematics and science in Korea. *Eurasia Journal of Mathematics, Science and Technology Education, 11*(1), 3–15. doi:10.12973/eurasia.2015.1301a

Kiray, S. A., & Kaptan, F. (2012). A new model for the integration of science and mathematics: The balance model. *Energy Education Science and Technology, Part B. Social and Educational Studies, 4*(3), 1181–1196.

Koirala, H. P., & Bowman, J. K. (2003). Preparing middle level preservice teachers to integrate mathematics and science: Problems and possibilities. *School Science and Mathematics, 103*(3), 145–154. doi:10.1111/j.1949-8594.2003.tb18231.x

Krajcik, J., & Czerniak, C. M. (2014). *Teaching science to children: A project-based science approach.* Routledge. doi:10.4324/9780203113660-7

Lau, M., & Multani, S. (2018). Engineering STEM teacher learning: Using a museum-based field experience to foster STEM teachers' pedagogical content knowledge for engineering. In S. M. Uzzo, S. B. Graves, E. Shay, M. Harford, & R. Thompson (Eds.), *Pedagogical content knowledge in STEM* (pp. 195–213). Springer. doi:10.1007/978-3-319-97475-0_11

Lederman, N. G., & Lederman, J. S. (2013). Is it STEM or S & M that we truly love? *Journal of Science Teacher Education, 24*(8), 1237–1240. doi:10.100710972-013-9370-z

Lee, M. M., Chauvot, J. B., Vowell, J., Culpepper, S. M., & Plankis, B. J. (2013). Stepping into iSMART: Understanding science–mathematics integration for middle school science and mathematics teachers. *School Science and Mathematics, 113*(4), 159–169. doi:10.1111sm.12015

Lehman, J. D., Kim, W., & Harris, C. (2014). Collaborations in a community of practice working to integrate engineering design in elementary science education. *Journal of STEM Education: Innovations and Research, 15*(3), 21–28.

Lesseig, K., Nelson, T. H., Slavit, D., & Seidel, R. A. (2016). Supporting middle school teachers' implementation of STEM design challenges. *School Science and Mathematics, 116*(4), 177–188. doi:10.1111sm.12172

Lonning, R. A., & DeFranco, T. C. (1997). Integration of science and mathematics: A theoretical model. *School Science and Mathematics, 97*(4), 212–215. doi:10.1111/j.1949-8594.1997.tb17369.x

Loughran, J., Mulhall, P., & Berry, A. (2004). In search of pedagogical content knowledge in science: Developing ways of articulating and documenting professional practice. *Journal of Research in Science Teaching, 41*(4), 370–391. doi:10.1002/tea.20007

MacMath, S., Wallace, J., & Chi, X. (2009). Curriculum integration: Opportunities to maximize assessment as, of, and for learning. *McGill Journal of Education, 44*(3), 451–466. doi:10.7202/039949ar

Magnusson, S., Krajcik, J., & Borko, H. (1999). Nature, sources and development of pedagogical content knowledge for science teaching. In J. Gess-Newsome & N. G. Lederman (Eds.), *Examining pedagogical content knowledge: The construct and its implications for science education* (pp. 95–132). Kluwer Academic Publishers.

Manning, M. L., & Bucher, K. T. (2012). *Teaching in the middle school* (4th ed.). Pearson.

Marble, S. (2007). Inquiring into teaching: Lesson study in elementary science methods. *Journal of Science Teacher Education, 18*(6), 935–953. doi:10.100710972-007-9071-6

Margot, K. C., & Kettler, T. (2019). Teachers' perception of STEM integration and education: A systematic literature review. *International Journal of STEM Education, 6*(1), 1–16. doi:10.118640594-018-0151-2

McHugh, L., Kelly, A. M., & Burghardt, M. D. (2018). Professional development for a middle school mathematics infused science curriculum. *Journal of Science Teacher Education, 29*(8), 804–828. doi:10.1080/1046560X.2018.1514825

Ministry of National Education. (2018a). *Fen bilimleri dersi öğretim programı (ilkokul ve ortaokul 3, 4, 5, 6, 7, ve 8. sınıflar)* [Science course curriculum (Primary and middle school 3, 4, 5, 6, 7, and 8th grades)]. Ministry of National Education.

Ministry of National Education. (2018b). *Matematik dersi öğretim programı (ilkokul ve ortaokul 3, 4, 5, 6, 7, ve 8. sınıflar)* [Mathematics course curriculum (Primary and middle school 3, 4, 5, 6, 7, and 8th grades)]. Ministry of National Education.

National Research Council. (2012). *A framework for K-12 science education: Practices, crosscutting concepts, and core ideas*. The National Academies Press.

Next Generation Science Standards Lead States. (2013). *Next generation science standards: For states, by states*. National Academies Press.

Ní Ríordáin, M., Johnston, J., & Walshe, G. (2016). Making mathematics and science integration happen: Key aspects of practice. *International Journal of Mathematical Education in Science and Technology, 46*(2), 233–255. doi:10.1080/0020739X.2015.1078001

Nilsson, P. (2008). Teaching for understanding: The complex nature of pedagogical content knowledge in preservice education. *International Journal of Science Education, 30*(10), 1281–1299. doi:10.1080/09500690802186993

Offer, J., & Vasquez-Mireles, S. (2009). Mix it up: Teachers' beliefs on mixing mathematics and science. *School Science and Mathematics, 109*(3), 146–152. doi:10.1111/j.1949-8594.2009.tb17950.x

Orton, T., & Roper, T. (2000). Science and mathematics: A relationship in need of counselling? *Studies in Science Education, 35*(1), 123–153. doi:10.1080/03057260008560157

Pang, J. S., & Good, R. (2000). A review of the integration of mathematics and science: Implications for further research. *School Science and Mathematics, 100*(2), 73–82. doi:10.1111/j.1949-8594.2000.tb17239.x

Park, M., Dimitrov, D. M., Patterson, L. G., & Park, D. (2017). Early childhood teachers' beliefs about readiness for teaching science, technology, engineering, and mathematics. *Journal of Early Childhood Research, 15*(3), 275–291. doi:10.1177/1476718X15614040

Radloff, J., & Guzey, S. (2017). Investigating changes in preservice teachers' conceptions of STEM education following video analysis and reflection. *School Science and Mathematics, 117*(3-4), 158–167. doi:10.1111sm.12218

Rennie, L., Venville, G., & Wallace, J. (2012). *Integrating science, technology, engineering, and mathematics: Issues, reflections, and ways forward*. Routledge. doi:10.4324/9780203803899

Ring, E. A., Dare, E. A., Crotty, E. A., & Roehrig, G. H. (2017). The evolution of teacher conceptions of STEM education throughout an intensive professional development experience. *Journal of Science Teacher Education, 28*(5), 444–467. doi:10.1080/1046560X.2017.1356671

Ríordáin, M. N., Johnston, J., & Walshe, G. (2016). Making mathematics and science integration happen: Key aspects of practice. *International Journal of Mathematical Education in Science and Technology, 47*(2), 233–255. doi:10.1080/0020739X.2015.1078001

Sackes, M., Flevares, L. M., Gonya, J., & Trundle, K. C. (2012). Preservice early childhood teachers' sense of efficacy for integrating mathematics and science: Impact of a methods course. *Journal of Early Childhood Teacher Education, 33*(4), 349–364. doi:10.1080/10901027.2012.732666

Schoenfeld, A. H. (2004). Multiple learning communities: Students, teachers, instructional designers, and researchers. *Journal of Curriculum Studies*, *36*(2), 237–255. doi:10.1080/0022027032000145561

Shulman, L. S. (1987). Knowledge and training: Foundations of the new reform. *Harvard Educational Review*, *57*(1), 1–22. doi:10.17763/haer.57.1.j463w79r56455411

Springer, M. (2013). Charting the course of curriculum integration. In P. G. Andrews (Ed.), *Research to guide practice in middle grades education* (pp. 187–216). Association for Middle Level Education.

Stinson, K., Harkness, S. S., Meyer, H., & Stallworth, J. (2009). Mathematics and science integration: Models and characterizations. *School Science and Mathematics*, *109*(3), 153–161. doi:10.1111/j.1949-8594.2009.tb17951.x

Treacy, P., & O'Donoghue, J. (2014). Authentic integration: A model for integrating mathematics and science in the classroom. *International Journal of Mathematical Education in Science and Technology*, *45*(5), 703–718. doi:10.1080/0020739X.2013.868543

Venville, G., Sheffield, R., Rennie, L., & Wallace, J. (2008). The writing on the classroom wall: The effect of school context on learning in integrated, community-based science projects. *Journal of Research in Science Teaching*, *45*, 857–880. doi:10.1002/tea.20245

West, S., Vásquez-Mireles, S., & Coker, C. (2006). Math and/or science education: Separate or integrate? *Journal of Mathematical Science and Mathematics Education*, *1*(2), 11–18.

ADDITIONAL READING

Bartels, S. L., Rupe, K. M., & Lederman, J. S. (2019). Shaping preservice teachers' understandings of STEM: A collaborative math and science methods approach. *Journal of Science Teacher Education*, *30*(6), 666–680. doi:10.1080/1046560X.2019.1602803

Capobianco, B. M., DeLisi, J., & Radloff, J. (2018). Characterizing elementary teachers' enactment of high-leverage practices through engineering design-based science instruction. *Science Education*, *102*(2), 342–376. doi:10.1002ce.21325

Carlson, J., & Daehler, K. R. (2019). The refined consensus model of pedagogical content knowledge in science education. In A. Hume, R. Cooper, & A. Borowski (Eds.), *Repositioning pedagogical content knowledge in teachers' knowledge for teaching science* (pp. 77–92). Springer. doi:10.1007/978-981-13-5898-2_2

Guzey, S. S., Tank, K., Wang, H. H., Roehrig, G., & Moore, T. (2014). A high-quality professional development for teachers of grades 3–6 for implementing engineering into classrooms. *School Science and Mathematics*, *114*(3), 139–149. doi:10.1111sm.12061

Hynes, M. M. (2012). Middle-school teachers' understanding and teaching of the engineering design process: A look at subject matter and pedagogical content knowledge. *International Journal of Technology and Design Education*, *22*(3), 345–360. doi:10.100710798-010-9142-4

Maiorca, C., & Mohr-Schroeder, M. J. (2020). Elementary preservice teachers' integration of engineering into STEM lesson plans. *School Science and Mathematics*, *120*(7), 402–412. doi:10.1111sm.12433

Peterman, K., Daugherty, J. L., Custer, R. L., & Ross, J. M. (2017). Analyzing the integration of engineering in science lessons with the Engineering-Infused Lesson Rubric. *International Journal of Science Education*, *39*(14), 1913–1931. doi:10.1080/09500693.2017.1359431

Srikoom, W., Faikhamta, C., & Hanuscin, D. L. (2018). Dimensions of effective STEM integrated teaching practice. *K-12 STEM Education*, *4*(2), 313-330.

KEY TERMS AND DEFINITIONS

Curriculum Integration: Using different disciplines to solve the real-world and complicated problems.

Engineering Design Process: An iterative process used to solve a problem from daily life.

Science and Mathematics Integration: Integrating the knowledge and practices of science and mathematics through the use of the engineering design process and two disciplines are harmonized equally.

STEM Education: Integrating science and engineering disciplines and supporting the integration with mathematics and technology disciplines.

Chapter 12
Mathematics in Environmental Issues:
Research-Based Interdisciplinary Practices in Preservice Teacher Education

Maria João Silva
https://orcid.org/0000-0003-1017-8315
CIED, Escola Superior de Educação de Lisboa, Instituto Politécnico de Lisboa, Portugal

Margarida Rodrigues
https://orcid.org/0000-0003-4658-6281
CIED, Escola Superior de Educação, Instituto Politécnico de Lisboa, Portugal & UIDEF, Instituto de Educação, Universidade de Lisboa, Portugal

Nuno Melo
CIED, Escola Superior de Educação, Instituto Politécnico de Lisboa, Portugal & UIDEF, Instituto de Educação, Universidade de Lisboa, Portugal

Pedro Sarreira
CIED, Escola Superior de Educação, Instituto Politécnico de Lisboa, Portugal & UIDEF, Instituto de Educação, Universidade de Lisboa, Portugal

ABSTRACT

In this chapter, the authors present and assess, in the context of initial teacher education, the process and results of the implementation of an interdisciplinary didactic sequence on environmental and mathematics education, developed to promote preservice teachers research-based interdisciplinarity. The data were gathered, using participant observation and the reports produced by the preservice teachers' groups. The collected data were treated using content analysis techniques. The results of the participant observation and the analysis of the preservice teachers report validated the research-based interdisciplinary didactic sequence on environmental and mathematics education. The critical and research-based activities, reflection, and assessment, developed by preservice teachers, allowed the production of knowledge to improve practices in higher and basic education contexts.

DOI: 10.4018/978-1-6684-5765-8.ch012

INTRODUCTION

Collaboration, communication, reasoning, critical thinking, problem-solving, creativity and innovation are internationally recognized competencies that should be developed by students to face the complex challenges of a world changing in an accelerated way (Biseth et al., 2022; Ng, 2019; World Economic Forum, 2016). Those competencies are also included in the Portuguese referential *Students' Profile by the End of Compulsory Schooling* (Martins et al., 2017). The referred challenges of our changing world include climate change, the recent COVID-19 pandemic, technological innovation, and ubiquity, as well as the threatening of fake news to a democratic and literate world.

To develop the abovementioned 21st century competences, and to achieve the United Nations Sustainable Development Goals (United Nations, 2022), there is a need for the integration of interdisciplinarity in the curriculum, making it possible the development of students' interdisciplinary thinking in solving real complex problems (Drake & Reid, 2020; Ng, 2019).

Accordingly, the importance of interdisciplinarity in preservice Teacher Education, namely between natural sciences and mathematics, is internationally acknowledged (Geiger et al., 2018; An, 2017). In this context, the need for interdisciplinary activities in preservice Teacher Education is twofold, in what concerns (Geiger et al., 2018): i) supporting preservice teachers in experiencing and evaluating authentic learning activities that cross diverse subject areas; and ii) developing preservice teachers' interdisciplinary pedagogical content knowledge.

The relevance of interdisciplinarity in preservice Teacher Education in Portugal is also supported by the Portuguese curriculum for basic education, which emphasizes essential knowledge and structuring skills for citizenship and sustainable development, with a focus on local and real-life contexts (OECD, 2018).

To develop students' competences for the 21st century, preservice teachers must have themselves competences for critical reflection, innovation, continuous learning, and change (Aspfors & Eklund, 2017). To support and mediate the development of those competences, Teacher Education should be research-based, integrating research-based knowledge, research-related activities, and practically oriented requirements associated to the practical nature of teacher´s daily work (Aspfors & Eklund, 2017).

In this chapter, the authors present and assess the investigative process and results of the implementation of a didactic sequence on environmental and mathematics education, in the context of Initial Teacher Education in Portugal. The following research questions guided the investigative process: i) How do the preservice teachers, throughout the didactic sequence, design, implement and assess authentic interdisciplinary learning activities for children? ii) What shared concepts are included by the preservice teachers in the authentic learning activities for children, to bring disciplinary subjects together? and iii) How does the didactic sequence contribute to a research-based education of preservice teachers?

BACKGROUND

Disciplinary Boundary Crossing

Interdisciplinarity between mathematics and natural sciences has a fundamental role in promoting a holistic and integrated view of reality and of its complexity (Beane, 2000; Morin, 2001). This integrated view is needed to develop cross-disciplinary 21st century competencies (Biseth et al., 2022; Rodrigues et al., 2019).

Disciplinary boundary crossing, with interactions between different types of knowledge and

transitions of individuals across communities and their interactions with new and different ideas and cultural norms (…) are of interest within educational contexts because of the potential for learning at intersection between communities who create and value different types of knowledge (Geiger et al., 2018, pp. 221-222).

In this chapter, the authors adopt the following working concepts, as presented by Ng (2019):

- **Multidisciplinary approach,** as the use of a theme to support teachers and students in connecting disciplinary subjects, which can be taught separately;
- **Interdisciplinary approach,** as the use of a shared concept to bring disciplinary subjects together, without separation of disciplinary knowledge and skills.

The relevance of disciplinary boundary crossing in curriculum was stressed, in the 1930's, by the progressive education movement, pioneered by Dewey, linking real-world problem solving with citizenship and democracy in schools and society (Beane, 1997; Drake & Reid, 2020). Since then, until today, there is a rich history of interdisciplinarity and integrative curriculum (Beane, 1997; Drake & Reid, 2020). However, an integrated curriculum has not yet been widely implemented at the national level, prevailing the disciplinary logic that compartmentalizes knowledge, showing the gap between different curriculum levels, namely the prescript curriculum and the implemented curriculum (Sacristán, 2000; Ponte, 2020). Nowadays, the need for disciplinary boundary crossing is emphasized by the recognition of the cross disciplinary 21st century competences.

Basic education in Portugal emphasizes an essential core curriculum, aiming "to ensure a mastery of core disciplinary subjects, while at the same time allowing space for interdisciplinary learning built on core subject mastery" (DGE, 2018, 2021; OECD, 2018, p. 9). To achieve the competences outlined in the *Students' Profile by the End of Compulsory Schooling* (Martins et al., 2017), the Portuguese Ministry of Education supports, namely by curricular autonomy and flexibility, the improvement of learning and teaching quality, including interdisciplinary teaching and development of projects that combine the cross-curriculum competences for different subjects (OECD, 2018).

In this context, Teacher Education must address interdisciplinary teaching and learning (Geiger et al., 2018; OECD, 2018; An, 2017). This way, there is a need to overcome the traditional isolation of subject didactics to the necessary integration in a curriculum with macro-purposes, namely the abovementioned knowledge and structuring skills for citizenship and sustainable development (Roldão, 2017).

An (2017) advocates that teachers need to develop an interdisciplinary pedagogical content knowledge (IPCK) to: i) approach themes that require boundary crossing, connecting disciplinary subjects; and ii) develop interdisciplinary themed activities, making it possible for students to link knowledge across curricula, which implies addressing content areas from multiple subjects simultaneously and identifying knowledge connections within and between particular subjects. IPCK is situated in the intersection of teachers' Pedagogical Knowledge, Pedagogical Content Knowledge of different subjects, and Interdisciplinary Content Knowledge, being specified as an explicit knowledge of interdisciplinary pedagogy (An, 2017).

The study conducted by An (2017), with 28 elementary preservice teachers mathematics–science, showed that their knowledge of using science-themed activities as instructional approaches for teaching mathematics had changed across several science content areas and revealed that the contextualized educational activities were used as resources for developing teachers' capacity for designing mathematics lessons. Accordingly, and consistently, the research on interdisciplinary teaching and learning has been emphasizing the importance of authentic didactical approaches, such as Inquiry Based Learning, Problem Based Learning, Project Based Learning, and Design Based Learning to effectively develop competences as problem-solving, critical thinking, collaboration, and creativity (Bean, 1997; Ng, 2019; Drake & Reid, 2020). Inquiry Based Learning is centred on the process of searching for the answer of one or more questions (Goldman et al., 2010). According to Yew and Goh (2016), Problem Based Learning is an iterative and collaborative process that includes problem analysis, self-directed learning and solutions reporting. Furthermore, the authors of this chapter adopt the vision of Merritt et al. (2017), which argue that Project Based Learning is to learn through projects focused on problems in their real scenarios and that it can, in this way, coincide with Problem Based Learning. Design Based Learning is learning centred on the creation of artifacts, systems and solutions, using innovation and processes of Inquiry and Problem Based Learning (Gómez Puente et al., 2013).

The stated didactical approaches are based on social constructivism theories and can be reified in multiple ways, namely depending on students' ages, and scientific domains (Goldman et al., 2010; Yew and Goh, 2016; Merritt et al., 2016; Gómez Puente et al., 2013). In this chapter, those approaches are centred on real problems, and situated in the local environment and scholar community.

The research on integration of natural sciences and mathematics in preservice Teacher Education in Portugal confirms and situates, at a national context, the abovementioned international findings (Rodrigues et al., 2019; Cavadas et al., 2022; Correia & Baptista, 2022). In the research presented in the present chapter, the integration of natural sciences and mathematics is used to address environmental issues in Initial Teacher Education, using multi and interdisciplinary approaches.

Research-based Teacher Education

To develop teachers' critical thinking and innovation competences, needed in nowadays changing contexts, the European Union (2015) recognizes and emphasizes the need for an Initial Teacher Education in which preservice teachers use and participate in research in their learning and teaching practices. Initial Teacher Education should create opportunities for "practice-oriented and research-based professional development that will strengthen the agency (capacity for action) of teachers for learner-oriented teaching and innovation" (European Commission, 2015, p. 3).

Research-based Teacher Education is needed to support preservice teachers in developing a reflective practice that requires: i) the use of previous theoretically and practical knowledge; ii) the problematizing theorization of the practical situation; and iii) the local production of knowledge to be used in other theoretical and practical contexts (Roldão, 2021). According to Aspfors and Eklund (2017), in Initial Teacher Education, the research-based approach "encourages the development of student teachers' critical reflection abilities and the systematic scrutiny of their daily work" (p.403) and uses research results in teaching and learning activities. These authors developed a longitudinal study aimed at investigating, over a 5-year period, newly qualified teachers' experiences in their school reality. The study was focused on 10 newly qualified primary school teachers, who were educated at a Teacher Education department at a university in Finland. The results of this study demonstrate that the research-based Teacher Education provided a solid basis for the future teacher profession, encouraging the participants to become innovative.

In this chapter, the authors adopt a research-based approach in Initial Teacher Education, using the processes, results, and products of previous research/intervention projects, to create a didactic sequence for a class of preservice teachers, in the context of an Initial Teacher Education master's degree. That didactic sequence uses and challenges preservice teachers to make use of those previous projects' products, as boundary objects that support bridging the disciplinary boundary crossing (Geiger et al., 2018) in interdisciplinary investigative teaching and learning activities.

METHODOLOGY

Purpose

The purpose of the research presented in this chapter was to assess the process and results of the implementation of an interdisciplinary didactic sequence on environmental and mathematical education, in the context of preservice Teacher Education.

Research Design

Participants

This research was developed in 2021/22 with a class of a Curricular Unit named Mathematics in Environmental Issues of an Initial Teacher Education master's degree running at Escola Superior de Educação de Lisboa (ESELx). This school is located on a hill of approximately 11ha in the city of Lisbon (Campus de Benfica), close to the Monsanto Forest Park, but surrounded by high-intensity traffic roads. On the Campus of Benfica there are two more higher education schools, parking lots and some green spaces, with multiple native and exotic species.

The aforementioned master's degree certifies for teaching in primary schools (grade 1 to 4) and for teaching Mathematics and Natural Sciences in grades 5 and 6. The participants were 17 preservice teachers, 16 women and 1 man, as well as a 5th grade class of a partner school with whom the activities designed by preservice teachers were implemented.

Preservice teachers' anonymity is maintained, in line with the ethical criterion of confidentiality.

Didactic Sequence

The implemented didactic sequence has been developed based on the teaching of the Mathematics in Environmental Issues Curricular Unit, in five consecutive school years, since 2017/18, by the first two authors of this chapter, which are the 2 Curricular Unit teachers (2 women), being one environmental education teacher/researcher and one didactics of mathematics teacher/researcher). It is an interdisciplinary didactic sequence on environmental and mathematics education, in the context of preservice Teacher Education.

The didactic sequence: i) is student-centred (preservice teachers are challenged to autonomously explore local and real-life environmental contexts, using sensory and instrumental approaches, as well as to critically reflect on related research bibliography); ii) adopts a design-based strategy (Plomp, 2013) that includes an action-research approach (Creswell, 2008), in which the groups of preservice teachers are asked to design interdisciplinary educational activities involving mathematics and natural sciences, to be implemented, assessed, and improved with 5th grade children; iii) uses resources and knowledge produced by previous intervention and research environmental projects, such as a toolkit that supports the use of sensors in environmental exploration activities, the native species planted on the Campus, the online information about those species, an eco-track that guides the exploration of plants in the neighbourhoods, and the school's recycling bins.

The **first stage** of the didactic sequence included the participatory problematization of the main current environmental issues, and the critical reflection on research papers reporting mathematics and natural sciences interdisciplinary educational activities. In line with Ng (2019), a multidisciplinary (the disciplinary subjects were analysed and integrated in teacher mediation and pre-service teachers' reflection) and interdisciplinary approach (integration of subjects to learn common concepts and processes, such as proportion and quantity, and systems' stability and change) was adopted.

In the autonomously exploration of local and real-life environmental contexts, the **second stage** of the didactic sequence, pre-service teachers investigated local biodiversity (native and exotic plant species from the Campus and from the nearby Forest Park) and environmental problems, such as air pollution, and noise pollution. In the mediation of that exploration, the teachers of the Curricular Unit used diverse strategies, such as problem solving with multisensory and inquiry tasks, and diverse resources, such as sensors and the Pl@ntnet app, as well as other resources produced in previous local research/intervention projects, such as the Eco-Sensors4Heath Toolkit (Eco-Sensors4Health Project, 2019), and the ESELx Eco-track (https://tinyurl.com/ecotrilhoESELx), which will be presented in the next section.

In this stage, the prospective teachers explored multisensory and quantitative environmental information. The quantitative data were organized, and processed, mobilizing statistical techniques. Teacher mediation supported the preservice teachers in sense-making, critical reflection, and problem solving.

The **third stage** of the didactic sequence was centred in the design, by the groups of preservice teachers, of environmental and mathematics education interdisciplinary educational activities, to be implemented and assessed with 5th grade children. The groups were challenged to situate such educational activities in the Campus environmental issues (indoors and outdoors), making use of knowledge of mathematics education and of research-based products (from previous local projects).

In the **fourth stage** of the didactic sequence, the preservice teachers implemented the planned activities, with a 5th grade class, in the Campus of ESELx. The implemented activities were centred on environmental issues, chosen by the groups, such as solid waste, energy waste, green spaces, and air quality. Afterwards, preservice teachers were invited to reflect critically on the implemented activities.

In the **fifth** and last **stage** of the didactic sequence, the groups of preservice teachers wrote a report on the creation, implementation, and assessment of such activities. The reports were also presented and discussed in class.

Previous Projects' Resources and Knowledge Used in the Didactic Sequence

Following the principles of research-based Teacher Education (Aspfors & Eklund, 2017), in the exploration of local environmental contexts, and afterwards in the design and implementation of the mathematics and natural sciences educational activities, preservice teachers used diverse products of previous intervention and research projects, in which the authors of this chapter were teachers/researchers. The main research/intervention projects involved are presented in this section, with an emphasis on the knowledge and resources used in the here described research.

The Eco-Sensors4Health project aimed at improving environmental health in schools, through the assessment of environmental health problems by children and through children's intervention on related environmental factors that may affect health, towards the creation of healthy and sustainable schools. A Toolkit was developed (Silva et al., 2018) to support teachers in implementing interdisciplinary school activities that make possible children's participation in the identification of environmental problems (using senses, eco-sensors, safety scales, and registration forms), and in the creation, implementation, and monitoring of solutions to the identified problems (Silva, 2020). This Project also integrated the use of eco-sensors to explore environmental problems, in different curricular units of school preservice Teacher Education courses. In this Project, environmental/natural sciences and mathematics were mobilized in interdisciplinarity activities to support children in organizing and processing quantitative data, and in making sense of quantities and systems' stability and change.

In the second stage of the didactic sequence presented in this chapter, the preservice teachers used eco-sensors, together with the registration forms and safety scales of the Toolkit (Eco-Sensors4Health Project, 2019), to autonomously explore environmental problems in the Campus, acquiring, statistically processing, and making sense of environmental quantitative data, in an interdisciplinary way. In the third and fourth stages of the didactic sequence, some preservice groups adapted this experienced strategy.

Despite the proximity of ESELx to the Monsanto Forest park, the Campus was very sparsely wooded. In 2017, the first intervention was carried out in the Campus with the planting of several dozen trees of native species, in the context of the GLOCAL-agir Project. However, the academic community still reveals little knowledge and interest about this process of afforestation and, in general, is unaware of the existing plant diversity. This indifference and little knowledge expressed by students and academic community in general, fits well in what has been defined as *Plant Blindness* (Wandersee & Schussler, 1999). This phenomenon has been described as the inability to notice plants in one's environment, recognize their importance or appreciate their unique biological features, and results in a limited interest in plant conservation. The GLOCAL-agir Project aimed to increase the Campus biodiversity, and to value those native species in the context of environmental education activities. Furthermore, this project promoted the exploration of local ecosystems in different curricular units of the Initial Teacher Education courses, as well as the exploration and valuing of the local biodiversity and geodiversity (Almeida et al., 2019). In this project, a preservice teachers' class of an environmental education Curricular Unit used carbon dioxide, temperature, and sound sensors to identify environmental problems in the Campus (Silva et al., 2017). Using those sensors, the students identified and characterized, in a quantitative way, problems such as thermal discomfort, sound pollution, and carbon dioxide pollution, recognized the causes, and explored the solutions, namely "to plant trees to create shadow (they used the sensors to prove the lower temperatures at the shadowed places) (Silva et al., 2017), and to create a barrier, «for instance of cork», to protect the Campus from the traffic noise" (Silva et al., 2017, p. 169).

The Plant@ESELx project, an intervention project in the school's outdoor spaces, identified, characterized, and mapped the Campus main plant species, with students' participation, to enhance the school's plant heritage and raise the academic community's awareness to its conservation and didactical potential (Melo et al., 2022). The project started with a comprehensive inventory of the existing plants (several dozen specimens were selected), their identification (a total of 40 different species were correctly identified) and characterization. The collection of information and photographs carried out by students included taxonomic information, geographic origin, conservation status and ecological and morphological characteristics of each species. All the information collected was organized on a webpage, easily accessible to the entire academic community (ESELx, 2021). At the same time, identification plates were created and placed next to the featured specimens. These plates contain a QR code allowing, through smartphones and other mobile devices, quick access to all the information available on the webpage (ESELx, 2021). Finally, individual sheets were developed in PDF format for each species, containing the essential information about the species and its location on the Campus.

The large amount of information and resources produced by this project can help to overcome indifference towards plants and increase the environmental awareness of preservice teachers and of the entire academic community. Also, the resources produced can be important didactic tools for approaching interdisciplinary activities involving natural sciences, environmental education, and mathematics.

Eco-Schools is a program "which encourages young people to engage in their environment by allowing them the opportunity to actively protect it" (FEE, 2022). It aims at: i) contributing to the achievement of the Sustainable Development Goals (2030-UN Agenda); ii) integrating Environmental Education / Education for Sustainable Development into formal, non-formal and informal education; iii) improving schools' environmental management, raising awareness, and involving the community; iv) guiding to action (change in attitude and behavior, commitment, participation and involvement, citizenship and governance); v) involving the entire school community with an emphasis on students (ABAE, 2018).

In ESELx Eco-Schools project, since 2019, there has been a progressive improvement in several areas related to sustainability, namely in the reduction of water and energy consumption, in waste management (mandatory topics), biodiversity, healthy and sustainable food, sustainable mobility, indoor air quality, and reduction of noise pollution. In waste management, there was an increase in the number of recycling bins for paper/cardboard, plastic/metal and glass. A recycle bin for Electrical and Electronic Equipment Waste was projected and built for collection of equipment and of light bulbs and batteries. There is a collaboration with the Social Action Services in the beginning of the composting of biodegradable kitchen waste.

Based on the knowledge developed by Plant@ESELx project, the ESELx Eco-track (https://tinyurl.com/ecotrilhoESELx) was developed with students, in the context of the ESELx Eco-Schools project. It is an interactive itinerary, between the Campus and the proximity area of Monsanto Forest Park. It was built with Google Maps, and includes stations, with plantmarks (plants that are landmarks). This Eco-track integrates a set of activities, developed by students, to guide the study and exploration of the species by the school community and visitors.

In the second stage of the didactic sequence presented in this chapter the preservice teachers were challenged to explore the Campus and the local Forest Park, using the ESELx Eco-track, to apply the results of the research/intervention projects (GLOCAL-agir, Plant@ESELx, and ESELx Eco-Schools) in informed environmental explorations. Furthermore, in the third stage of the didactic sequence, the groups of preservice teachers used the recycling bins, the container for electrical and electronic equipment waste, light bulbs and batteries, as well as related posters to raise awareness of sustainability issues, to design the activities for elementary school children.

Techniques of Data Collection and Analysis

The data were gathered through participant observation of the lessons and through the collection of all the group reports. The content analysis technique, with inductive categorization, was used to treat the data (Creswell, 2008). The analytic categories included interdisciplinary environmental problems, type of strategies (proposed in the activities designed by prospective teachers) and used resources.

Following Ng (2019), in this chapter the authors use the classification of environmental/natural sciences and mathematics shared concepts to identify interdisciplinarity in the educational activities designed, implemented, and assessed by preservice teachers.

RESULTS

Exploratory Activities With Pre-service Teachers

The 17 preservice teachers worked in 4 groups (3 with 4 elements, and 1 with 5), in all the stages of the Mathematics in Environmental Issues didactic sequence. In the second stage of the didactic sequence, the four groups used human senses, the ESELx Eco-track, and the Pl@ntnet app, to explore local biodiversity in the Campus and the local Forest Park.

Furthermore, the groups used the eco-sensors and the Eco-Sensors4Health Toolkit to identify authentic environmental problems in the school. Following a 2-hour class, with the environmental education teacher, which supported the familiarization with those resources and potentialities, the groups developed field work and a short report on the diverse problems. One group addressed noise pollution, another group assessed thermal discomfort, another one assessed thermal discomfort together with noise pollution, and another one assessed all the three problems. The experimental reports were assessed by the mathematics didactics teacher and discussed with the groups. This way, it was possible to overcome the groups' difficulties, namely in what concerns the different magnitudes, units and scales, and their meaning in the context of the environmental problems addressed, fulfilling an interdisciplinary approach. This second stage allowed preservice teachers to experience environmental problems, contributing to the third stage in which they were challenged to plan interdisciplinary activities based on real environmental problems. Nevertheless, in this second stage, a multidisciplinary approach was also used, with the two teachers making explicit the specific contributions of the two subjects to address the environmental problems.

Problems, Strategies, and Use of Products of Previous Projects in the Designed Educational Activities

Participant observation, together with the analysis of the final group reports, allowed the collection of data related to the activities designed, implemented with a 5th grade class and assessed by the preservice teachers. The analysis of such data produced the results that are subsequently presented in Table 1.

All the preservice teachers' groups centered the educational activities on situated environmental issues: energy consumption in the school (Group 1 – G1); plant diversity in the Campus (Group 2 – G2); air quality in school and Campus (Group 3 – G3); solid waste management in school (Group 4 – G4). These are authentic (real-life) situated issues that the groups addressed, creating authentic problems, and subproblems, to guide the educational activities with children (Table 1). This way, all the groups used an authentic problem-based approach to learning, making it possible interdisciplinarity in such activities.

Table 1. Authentic problems, multisensory strategies, and research/intervention products used in the activities designed, implemented, and assessed by the preservice teachers' group

Preservice Teachers' Groups	Authentic (Real-Life) (Sub)Problems	Situated Embodied Multisensory Strategies	Products of Previous Research and Intervention Projects
G1	- What should we do to not waste energy in school? (What lamps should we use? Should we maintain the Electrical and Electronic Equipment (EEE) in standby?) - Should we use electric or fossil fuel cars? - Where should we put the diverse Waste from EEE (WEEE)?	- Spatial orientation and itineraries for discovering electric vehicle charging station and WEEE waste bins - Transport of water in a one-legged race, minimizing losses and simulating the transport of electrical energy	The WEEE recycle bin was a product of the Eco-School Program, as well as the suggestions to reduce energy consumption in school
G2	What are the characteristics of the plants located on the map? Where are the regions of origin of those plants? What is the influence of plants on the concentration of carbon dioxide in the air? How can I create the maximum number of flowerbeds with the same number of species?	- Spatial orientation and itineraries for discovering the plants marked on the map - Sensing (with vision, touch, and smell), plants' textures, shapes, and colours - Seeding and planting	Suggestions and materials of Eco-sensors4Health (2019) Project to support experimental work in air pollution inquiries. ESELx Eco-Track, produced by the Eco-Schools Program and Plant@ESELx Project website
G3	-What can we do to decrease CO_2 concentrations in our environment? What are the sources and sinks of carbon dioxide in the air? - What plants are there in the Campus? What are the dimensions of those plants?	Sensing (with vision, touch, and smell), drawing and describing verbally plants' textures, shapes, and colours	Suggestions and materials of Eco-sensors4Health (2019) Project to support experimental work in air pollution inquiries. ESELx Eco-Track, produced by the Eco-Schools Program and Plant@ESELx Project website
G4	- Where should I put snack waste? - Where are the recycle bins? - How far does each product travel in its life cycle? What are the costs? - How to optimize the amount of waste in the recycle bin?	Spatial orientation and itineraries for discovering recycle bins inside and outside the school	The recycle bins inside the school were products of the Eco-School Program, as well as the suggestions to improve recycling awareness

To solve the (sub)problems proposed, all the preservice teachers' groups designed situated embodied multisensory strategies, challenging, and supporting children in tasks of spatial orientation, itineraries, and of visual, tactile, and proprioception sensing of the environment (Table 1). These are embodied multisensory tasks that resulted from the authentic and situated learning approach, adopted by the preservice teachers. The holism, and the diversity, of the proposed field work tasks trigger interdisciplinarity, not only between natural sciences and mathematics, but also visual arts, and Portuguese language, developing communicational competences (see Table 1).

In the activities designed by G1 and G4, the participant children were challenged and scaffolded in locating and finding different environmental management equipment, such as outdoors and indoors recycle bins, and the electric and electronic waste recycle bin, created by the ESELx Eco-Schools project. In the activities created by G3 and G2, children were challenged to locate, find, and explore plant species in the Campus. The native explored species were planted in the GLOCAL-agir project. To design these activities, the groups used the ESELx Eco-Track, an Eco-Schools' product, based on the knowledge and resources produced in the Plant@ESELx Project.

Furthermore, G2 and G3 used eco-sensors to allow children to explore sources and sinks of carbon dioxide in the air, using experimental work, and an inquiry strategy. These activities were supported by the suggestions and materials of the Eco-sensors4Health Project (2019).

In this way, all the groups mobilized products from previous research/intervention projects (Table 1), in the context of problem-solving active strategies, such as field work, inquiry, and experimental work.

Interdisciplinarity in the Designed and Implemented Educational Activities

In this section, the data collected, through participant observation and analysis of the final reports developed by preservice teachers' groups, are presented and interpreted to find out if the designed educational activities were interdisciplinary, analysing if those activities included cross-cutting concepts shared by environmental/natural sciences and mathematics.

Crosscutting Shared Concepts

Following Ng (2019), the authors of this chapter searched for concepts that are shared and crosscut different subjects, namely environmental/natural sciences, and mathematics to identify interdisciplinary educational activities. The crosscutting concepts, used as categories in the content analysis of the final reports developed by preservice teachers' groups, were selected and adapted from the ones proposed by National Research Council (2012) to Sciences and Engineering.

It was possible to identify four categories of shared crosscutting concepts used in the educational activities designed by preservice teachers (see Table 2): Spatial orientation and location (used by the 4 preservice teachers' groups); Investigation of patterns and symmetries (used by one preservice teachers' group); Analysis of cause-and-effect processes (used by the 4 preservice teachers' groups); Sense making of scale, proportion, and quantity (used by the 4 preservice teachers' groups). All these concepts are worked in Natural Sciences and Mathematics' Curricular Units of ESELx Initial Teacher Education.

Spatial orientation and location are shared in environmental/natural sciences and mathematics' elementary school Portuguese curricula (DGE, 2018; 2021), and in scientific and educational activities. In the educational activities designed by preservice teachers, children are asked to use instructions and maps to find diverse plants and environmental management facilities outdoors and indoors (Table 2, Figure 1). Furthermore, children were also asked by G2 to communicate locations using coordinate identification, namely the location of the plants in the Campus, and of the region of origin of the plants in the world map (Table 2, Figure 2).

Table 2. Crosscutting shared concepts in the educational activities designed by preservice teachers

Shared Concepts	Tasks Designed by Preservice Groups
Spatial orientation and location	G1 - Using instructions to find electric vehicle charging station and the WEEE recycle bin G2 – Using a map to find plants in the campus; Communicate locations using coordinate identification (of the plants in the Campus, and of the region of origin of the plants in the world map) G3 – Using a map to find plants in the Campus G4 – Using instructions and maps to find recycle bins inside and outside the school
Symmetries	G2 – Identification of the reflection symmetry of the leaves; Symmetry analysis of the plants and fruits
Cause-and-effect	G1 – Study of the effect of the use of different types of lamps on the electric consumption and cost; study of the effect of the choice of an electric or fossil fuel car on costs and carbon dioxide emissions; study of the effect of the shut down or standby on electricity consume and costs G2 – Study of the effect of plants on carbon dioxide concentration in air G3 – Study of the effect of a burning candle and of a plant on carbon dioxide concentration in air G4 – Study of the effect of the consumption of nonlocal, or local, products on fossil fuel costs; relation between the volume of packaging waste, compaction, and the amount of waste in the recycling bin
Proportion and quantity	G1 – Analysis and sense making of the proportion of the electricity bill due to maintain the EEE in standby G2 – Multiplicative reasoning in creating the maximum number of flowerbeds with the same number of plant species G3 – Estimation, analysis and sense making of the variation of carbon dioxide concentration in air, as a function of the presence of sources and sinks G4 – Estimation, measurement, and sense making of distances covered to the outside recycling bins; estimation, measurement, and sense making of distances travelled by each product in its life cycle; relation between the volume of packaging waste, compaction, and the amount of waste that can be put in the recycling bin

Only one group (G2) created tasks to challenge children to investigate symmetries (Table 2), to explore and characterize different trees (Figure 3). Although symmetry is important in characterizing living beings, contributing to their classification (as happens with the classification of leaves), it is a topic that is only mentioned as an essential learning in the Portuguese elementary school curriculum of mathematics (DGE, 2018; 2021), and not in natural sciences.

Cause-and-effect was a crosscutting shared concept used in the educational activities designed by all the groups of preservice teachers. The study of cause-and-effect processes was present in diversified tasks (Table 2), needed to solve the authentic environmental problems designed to children. In those tasks, children were invited to investigate cause-and-effect processes with different strategies, namely experimental work with interpretation of data presented graphically (Figure 4), as well as numerical and manipulative simulations (Figure 5).

Figure 1. Map created by G4 to guide children to find recycle bins inside the school
Source: Authors

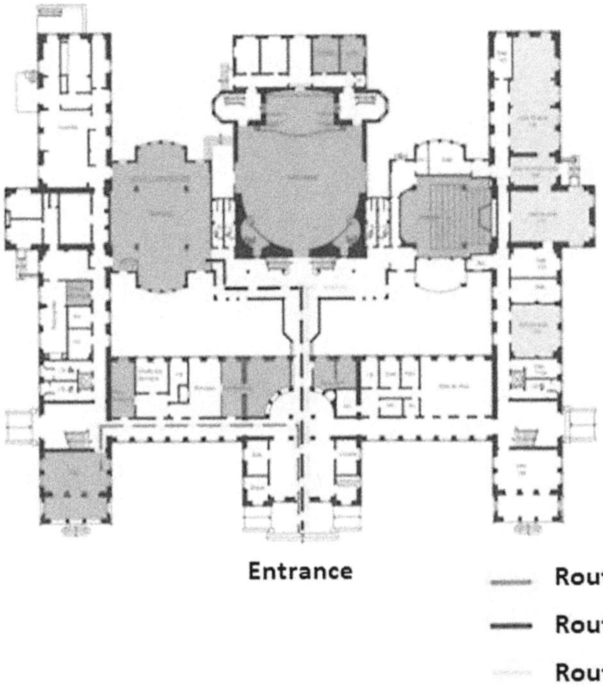

Figure 2. Map created by G2 to support children in communicating the location of diverse plant species, using coordinates
Source: Authors

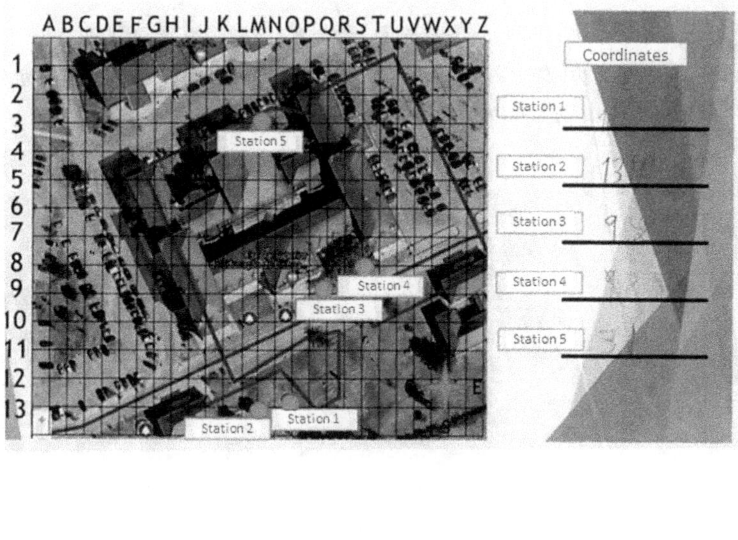

Figure 3. Registration forms created by G2 to support children in exploring symmetries to characterize different plants
Source: Authors

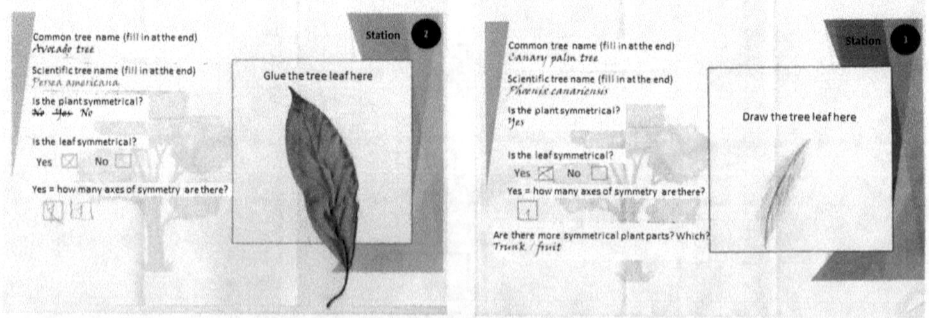

Figure 4. G3 challenged children to match different data, acquired by eco-sensors and visualized in line charts to different experimental conditions (sinks and sources of carbon dioxide in air)
Source: Authors

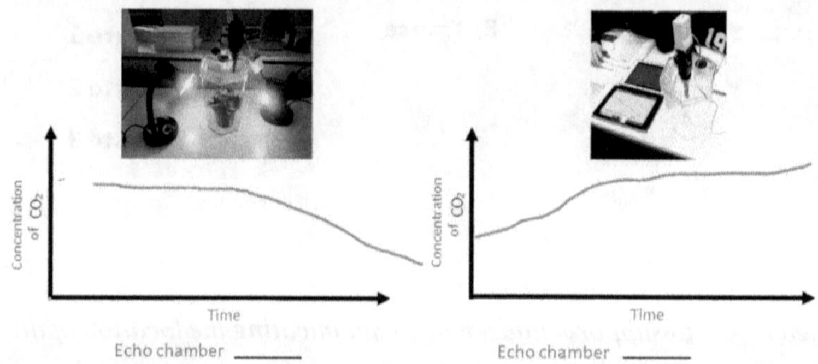

Figure 5. Manipulative simulation by G4 to investigate the relation between compaction, volume of packaging waste, and the amount of waste that can be put in the recycling bin
Source: Authors

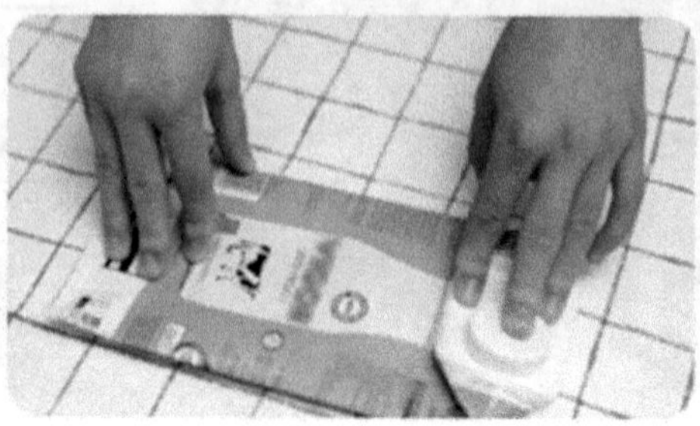

Proportion and quantity were used as environmental/natural sciences and mathematics' shared concepts by all the preservice teachers' groups in the design of the educational activities. The designed tasks asked children to address proportion and physical and chemical quantities in diverse ways, such as (Table 2): i) paper and pencil problem solving, in the context of an energy peddy paper; ii) interpretation of environmental data line charts, considering experimental conditions within experimental work; iii) estimation, measurement, and sense making of quantities in embodied itineraries for waste recycling; and iv) sense making of quantity changes and relations in waste separation.

Considering the results presented, it is possible to state that the four sets of shared crosscutting concepts were used by the preservice teachers to bring disciplinary subjects together in the designed tasks and activities, without separation of disciplinary knowledge and skills.

Integration of Mathematics and Environmental/Natural Sciences

As explained in the previous section, preservice teachers were able to design environmental activities that scaffold the use and development of mathematical and natural sciences' competences, using shared crosscutting concepts to implement interdisciplinarity. Nevertheless, not all the designed activities and not all the tasks were interdisciplinary (Table 3).

The activities of 2 groups (G2 and G4), and all the tasks in these activities, were interdisciplinary. However, some shared crosscutting concepts, although present in problem solving, were not explicitly addressed when dealing with maps and carbon dioxide concentrations in air, as for instance scale, quantities, and proportion.

All the activities designed by G3 made use of shared crosscutting concepts in the context of an air pollution problem solving inquiry strategy. Still, while in some tasks the request for organizing data in ascending order is part of the problem-solving strategy (as in ordering carbon dioxide concentrations of different locations in the Campus and school), in other tasks a similar request does not directly contribute to the goal of the inquiry activity (as in ordering the length of five plant's leaves). In the first case, the tasks are interdisciplinary, while in the last case they are multidisciplinary. Furthermore, some tasks are only centered on environmental education goals, such as the ones related to multisensory observation of the plants.

G1 was the only group that designed some activities that were only centered in environmental education, and in which mathematical learning was not identified. However, one of those activities included a task that asked students to assemble the pieces of a puzzle to discover the container for electric and electronic waste, this way involving spatial reasoning.

Table 3. Levels of integration of Mathematics and Environmental/Natural Sciences in the educational activities designed by preservice teachers

Groups	Integration of Mathematics and Environmental/Natural Sciences
G1	Three of the five designed activities are interdisciplinary. The other two are environmental education activities without the contribution of mathematics
G2	All the four designed activities are interdisciplinary, despite not all the potential of this interdisciplinarity is explicit
G3	Although some tasks of the five activities are only centred on environmental sciences, and some others are multidisciplinary, the five activities are globally interdisciplinary, as well as most of the tasks
G4	All the four designed activities are interdisciplinary, despite not all the potential of this interdisciplinarity is explicit

Critical Reflection on the Implementation of Educational Activities With Children

The analysis of the critical reflections of the four preservice teachers' groups is presented in this section to understand how the different groups used the participant observation of the implementation of the educational activities with children to improve such activities, including their mediation. Not all the activities designed were implemented, since the 5th grade class could only participate for 90 minutes. Each preservice teachers' group worked with a group of six children and selected the adequate sequence of activities to fit the time available. The activities were implemented with children in the Campus, mainly outdoors, but G1, G3 and G4 also developed some tasks indoors.

All the groups mentioned a noticeable engagement of children in the activities implemented and highlighted children's motivation and engagement as positive feedback to the developed activities. G1 and G2 specified that children were specially motivated to outdoors environmental and physical tasks (for instance, the task in which children were invited to carry water in a one-legged race, trying to minimize losses, while simulating the transport of electrical energy), and not so motivated when mathematics was involved. Nevertheless, it was also noticed by G3 that interdisciplinarity made it possible for children to apply mathematics in real-world tasks and problems, increasing learning motivation.

Despite the reported global success, in the critical reflections, G1, G2, and G3 stated that the implementation of the activities with children reinforced their awareness to the importance of the assessment of children's prior knowledge to design adequate activities and tasks. G2 referred the need to adapt the activities plans, after the assessment of children's prior knowledge. This aspect can be illustrated with the following excerpt of G1´s critical reflection:

Among the mathematical problems, we can highlight the problem of the "Illuminated Idea" station, which was the task in which the children needed greater support from us. We can even say that the students, without our help, would hardly be able to solve the problem. Thus, in our opinion, we consider that this fact is due to the prior lack of knowledge of the target audience, otherwise, we would have adapted the tasks to their characteristics and skills.

In this way, after carrying out the peddy-paper, we had the opportunity to reformulate some aspects considering what was previously mentioned and, thus, they were considered in the realization of the plans that we present below. (G1 report, p. 17)

Furthermore, G1 highlighted children's diversity in attitudes and competences, and reported having adjusted the activities' plans, after the observation of children's performances and difficulties, during the activities' implementation.

Likewise, G3 acknowledged the importance of the activities' implementation to learn about teaching and learning time management. In addition, G4 described and reflected on children's strategies in performing measurements and measurements estimations. The following excerpt describes children's strategies in estimating the distance from where they were to the nearest recycle bin:

One of the groups used a reference item, which was the distance they had to walk in their school to get to the gym, realizing that the distance from the point where they were to the nearest recycle bin was equal to the distance they had to walk to the gym or a little more. Thus, this group estimated the distance to be 20 meters. On the other hand, the other group thought about the number of steps needed to get to the nearest rubbish bin. Then they estimated the measure of a step and multiplied it by the number of steps they thought it would take to get to the rubbish bin. Thus, the group concluded that the distance was 30 meters. (G4 report, p. 29)

Two of the groups (G2, G3) emphasized the gratification and encouragement felt with explicit positive feedback children offered:

Nevertheless, the children also expressed a very positive feedback, since it was a pleasant and fruitful task from the point of view of new learning. Some students expressed their satisfaction with sentences such as: "you are going to be good teachers", "I prefer to be here than to be closed in a classroom", "we liked it a lot" or "I had no idea it was going to be like this, I am enjoying it a lot". (G2 report, p. 36)

G2 linked the positive feedback to the potentialities of the didactic resource produced:

This clearly manifests the potential of activities of this nature, which calls for an exploration of the environment and encourage closer contact with the environment and the elements that compose it, breaking with more traditional approaches. Finally, the importance of promoting environmental education is reinforced, in this case, allied to mathematics, highlighting its transversal character and the potential of this articulation, which, by nature, is intrinsic. (G2 report, p. 36)

Also, G3 emphasized the way mathematical topics were connected to children's reality:

We believe that it was much more beneficial and interesting for the children that these mathematical contents had emerged in a transversal way, applied to situations in the children's reality. In this way, consciously or unconsciously, the child understands the importance of mathematics for the study of the reality that surrounds us and the multiplicity of ways in which it can arise. (G3 report, p. 38)

In this way, the critical reflections on the implementation of educational activities with children made it possible to preservice teachers to assess their design and mediation performance, based on the learning practices observed. Nevertheless, only two groups used bibliographic references to support that assessment and the identification of needed improvements.

CONCLUSION

The research presented in this chapter is centered on the implementation and assessment of a didactic sequence and was developed in the context of the Mathematics in Environmental Issues Curricular Unit in Initial Teacher Education in Portugal. The didactic sequence aimed at the research-based design, implementation (with a 5th grade class), and assessment of interdisciplinary authentic activities by the participant preservice teachers.

In what regards the first research question, "How do the preservice teachers, throughout the didactic sequence, design, implement and assess interdisciplinary authentic learning activities for children?", the outcomes obtained showed that the activities designed, implemented, and assessed by the groups of preservice teachers: i) were authentic activities since were based on situated environmental issues and real-world problems, and used active educational strategies, as problem-based learning, inquiry-based learning, embodied learning, experimental work, and field work; ii) were globally multidisciplinary, since disciplinary knowledge and skills were used together, without separation (Ng, 2019), to explore the environmental systems and facilities, and solve the proposed environmental problems. The development of activities and tasks exclusively interdisciplinary was not a goal of the didactic sequence.

In this way, the didactic sequence presented and assessed in this chapter, is in line with international recommendations (Ng, 2019), and it is a contribution to the interdisciplinary approach, crossing the boundaries between mathematics and natural sciences, which has been lately developed in Teachers Initial Training in Portugal (Cavadas et al., 2022; Correia & Baptista, 2022; Rodrigues et al., 2019).

In what concerns the second research question, "What shared concepts were included in the authentic learning activities for children, to bring disciplinary subjects together?", the analysis of the collected data allowed, based on the categories proposed by National Research Council (2012) and Ng (2019), the identification of the following crosscutting shared concepts, used to bring together mathematics and environmental/natural sciences: spatial orientation and location, symmetries, cause-and-effect, proportion, and quantity. Only one group of preservice teachers included symmetries in the developed activities, but all the other crosscutting shared concepts were integrated by all the preservice teachers' groups in the developed activities.

This research only analyzed the results of the didactic sequence in the 2021-2022 academic year. In this year, "patterns" were not explored as a boundary crossing concept in the activities developed by the preservice teachers. However, in previous academic years, some of the designed tasks asked children to build and explore patterns, using parts of plants, or plants texture decals. In the same way, scales were not explored in the academic year analyzed in this research. Nevertheless, in preceding academic years, preservice teachers developed tasks that challenge students to map a green space, considering its scale and the existing plants.

Regarding the third research question, "How does the didactic sequence contribute to a research-based education of preservice teachers?", based on the participant observation of the didactic sequence and on the analysis of the preservice teachers' final reports, and using the approach of Aspfors and Eklund (2017), it is possible to conclude that preservice teachers: i) experienced exploratory activities created, using knowledge and resources produced in former research/intervention projects, in which the authors of this chapter were involved; ii) developed innovative, authentic and local interdisciplinary activities for children, also using situated knowledge and resources produced in research/intervention projects; and iii) developed reflective practices, while implementing the designed activities with children, with subsequent practice-based critical reflection and consequent improvements.

The critical and research-based activities, reflection, and assessment developed by preservice teachers allowed the production of knowledge to improve teachers and preservice practices in higher and basic education contexts. In this context, the analysis of the results of the present research also highlights the need for continuous improving in the didactic sequence of Mathematics in Environmental Issues Curricular Unit, namely in what concerns deepening interdisciplinary pedagogical content knowledge of preservice teachers (An, 2017), improving their awareness of shared concepts and respective elicitation, and developing their practice-based research competences (Aspfors & Eklund, 2017; Roldão, 2021).

ACKNOWLEDGMENT

This research received no specific grant from any funding agency in the public, commercial, or not-for-profit sectors. The authors acknowledge all the participant preservice teachers and children.

REFERENCES

Almeida, A., Valente, B., Silva, M. J., Rodrigues, M., & Manteigas, V. (2019). O projeto GLOCAL – AGIR: Conhecer o território para o valorizar [The Project GLOCAL - AGIR: Know the territory to value it]. In J. Pinhal, C. Cavaco, M. J. Cardona, F. A. Costa, J. Marques, A., & R. Faria (Eds.), Atas do XXV Colóquio da AFIRSE Portugal: A investigação, a formação, as políticas e as práticas em educação – 30 Anos de AFIRSE em Portugal (pp. 92-102). Instituto de Educação da Universidade de Lisboa.

An, S. A. (2017). Preservice teachers' knowledge of interdisciplinary pedagogy: The case of elementary mathematics–science integrated lessons. *ZDM Mathematics Education*, *49*(2), 237–248. doi:10.100711858-016-0821-9

Aspfors, J., & Eklund, G. (2017). Explicit and implicit perspectives on research-based teacher education: Newly qualified teachers' experiences in Finland. *Journal of Education for Teaching*, *43*(4), 400–413. doi:10.1080/02607476.2017.1297042

Associação Bandeira Azul da Europa (ABAE). (2018). *Programa Eco-Escolas* [Eco-Schools Program]. https://ecoescolas.abae.pt/wp-content/uploads/sites/3/2018/09/Programa-Eco-Escolas-SINTESE.pdf

Beane, J. A. (1997). *Curriculum Integration. Designing the Core of Democratic Education*. Teachers College Press.

Beane, J. A. (2000). O que é um currículo coerente? [What is a coherent curriculum?]. In J. A. Pacheco (Org.), Políticas de integração curricular (pp. 39-58). Porto Editora.

Biseth, H., Svenkerud, S. W., Magerøy, S. M., & Rubilar, K. H. (2022). Relevant transformative teacher education for future generations. *Frontiers in Education*, *7*(806495). Advance online publication. doi:10.3389/feduc.2022.806495

Cavadas, B., Rézio, S., Nogueira, J. R., & Branco, N. A. (2022). Framework and a research design proposal to identify preservice teachers' integration performance of science and mathematics. *Can. J. Sci. Math. Techn. Educ.*, *22*(1), 101–129. doi:10.100742330-022-00198-2

Correia, M., & Baptista, M. (2022). Supporting the development of pre-service primary teachers PCK and CK through a STEM Program. *Education Sciences*, *12*(4), 258. doi:10.3390/educsci12040258

Creswell, J. W. (2008). *Educational research: Planning, conducting, and evaluating quantitative and qualitative research* (3rd ed.). Merrill Prentice Hall.

Direção Geral de Educação. (2018). *Aprendizagens Essenciais - Ensino Básico* [Essential Learning - Basic Education]. Ministério da Educação. https://www.dge.mec.pt/aprendizagens-essenciais-ensino-basico

Direção Geral de Educação (DGE). (2021). *Novas Aprendizagens Essenciais de Matemática - Ensino Básico* [New Essential Learning of Mathematics - Basic Education]. Ministério da Educação. https://www.dge.mec.pt/aprendizagens-essenciais-ensino-basico

Drake, S. M., & Reid, J. L. (2020). 21st Century competencies in light of the history of integrated curriculum. *Frontiers in Education*, *5*(122), 122. Advance online publication. doi:10.3389/feduc.2020.00122

Eco-Sensors4Health Project. (2019). *Eco-Sensors4Health Toolkit: Guia Eco-sensores para a saúde* [Eco-Sensors4Health Toolkit: Eco-Sensors Guide to Health]. Instituto Politécnico de Lisboa, Instituto Politécnico de Viseu, Ciência Viva – ANCCT, Município de Viseu. https://ecosensors4health.files.wordpress.com/2019/03/toolkit-eco-sensors4health_cv_vf.pdf

Escola Superior de Educação de Lisboa (ESELx). (2021). *Projeto Plant@Eselx* [Project Plant@Eselx]. https://www.eselx.ipl.pt/plantas-eselx

European Commission. (2015). *Shaping career-long perspectives on teaching. A guide on policies to improve initial teacher education*. Publications Office of the European Union.

Foundation for Environmental Education (FEE). (2022). *About Eco-Schools*. https://www.ecoschools.global/how-does-it-work

Geiger, V., Mulligan, J., Date-Huxtable, L., Ahlip, R., Jones, D. H., May, E. J., Rylands, L., & Wright, I. (2018). An interdisciplinary approach to designing online learning: Fostering pre-service mathematics teachers' capabilities in mathematical modelling. *ZDM Mathematics Education*, *50*(1-2), 217–232. doi:10.100711858-018-0920-x

Goldman, S. R., Radinsky, J., Tozer, S., & Wink, D. (2010). Learning as Inquiry. In P. Peterson, E. Baker, & B. McGaw (Eds.), *International encyclopedia of education* (3rd ed., pp. 297–302). Elsevier., doi:10.1016/B978-0-08-044894-7.00495-4

Gómez Puente, S. M., van Eijck, M., & Jochems, W. (2013). A sampled literature review of design-based learning approaches: A search for key characteristics. *International Journal of Technology and Design Education*, *23*(3), 717–732. doi:10.100710798-012-9212-x

Martins, G., Gomes, C., Brocardo, J., Pedroso, J., Carrillo, J., Silva, L., ... Rodrigues, S. (2017). *Perfil dos alunos à saída da Escolaridade Obrigatória* [Profile of students leaving compulsory schooling]. Ministério da Educação/Direção-Geral da Educação (DGE).

Melo, N., Telo, C., & Sarreira, P. (2022). O Projeto Plant@ESELx e a participação dos estudantes na caracterização do património vegetal da ESELx [The Project Plant@ESELx and the participation of students in the characterization of eselx's plant heritage]. *Da Investigação às Práticas. Estudos de Natureza Educacional*, *12*(1), 188–208. doi:10.25757/invep.v12i1.321

Merritt, J., Lee, M., Rillero, P., & Kinach, B. M. (2017). Problem-Based Learning in K–8 Mathematics and Science Education: A Literature Review. *The Interdisciplinary Journal of Problem-Based Learning*, *11*(2). Advance online publication. doi:10.7771/1541-5015.1674

Morin, E. (2001). L' enseignement des connaissances [The Project Plant@ESELx and the participation of students in the characterization of eselx's plant heritage]. In F. C. Gulbenkian (Ed.), *Novo conhecimento, nova aprendizagem* (pp. 25–33). Fundação Calouste Gulbenkian.

National Research Council. (2012). *A Framework for K-12 Science Education: Practices, crosscutting concepts, and core ideas*. The National Academies Press. https://www.nap.edu/catalog/13165/a-framework-for-k-12-science-education-practices-crosscutting-concepts

Ng, S. B. (2019). Exploring STEM competences for the 21st century. *Current and Critical Issues in Curriculum, Learning and Assessment, 30*. https://unesdoc.unesco.org/ark:/48223/pf0000368485

OECD. (2018). *Curriculum Flexibility and Autonomy in Portugal: An OECD Review*. http://www.oecd.org/ education/2030/Curriculum-Flexibility-and-Autonomy-in-Portugal-an-OECD-Review.pdf

Plomp, T. (2013). Educational design research: An introduction. In T. Plomp & N. Nieveen (Eds.), Educational design research - part a: an introduction (pp. 10 – 51). Netherlands Institute for Curriculum Development (SLO).

Ponte. (2020). A didática da matemática e o trabalho do professor [The didactics of mathematics and the work of the teacher]. *RBECM, Passo Fundo, 3*(3), 809-826.

Rodrigues, M., Caseiro, A., Silva, M. J., Loução, A., & Monteiro, S. (2019). Integração curricular de matemática e ciências naturais na abordagem de problemas de saúde ambiental na formação para a docência [Curricular integration of mathematics and natural sciences in the approach of environmental health problems in teaching training]. In M. J. Silva & R. Brito (Coords.), *Utilização pedagógica de sensores eletrónicos para a participação na saúde ambiental das escolas* (pp. 30-50). Centro Interdisciplinar de Estudos Educacionais.

Roldão, M. C. (2017). Conhecimento, didáctica e compromisso: O triângulo virtuoso de uma profissionalidade em risco [Knowledge, didactics and commitment: the virtuous triangle of a professionality at risk]. *Cadernos de Pesquisa, 47*(166), 1134–1149. doi:10.1590/198053144367

Roldão, M. C. (2021). Investigação como instrumento da formação profissional de docentes. [Research as an instrument of professional training of teachers]. *Revista Formação Docente – Revista Brasileira de Pesquisa sobre Formação de Professores, 13*(28), 79-90. doi:10.31639/rbpfp.v13i28.545

Sacristán, J. (2000). O currículo: Uma reflexão sobre a prática [The curriculum: A reflection on the practice] (3rd ed.). Artmed. (Original work in Spanish published in 1991)

Silva, M. J. (2020). Children using electronic sensors to create and use knowledge on environmental health. *First Monday, 25*(3). Advance online publication. doi:10.5210/fm.v25i3.9646

Silva, M. J., Almeida, A., Valente, B., Rodrigues, M., & Manteigas, V. (2017). Sensing locally in the global environment: Using sensors in teachers' education. In *Proceedings of IADIS International Conference Educational Technologies 2017 - ICEduTech: IADIS International Conference Educational Technologies 2017 - ICEduTech* (pp. 166-170). IADIS.

Silva, M. J., Ferreira, E., Souza, A., & Alves, A. R. (2018). Eco-Sensors4Health Toolkit: Scaffolding children participation in schools' environmental health. In *Proceedings of Global Health 2018, The Seventh International Conference on Global Health Challenges.: The Seventh International Conference on Global Health Challenges* (pp. 47-52). https://thinkmind.org/index.php?view=article&articleid=global_health_2018_4_10_70026

United Nations. (2022). *Make the SDGS a Reality*. United Nations Department of Economic and Social Affairs, Sustainable Development. https://sdgs.un.org/#goal_section

World Economic Forum. (2016). *New vision for education: Fostering social and emotional learning through technology*. https://www3.weforum.org/docs/WEF_New_Vision_for_Education.pdf

Yew, E. H. J., & Gohb, K. (2016). Problem-Based Learning: An Overview of its Process and Impact on Learning. *Health Profession Education*, *2*(2), 75–79. doi:10.1016/j.hpe.2016.01.004

ADDITIONAL READING

Amprazis, A., & Papadopoulou, P. (2020). Plant blindness: A faddish research interest or a substantive impediment to achieve sustainable development goals? *Environmental Education Research*, *26*(8), 1065–1087. doi:10.1080/13504622.2020.1768225

Cavadas, B., Rézio, S., Nogueira, J. R., & Branco, N. (2022). A Framework and a Research Design Proposal to Identify Preservice Teachers' Integration Performance of Science and Mathematics. *Can. J. Sci. Math. Techn. Educ.*, *22*(1), 101–129. doi:10.100742330-022-00198-2

Davis, B.Spatial Reasoning Study Group. (2015). *Spatial reasoning in the early years: Principles, assertions, and speculations*. Routledge. doi:10.4324/9781315762371

Marchand, T. H. J. (2010). Introduction: Making knowledge: explorations of the indissoluble relation between mind, body, and environment. In T. H. J. Marchand (Ed.), Making knowledge: explorations of the indissoluble relation between mind, body and environment (pp. 01–20). Wiley-Blackwell.

Ng, O. L., & Sinclair, N. (2015). Young children reasoning about symmetry in a dynamic geometry environment. *ZDM Mathematics Education*, *47*(3), 421–434. doi:10.100711858-014-0660-5

Sauvé, L. (2005). Currents in environmental education: Mapping a complex and evolving pedagogical field. *Canadian Journal of Environmental Education*, *10*(1), 11–37.

Thompson, P. W. (1993). Quantitative reasoning, complexity, and additive structures. *Educational Studies in Mathematics Education*, *25*(3), 165–208. doi:10.1007/BF01273861

Wals, A. E. J., Brody, M., Dillon, J., & Stevenson, R. B. (2014). Convergence between science and environmental education. *Science*, *344*(6184), 583–584. doi:10.1126cience.1250515 PMID:24812386

KEY TERMS AND DEFINITIONS

Authentic Learning: A constructivist approach, in which students construct their own knowledge, based on their previous representations and experiences, while developing new and complex inquiry and real-world problem-solving activities, using multiple resources, such as digital technologies.

Cause-and-Effect: In Natural Sciences, it is a relationship between two events in which the second is a consequence of the first. Mathematically, it is a relationship between two variables in which a change in one variable produces a change in the other variable.

Education for Sustainable Development: An interdisciplinary process aiming at developing critical citizenship to achieve a social, economic, and environmental development, with social justice, and quality of life for present and future generations, without endangering ecological systems, services, and resources.

Environmental Education: An interdisciplinary process aiming at developing experiences, awareness, knowledge, skills, values, attitudes, practices, motivations, and critical participation in and for the environment, promoting, in this way, environmental literacy and citizenship.

Exotic Specie: Specie introduced, through human activity, into an area where it does not occur naturally.

Multisensory: Integration of sensations and information from different sensory modalities, such as visual, auditory, gustatory olfactory, as well as vestibular and somatic senses.

Native Specie: Specie that have become part of a particular region, ecosystem, or habitat through natural processes, without human intervention.

Proportion: A comparative relationship between quantities.

Quantity: The amount of something that can be measured.

Spatial Location: The ability to identify and describe where an object is physically located with the help of a space coordinate system.

Spatial Orientation: The ability of an individual to determine his/her position or relative motion in relation to the surrounding environment.

Sustainable Development Goals: Set of 17 goals, subdivided into 169 targets, from the United Nations (UN) 2030 Agenda ("Transforming our World: the 2030 Agenda for Sustainable Development"), which was adopted by the UN General Assembly in 2015 by the 193 Member States of the United Nations. Those goals and targets aims to stimulate actions to build a more sustainable world by 2030.

Symmetry: A geometric transformation that preserves the distances and simultaneously leaves the figure globally invariant.

Chapter 13
Adaptation of the Math-Biology Values Instrument to Preservice Teachers: German, Portuguese, and Spanish Versions

José María Marcos-Merino
Universidad de Extremadura, Spain

Nora Rodilla-Martín
Universidad de Extremadura, Spain

María Rocío Esteban-Gallego
Universidad de Extremadura, Spain

Jesús A. G. Ochoa de Alda
Universidad de Extremadura, Spain

Nelson Mestrinho
Polythecnic Institute of Santarém, Portugal

Bento Cavadas
https://orcid.org/0000-0001-6021-6581
Lusófona University, Portugal & Polytechnic Institute of Santarém, Portugal

ABSTRACT

Primary education curricula highlight the need for integrating STEM subjects in the classroom. Thus, preservice teachers' training programs must involve subject integration. Students' personal value to this integration is a crucial component of their motivation, influencing their outcomes. However, practitioners lack validated instruments to monitor student task-value changes to specific subject combinations. Using math in a biological context exemplifies STEM subject integration. This chapter shows the adaptation of the Math-Biology Values Instrument to preservice teachers, a tool to measure task values of using math in biology. Structural and construct validities were obtained from a sample of Spanish and an independent multicultural sample of preservice teachers, respectively. Multigroup confirmatory factor analysis supported partial measurement invariance across sexes, backgrounds, and countries. The instrument provides enough validity to monitor task-value in preservice teachers.

DOI: 10.4018/978-1-6684-5765-8.ch013

INTRODUCTION

Fundamental problems of 21st-century society cannot be solved through single disciplines but require a global and integrative perspective. Current education fails to provide this perspective because, following a classical reductionist approach, educational systems are arranged into strict compartments of separate subjects (Brewer & Smith, 2011; Labov et al., 2010; McCright et al., 2013). From primary education, disciplines divide into different subjects, moving away from a global view of science (Stone, 2014). Acquiring a systemic perspective is significant for primary (elementary) education preservice teachers. If appropriately trained, they can transmit a global picture of science, technology, engineering, and mathematics (STEM) disciplines since primary education. As in other European curricula, the Spanish elementary education curriculum includes a cross-sectional block of contents for integrating natural science subjects named "Initiation to research activity," where sciences, technology, engineering, and mathematics are fully integrated (Spanish Government, 2022). Thus, as previously noticed (Šorgo, 2010), preservice teachers' science instruction requires an interdisciplinary perspective reflecting the current science and engineering practice. Mathematics plays a pivotal role in achieving this perspective because it allows students to model, represent, and interpret experimental and observational results from various disciplines, such as biology, a science with a long descriptive tradition (May, 2004; Oltvai & Barabási, 2002). According to current educational research, integrating math into biology curricula results in increased coherence in the undergraduate biological sciences curriculum, improves quantitative skills in students, and a greater appreciation among graduates for the essential relationship between mathematics and modern biology (Thompson et al., 2013).

The barriers to establishing math-science connections in the classroom are mainly grounded in prior experiences and inadequate training. In preservice teachers, previous student experiences determine their preconceptions about teaching mathematics. They make decisions based on an external locus of control (such as the university professor, the textbook, or other sources) and harbor higher anxiety about teaching math than science (Cady & Rearden, 2007; Mellado et al., 2014; Hernández-Barco et al., 2021). Content integration is more challenging for many preservice teachers than traditional content (Leszczynski, 2014). Teacher training programs generally lack a focus on integrated teaching and do not devote time to building connections (DeLuca et al., 2015; Kim et al., 2004). As a result, preservice teachers possess weak content knowledge in both disciplines and inadequate integration models. These teachers frequently ignore fundamental differences in mathematics and science knowledge organization. Past efforts to integrate math and science reflected an insufficient understanding of negative attitudes about integration among teachers (Bursal & Paznokas, 2006; Leszczynski, 2014). This attitude, defined as the enduring tendency to evaluate this integration such as positive/negative or good/bad, can be divided into three dimensions: 1) perceived control, which includes elements such as self-efficacy (subjective perception of own capability); 2) affective states, which provides for emotions such as enjoyment and anxiety and 3) cognitive beliefs, such as perceived relevance or subjective value (Bursal & Paznokas, 2006; Buss, 2010; Cady & Rearden, 2007; Leszczynski, 2014; Riegle-Crumb et al., 2015). Subjective value is the extent to which an outcome (math-science integration in this case) is desired or valued concerning the probability of success. Psychology and neuroscience recognize this subjective value as critical to achieving motivation (Eccles & Wigfield, 2020). The subjective value of preservice teachers to math-biology integration is unexplored, mainly due to de absence of a specific, reliable, and validated instrument to monitor changes in these subjective values. The availability of this instrument would pave the way to understanding attitudes toward math-biology integration. This chapter presents the development and psychometric properties of an instrument to monitor changes in the subjective value that preservice teachers ascribe to using mathematics in biology.

BACKGROUND

Preservice teachers assign a subjective value to particular teaching-learning activities and outcomes (Torsney et al., 2017). Psychology conceptualizes the relationship between this personal value and learning into Expectancy-Value (Eccles & Wigfield, 2020), Control-Value (Pekrun, 2006), and Information-Seeking (Sharot & Sunstein, 2020) theories (EVT, CVT, IST, respectively). According to EVT, subjective task value and perceived competence (e.g., expectations for success) drive achievement-related choices and performances, personal behaviors, and academic achievement. Additionally, in CVT, the role of task value in motivation and achievement is mediated by achievement emotions (Forsblom et al., 2021; Pekrun, 2006), in agreement with the current view of emotions as subjective valuation processes. Finally, IST rests on the idea that people calculate the value of knowledge before deciding on information-seeking or avoidance, which is crucial in education. In the context of current neurophysiological theories, the IST framework ascribes the value of knowledge to the brain's reward circuits, a complex network involving different aspects of incentive learning, adaptive behaviors, emotional learning, and good decision making (Sharot & Sunstein, 2020). Reward circuits control the decisions to obtain reward and avoid harm (Etkin et al., 2015; Lebreton et al., 2015). Thus, the value preservice teachers expect after succeeding would determine their motivation to perform achievement tasks. Together, these considerations support task-value as an essential construct to monitor during teaching-learning.

To understand the motivation for math-science integration, educational researchers and practitioners require specific and reliable instruments to monitor students' subjective values towards classroom activities involving math-biology integration. Even though various instruments measure different personal values, developing an instrument to measure specifically math-biology integration is very recent. The Math-Biology Values Instrument (MBVI) was initially developed for life science students (Andrews et al., 2017). This instrument consists of an 11-item self-report scale measuring three subscales: interest (4 items), utility (4 items), and cost (3 items). The interest subscale refers to the enjoyment a student gets from using mathematics to understand biology or the personal interest rather than external pressures or rewards that a student has in this task. The utility subscale refers to the utility a student gives to mathematics in their professional future as a biologist. The cost subscale refers to negative aspects of using mathematics in biology and includes extra effort, loss of opportunities, and emotional toll (such as performance anxiety and fear of failure). Integrating positive values (interest and utility) and negative (cost) is also an uncommon feature of this instrument. Instruments including negative subscales of value are relatively new (Andrews et al., 2017; Luttrell et al., 2010). The development of MBVI integrates multiple forms of validity evidence to support the use of the instrument to measure undergraduate life science majors' math–biology task values (Andrews et al., 2017). However, there is still room for improvement in the instrument: measurement invariance has never been explored, nor has the validation of MBVI in different contexts and languages.

MAIN FOCUS OF THE CHAPTER

This chapter focuses on developing a reliable instrument to measure the subjective value of a task (task-value) in preservice teachers. This task is the use of mathematics in the context of biology (or science). Instrument validation is required to ensure that it measures what it is intended to measure, and that the information yield is appropriate, meaningful, and useful for decision-making. In order to validate the instrument, it is necessary to accumulate various sources of validity, scaffolding an argument to support the instrument's intended use. By translating and adapting MBVI to preservice teachers, it is assumed that the content is likely adequate and represents the target construct (the task-value of using mathematics in biology) as the original MBVI does in the frame of EVT (Andrews et al., 2017).

The ability of MBVI to capture changes in subjective values is under discussion. After a research experience using mathematical modeling to quantify microbial growth, the Likert-scale assessment of values surrounding the role of math in biology found limited evidence of gains (Furrow et al., 2020). In contrast, a previous (still not validated) adaptation of MBVI to assess preservice teachers' attitudes towards the use of chemistry in biology showed the ability of this instrument to capture changes in interest and utility after a research experience (Marcos-Merino et al., 2020). Thus, adaptation and validations of MBVI are required not only to show the adequateness of a factor structure but also to show the ability of the instrument to capture construct changes,

The adaptation and translation of an instrument requires exploring if the original factorial structure can be recovered (structural validity) and verifying this structure using an independent sample (construct validity). Equivalence of the structural and construct validities of the adapted instrument with the original provides additional external validity.

Psychometric instruments are commonly used to compare different groups: Control vs. experimental group, female vs. males, different countries, and different conditions. One crucial feature of psychometric instruments is the equivalence of a construct across groups being both meaningful and valid. Measurement invariance across groups allows researchers to examine whether respondents from different groups interpret the same measure conceptually similarly. This equivalence is required to produce more comprehensive and broadly applicable results. Thus, measurement invariance increases the robustness and validity of the instrument. The original version of MBVI did not analyze measurement invariance

Although construct validity for MBVI has been provided (Andrews et al., 2017), this confirmation of latent factor structure in different contexts is still required for intercultural studies. This chapter validates a version of MBVI for preservice teachers (hereafter MBVI-PT) in three different cultural contexts: Germany, Portugal, and Spain. As in the original version, adaptation results in an 11-item self-reported questionnaire that measures three critical subscales of subjective value (interest, utility, and cost) of the task of using mathematics in biology. In the original MBVI (designed for life science students): We have slightly changed MBVI by extending the target to preservice elementary teachers (instead of life sciences students), ensuring that it was suitable and consistent with the original version. Then, we carefully translated this adaptation (firstly to Spanish, and after that to German and Portuguese). This chapter provides evidence of structural, construct, and external validities together with a study of measurement invariance of construct across countries, sexes, and prior learning.

Research Questions

This chapter analyzes if items of adapted Spanish MBVI-PT load into three factors (interest, utility, and cost values) according to an acceptable model fit (research question 1). Secondly, we analyze if this model is adequate for an independent sample of Spanish, German, and Portuguese preservice elementary teachers (research question 2). Further, we explored measurement invariance across countries (Germany, Portugal, and Spain), prior learning (STEM or non-STEM), and sexes (male or female) (research question 3). This approach serves us to establish if differences among subgroups are valid (research question 4).

METHODOLOGY

Participants

Participants were 643 university students (primary preservice teachers) from 3 European countries: Germany, Portugal, and Spain, all enrolled in a Degree in Primary Education and attending science or mathematics education subjects (Table 1). German preservice teachers (96 participants) come from two nearby universities in the Land of Schleswig-Holstein: the Europe University of Flensburg and the University Christian Albrecht of Kiel. Their students come from the surrounding area and belong to similar social and cultural contexts. Portuguese preservice teachers (72 participants) come from the Higher School of Education of Santarém. Spanish preservice teachers (475 participants) come from the Faculty of Education and Psychology (Badajoz) and from the Teacher Training School (Cáceres), faculties of the University of Extremadura. Both faculties are 100 km apart in the region of Extremadura, and their students come from the surrounding area and belong to similar social and cultural contexts. We split the Spanish sample into two, one of 350 students for Exploratory Factor Analysis (EFA) and the other of 125 students for Confirmatory Factor Analysis (CFA).

One important difference among participants is that Portuguese preservice teachers were enrolled in general degrees in childhood and primary education, and Spanish preservice teachers were enrolled in general degrees in primary education. In contrast, German participants were enrolled in a degree in primary education specializing in science and mathematics. Also, Portuguese students attended science education courses, whereas German and Spanish participants had classes focused on biology education. These differences were considered during instrument content adaptation.

Table 1. Description of participants according to their country (Germany, Portugal, or Spain), prior learning (STEM or non-STEM during Upper Secondary Education), sex (male or female), and referencing the factor analysis in which they are used (EFA/CFA).

Country	CFA Dataset			EFA Dataset
	Germany	Portugal	Spain	Spain
Sample size	96	72	125	350
Sex				
Female	80	68	84	232
Male	16	4	41	118
Prior learning				
STEM	86	12	22	88
Non-STEM	10	60	123	262
Total sample size	643			

Instrument Adaptation

We followed the International Test Commission's Guidelines for Translating and Adapting Tests (International Test Commission 2017) and the Guidelines on Computer-Based and Internet-Delivered Testing (International Test Commission 2005). In particular, we gave special attention to the following points: i) permission requested and obtained from author Melissa Aikens for mimicking the original test; ii) we analyze the legitimacy of subscales in the three different preservice teachers' groups, German, Portuguese, and Spanish; iii) we discuss translations to fit linguistic and training program differences (the Portuguese version uses science instead biology and German version focus interest items on teaching instead of understanding); iv) German and Portuguese collaborators assess items were understood in the same way across language ensuring construct makes sense in the cultures of both groups; v) we verify the suitability of internet-delivered testing and procedure; vi) test was administered in a controlled mode (no direct human supervision of the assessment session is involved, but the test is made available only to known test-takers); vii) participants acted informed, voluntarily, and anonymously, viii) the study was reviewed and approved by the ethical committee of each institution.

The instrument consisted of an adaptation of MBVI to measure subjective task values in preservice teachers. MBVI is based on three subscales containing 11 Likert-type response items ranging from 1 "strongly disagree" to 7 "strongly agree." This instrument was initially developed and validated to measure life science students' interest in using mathematics to understand biology, their perceived usefulness to their life science careers, and the cost of using mathematics in biology courses (Andrews *et al.*, 2017). The content was adapted to measure preservice teachers' interest in using mathematics to understand (or to teach, in the German version) biology (science in the Portuguese version), their perceived usefulness to their career as primary education teachers, and the cost of using mathematics in biology (science in the Portuguese version) courses (Table 2).

Instrument Description

In our translated versions, interest and utility items were adapted to a future as a primary education teacher of biology instead of a life science career (Table 2). Minor differences in the instrument were due to differences in preservice teacher training programs. The Portuguese version analyzes subjective task values for using mathematics in science understanding and teaching (Portuguese participants attended science and mathematics content and education courses). The Spanish version analyzes the use of mathematics in the context of biology understanding and teaching (these participants were attending biology education courses). In contrast, the German version analyzes the use of mathematics in the context of biology teaching (these participants were attending biology education courses). In this regard, the Portuguese and Spanish versions were close to the original MBVI. The German version differs from the Portuguese and Spanish versions in the interest items. In the German version, interest items focus on using mathematics to teach biology, whereas in other versions focus on interest in using mathematics to understand biology. Items were presented randomly to students through a Google Form online survey (Appendix I).

Table 2. Items of the original MBVI for life science students (left column), the adapted and translated versions to German, Portuguese, and Spanish for preservice teachers (middle columns), and the English back translation (right column).

Items	Biology students	Preservice Teachers			
	English	German	Portuguese	Spanish	English
Interest					
Int1	Using math to understand biology intrigues/would intrigue me.	Es würde mich faszinieren, Mathematik zu benutzen, um Biologie zu unterrichten.	Tenho curiosidade em usar a Matemática para compreender melhor os conceitos em Ciências.	Tengo curiosidad por utilizar las matemáticas para entender la biología	Using math to teach/ understand biology would intrigue me.
Int2	It would be fun to use math to understand biology.	Es würde mir Spaß machen, Mathematik zu benutzen, um Biologie zu unterrichten	Poderá ser divertido utilizar a Matemática para compreender melhor as Ciências.	Podría ser entretenido utilizar las matemáticas para entender la biología	It would be fun to use math to teach/ understand biology.
Int3	Using math to understand biology appeals/would appeal to me.	Mit Mathe Biologie zu unterrichten, spricht mich an	Usar a Matemática para compreender as Ciências é algo que me atrai.	Me resulta atractivo utilizar las matemáticas para entender la biología	Using math to teach/ understand biology appeals to me.
Int4	Using math to understand biology is/ would be interesting to me.	Es ist für mich interessant, mit Hilfe von Mathematik, um Biologie zu unterrichten.	Utilizar a Matemática para compreender as Ciências é algo que considero interessante.	Podría ser interesante utilizar las matemáticas para entender la biología	Using math to teach/ understand biology is interesting to me.
Utility value					
Uti1	Math is valuable for me for my life science career	Mathematik ist eine wertvolle Ressource für meine Zukunft als Biologielehrer	A Matemática é um recurso valioso para o meu futuro enquanto professor/a que vai ensinar conteúdos de Ciências.	Las matemáticas son un recurso valioso para mi futuro como maestro de contenidos de biología	Math is valuable for me for my future as a biology teacher.
Uti2	It is important for me to be able to solve math questions for my career in the life sciences	Es ist wichtig für meine Zukunft als Biologielehrerin, mathematische fragen lösen zu können	É importante, para o meu futuro enquanto professor/a que vai ensinar conteúdos de Ciências, ser capaz de resolver situações matemática	Es importante ser capaz de resolver cuestiones matemáticas para mi futuro como maestro de contenidos de biología	It is important for me to be able to solve math questions for my future as biology teacher.
Uti3	An understanding of math is essential for me for my life science career	Mathematik zu verstehen ist essentiell für meine Zukunft als Biologielehrerin	Entender a Matemática é essencial para mim, enquanto professor/a que irá ensinar conteúdos de Ciências.	Entender las matemáticas es esencial para mi futuro como maestro de contenidos de biología	An understanding of math is essential for me for my future as a biology teacher.

Continued on following page

Table 2. Continued

Items	Biology students	Preservice Teachers			
	English	German	Portuguese	Spanish	English
Uti4	Math will be useful to me in my life science career	Mathematik wird für meine Zukunft als Biologielehrer nützlich sein	A Matemática ser-me-à útil na minha carreira enquanto professor/a que irá ensinar conteúdos de Ciências.	Las matemáticas serán útiles para mi futuro como maestro de contenidos de biología	Math will be useful to me in my future as a biology teacher.
Perceived cost					
Cos1	I have/would have to work harder for a biology course that incorporates math than for one that does not.	Ich müsste härter für einen Biologiekurs arbeiten, der Mathematik beinhaltet, als für einen, der das nicht tut	Terei de trabalhar mais para uma unidade curricular de Ciências se esta integrar Matemática do que teria se tal não acontecesse.	Incorporar las matemáticas a los contenidos de biología me supone un esfuerzo adicional	I would have to work harder for a biology course that incorporates math than for one that does not.
Cos2	I worry/would worry about getting worse grades in a biology course that incorporates math than one that does not.	Ich würde mir Sorgen machen, schlechtere Noten in einem Biologiekurs zu bekommen, der Mathe enthält, als einen, der das nicht tut	Enquanto estudante, preocupa-me que, por incorporar Matemática, possa tirar classificações piores em unidades curriculares da área das Ciência.	Como alumno me preocupa que, por incorporar las matemáticas, pueda sacar peores notas en contenidos de biología	I would worry about getting worse grades in a biology course that incorporates math than one that does not.
Cos3	Taking a biology course that incorporates math intimidates/would intimidate me.	Einen Mathematikkurs zu belegen, würde mich einschüchtern	Assusta-me um pouco frequentar uma unidade curricular de Ciências que integra Matemática.	Me asusta un poco tener que incorporar las matemáticas a mis futuras clases de biología	Taking a biology course that incorporates math would intimidate me.

Procedure

These students completed voluntarily and anonymously an online survey about their subjective value toward math-biology integration for their future as primary education teachers. Participants also provided information about their sex (male or female) and the academic itinerary they elected during Upper Secondary Education (Appendix I). We used this information to proxy their prior learning, and we categorized it as STEM (including science and technology itineraries) or non-STEM (including humanities, social sciences, and arts itineraries, or higher-level professional education). Participants were previously informed about the anonymity of their data and the aim of the research (their attitudes towards the integration of biology and mathematics in its future teaching), the definition of biology (the study of living beings from biomolecules to ecosystems), and mathematics (the study of properties and relationships between abstract entities such as numbers, graphs, geometric figures, or symbols). The average time taken to complete the test was about 5 minutes.

Data Analysis

Statistical analyses were performed using R-based free software Jamovi 1.8 (jamovi.org) and JASP 0.16.3 (jasp-stats.org). Since variables did not follow a normal distribution (Shapiro-Wilk test p-value <0.001), we used non-parametric statistics such as robust ANOVA (R-package' walrus') to identify sources of variability in the observed variances.

Subjective values were modeled as latent constructs. EFA was conducted to ascertain the structural consistency of the versions of the instrument with respect to the original (a three-factor model for the 11 items of MBVI). EFA was run using principal axis factoring (idoneous for the absence of normality) with an oblimin rotation (allowing factors to correlate). We chose the optimal number of factors guided by three different model-fit indices: The root mean square error of approximation (RMSEA), the Tucker-Lewis Index (TLI), and the Bayesian Information Criteria (BIC). These three indexes account for and penalize model complexity. RMSEA and TLI penalize models with more than the necessary number of factors by including degrees of freedom in their calculation (an adjustment for complexity in terms of the number of free parameters), rewarding more parsimonious models. RMSEA values ≤ 0.08 and TLI ≥ 0.9 are typically taken as indicators of good model fit (Finch 2020). This procedure results in a set of good models from which we select the simplest using BIC to measure comparative fit. BIC introduces a penalty term for the number of parameters in the model. The use of information criteria like the AIC or BIC in model selection is common, and its use to determine the number of factors to retain in EFA has been discussed before (Akaike, 1987). This procedure prevents selecting more parameters (factors) than those that can be justified by the data (compromising the model's usefulness for future predictions). We used Bartlett's test of Sphericity, KMO measurement for sampling adequacy of each variable, and Cronbach's alpha to assess the internal consistency of each subscale. Bartlett's test values <0,001 and KMO values closer to 1.0 are considered ideal. Acceptable alpha values range from 0.7 to 0.95.

CFA was used to confirm the construct validity of the three-factor model structure in a multicultural dataset. This dataset contained an independent sample of Spanish preservice teachers together with samples of German and Portuguese preservice teachers (Table 1). We estimated the model fit of CFA by multiple indices: χ^2 statistic, RMSEA, TLI, and the Comparative Fit Index (CFI). The χ^2 is shown to allow comparison with previous works. This index has reduced usefulness because it assumes multivariate normality and is sensitive to sample size (nearly always rejects the model when large samples are used). We established a threshold for good/acceptable values according to Sun (2005) and Kline (2011), where: CFI and TLI $\geq 0.95/0.90$ and RMSEA $\leq 0.06/0.1$.

Measurement invariance was tested using Multigroup Confirmatory Factor Analysis (MCFA), which examines the change in model fit indexes in a series of nested CFA models in which parameters are progressively constrained to be equal across groups. Measurement invariance was tested across countries (German, Portugal, and Spain), sex (males and females), and secondary education background (STEM and non-STEM). The fit of each nested model is compared to its less-constrained antecedent by observing differences in model fit values. First, we tested a 3 factor model with unconstrained parameters (configural invariance or invariance of the model form) across countries, sex, and backgrounds. Second, we tested this model with factor loadings constrained to be equal across groups (metric invariance). By subtracting, we compared its model's fit to the fit obtained on the configural model. Third, we constrained observed intercepts to be equal across the groups (scalar invariance), and we compared its fit to the fit obtained on the invariant model. Fourth, we additionally constrained residual variances and compared the fit of the resulting model to the scalar invariant model (strict invariance). The changes in CFI (ΔCFI ≤ -0.01) and in McDonald's fit index (ΔMFI) ≤ -0.02 across the models were used to evaluate whether the construct was invariant at each of the stages of the multigroup analysis mentioned above. These ΔCFI and ΔMFI indexes are minimally affected by sample size and model complexity (Cheung & Rensvold, 2002, Kang *et al.*, 2016).

RESULTS AND DISCUSSION

Raw Data Analysis

As shown in Table 1, the sample of preservice teachers was characterized by a high proportion of females, a common characteristic of all primary education degrees. This imbalance in sex distribution among participants was also observed in the study used to validate MBVI (Andrews et al., 2017), making results more comparable. Uneven distribution in the prior learning of preservice teachers was also observed. Although most preservice teachers enter the primary education degree from secondary higher education, Portuguese and Spanish preservice teachers enter the teacher training program generally from a social science background, whereas Germans enter preferentially from an experimental science background (Table 1).

Table 3 shows the mean and standard deviations of raw scores self-reported by preservice teachers to different items. A first look suggested intercultural differences. Although students of the three countries reported comparable cost values, Portuguese reported more interest and utility values. Portuguese students' high interest and utility value levels could be explained by their involvement in interdisciplinary activities between science and mathematics courses since the beginning of their program. To explore sources of data variability, a robust ANOVA was performed. This robust analysis is adequate against the unbalance in the sample size and the violation of the two ANOVA assumptions (normality and homoscedasticity). According to rANOVA results, no significant differences (p-value$\leq$0.001) associated with sex or prior learning were observed in any of the variables. In contrast, the country was a predictor of changes in item raw scores of interest and utility (all p-values$\leq$0.001) but not cost.

Table 3. Means and standard deviations (in parenthesis) from preservice teachers' responses to MBVI-PT in Germany, Portugal, and Spain (CFA dataset of Table 2).

Items	German	Portuguese	Spanish
Interest			
Int1	3.19 (1.57)	5.72 (1.05)	4.01 (1.68)
Int2	3.54 (1.54)	5.96 (0.99)	4.40 (1.57)
Int3	3.47 (1.67)	5.51 (1.79)	3.78 (1.76)
Int4	3.83 (1.80)	5.76 (1.14)	4.54 (1.59)
Utility			
Uti1	3.90 (1.62)	5.79 (1.01)	4.64 (1.00)
Uti2	4.21 (1.72)	6.11 (0.87)	4.69 (1.49)
Uti3	3.94 (1.63)	6.04 (0.90)	4.84 (1.49)
Uti4	4.36 (1.55)	5.99 (1.04)	4.77 (1.41)
Perceived cost			
Cos1	4.70 (1.83)	4.85 (1.64)	4.80 (1.64)
Cos2	4.20 (1.96)	4.31 (1.93)	4.59 (1.84)
Cos3	3.99 (1.87)	3.84 (1.93)	4.53 (1.67)

Content Validity

To add content validity, MBVI-PT was presented and discussed with experienced practitioners in science education from Germany, Portugal, and Spain. All agree that their students would recognize the importance of mathematics in their role as biology teachers to a greater or lesser extent. However, the Portuguese version was slightly modified to account for Portugal's undergraduate program differences because Portuguese preservice teachers are trained to teach science as a whole instead of its disciplines. The MBVI-PT German version was also modified to account for its undergraduate program. Since their preservice teacher training focused on math and biology, it was considered that their interest would be more related to teaching than understanding. These changes were deemed that add content validity and were appropriate until construct validity across subgroups was analyzed.

Structural Validity

Regarding structural validity, a series of EFAs (with an initial sample of Spanish participants) were conducted to study the relationship among items and select the number of factors to retain. We ran models with 2 to 4 factors and compared their fit indices (Table 4). According to RMSEA and TLI model fits (Table 4), a factorial model with three or four factors provides an adequate representation of the data (RMSEA ≤ 0.08 and TLI ≥ 0.9). BIC was used to choose between these two competing models. The three-factor model showed the lowest BIC value (−75,38) and was finally selected.

Table 4. Fit indexes from three EFAs using a different number of factors (left column) to select the most appropriate. The EFA dataset of Table 2 was used.

Number of Factors	% Variance Explained	RMSEA	CI 90% RMSEA		TLI	BIC	Test Model		
			Bottom	Upper			χ^2	df	p-Value
2	63.5	0.113	0.100	0.132	0.898	−5.59	193.58	34	< 0.001
3	67.5	0.073	0.053	0.093	0.960	−75.38	71.07	25	< 0.001
4	69.7	0.045	0.013	0.073	0.984	−70.31	29.27	17	0.032

The solution for the three-factor model showed high loadings for the three factors (0.78 on average across the 11 items), low cross-factor loadings, and a 67.5% variance explained. Moreover, all factors demonstrated good reliability. The three Cronbach's alpha values were higher than 0.77, and the sample size was acceptable according to KMO (Table 5). The 11 items of MBVI-PT fit the original instrument's expected three-factor model (Andrews *et al.*, 2017). Noteworthy, the correlations between interest and utility, interest and cost, and utility and cost (0.78, -0,38, and -0.06, respectively) showed a similar trend to those reported for MBVI (0.62, -0,58, and -0.27, respectively) and other for similar instrument (0.31, -0,28, and -0.04, respectively) (Lohbeck & Retelsdorf, 2021). However, in the Spanish preservice teachers' case, the cost is not associated with utility value. The variance explained by the model (67%) was close to the original MBVI value (69%), providing convergent validity. Results show that the Spanish version of MBVI-PT showed structural and convergent validity comparable to the original instrument.

Table 5. Results of EFA performed with a sample of 350 Spanish participants (Dataset is described in Table 1).

Items	Factor			KMO
	I	II	III	
I. Interest				
Int1	**0.864**			0.931
Int2	**0.844**			0.916
Int3	**0.837**			0.924
Int4	**0.701**			0.924
II. Utility value				
Uti1		**0.927**		0.933
Uti2		**0.845**		0.915
Uti3		**0.795**		0.909
Uti4		**0.589**		0.913
III: Perceived cost				
Cos1			**0.753**	0.765
Cos2			**0.746**	0.798
Cos3			**0.688**	0.770
Characteristics of factors				
% Variance explained	27.3	25.4	14.8	
% Cumulative variance explained	27.3	52.7	67.5	
Cronbach's alpha	0.906	0.914	0.772	
Factor correlations		**II**		
I	–			
II	**0.78**	–		
III	**- 0.38**	**- 0.06**	–	

Construct Validity

A CFA was conducted on independent participants of multicultural origin German, Portuguese and Spanish). Furthermore, this CFA was used to test the hypothesized three-factor model (Table 6). As often described, the $\chi 2$ test rejected the model data fit. However, three supplemental fit indices (RMSEA, TLI, and CFI) supported the adequacy of the three-factor model as a good model for the independent dataset. RMSEA was ≤ 0.06, and CFI and TLI were ≥ 0.95. Figure 1 represents the resulting factorial model.

Table 6. Results of CFA performed with an independent sample of 293 preservice teachers (125 Spanish, 96 German, and 72 Portuguese).

RMSEA	CI 90% RMSEA		TLI	CFI	Test Model		
	Bottom	Upper			χ^2	df	p-Value
0.056	0.037	0.075	0.981	0.986	77.58	41	< 0.001

Figure 1. Three-factor model for interest (Int), utility (Uti), and cost (Cst) subscales. Model results from CFA of Table 6.

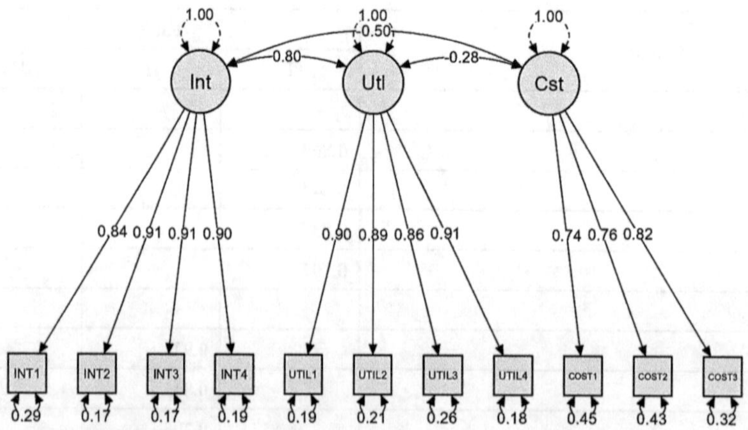

Measurement Invariance

MCFA was conducted to assess the comparability of the scores (Table 7). Configural invariance analysis showed that the three-factor model was adequate regardless of the country, sex, or prior learning (CFI≥0.95). Metric invariance analysis showed non-significant differences (ΔCFI ≤ −0.01 and ΔMFI ≤ −0.02) in factor loading regardless of country, sex, or prior learning. This invariance means that correlations among variables can be confidentially compared across groups. Scalar invariance analysis indicated non-significant differences (ΔCFI ≤ −0.01 and ΔMFI ≤ −0.02) in intercepts between STEM and non-STEM participants (prior learning group) but significant across sex and countries. This invariance means that mean scores can be confidentially compared across prior learning, not sex or countries. Residual invariance analysis (strict invariance) indicated non-significant differences (ΔCFI ≤ −0.01 and ΔMFI ≤ −0.02) between STEM and non-STEM participants, any comparison can be confidentially made. Thus, as typical during instrument development (Putnick & Bornstein, 2016), MBVI-PT does not provide a complete measurement invariance.

Table 7. Measurement invariance for the three-factor model across the country, sex, and prior learning.

Groups	Model	χ^2	df	p-Value	CFI	MFI	Δ CFI	Δ MFI
Country	*Configural invariance*	247.78	123	< 0.001	0.948	0.808	-	-
	Metric invariance	281.55	139	< 0.001	0.941	0.784	- 0.007	-0.012
	Scalar invariance	269.73	155	< 0.001	0.952	0.822	0.011	0.038
	Residual invariance	324.11	177	< 0.001	0.939	0.778	-0.013	-0.044
Sex	*Configural invariance*	129.93	82	< 0.001	0.982	0.921	-	-
	Metric invariance	137.46	90	< 0.001	0.982	0.922	0	0.001
	Scalar invariance	133.25	98	0.01	0.960	0.942	- 0.022	0.020
	Residual invariance	216.28	109	< 0.001	0.960	0.833	0	-0.109
Prior learning	*Configural invariance*	154.14	82	< 0.001	0.973	0.884	-	-
	Metric invariance	158.10	90	< 0.001	0.975	0.890	0.002	0.006
	Scalar invariance	157.20	98	< 0.001	0.978	0.904	0.005	0.014
	Residual invariance	216.28	109	< 0.001	0.960	0.833	-0.013	-0.071

CONCLUSION

The work presented in this chapter responds to the initial research questions. Items of MBVI-PT load into three factors (interest, utility, and cost values) according to an acceptable model fit (research question 1). This three-factor model is adequate to fit data obtained from Spanish, German, and Portuguese preservice elementary teachers (research question 2). Measurement remains fully invariant concerning prior learning (STEM or non-STEM) and partially invariant across countries (Germany, Portugal, and Spain) and sexes (male or female) (research question 3). Although the country of origin is the primary source of variability in the data, the lack of scalar invariance across countries did not permit the conclusion that differences among countries are significant (research question 4).

Limitations and Educational Implications

This study was conducted by sampling preservice teachers within particular cultural contexts. Although our findings are consistent with previous research on subjective task-value, this study requires replication before generalizing (more diverse settings and students). This limitation entails observing the necessary caution while generalizing its findings to students of other training programs, secondary education backgrounds, and cultures. Moreover, our instrument captures task values driving motivation in a domain-specific manner (math-biology/science integration at large) rather than specific tasks (e.g., using mathematical representations to comprehend a biological/scientific process such as germination). Success in math-biology integration depends on students completing individual tasks and assignments daily and weekly. Little is known about how students' study intentions and motivation operate on a task-specific level. Adapting the instrument to different classroom activities, settings, and methodologies will inform situational task values, as recently pointed out (Eccles & Wigfield, 2020). Finally, it is necessary to note that MBIV-PT is a self-reported questionary. Although this instrument is commonly used in psychology and correlates with neurobiology findings, it operates on the questionable assumption that participants have perfect self-knowledge about their subjective values. Responses can be biased by answering style, social desirability, interpretation of items, and the lack of incentives to participate.

Since task-value determines motivation, attitude, and academic achievement, and this value is malleable with appropriate educational interventions, MBVI-PT can monitor students' attitudes toward subject integration.

ACKNOWLEDGMENT

J.M.M.M., M.R.E.G., and J.A.G.O. research was supported by "ERDF A way of making Europe" Project GR21047 and IB18004; J.M.M.M. and M.R.E.G. research was also supported by the Spanish Ministry of Science and Innovation and the State Research Agency through grant PID2020-115214RB-I00 funded by MCIN/AEI/10.13039/501100011033. J.M.M.M. collected the data of German participants during a research internship at the Leibniz Institute for Science and Mathematics Education of the University Christian Albrecht of Kiel under the direction of Dr. Ute Harms. This internship was supported by the grant FPU-15/02737 of the Spanish Ministry of Education, Culture, and Sports. The authors acknowledge all those collaborating in tuning the instrument's content to different languages and participants.

REFERENCES

Akaike, H. (1987). Factor Analysis and AIC. *Psychometrika, 52*(3), 317–332. doi:10.1007/BF02294359

Andrews, S. E., Runyon, C., & Aikens, M. L. (2017). The math-biology values instrument: Development of a tool to measure life science majors' task values of using math in the context of biology. *CBE Life Sciences Education, 16*(3), 1–12. doi:10.1187/cbe.17-03-0043 PMID:28747355

Brewer, C., & Smith, D. (2011). *Vision and change in undergraduate biology education: a call to action.* American Association for the Advancement of Science.

Bursal, M., & Paznokas, L. (2006). Mathematics Anxiety and Pre-service Elementary Teachers' Confidence to Teach Mathematics and Science. *School Science and Mathematics, 106*(4), 173–180. doi:10.1111/j.1949-8594.2006.tb18073.x

Buss, R. R. (2010). Efficacy for Teaching Elementary Science and Mathematics Compared to Other Content. *School Science and Mathematics, 110*(6), 290–297. doi:10.1111/j.1949-8594.2010.00037.x

Cady, J. A., & Rearden, K. (2007). Preservice Teachers' Beliefs about Knowledge, Mathematics, and Science. *School Science and Mathematics, 107*(6), 237–245. doi:10.1111/j.1949-8594.2007.tb18285.x

Cheung, G. W., & Rensvold, R. B. (2002). Evaluating goodness-of-fit indexes for testing measurement invariance. *Structural Equation Modeling, 9*(2), 233–255. doi:10.1207/S15328007SEM0902_5

DeLuca, C., Ogden, H., & Pero, E. (2015). Reconceptualising Elementary Preservice Teacher Education: Examining an Integrated-Curriculum Approach. *New Educator, 11*(3), 227–250. doi:10.1080/1547688X.2014.960986

Eccles, J. S., & Wigfield, A. (2020). From expectancy-value theory to situated expectancy-value theory: A developmental, social cognitive, and sociocultural perspective on motivation. *Contemporary Educational Psychology, 61*, 101859. doi:10.1016/j.cedpsych.2020.101859

Etkin, A., Büchel, C., & Gross, J. J. (2015). The neural bases of emotion regulation. In Nature Reviews Neuroscience (Vol. 16, Issue 11, pp. 693–700). doi:10.1038/nrn4044

Finch, W. H. (2020). Using fit statistic differences to determine the optimal number of factors to retain in an exploratory factor analysis. *Educational and Psychological Measurement, 80*(2), 217–241. doi:10.1177/0013164419865769 PMID:32158020

Forsblom, L., Pekrun, R., Loderer, K., & Peixoto, F. (2022). Cognitive appraisals, achievement emotions, and students' math achievement: A longitudinal analysis. *Journal of Educational Psychology, 114*(2), 346–367. doi:10.1037/edu0000671

Furrow, R. E., Kim, H. G., Abdelrazek, S. M. R., Dahlhausen, K., Yao, A. I., Eisen, J. A., Goldman, M. S., Albeck, J. G., & Facciotti, M. T. (2020). Combining Microbial Culturing With Mathematical Modeling in an Introductory Course-Based Undergraduate Research Experience. *Frontiers in Microbiology, 11*, 581903. doi:10.3389/fmicb.2020.581903 PMID:33250873

Hernández-Barco, M., Cañada-Cañada, F., Corbacho-Cuello, I., & Sánchez-Martín, J. (2021). An Exploratory Study Interrelating Emotion, Self-Efficacy and Multiple Intelligence of Prospective Science Teachers. *Frontiers in Education*, *6*, 1–12. doi:10.3389/feduc.2021.604791

International Test Commission. (2005). International guidelines on computer-based and internet-delivered testing. *International Journal of Testing*, *6*(2), 143–171.

International Test Commission. (2017). *The ITC Guidelines for Translating and Adapting Tests*. https://www.intestcom.org

Kang, Y., McNeish, D. M., & Hancock, G. R. (2016). The role of measurement quality on practical guidelines for assessing measurement and structural invariance. *Educational and Psychological Measurement*, *76*(4), 533–561. doi:10.1177/0013164415603764 PMID:29795877

Kim, M. M., Andrews, R. L., & Carr, D. L. (2004). Traditional versus integrated preservice teacher education curriculum: A case study. *Journal of Teacher Education*, *55*(4), 341–356. doi:10.1177/0022487104266778

Kline, R. B. (2011). *Principles and practice of structural equation modeling*. Guilford.

Labov, J. B., Reid, A. H., & Yamamoto, K. R. (2010). Integrated biology and undergraduate science education: A new biology education for the twenty-first century? *CBE Life Sciences Education*, *9*(1), 10–16. doi:10.1187/cbe.09-12-0092 PMID:20194802

Lebreton, M., Abitbol, R., Daunizeau, J., & Pessiglione, M. (2015). Automatic integration of confidence in the brain valuation signal. *Nature Neuroscience*, *18*(8), 1159–1167. doi:10.1038/nn.4064 PMID:26192748

Leszczynski, E. (2014). *The Study of Middle School Mathematics and Science Teachers' Practices, Perceptions, and Attitudes Related to Mathematics and Science Integration* [Unpublished doctoral dissertation]. University of Montclair, Canada.

Lohbeck, A., & Retelsdorf, J. (2021). Assessing value beliefs among university students: Validation of the Value Beliefs Questionnaire for University Students (VBQU). *Studies in Educational Evaluation*, *70*, 101052. doi:10.1016/j.stueduc.2021.101052

Luttrell, V. R., Callen, B. W., Allen, C. S., Wood, M. D., Deeds, D. G., & Richard, D. C. S. (2010). The Mathematics Value Inventory for General Education Students: Development and Initial Validation. *Educational and Psychological Measurement*, *70*(1), 142–160. doi:10.1177/0013164409344526

Marcos-Merino, J. M., Gallego, R. E., & Ochoa de Alda, J. A. G. (2020). Valor subjetivo y emociones hacia el uso de Química en una práctica activa interdisciplinar. *Educación en la Química*, *31*(4), 101–111. doi:10.22201/fq.18708404e.2020.4.76221

May, R. (2004). Uses and abuses of mathematics in biology. *Science*, *303*(5659), 790–793. doi:10.1126cience.1094442 PMID:14764866

McCright, A., O'Shea, B., Sweeder, R., Urquhart, G., & Zeleke, A. (2013). Promoting interdisciplinarity through climate change education. *Nature Climate Change*, *3*(8), 713–716. doi:10.1038/nclimate1844

Mellado, V., Borrachero, A. B., Brígido, M., Melo, L. V., Dávila, M. A., Cañada, F., Conde, M. C., Costillo, E., Cubero, J., Esteban, R., Martínez-Borreguero, G., Ruiz, C., Sánchez, J., Garritz, A., Mellado, L., Vázquez, B., Jiménez, R., & Bermejo, M. L. (2014). Las emociones en la enseñanza delas ciencias. *Enseñanza de las Ciencias*, *32*(3), 11–36. doi:10.5565/rev/ensciencias.1478

Oltvai, Z., & Barabási, A. (2002). Life's complexity pyramid. *Science*, *298*(5594), 763–764. doi:10.1126cience.1078563 PMID:12399572

Pekrun, R. (2006). The control-value theory of achievement emotions: Assumptions, corollaries, and implications for educational research and practice. *Educational Psychology Review*, *18*(4), 315–341. doi:10.100710648-006-9029-9

Putnick, D. L., & Bornstein, M. H. (2016). Measurement invariance conventions and reporting: The state of the art and future directions for psychological research. *Developmental Review*, *41*, 71–90. doi:10.1016/j.dr.2016.06.004 PMID:27942093

Riegle-Crumb, C., Morton, K., Moore, C., Chimonidou, A., Labrake, C., & Kopp, S. (2015). Do Inquiring Minds Have Positive Attitudes? The Science Education of Pre-service Elementary Teachers. *Science Education*, *99*(5), 819–836. doi:10.1002ce.21177 PMID:27667862

Sharot, T., & Sunstein, C. R. (2020). How people decide what they want to know. *Nature Human Behaviour*, *4*(1), 14–19. doi:10.103841562-019-0793-1 PMID:31932690

Šorgo, A. (2010). Connecting biology and mathematics: First prepare the teachers. *CBE Life Sciences Education*, *9*(3), 196–200. doi:10.1187/cbe.10-03-0014 PMID:20810951

Spanish Government. (2022). *Real Decreto 157/2022, de 1 de marzo, por el que se establecen la ordenación y las enseñanzas mínimas de la Educación Primaria* [Royal Decree 157/2022, of March 1, establishing the organization and minimum standards for primary education].

Stone, E. M. (2014). Guiding students to develop an understanding of scientific inquiry: A science skills approach to instruction and assessment. *CBE Life Sciences Education*, *13*(1), 90–101. doi:10.1187/cbe-12-11-0198 PMID:24591508

Sun, J. (2005). *Assessing goodness of fit in confirmatory factor analysis. Measurement and Evaluation in Counseling and Development*. SAGE Publications Inc.

Thompson, K. V., Cooke, T. J., Fagan, W. F., Gulick, D., Levy, D., Nelson, K. C., Redish, E. F., Smith, R. F., & Presson, J. (2013). Infusing quantitative approaches throughout the biological sciences curriculum. *International Journal of Mathematical Education in Science and Technology*, *44*(6), 817–833. doi:10.1080/0020739X.2013.812754

Torsney, B. M., Ponnock, A. R., & Lombardi, D. (2017). The Role of Values in Preservice Teachers' Decision to Teach. *Teacher Educator*, *52*(1), 39–56. doi:10.1080/08878730.2016.1196986

KEY TERMS AND DEFINITIONS

Confirmatory Factor Analysis: A statistical method used to explore the adequateness of data to a given factorial model.

Construct (or Psychological Constructs): A mental construction of a psychological dimension (the underlying cause of covaried human behaviors) that scales and allows understanding of behavior.

Construct Validity: Sum of pieces of evidence to support that a construct adequately represents what is intended by the theoretical framework that supported its construction.

Exploratory Factor Analysis: A statistical method used to reduce data by extracting from a group of correlated items a common latent factor associated with all group items and exploring their underlying theoretical structure.

Factorial Model: Representation of the variance of a set of items into two or more factors, resulting in new variables derived from the common variance groups of items.

Measurement Invariance: Statistical property of a construct to scale and to be measured equivalently across different groups.

Psychometric Instrument: A tool to measure psychological constructs such as subjective value, attitudes, or motivation.

Robust ANOVA: A tool of non-parametric statistics to explore the source of variance in the data and robust against differences in sample sizes of groups and violations of normality and homoscedasticity of variables.

Task-Value: A psychological construct that captures the student's subjective perceptions of positive (interest, utility, relevance) and negative (physical cost, intellectual cost, affective cost, economical cost) incentives to be involved in a task.

APPENDIX

In this contribution we use different sources to validate the Math-Biology Values Instrument for Preservice Teachers (MBVI-PT). Here it follows a complete description of the instrument as administered to participants. This English version of the instrument can be adapted to different countries, as indicated in Table 2.

Table 8.

Math-Biology Values Instrument for Preservice Teachers (MBVI-PT)								
This survey is part of an investigation into your attitude towards teaching biology and mathematics and the importance you ascribe to this teaching for your future as a practitioner. Biology refers to the study of living beings, from ecosystems to biomolecules. Mathematics studies the properties and relationships between abstract entities such as numbers, graphs, geometric figures, or symbols. The collected data will be presented in aggregated anonymous form in future publications. Participation is voluntary								
Personal data								
Sex: ❑ Male ❑ Female								
Academic itinerary during Upper Secondary Education: ❑ Science and technology ❑ Humanities or social sciences ❑ Arts ❑ Higher-level professional education ❑ Other								
Anonymous identifier (only for longitudinal studies) Write down the first letter of your mother's name, the first letter of your father's name, last two numbers and letter of your identity cart (for example: MM54S) ID:								
Scale	**Item*	1 strongly disagree-7 totally agree						
Interest	Using Math to teach Biology intrigue me	1	2	3	4	5	6	7
	It would be fun to use Math to teach Biology	1	2	3	4	5	6	7
	Using Math to teach Biology appeals me	1	2	3	4	5	6	7
	Using Math to teach Biology would be interesting to me	1	2	3	4	5	6	7
Utility	Math is a valuable resource for my future as a Biology teacher	1	2	3	4	5	6	7
	It is important to be able to solve mathematical questions for my future as a Biology teacher	1	2	3	4	5	6	7
	Understanding Math is essential for my future as a Biology teacher	1	2	3	4	5	6	7
	Math will be useful for my future as a Biology teacher	1	2	3	4	5	6	7

Continued on following page

Table 8. Continued

Math-Biology Values Instrument for Preservice Teachers (MBVI-PT)								
Cost	Incorporating Math into Biology lessons require an additional effort	1	2	3	4	5	6	7
	I worry that, by incorporating Math, I can get worse grades in Biology	1	2	3	4	5	6	7
	It scares me to incorporate Math into my future Biology lessons	1	2	3	4	5	6	7

*Here, items are grouped by convenience, but items are randomly ordered when the instrument is used.

Chapter 14
Primary Pre-Service Teacher Changing Perspectives About Integrated STEAM Approach

Marisa Correia
https://orcid.org/0000-0001-6205-4475
Life Quality Research Centre (CIEQV), Polytechnic Institute of Santarém, Portugal

Maria Clara Martins
https://orcid.org/0000-0002-0689-8099
Polytechnic Institute of Santarém, Portugal

ABSTRACT

In recent years, there is a growing interest in educational approaches that include the integration of STEAM disciplines. However, to contribute to increase students' interest in these subjects from an early age, it is crucial to prepare primary teachers to carry out an integrated STEAM approach in schools. This study is aimed to demonstrate the potential of engaging primary pre-service teachers in a STEAM program within a science and a mathematics course. The results of the study emerged from the analysis of the participants' responses to a pre-post questionnaire, STEAM lesson plans they have developed, and reflective writing assignments they carried out at the end of the STEAM program. Findings show that primary pre-service teachers' attitudes toward a STEAM-integrated approach evolved positively through developing lesson plans. The results revealed benefits and challenges of planning STEAM activities according to the participants and their confidence to implement a STEAM-integrated approach in the context of practice. This work was supported by Portuguese Foundation for Science and Technology, I.P., Grant/Award Number UIDP/04748/2020.

DOI: 10.4018/978-1-6684-5765-8.ch014

INTRODUCTION

STEM (Science, Technology, Engineering, Mathematics) education movement emerged about 30 years ago in the United States to promote students' interest in these disciplines, driven by the need to prepare the STEM workforce to address urgent contemporary challenges, such as climate change and public health (Kelley & Knowles, 2016). The need to encourage young people to pursue careers in these areas has been widely pointed out in several international reports (e.g., Hurley & Hallissy, 2017), which, as a result, have recommended the integration of the STEM approach into school curricula at all levels of education and investment in teacher education. Research suggests that students should be involved in carrying out STEM activities, according to an integrated approach, from the earliest levels of schooling (Dejarnette, 2012). In recent years, a more comprehensive approach has been advocated, including articulating STEM with other curricular areas (arts, languages, history, and others), commonly known as STEAM (Yakman, 2012; Quigley et al., 2017). In fact, Martínez-Borreguero et al. (2022) conclude that an integrative STEM approach, based on active methodologies, contributes positively to the development of content knowledge of pre-service teachers in the initial levels of schooling.

In this context, it is imperative to study ways to support teachers and pre-service teachers to conceptualize and implement an integrated STEM approach (Estapa & Tank, 2017). In this sense, a STEAM program was developed in teacher education to prepare primary pre-service teachers (PSTs) to create and use STEAM lessons that take full advantage of the interdisciplinary connections between science and mathematics topics. This study investigates how primary PSTs develop STEAM lessons during a science and mathematics methods course and analyses their attitudes and perspectives about the integrated STEAM approach.

THEORETICAL BACKGROUND

Kelley and Knowles (2016) argue that the need to train more people in STEM areas comes from the social and environmental impacts of our century that jeopardize global security and economic stability. Such challenges imply the development of STEM literacy, which Mohr-Schroeder et al. (2015) define as the ability to apply concepts from science, technology, engineering, and mathematics in solving problems that cannot be solved by using a single area.

However, multiple definitions of STEM education coexist in the literature, which has led to ambiguity in the interpretation of the term (English, 2016; Perales-Palacios, 2020) and its transposition to the classroom (Martín-Páez et al., 2019). According to Hsu and Fang (2019), STEM education can be viewed as a set of STEM areas – a multidisciplinary or an integrated perspective – an inter and transdisciplinary perspective. According to Breiner et al. (2012), the latter foresees teaching these disciplines as a single and cohesive entity. Also sharing this integrated view of STEM education, Brown and Bogiages (2019) and Sanders (2009) support a model that integrates two or more STEM disciplines. Aguilera et al. (2021) assert that adding the term "integrated" to STEM education is redundant, as the acronym already alludes to disciplinary integration. Nevertheless, Wang et al. (2012) define STEM integration as an interdisciplinary pedagogical approach that removes barriers between disciplines to 1) deepen students' knowledge of each STEM area, 2) expand students' understanding of STEM areas through contact with relevant social and cultural contexts, and 3) increase students' interest in STEM areas. According to Roehrig et al. (2012), the way to operationalize STEM integration can be understood in two ways – content or context integration. The first model starts with the fusion of contents from different areas in a single activity or didactic sequence, allowing the teacher to explore contents from different areas and demonstrate how they are necessary to solve a given problem. The second type of integration focuses primarily on the content of one STEM area and uses the context of others to make the content more relevant.

Thibaut et al. (2018), based on an extensive literature review on STEM integration (iSTEM) and the social constructivist view on learning, propose a framework containing five key principles (integration of STEM content, problem-centred learning, inquiry-based learning, design-based learning, and cooperative learning) that describe the practices underlying integrated STEM education. First, STEM content integration must be explicit to help students develop their knowledge and skills in the different STEM disciplines. The model proposed by Thibaut et al. (2018) highlights inquiry-based learning as one of the critical principles of the STEM integrative approach, a methodology that has long been advocated in science education (e.g., Bybee, 2006). This instructional practice is strongly associated with the first levels of education, as children are naturally curious and constantly ask questions about the world around them (Greenwood et al., 2022). The iSTEM model also includes design-based learning, which is an approach that includes engineering design as a basis for creating connections with mathematics and/or science content and that should be encouraged early on in students (Difrancesca et al., 2014; Guzey et al., 2016; Shalali et al., 2017). This methodology implies formulating or identifying a problem, planning solutions, and creating models, prototypes or other products that respond to the problem (Guzey et al., 2016). The design process also opens the possibility of integrating the arts (Bequette & Bequette, 2012). Another aspect highlighted by Thibaut et al. (2018) is that the STEM approach implies that students should be involved in solving open-ended and real-world problems in a motivating context (Shahali et al., 2017). Finally, this model comprises collaborative learning as a feature of the STEM approach. Indeed, research has shown that collaboration plays a decisive role in children's learning (Howe et al., 2007).

In recent years, a broader pedagogical approach that includes non-stem disciplines has been widely highlighted in the literature (Yakman, 2012; Quigley et al., 2017; Aguilera & Ortiz-Revilla, 2021). This approach derives from the lack of creativity and innovation in recent college graduates in the United States (Land, 2013). However, some authors have pointed out the diversity of existing conceptions in the literature about the meaning of the letter "A" of the acronym (Colucci et al., 2017) as one of the main problems in implementing it. Quigley et al. (2017), among other authors, call STEAM the natural articulation of STEM disciplines with other areas, such as languages, social sciences, and arts (Sanders, 2009). Those authors define it as a transdisciplinary learning process in which the "A" represents the arts and humanities and aims to increase participation in STEM disciplines.

These ideas underlie Yakman's (2012) definition, which includes: language; the fine arts; the physical and manual arts; and the liberal or social arts. Quigley et al. (2017) say that the difference between the STEM and STEAM approaches is that the former argues in favour of including the arts while the latter intentionally includes them. Miller and Knezek (2013) contend that the curricular integration of technologies and the arts creates the appropriate environment for developing creativity and learning of science and mathematics topics and better integrating the disciplines. The idea that the insertion of the arts enhances a transdisciplinary learning experience has been upheld in the literature (Henriksen, 2014; Guyotte et al., 2015).

Efforts to integrate STEM education into curricula initially had a greater expression in secondary education (Hourigan et al., 2021), although several studies (e.g., Dejarnette, 2012) revealed that the introduction of the STEM approach in the early years of schooling has a positive effect on children's attitudes and learning. Primary education presents important conditions for implementing STEM education, among which are capitalizing on children's natural enthusiasm and interest in exploring the contents of STEM areas, developing basic knowledge in them, and fostering curriculum flexibility that more easily supports innovative approaches (Nadelson et al., 2013). Studies focused on the impact of the STEAM approach in primary education have shown very positive results in terms of students' competencies in the subjects covered (Bassachs et al., 2020; Greca et al., 2021) and their motivation (Lage-Gómez & Ros, 2021). However, according to Garcia-Carmona and Toma (2021), implementing in the classroom an integrated pedagogical approach to STEM brings many challenges, even in primary education, where the fact that the same teacher teaches different subjects could be a facilitating aspect. These authors emphasize that pre-service preparation of primary teachers has not focused on STEM areas (as is the case with engineering). Other authors consider it necessary for the success of STEM or STEAM education in primary education to develop teacher education programs that promote knowledge about inquiry and the confidence to use this methodology (Kim & Bolger, 2017; Nadelson et al., 2013). The lack of consensus on a common definition of STEM integration makes it urgent to develop a conceptual framework for STEM education in primary teacher education (Hourigan et al., 2021). Hence, it is recommended that teacher education is not limited to focusing on STEM areas in isolation, allowing pre-service teachers (PSTs) to carry out and implement activities according to an integrated STEM approach (Madden et al., 2016). In addition, Kim and Bolger (2017) emphasized that it is essential to prepare primary teachers to develop STEAM activities and, thus, contribute from an early age to increasing confidence in their planning and implementation skills following the principles of an integrated STEAM approach.

METHODS

This study used a mixed-methods research design that combined both quantitate and qualitative data to investigate the effect of a STEAM program on PSTs attitudes and perspectives toward STEAM pedagogy. Additionally, the study focused on their ability to develop STEAM lessons according to an integrated approach.

Participants

The research participants were pre-service primary school teachers studying in a Portuguese higher education institution, attending the last year of a teacher education program. The study sample was composed mainly of females (23 – female; 1 – male; 23-28 years old) enrolled in a science and mathematics methods course, who had no prior experience planning STEAM lessons. During the program, the PSTs worked in self-selected groups (Table 1).

Table 1. List of PSTs groups

Participants	Groups
PST1	G1
PST2	
PST3	G2
PST4	
PST5	
PST6	
PST7	G3
PST8	
PST9	
PST10	G4
PST11	
PST12	
PST13	
PST14	G5
PST15	
PST16	
PST17	
PST18	G6
PST19	
PST20	
PST21	G7
PST22	
PST23	
PST24	

STEAM Program

For eight weeks (2,5 hours per week), the PSTs learned about the theoretical background behind STEAM education. During the first part of the program, the PSTs performed several STEAM activities (Table 2), reflected on underlying instructional practices, evaluated the teaching materials, and identified the math and science curriculum topics covered.

Table 2. STEAM activities

Activities	Description
The measure of the solar system (Correia & Martins, 2020)[1]	Creating a representation of the Solar System at scale (in terms of distance between planets), using the representation of unit fractions, not units on the number line, and determining the relative and absolute distances.
Science at Christmas time (Dias et al., 2017)[2]	The activity involves exploring electrical circuits (physics), solving mathematical problems (mathematics) and writing informative texts (language).
My itineraries (Camacho et al., 2022)[3]	This activity focuses on mathematics, physics and geography topics, comprising tasks that involve using digital resources (e.g., Google Maps®) and building a compass.

The PSTs were introduced to the topics listed in Table 3 during the Methods course over 40 hours.

Table 3. Topics covered in the Science and Mathematics Methods course

Topics	Number of hours
Science and mathematics curriculum in primary education The benefits of science education in primary education Social constructivism in science education	5
Development of 21st-century skills Collaborative learning Problem and project-based learning	5
Hands-on activities Inquiry-based learning 5E-instructional model (Bybee et al., 2006)	10
Assessment in mathematics and science education	5
Integration of technology	5
Interdisciplinary approach STEAM education	10

In the last eight-week period of the course, students developed STEAM lesson plans about science and mathematics topics, compared and shared ideas by presenting their STEAM materials to each other, and reflected on the potential and challenges of the integrated STEAM approach. After completing the Science and Mathematics Methods course, PSTs had the opportunity to develop STEAM lesson plans during five weeks of teaching practice.

Data Collection and Analysis

The participants signed the informed consent forms, and could choose to withdraw at any stage of the research. In addition, each participant received a code to replace their names, thus ensuring their anonymity.

All 24 PSTs agreed to complete an initial questionnaire to attend the Science and Mathematics Methods course. The instrument was adapted from a validated instrument developed by Mahoney (2010) and later modified by Kim and Bolger (2017). The questionnaires were completed anonymously; nevertheless, each answer was identified by a code.

The prior (QA) and post (QB) questionnaire consists of several statements (Table 4) which students had to place on a Likert scale from 0 to 4 (0 - "I don't know"; 1 - "I totally disagree"; 2 - "I strongly disagree"; 3 - "I agree in large part"; 4 – "I totally agree").

The post-questionnaire had, as distinct from the pre-questionnaire, a group of other open-response questions for PSTs to express their opinion on the development of STEAM activities and the experiences they had in the context of teaching practice. Twenty-four participants answered the first questionnaire, but only 20 answered the second. The questionnaire was answered anonymously despite each student indicating the last four digits of the identification citizen card. Thus, it was possible to preserve the anonymity of the participants, as it was possible to compare the answers they gave in the first and second questionnaires. For quantitative data treatment, the sample size considered the 20 participants who answered both questionnaires.

During the PSTs attendance in the methods course, they planned a STEAM activity for the 3rd and 4th grades of primary school. Content analysis also included the STEAM lessons planned by the participants that were examined according to the iSTEM model proposed by Thibaut and colleagues (2018): 1) integration of STEM content; 2) problem-centered learning (focus on authentic, open-ended, and real-world problems); 3) inquiry-based learning (hands-on activities; posing questions; planning and carrying out investigations; collecting, analyzing, and interpreting data/information); 4) design-based learning; 5) collaborative learning (learning through design; developing and using models; designing solutions; opportunities to learn from failure and to redesign based on that learning). To the first category, "integration of STEM content", was added dimension "A", which represents the arts and humanities (Quigley et al., 2017).

Table 4. Dimensions of prior and post questionnaires

Dimensions	Statements
Interest in STEAM activities	1. «To me, STEM activities are appealing.» 2. «I like to learn/develop STEM activities.» 3. «I am interested in courses, programs, subjects or others that address STEM activities.» 4. «I like the challenges of the STEM approach.»
Motivation and perceived ability to develop STEAM activities	1. «I do well in STEAM activities.» 2. «I am not very confident about carrying out STEM activities.» 3. «I have a lot of difficulties in carrying out STEM activities.» 4. «Performing STEM activities is easy for me.»
Valuing STEAM education	1. «STEM activities are important to me.» 2. «I don't value what I learn about STEM activities.» 3. I believe in the need to promote STEM education.» 4. «Learning about STEM activities will not help me.»
Deepening STEAM knowledge	1. «I don't like to deepen my knowledge about STEAM activities by participating in courses, subjects, etc.» 2. «I am interested in discovering new ways to implement STEM activities.» 3. «STEM education is not a vital part of lesson planning.» 4. «I intend to continue implementing STEM activities.»

At the completion of the program, PSTs developed a reflective writing assignment, having been given some questions that they would have to try to answer. Students were asked to identify the essential features of a STEAM activity and the potential benefits and challenges of developing STEAM lesson plans. And if they had the opportunity to implement this approach in the context of teaching practice, they should seek to integrate the analysis of that experience into this assignment.

To further investigate participants' views about developing STEAM lesson plans, a few open questions were added to the post-questionnaire adapted from Kim and Bolger (2017). Qualitative coding analyses of open-ended questions and reflective writing assignments were performed by developing a set of categories to describe the themes present in the responses to each question. Lastly, each answer was coded with one or more categories since some answers contained multiple ideas that suited distinct categories.

FINDINGS AND DISCUSSION

Pre-Service Teachers' STEAM Lesson Plans

Each group of students had to design a STEAM lesson plan after they were introduced to the principles underlying an integrated approach within the methods course (Table 3) and had the opportunity to perform three STEAM activities (Table 2). During the process, the PSTs received feedback on their developing lesson plans, to introduce improvements and reach a final version that effectively integrates the principles of a STEAM-integrated approach. These discussions offered insight to the instructors (the authors), who reoriented the support provided accordingly. The results of the analysis of lesson plans were summarized according to the predetermined categories in Table 5.

Table 5. Analysis of PSTs' lesson plans according to the iSTEAM model (Thibaut et al., 2018)

Category	Instructional Practices				
	S	T	E	A	M
Integration of STEAM content	G1, G2, G3, G4, G5, G6, G7	G1, G2, G3, G4, G5, G6, G7	G3, G5, G6, G7	G2, G4, G5, G7	G2, G4, G5, G6, G7
Problem-centered learning	G1, G3, G4, G5, G6, G7				
Inquiry-based learning	G1, G2, G3, G5, G6				
Design-based learning	G3, G5, G6, G7				
Collaborative learning	G1, G2, G3, G4, G5, G6				

The STEAM activity proposed by G1 starts with a problem related to the buoyancy of objects in water and a video presenting the topic. Next, students should explore floating and sinking with a PhET® simulation by changing variables such as the mass and volume of the objects, registering their behaviour in water. The virtual experiment is followed by a real one, by throwing different objects into a container with water. Initially, the group planned to integrate the mathematics content into their lesson. However, they concluded that children in primary education would hardly understand the formula for calculating density and had no more ideas beyond these.

G2 presented a starting context (planning a visit to the zoo) instead of an open-ended and real-world problem. In this sense, students had to explore the zoo webpage to answer a few questions. The following task involves programming by drawing a path in the map from the zoo entrance to the giraffe "habitat" so that Ozobot® robot can follow the drawn line on the map. For the robot to run the path successfully, lines with colored sequences must be used in order to produce the necessary changes in speed and rotation. Students continue to plan paths for the robot to encounter other animals through the clues presented (diet, camouflage, etc.). The activity ends with the painting of objects to camouflage them in the classroom. This is the only lesson plan that presented a scoring rubric, which, according to Satchwell and Loepp (2002), is consistent with an authentic assessment that promotes connections of all key concepts studied in mathematics, science, and technology.

G3 planned a STEAM lesson focusing on the subject of sound production. First, students are engaged in a hands-on task to explore the concepts of pitch and sound intensity by manipulating several objects and sound instruments (e.g., ruler, balloon, whistle, flute, xylophone). Exploring these sound attributes should allow students to relate them to examples of undesirable sounds (noise). After that, students are challenged to build a device that produces different tones to distinguish the attributes of sound (pitch and sound intensity). So, they build an electrical circuit with a micro:bit® and earphones and use a code available at Microsoft Makecode®[4]. The micro:bit® can detect external light level intensity reaching the LEDs; thus, different exposure to light produces sounds with distinct frequencies.

G4 developed an activity about healthy eating that combines science content with arts and math. First, it requires students to sketch the food wheel to understand their previous conceptions. After that, they explore some online interactive resources (infographics, games, etc.) about the topics and make a 2D or 3D representation of the food wheel, to confront the product with the previous one. The second part of the activity is centered on food safety, particularly preventing the growth of microorganisms. For that, students watch a video and answer a few questions. Finally, the third part consists of creating a healthy food menu and calculating its food cost, reflecting on the criteria that guided their options.

Only G5 proposed a lesson that integrates the content of all STEAM disciplines, although not so relevant as science learning goals. Students have to pose questions, make predictions, and plan and carry out one investigation about how lung capacity affects the amount of air we exhale. Lastly, they have to build a model of the respiratory system. This lesson provided the opportunity to engage students in a hands-on design challenge, but that is not sufficiently emphasized. PSTs also mentioned arts integration in the design of the model.

The STEAM lesson proposed by G6 engages students on friction force on a flat surface by watching a video followed by a discussion of their previous ideas about the subject. The next task comprises a hands-on activity about friction force on a slanted surface, measuring the stopping distance and time of a vehicle on a slope with different types of surfaces. Furthermore, students create a graph based on the data collected throughout the experiment and discuss the factors affecting the stopping distance. A second activity consists of designing a balloon car from recycled materials to discuss some concepts also involved in the first part of the lesson, such as frictional force and Newton's third law of motion. Lastly, students are challenged to race their cars against each other and film the action.

G7 created a learning scenario focused on ocean pollution that integrates mathematical concepts and computational thinking. First, students analyze and discuss some news about the subject; according to the group, it promotes literacy and critical thinking. After that, they code the DOC robot by Clementoni® to complete some paths to collect plastic garbage on a board representing the ocean. Next, students reflect on the best way to complete the paths, having covered the shortest possible distance to clean up the ocean. This experience intends to get the students to think of possible solutions to the garbage islands in the Pacific using robotics. Like G2, this lesson plan does not include building a robot, so it does not involve engineering (National Academy of Sciences, 2014). However, robotics is a technology that provides a very rich and attractive learning environment for STEM education (Barak & Assal, 2018). Engineering is present by involving students in the design of a tool to clean plastic from the sea. This task does not include all design-based learning steps because students only have to design the tool; they do not have to build it.

As in the study developed by Correia and Baptista (2022), some features of the iSTEM model are more common in primary PSTs' lesson plans (Thibaut et al., 2018). Indeed, in most of the lesson plans, students work collaboratively to solve open-ended and real-world problems in inquiry based-learning activities, and curricular integration between science and technology is more frequent in the lesson plans. However, contrary to the study mentioned above, in most lesson plans, the role of technology was not limited to watching a video. Despite being less present, the interdisciplinarity between science and mathematics is more explicit in the lesson plans in terms of learning objectives and the contents covered.

The lesson plan delivered by G1 was the only one that focused uniquely on two STEM dimensions (science and technology) which, for Brown and Bogiages (2019) and Sanders (2009), is enough to be considered an integrated view of STEM education. Another aspect that emerged was that in all the lesson plans, the same attention was not applied to two or more STEM disciplines (Satchwell & Loepp, 2002; Shahali et al., 2017), although in most cases, there are strong connections between the contents of different disciplines. Moreover, according to Roehrig et al. (2012) definition, there is not a real content integration but a context integration because the lesson plans focused on science content and used other disciplines to make that content more relevant.

Still on the integration of content from different STEAM areas, there was an emphasis on science and mathematics, which is usual in primary education (Gresnigt et al., 2014), but it can lead to a disregard for the component of engineering and design process (DiFrancesca, 2014). In fact, it was found that only four groups took advantage of the context provided by engineering to learn science and/or mathematics content (Roehrig et al., 2012; Shahali et al., 2017). Another aspect that stood out in the results was that only four lesson plans included arts and, similarly to Quigley and Herro's study (2016), most of those connections did not go beyond design arts.

Effect of Developing Lessons Plans on Pre-Service Teachers' Attitudes Toward an Integrated STEAM Approach

The attitudes of 20 PSTs (23-28 years old) towards the importance and relevance of STEM education in their future role as teachers were analyzed with a questionnaire with four dimensions, each with four questions. The PSTs were asked the same set of questions in two different moments (before and after the course) to assess the differences in their conceptions about STEAM education. Summaries of interest variables are presented as mean ± standard and available in Table 6, as well as more detailed results like minimum and maximum values of the sums of the variables under study.

The extreme and quartile diagrams (Figure 1) allow visualizing data behavior and its variability in the two moments (QA and QB).

PSTs reveal a growing identification with the STEAM approach. In fact, regarding interest in STEAM activities, it appears that the variable under study assumes, on average, after the training course, values very close to the maximum value (Mean - 15.35) with a smaller deviation (SD – 1.46) compared to the pre-questionnaire.

The motivation to perform STEAM activities was the dimension with the most significant impact (Mean 9.1 and SD 2.53 in QA; Mean 11.3 and SD 0.98 in QB). These measurements are consistent with the measurements of the extreme and quartile diagram (figure 1) since in QB there was very little variability in the data. Descriptive data measures suggest that participants have come to value the STEAM approach much more. In fact, although the average sum of the indicated levels is slightly higher in QB (Mean 15.55), there is a lower deviation after the STEM program (SD 0.68).

Table 6. Summary measures of four variables analyzed in pre-questionnaire (QA) and post-questionnaire (QB)

Categories		N	Minimum	Maximum	Mean	Std. Deviation
Interest in STEAM activities	SumQA1_4	20	10,0	16,0	14,800	1,5079
	SumQB1_4	20	10,0	16,0	15,350	1,4609
Motivation to develop STEAM activities	SumQA5_8	20	4,0	12,0	9,100	2,5319
	SumQB5_8	20	10,0	15,0	11,300	,9787
Valuing education with STEAM	SumQA9_12	20	14,0	16,0	15,150	,7452
	SumQB9_12	20	14,0	16,0	15,550	,6863
Deepening knowledge about STEAM	SumQA13_16	20	9,0	16,0	14,200	1,8806
	SumQB13_16	20	12,0	16,0	14,800	1,0052

Regarding the deepening of the STEM approach, after the course, the data suggest that future teachers showed a greater willingness to continue learning about this approach and to value it in their planning (Cf. Table 6). Furthermore, the variability of the data is smaller and tends to be higher (Cf. Figure 1).

Figure 1. Comparative diagram of extremes and quartiles for the four dimensions

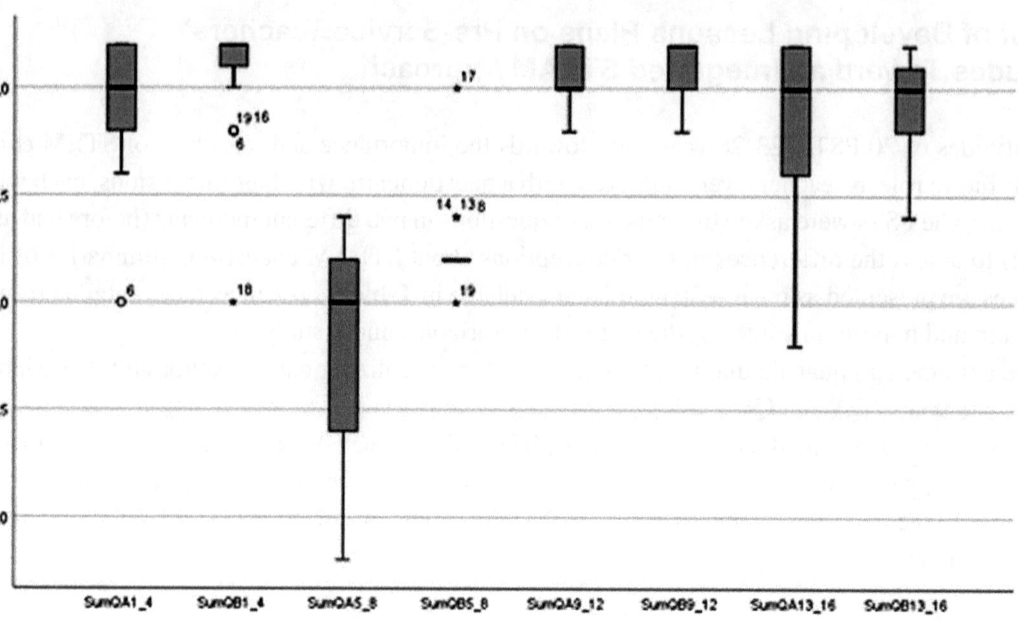

To assess the evolution of the students, we considered the sum of the scores for each question within each of the four dimensions of the questionnaire.

The comparison of the variables in the two moments (QA and QB) was performed through a Wilcoxon test, used for paired samples (Table 7). Several Kolmogorv-Smirnov tests were considered to assess the normality assumption for each summation, but the results showed that such an assumption does not hold.

The first three dimensions showed a positive evolution with a significant Wilcoxon test (p-value<0.05). However, in the fourth dimension, students did not show statistical differences between the two moments (p-value=0.227), meaning that the developing lessons positively influenced pre-service teacher attitudes about the STEAM approach mainly in what concerns motivation and valuing, being inconclusive about deepening PSTs' STEAM knowledge. Overall, the experience of developing STEAM lesson plans within the Science and Mathematics Methods course impacted PSTs' attitudes positively toward the STEAM approach (Kim & Bolger, 2017).

Table 7. Test statistics used to compare samples

Test Statistics[a]				
	SumQB1_4 - SumQA1_4	**SumQB5_8 - SumQA5_8**	**SumQB9_12 - SumQA9_12**	**SumQB13_16 - SumQA13_16**
Z	-1,992[b]	-3,124[b]	-2,309[b]	-1,208[b]
Asymp. Sig. (2-tailed)	,046	,002	,021	,227
a. Wilcoxon Signed Ranks Test | b. Based on negative ranks.				

Pre-Service Perspectives About an Integrated STEAM Approach

In the final questionnaire, the PSTs answered open-ended questions about the STEAM approach and the development of STEAM activities according to their experience. A set of categories emerged from the analysis of the PSTs' answers to this open questionnaire that integrated PSTs' attitudes post-questionnaire.

Of the 24 participants, only 20 answered the final questionnaire after the Methods course, so only these pre-service teachers were considered for the analysis of the questionnaires. The PSTs filled out the questionnaire anonymously (APST1 – APST20), having placed the last four digits of their identification card so that it was possible to make an association between their answers in the first questionnaire, before the course, and in the second questionnaire, after the course.

Regarding what the PSTs considered to be more difficult in developing a STEAM activity, the participants identified more than one aspect (Table 8).

Table 8. PSTs' perspectives, after the training course, about what they consider to be more difficult in developing a STEAM activity

What do you consider most difficult in developing a STEAM activity?	Number of responses (n=20)
Interdisciplinarity	11
Preparation of an activity guide	7
Including technology	2
Include engineering	2
Other responses (vague or imperceptible)	2
Classroom practice	1

Fifty-five per cent of the PSTs refer to interdisciplinarity as something challenging in the development of a STEAM activity ("Create an activity that brings together all the desired areas." (APST14, QB)).

Thirty-five per cent of the participants say that the elaboration of an activity guide following the 6E model and inquiry-based learning is a demanding process, namely in elaborating and identifying different stages. ("For me, the most difficult thing was to distinguish between the exploration phase and the elaboration phase." (APSST18, QB)).

Ten per cent of the participants mention that the inclusion of technology is a difficulty in the process of developing a STEAM activity ("Sometimes, understanding the use of certain digital resources." (APST11, QB)).

Ten per cent specifically identify the difficulty in integrating engineering in the development of these activities ("I consider engineering integration more difficult in the approach, since I do not have much knowledge about it." (APST12, QB)).

The PSTs could reflect on the benefits and challenges of implementing STEAM activities.

Regarding the challenges or difficulties inherent in the implementation of STEAM activities (Table 9), participants (20%) mention that some activities involve more time, so this fact can represent a challenge compared to the option of other types of tasks "I highlight the lack of time to implement this type of activities." (APST20, QB)). Future teachers (10%) see the autonomy that STEAM activities demand from students as a difficulty or challenge in implementation ("(...) it is necessary for the teacher to practice frequently so that students become familiar with this type of activities"). 30% of PSTs refer, as a challenge, the fact that this kind of work requires a greater ability from the teacher to promote interdisciplinarity and integrate several STEAM areas, which can represent a challenge for the teacher ("As a challenge, I identify the need to interconnect all areas of knowledge, including engineering." (APST6, QB)). Twenty per cent refer to the lack of resources in schools, making it hard to implement this type of activities and demanding more effective and demanding management of resources ("(...) I emphasize the fact that sometimes there are no materials needed to carry them out in institutions." (APST7), QB)). Finally, 25% of the participants do not see any challenges.

Table 9. Challenges associated with STEAM activities classroom practice according to the PSTs

Challenges	Number of answers (n=20)
Require from the teacher greater ability to promote interdisciplinarity	6
It is not or they are not indicated	5
Need to spend more time	4
More demanding resource management	4
Require greater autonomy from students	2

Regarding the inherent advantages of implementing STEAM activities, all participants identify several gains (Table 10). Fifty per cent refer to students' more active role ("I believe that STEAM activities have several advantages such as the approach of several contents from different areas at the same time, the fact that they favor learning, since students learn through more practical, significant activities, as it starts from everyday situations and issues." (APST12, QB)), 40% of future teachers refer to interdisciplinarity as a great advantage of STEM activities providing more meaningful contexts for students, 10% identify the ability to work collaboratively ("The main advantage is the combination of several areas. For example, in the class where I implemented, I felt that everyone felt excited, they were divided into groups, they set out to discover, among others" (APST4, QB)), in particular, and 25% refer to the learning of knowledge and the development of competences. ("The big advantage is the development of skills such as scientific and critical thinking." (APST17, QB)).

Table 10. Advantages associated with STEAM activities classroom practice according to the PSTs

Benefits	Number of answers (n=20)
More active role of students	10
Interdisciplinarity/ Meaningful Contexts for Students	8
Knowledge learning and skills development	5
Collaborative work	2

Future teachers also mention how much they value the opportunity to experience planning and, in some cases, implement STEM activities in their initial education. When asked about the most positive aspect of being involved as PSTs in the development of STEAM activities, they mention the knowledge they have of some of the characteristics of the STEAM approach ("Learning a new methodology in the classroom and designing a way of collaborative work involving several areas STEAM." (APST2, QB)). They also refer as an added value to the symbiosis that enriches the school itself, which is often not familiar with this approach, as well as the development of their self-confidence to implement STEAM activities in their future practice ("It was great to give us security in developing STEAM activities in the future." (APST15, QB)).

All the PSTs planned STE(A)M activities during the training course. However, about 35% did not experience the implementation of a STEAM activity. When asked about the reasons, they mentioned that they had to comply with the planning of the cooperating head teacher, which did not include any activities of this nature, either due to time management issues or other existing projects in the school. These data suggest that, although the PSTs have developed a positive attitude towards the STEAM approach, the experience of planning activities in the training course is not always enough to lead them to implement these activities in their classroom practice. It is, therefore, necessary to reinforce the work developed, so that PSTs are prepared for the challenges of implementing STEAM activities and closely monitoring their practices.

Additionally, the analysis of reflective writing assignments (RWA) developed by the PSTs provided a set of categories (Table 11). Results concerning the benefits of STEAM approach demonstrated clearly that PSTs found that the integration of content from different areas was the most important feature. One PST emphasized interdisciplinarity, giving her lesson plan as an example:

A topic from one of the content areas can be interconnected with others, for example, in our lesson plan about the respiratory system, we were able to carry out the activity of measuring children's lung capacity. At the same time, we were able to combine science with content from math. (PST14, RWA)

One participant reflected on what she learned from this course that she plans to apply in the future in teaching practice, saying that

I learned, once again, that teaching should not be structured and limited by areas, and the interconnection between several or even all areas in a single activity allows the acquisition of more structured and meaningful knowledge for students. (PST9, RWA)

Many PSTs' answers not only highlighted the importance of including technology, but also the use of it to articulate content from different domains, as illustrated by the following extracts. Furthermore, some participants mentioned the role of technology, through robotics, in linking content from other disciplines, namely science and mathematics.

When it comes to integrating technology with science, engineering and mathematics, it is crucial, not only because we are moving towards an even more technological future but it is also a dimension in which children show interest. In the future, there will be more and more professions that involve technologies, so children from an early age must have contact and learn with them (from computers, tablets, and robots, among others). In fact, it is crucial to show children the risks of excessive use of technologies, teaching them some precautions they should take. Therefore, using technologies facilitates the teaching of various subjects, the development of new knowledge (e.g., when we want to teach a certain content, it is possible to use technologies for children to acquire knowledge) and competencies (e.g., how to work with a certain game, how to program a certain application, and how to create a game, among others). (PST22, RWA)

Technology is a great starting point to connect diverse domains such as science, engineering and mathematics in the sense that it is an area of high interest to students and has many complex and rich tools and resources that can provide the acquisition of knowledge and learning in a meaningful and stimulating way. (PST9, RWA)

Table 11. Pre-service teachers' perspectives on STEAM approach

Categories of answers	Subcategories	Number of answers
Benefits of STEAM approach	Interdisciplinarity	21
	Collaborative learning	19
	Inquiry-based learning	15
	Focus on real-world problems	14
	Critical thinking	13
	Creativity	9
	Integration of non-STEM content	8
	Design-based learning	8
	Students' interests and needs	7
	Computational thinking	5
	Students' positive emotions	1
Challenges of STEAM approach	Integration of STEM content	11
	Students' interests and difficulties	10
	Inquiry-based learning	6
	Assessment	1

N = 23; one (PST1) of 24 participants did not complete the assignment

The second most common benefit of STEAM approach the PSTs focused on was collaborative learning. This is also considered a strength in Correia and Baptista's (2022) study, however, unlike them, the PSTs in the present study did not see it as a challenge, as it is well demonstrated in the following statement

The main lessons I learn from planning STEAM activities is the ability to integrate various components of the curriculum and thus facilitate student learning and, with this, promote group work, which is not very common in the classroom. From my experience in teaching practice in schools, students do not work in groups, and to develop 21st-century skills, this methodology is essential, which is in line with the guiding documents of the curriculum. In fact, it is possible to identify some common principles with STEM activities, such as critical thinking and creative thinking, reasoning and problem solving, scientific, technical and technological knowledge, and personal development and autonomy. (PST13, RWA)

Whether by mentioning explicitly inquiry-based learning and/or some aspects of it (e.g., planning and performing hands-on experiments and collecting data), most PSTs demonstrated to value these instructional practices. For instance, some PSTs mentioned this instructional practice as a core feature of teaching integrated STEM (Thibaut et al., 2018), others just mentioned some instructional practices associated.

The fourth category was that PSTs found that using real-world problems in a motivating context is crucial to promote meaningful learning, as in the study developed by Correia and Baptista (2022). About this, PST10 gives an interesting example in the following description.

Another learning in this program, which I later used, in the design of the lesson plan, was the use of real situations to explore the contents and reach the intended conclusions. I think it makes the activity/task much richer and closer to the student's context, which, in a way, serves as a motivation for carrying it out. I believe that with the design of the activity, I learned to use questions/comments/contributions from the students, creating a task from that, in order to work with the contents that are being addressed. (PST10, RWA)

The following categories regarding the benefits of STEAM approach comprise two "21st-century skills". First, by promoting STEAM activities "I will be able to train future active citizens in the world of work, developing their critical thinking, along with other relevant skills" (PST24, RWA). Second, this approach also develops creativity related, for example, with engineering. Kim and Bolger (2017) also found that "many pre-service teachers also saw the process as highly educative for themselves and saw the potential benefit of creativity in lesson planning for the future" (p. 598). In this respect, PST23 described their lesson plan in which they asked students to design a machine to clean de ocean.

Some PSTs acknowledged that this approach has the benefit of promoting the integration of non-STEM disciplines, as described in the following assertions.

In addition to the STEM areas, it also involves literacy (...), the development of critical thinking about the news and about what is analyzed in it. (PST23, RWA)

There is also a point that I consider very interesting in our proposed activity, the use of children's literature. The children's book approach is part of the letter "A" of the STEAM activity, as it belongs to the Arts component, with regard to the language component, Languages and Humanities. (PST24, RWA).

Engaging students in engineering design challenges was emphasized by a group of PSTs that integrated this methodology into their lesson plans. Likewise, in Correia and Baptista's study, only a few PSTs' involve children in a prototype design task. This extract provides an example of that

Insofar as it is proposed that children devise a tool to clean up the garbage from the sea. (...) I consider that this methodology is an asset for students to develop the skills above in an interactive and autonomous way, allowing them to build, create, test, and solve problems. (PST23, RWA)

To seven PSTs, STEAM approach enhances students' interest and motivation for learning, especially because students are engaged in active learning activities. In Kim and Bolger's (2017) findings, the most common answer was the opportunity to think about students' interests. Moreover, technologies, such as robotics, "are crucial for motivating students and integrating the various content areas" (PST9, RWA). Examples of this are evident in the following extracts

I think about using these activities to make students feel more motivated, more involved and more creative in doing them and to encourage students to create through diverse experiences. (PST17, RWA)

(...) planning this activity brought me significant learning, not only the perception of the difficulty that it is to interconnect several curricular areas, but also the importance that these activities have for students' learning. These activities are innovative and most of them use resources that are not usual

for children to handle or have contact with in their daily lives, and this arouses the curiosity, attention, and interest of children (…). With STEM activities, students are often learning or deepening knowledge without realizing it. (PST8, RWA)

It is something that captivates the attention of students, as it is something different from the usual classroom routine, which motivates students to perform tasks. Thus, this model of activities will be something that I will use as a future professional in the field of education to promote the learning of my students, since it is an excellent way to teach the contents in a different and more playful way. (PST10, RWA)

Additionally, to stimulate students' interest, the PSTs mentioned that "applications, simulators, games or others are ways to use technology and, in some cases, stimulate computational thinking" (PST5, RWA). Three lesson plans integrated robotics, therefore, naturally, it was an aspect focused on by those PSTs as the following statement demonstrates:

The theme of our activity focused, mostly, on the theme of science (characteristics of living things), as well as on mathematics knowledge, associated with technology, computational thinking, more precisely, the programming of a robot. (PST6, RWA)

Only one PST explicitly mentioned that STEAM activities induce positive emotions in students toward learning STEM content.

Through STEM activities and even interdisciplinarity, children will develop skills to work in groups, to evolve or acquire knowledge, to explore various materials that contribute to their learning, to develop positive feelings/emotions, to be prepared for the future. (...) Not only would we [PSTs] develop a variety of learning, knowledge and skills, but it also enhances our emotional side, especially in children, since they develop attitudes and positive feelings/emotions, which promotes the desire to learn more, to discover more and to explore more. (PST22, RWA)

The most common difficulty experienced by the PSTs during the design of a STEAM lesson was how to establish connections between the different STEAM disciplines. Some examples of PSTs answers include

When we planned this activity, the great challenge we faced was articulating all [STEAM] areas so that the whole activity was coherent and that the main theme was the guiding thread for all the proposed tasks. (PST23, RWA)

While planning the STEAM activity, some difficulties arose, but the one that I consider most relevant was how we would involve mathematics in it. (PST12, RWA)

There were at least three curriculum areas that were included in the STEAM activity, however, of these three areas, mathematics did not play a very important role. Thus, one of the great challenges that the group felt is related to the fact that it is difficult to interconnect several areas in the same activity without losing focus on the initial theme. (PST8, RWA)

According to some PSTs, the challenges of developing integrated lesson plans are due to their lack of experience with STEAM approach, particularly in technology and engineering, as well demonstrated in the following citation:

Personally, it was a challenge for me to think about the questions to be asked in the activity worksheet involving engineering. This is a discipline that I personally don't have much contact with, so initially, it was not easy to think of questions that would contribute to the activity and address the contents of this discipline. (PST24, RWA)

Overall, despite the challenges to integrate different disciplines, PSTs intend to implement this form of education in the future classroom practice. Similarly, in Kim and Bolger's (2017) study, the PSTs found it difficult to relate subjects from different disciplines, particularly technology and engineering. Another interesting aspect that emerges in this study is that the main challenge for the PSTs in planning STEAM activities is also, from their perspectives, the main benefit of this approach, which is similar to other studies (Correia & Baptista, 2022; Kim & Bolger, 2017). Particularly, the PSTs expressed that the most challenging area to include in an integrated approach is engineering, such as in Correia and Baptista's (2022) study.

Students' interests and difficulties also obtained a considerable amount of PSTs' responses. During the lesson planning, the teachers were concerned with the adequacy of the activity worksheet for the students, so that it was not too complex for the children and that this could lead to their lack of interest and motivation. Another challenge was the extension of the activity worksheet, to not discourage students from answering it.

Regarding the challenges experienced in lesson planning, the most difficult was the definition of a common thread between the various tasks present in the activity sheet and, mainly, how to bring students closer to the topic in question. Once this initial point was made, numerous ways of exploring the contents to work quickly emerged. (PST19, RWA)

However, I came across some challenges during the elaboration of the STEAM activity sheet, the main thing was to organize the various elements that we wanted to include in the sheet and to write correctly and clearly the questions in order to make sense to everyone. (PST21, RWA)

PSTs that had the opportunity to implement the STEAM activity during teaching practice mentioned some problems in managing collaborative work, supporting children to use technologies. Some examples of these difficulties arouse in the following extract

During teaching practice, I had the chance to implement the STEAM activity, however, although I felt that the students liked it and that they were able to carry out the tasks described in the worksheet, I consider that the degree of autonomy was low. I realized that contrary to what we think, children are not able to use digital educational tools, because they essentially use tools or applications for online games. And when I implemented the activity, I felt that, although it was described in the worksheet, they really needed the adult's help. (PST16, RWA)

Even though the PSTs considered inquiry-based learning an important feature of STEAM approach, they faced numerous difficulties in planning it. Mainly, "the selection of tasks to include in each point of the 5E model, mainly the 'Elaborate' phase" (PST9, RWA) and the "Organization and sequence of tasks" (PST13, RWA). Although the PSTs have performed inquiry activities before, they had never had the experience of planning it, and for that, they had to be acquainted with all stages of the 5E- instructional model, similar to Correia and Baptista (2022), as the following extract demonstrates:

In planning these activities, the main challenge I faced was distinguishing the Explore phase from the Elaborate phase (…) With the development of the activity sheet, I was able to perceive that the explore phase, although there is the elaboration of something, is always based on guidelines presented in the activity sheet and the elaboration phase is a stage of application of the acquired knowledge. I consider that this was also a learning experience for me, in addition to the fact that I also realized the high level of demand of creating a STEAM activity. With this, I developed my skills to create this type of activities, as I had never had that experience. I had only performed these activities as a student and had never created one. (PST16, RWA)

Only PST9 mentioned difficulties dealing with assessment when planning the STEAM activity, especially "because I am not used to asking children to evaluate the activity, and I was not sure, in practice, how this part would go" (PST9, RWA).

Comparing the PSTs' answers in the final questionnaire with their reflections regarding the benefits and challenges associated with STEAM approach, there are many similarities. For instance, the PSTs highlighted interdisciplinarity as the main benefit and challenge of STEAM activities. Another interesting result is the fact that the teachers, perhaps due to the anonymity of the questionnaire and the absence of guidelines, unlike the reflective writing assignment, distanced themselves from STEAM approach features and pointed out difficulties related to the amount of effort, expertise, and time required for the preparation of this type of learning activities. These findings are close to those obtained by Kim and Bolger (2017), specifically that the PSTs remained concerned about their ability to implement STEAM lessons.

CONCLUSION

At the end of the semester, after the Science and Mathematics Methods course and classroom practice, the PSTs were given a questionnaire to provide insight into their perspectives about STEAM lesson plans. Findings suggest that as the PSTs engaged in planning STEAM lessons, they began to see more of the advantages of this integrated approach. The content analysis of reflective written assignments exhibits a solid knowledge of the participants of the features of iSTEM model (Thibaut et al., 2018). However, planning STEAM activities was a challenging task for the PSTs, especially the difficult task of integrating content from different domains. Despite the growing confidence of the PSTs to develop STEAM lesson plans as a result of their participation in a Methods course, supported by the statistical analysis of the questionnaires, qualitative analysis also revealed that they were still concerned about their ability to design and implement STEAM activities.

The analysis of the lesson plans designed by the PSTs focused on how the iSTEM model was incorporated. They made a real effort to foresee almost all model features; nevertheless, engineering and process design played a very small role or were completely absent. The same applies to the integration of the arts. Therefore, more attention should be paid in preparing teacher candidates at the primary level to integrate engineering (DiFrancesca et al., 2014) and to go beyond the integration of digital/design arts (Quigley & Herro, 2016).

Although students expressed confidence in implementing STEM activities in primary school, not all PSTs could implement their lesson plan in practice, so it is imperative to progress with this experience to another stage that involves PSTs in planning STEAM activities and their implementation in the context of professional practice. Teacher education should, therefore, continue to move towards providing more practice situations where future teachers implement STEAM activities and acquire experience and in contexts where this type of work is a reality.

REFERENCES

Aguilera, D., & Ortiz-Revilla, J. (2021). STEM vs. STEAM Education and Student Creativity: A Systematic Literature Review. *Education Sciences, 11*(7), 331. doi:10.3390/educsci11070331

Barak, M., & Assal, M. (2018). Robotics and STEM learning: Students' achievements in assignments according to the P3 Task Taxonomy—practice, problem solving, and projects. *International Journal of Technology and Design Education, 28*(1), 121–144. doi:10.100710798-016-9385-9

Bassachs, M., Cañabate, D., Nogué, L., Serra, T., Bubnys, R., & Colomer, J. (2020). Fostering critical reflection in primary education through STEAM approaches. *Education in Science, 10*(12), 1–14. doi:10.3390/educsci10120384

Bequette, J., & Bequette, M. (2012). A Place for Art and Design Education in the STEM Conversation. *Art Education, 65*(2), 40–47. doi:10.1080/00043125.2012.11519167

Breiner, J., Harkness, S., Johnson, C., & Koehler, C. (2012). What is STEM? A discussion about conceptions of STEM in education and partnerships. *School Science and Mathematics, 112*(1), 3–11. doi:10.1111/j.1949-8594.2011.00109.x

Brown, R., & Bogiages, C. (2019). Professional Development Through STEM Integration: How Early Career Math and Science Teachers Respond to Experiencing Integrated STEM Tasks. *International Journal of Science and Mathematics Education, 17*(1), 111–128. doi:10.100710763-017-9863-x

Bybee, R. (2006). Scientific Inquiry and Science Teaching. In L. B. Flick & N. G. Lederman (Eds.), *Scientific inquiry and nature of science: Implications for teaching, learning, and teacher education*. Springer. doi:10.1007/978-1-4020-5814-1_1

Bybee, R., Taylor, J., Gardner, A., Van Scotter, P., Powell, J., Westbrook, A., & Landes, N. (2006). *The BSCS 5E Instructional Model: Origins, Effectiveness and Applications*. BSCS.

Camacho, G., Martins, M. C., & Correia, M. (2022). *Os meus itinerários* [My itineraries]. Casa das Ciências. https://www.casadasciencias.org/recurso/8998

Colucci-Gray, L., Burnard, P., Cooke, C., Davies, R., Gray, D., & Trowsdale, J. (2017). *BERA Research Commission: Reviewing the potential and challenges of developing STEAM education through creative pedagogies for 21st learning: how can school curricula be broadened towards a more responsive, dynamic, and inclusive form of education?* BERA.

Correia, M., & Baptista, M. (2022). Supporting the Development of Pre-Service Primary Teachers PCK and CK through a STEM Program. *Education Sciences, 12*(4), 258. doi:10.3390/educsci12040258

Correia, M., & Martins, M. C. (2020). *CreativeLab_SciMath A Medida do Sistema Solar* [CreativeLab_SciMath The Measurement of the Solar System]. Casa das Ciências. https://www.casadasciencias.org/recurso/8867

Daugherty, M., Carter, V., & Swagerty, L. (2014). Elementary STEM education: The future for technology and engineering education? *The Journal of STEM Teacher Education, 49*(1), 45–56. doi:10.30707/JSTE49.1Daugherty

Dejarnette, N. (2012). America's children: Providing early exposure to STEM (science, technology, engineering, and math) initiatives. *Education, 133*(1), 77–84.

Dias, D., Correia, M., & Martins, C. (2017). *As ciências em tempo de Natal* [The sciences in Christmas time]. Uma proposta didática interdisciplinar para o 1.º Ciclo. Casa das Ciências. https://www.casadasciencias.org/recurso/8629

DiFrancesca, D., Lee, C., & McIntyre, E. (2014). Where is the 'E' in STEM for Young Children? Engineering Design Education in an Elementary Teacher Preparation Program. *Issues in Teacher Education, 23*(1), 49–64.

English, L. (2016). STEM education K-12: Perspectives on integration. *International Journal of STEM Education, 3*(3), 1–8. doi:10.118640594-016-0036-1

Estapa, A., & Tank, K. (2017). Supporting integrated STEM in the elementary classroom: A professional development approach centered on an engineering design challenge. *International Journal of STEM Education, 4*(6), 6. Advance online publication. doi:10.118640594-017-0058-3

García-Carmona, A., & Toma, R. (2021, Fevereiro). *STEM integrado: significados, dificultades y perspectivas en el contexto educativo español* [Integrated STEM: meanings, difficulties and perspectives in the Spanish educational context]. Comunicação apresentada em 29 Encuentros de Didáctica de las Ciencias Experimentales y 5ª Escuelas de Doctorado. Universidad de Córdoba y APICE (Asociación Española de Profesores e Investigadores en Didáctica de las Ciencias Experimentales).

Greca, I., Ortiz-Revilla, J., & Arriassecq, I. (2021). Diseño y evaluación de una secuencia de enseñanza aprendizaje STEAM para educación primaria [Design and evaluation of a STEAM teaching-learning sequence for primary education]. *Revista Eureka sobre Enseñanza y Divulgación de las Ciencias, 18*(1), 1802–1819. doi:10.25267/Rev_Eureka_ensen_divulg_cienc.2021.v18.i1.1802

Greenwood, R., Austin, S., Bacon, K., & Pike, S. (2022). Enquiry-based learning in the primary classroom: student teachers' perceptions. *Education 3-13, 50*(3), 404-418. doi:10.1080/03004279.2020.1853788

Gresnigt, R., Taconis, R., van Keulen, H., Gravemeijer, K., & Baartman, L. (2014). Promoting science and technology in primary education: A review of integrated curricula. *Studies in Science Education*, *50*(1), 47–84. doi:10.1080/03057267.2013.877694

Guyotte, K. W., Sochacka, N. W., Costantino, T. E., Kellam, N., Kellam, N. N., & Walther, J. (2015). Collaborative creativity in STEAM: Narratives of art education students' experiences in transdisciplinary spaces. *International Journal of Education & the Arts*, *16*(15). http://www.ijea.org/v16n15/

Guzey, S., Moore, T., & Harwell, M. (2016). Building up STEM: An analysis of teacher-developed engineering design-based STEM integration curricular materials. *Journal of Pre-College Engineering Education Research*, *6*(1), 11–29. doi:10.7771/2157-9288.1129

Henriksen, D. (2014). Full STEAM ahead: Creativity in excellent STEM teaching practices. *The STEAM Journal*, *1*(2), 1–9. doi:10.5642team.20140102.15

Hourigan, M., O'Dwyer, A., Leavy, A., & Corry, E. (2021). Integrated STEM – a step too far in primary education contexts? *Irish Educational Studies*. Advance online publication. doi:10.1080/03323315.2021.1899027

Howe, C., Tolmie, A., Thurston, A., Topping, K., Christie, D., Livingston, K., Jessiman, E., & Donaldson, C. (2007). Group work in elementary science: Organisational principles for supporting pupil learning. *Learning and Instruction*, *17*(5), 549–563. doi:10.1016/j.learninstruc.2007.09.004

Hsu, Y.-S., & Fang, S.-C. (2019). Opportunities and Challenges of STEM Education. In Y.-S. Hsu & S.-C. Fang (Eds.), Asia-Pacific STEM Teaching Practices (pp. 1-16). Springer International Publishing. doi:10.1007/978-981-15-0768-7_1

Hurley, J., & Hallissy, M. (2017). *STEM Education Policy Statement 2017–2026*. Department of Education and Skills.

Kelley, T., & Knowles, J. (2016). A conceptual framework for integrated STEM education. *International Journal of STEM Education*, *3*(1), 11. doi:10.118640594-016-0046-z

Kim, D., & Bolger, M. (2017). Analysis of Korean elementary pre-service teachers' changing attitudes about integrated STEAM pedagogy through developing lesson plans. *International Journal of Science and Mathematics Education*, *15*(4), 587–605. doi:10.100710763-015-9709-3

Lage-Gómez, C., & Ros, G. (2021). Transdisciplinary integration and its implementation in primary education through two STEAM projects. *Journal for the Study of Education and Development*. doi:10.1080/02103702.2021.1925474

Land, M. (2013). Full STEAM ahead: The benefits of integrating the arts into STEM. *Procedia Computer Science*, *20*, 547–552. doi:10.1016/j.procs.2013.09.317

Madden, L., Beyers, J., & O'Brien, S. (2016). The Importance of STEM Education in the Elementary Grades: Learning from Pre-service and Novice Teachers' Perspectives. *The Electronic Journal of Science Education*, *20*, 1–18.

Mahoney, M. (2010). Student attitudes toward STEM: Development of an instrument for high school STEM-based programs. *The Journal of Technology Studies*, *36*(1), 24–34. doi:10.21061/jots.v36i1.a.4

Martín-Páez, T., Aguilera, D., Perales-Palacios, F., & Vílchez-González, J. M. (2019). What are we talking about when we talk about STEM education? A review of literature. *Science Education, 103*(4), 799–822. doi:10.1002ce.21522

Martínez-Borreguero, G., Naranjo-Correa, F., & Mateos-Núñez, M. (2022). Cognitive and Emotional Development of STEM Skills in Primary School Teacher Training through Practical Work. *Education Sciences, 12*(7), 470. doi:10.3390/educsci12070470

Microsoft Makecode. (2022). *Light Sensor Tone control. Version 5.0.12.* https://makecode.microbit.org/v0/projects/guitar/lightsensor

Miller, J., & Knezek, G. (2013). STEAM for Student Engagement. In R. McBride, & M. Searson (Eds.), *Proceedings of SITE 2013--Society for Information Technology & Teacher Education International Conference* (pp. 3288-3298). Association for the Advancement of Computing in Education (AACE).

Mohr-Schroeder, M., Cavalcanti, M., & Blyman, K. (2015). STEM education: understanding the changing landscape. In A. Sahin (Ed.), *A practice-based model of STEM teaching* (pp. 3–14). Sense. doi:10.1007/978-94-6300-019-2_1

Nadelson, L. S., Callahan, J., Pyke, P., Hay, A., Dance, M., & Pfiester, J. (2013). Teacher STEM Perception and Preparation: Inquiry-Based STEM Professional Development for Elementary Teachers. *The Journal of Educational Research, 106*(2), 157–168. doi:10.1080/00220671.2012.667014

National Academy of Sciences. (2014). *STEM integration in K-12 education: Status, prospects, and an agenda for research.* National Academies Press.

Perales-Palacios, F., & Aguilera, D. (2020). Ciencia-Tecnología-Sociedad vs. STEM: ¿evolución, revolución o disyunción? [Science-Technology-Society vs. STEM: evolution, revolution or disjunction?]. *Ápice. Revista de Educación Científica, 4*(1), 1–15. doi:10.17979/arec.2020.4.1.5826

Quigley, C., & Herro, D. (2016). "Finding the Joy in the Unknown": Implementation of STEAM Teaching Practices in Middle School Science and Math Classrooms. *Journal of Science Education and Technology, 25*(3), 410–426. doi:10.100710956-016-9602-z

Quigley, C., Herro, D., & Jamil, F. (2017). Developing a conceptual model of STEAM teaching practices. *School Science and Mathematics, 117*(1–2), 1–12. doi:10.1111sm.12201

Roehrig, G., Moore, T., Wang, H.-H., & Park, M. (2012). Is Adding the E Enough? In-vestigating the Impact of K-12 Engineering Standards on the Implementation of STEM Integration. *School Science and Mathematics, 112*(1), 31–44. doi:10.1111/j.1949-8594.2011.00112.x

Sanders, M. (2009). STEM, STEM education, STEM mania. *Technology Teacher, 68*(4), 20–26.

Satchwell, R., & Loepp, F. (2002). Designing and Implementing an Integrated Mathematics, Science, and Technology Curriculum for the Middle School. *Journal of Industrial Teacher Education, 39*(3), 41–66.

Shahali, E., Halim, L., Rasul, M., Osman, K., & Zulkifeli, M. (2017). STEM learning through engineering design: Impact on middle secondary students' interest towards STEM. *Eurasia Journal of Mathematics, Science and Technology Education, 13*(5), 1189–1211. doi:10.12973/eurasia.2017.00667a

Thibaut, L., Ceuppens, S., De Loof, H., De Meester, J., Goovaerts, L., Struyf, A., Boeve-de Pauw, J., Dehaene, W., Deprez, J., De Cock, M., Hellinckx, L., Knipprath, H., Langie, G., Struyven, K., Van de Velde, D., Van Petegem, P., & Depaepe, F. (2018). Integrated STEM Education: A Systematic Review of Instructional Practices in Secondary Education. *European Journal of STEM Education, 3*(1), 2. doi:10.20897/ejsteme/85525

Wang, H., Moore, T., Roehrig, G., & Park, M. (2011). STEM Integration: Teacher Perceptions and Practice. *Journal of Pre-College Engineering Education Research, 1*(2), Article 2. doi:10.5703/1288284314636

Yakman, G. (2012). Recognizing the A in STEM Education. *Middle Ground Magazine, 16*(1), 15–16.

ADDITIONAL READING

Baptista, M., & Martins, I. (2019). STEM Approach as a Means for Students' Science Learning. *Acta Scientiae, 21*(6), 98–115. doi:10.17648/acta.scientiae.5421

Boice, K. L., Jackson, J. R., Alemdar, M., Rao, A. E., Grossman, S., & Usselman, M. (2021). Supporting Teachers on Their STEAM Journey: A Collaborative STEAM Teacher Training Program. *Education Sciences, 11*(3), 105. doi:10.3390/educsci11030105

Byrd, K. (2019). Elementary Preservice Teacher Preparation to Teach Mathematics and Science in an Integrated STEM Framework. *Dissertations.* 1619. https://aquila.usm.edu/dissertations/1619

Correia, M., & Baptista, M. (2021). The effects of a STEM approach on Pre-service elementary teachers' Subject Matter Knowledge about Sound. *Acta Scientae, 23*(6), 179–208. doi:10.17648/acta.scientiae.6246

Herro, D., Quigley, C., & Cian, H. (2019). The Challenges of STEAM Instruction: Lessons from the Field. *Action in Teacher Education, 41*(2), 172–190. doi:10.1080/01626620.2018.1551159

Morais, C., Moreira, L., Baptista, M., & Martins, I. (2021). Digital Tools Entering the Scene in STEM Activities for Physics Teaching. In A. Reis, J. Barroso, J. B. Lopes, T. Mikropoulos, & C. W. Fan (Eds.), *Technology and Innovation in Learning, Teaching and Education. TECH-EDU 2020. Communications in Computer and Information Science* (Vol. 1384, pp. 124–137). Springer International Publishing. doi:10.1007/978-3-030-73988-1_9

Ortiz-Revilla, J., Greca, I., & Arriassecq, I. (2022). A Theoretical Framework for Integrated STEM Education. *Science & Education, 31*(2), 383–404. doi:10.100711191-021-00242-x

Ríordáin, M., Johnston, J., & Walshe, G. (2016). Making mathematics and science integration happen: Key aspects of practice. *International Journal of Mathematical Education in Science and Technology, 47*(2), 233–255. doi:10.1080/0020739X.2015.1078001

Romero-Ariza, M., Quesada, A., Abril, A. M., & Cobo, C. (2021). Changing teachers' self-efficacy, beliefs and practices through STEAM teacher professional development. *Journal for the Study of Education and Development, 44*(4), 942–969. doi:10.1080/02103702.2021.1926164

KEY TERMS AND DEFINITIONS

21st-century skills: Skills that are deemed necessary to effectively function as citizens in the 21st-century workplace, that include: creativity and innovation; critical thinking and problem solving; communication and collaboration; information, media and technology skills.

Cooperative learning: Teamwork and collaboration with others strongly guided by the teacher, who guides moves from team to team, observes the interactions, and intervenes whenever it is appropriate. Moreover, teacher promotes shared responsibility, so that students assume different roles and learn to share knowledge, tasks, and strategies.

Design-based learning: Implies engineering design processes and engineering practices to develop a model, a product or a prototype to solve a real-life problem, but also deepen their knowledge of related content.

Inquiry-based learning: Students act as scientists by planning and designing experiments, making predictions, carrying out experiments, collecting and critically analyzing data, discussing with their peers and devising evidence-based explanations to answer the questions initially posed. Although it is a methodology used in the teaching of physical and natural sciences, it is not limited to this field.

Problem-based learning: Starting from a real open problem or from a context provided by the teacher, which implies that students work collaboratively and present a solution to the problem. Unlike project-based learning, this approach focuses on the process and not achieving a predetermined product.

Project-based learning: Students are engaged on a project for an extended time – from a week to a semester – that involves solving a real-world problem or answering a complex question. They demonstrate the knowledge and skills developed by creating a public product or a presentation for a real audience.

Social Constructivism: Learning is socially situated, and knowledge is constructed based on previous experiences through interaction with others.

STEM: Science, technology, engineering and mathematics.

STEAM: Science, technology, engineering, arts (English language, arts, social studies, etc.), and mathematics.

1 Correia, M., & Martins, M. C. (2020). *CreativeLab_SciMath A Medida do Sistema Solar*. Casa das Ciências. https://www.casadasciencias.org/recurso/8867

2 Dias, D., Correia, M. & Martins, C. (2017). *As ciências em tempo de Natal. Uma proposta didática interdisciplinar para o 1.º Ciclo*. Casa das Ciências. https://www.casadasciencias.org/recurso/8629

3 Camacho, G., Martins, M. C., & Correia, M. (2022). *Os meus itinerários*. Casa das Ciências. https://www.casadasciencias.org/recurso/8998

4 Microsoft Makecode (2022). *Light Sensor Tone control*. Version 5.0.12. https://makecode.microbit.org/v0/projects/guitar/lightsensor

Chapter 15
How Did the Egyptians Transport the Stone Blocks of the Pyramids?
Design of a STEM Practice to Address the Simple Machines in Initial Teacher Training

Aurora Muñoz-Losa
Universidad de Extremadura, Spain

José María Marcos-Merino
Universidad de Extremadura, Spain

ABSTRACT

The new trend in the Spanish educational system towards the teaching-learning of STEM skills makes it necessary to reinforce both didactic and contents of the different STEM areas. A STEM experience with simple machines was designed for 115 pre-service elementary teachers, being all students of the 4th grade in primary education in the context of a subject of didactics of experimental sciences. After the completion of the intervention, hard difficulties with several concepts were identified in participants, especially those related to mathematics (measurement of angles, concepts of trigonometry, graphic representations). These difficulties should be addressed in initial teacher training with the aim to improve their future professional performance as STEM teachers.

INTRODUCTION AND BACKGROUND

Research in Science Education widely highlighted those pre-service elementary teachers need to acquire a systemic perspective to improve science teaching in their future professional performance. This systemic perspective is needed to reflect how science is practiced in the real world and to link school science to significant problems of current society (Brewer & Smith, 2011; Labov *et al.*, 2010; McCright *et al.*, 2013; NGSS Lead States, 2013; Timmis *et al.*, 2019). However, educational curricula

DOI: 10.4018/978-1-6684-5765-8.ch015

place science (as the rest of other subjects) in separate compartments. As a results of this reductionist approach, integration of subjects is perceived as difficult and costly by most teachers (both in-service and pre-service) (Leszczynski, 2014). Considering this situation, it is necessary to train future teachers to integrate science with other related subjects (such as mathematics or technology). STEM (Science, Technology, Engineering and Mathematics) approach could be a chance to strengthen these interactions in science lessons with pre-service teachers (Bybee, 2010). This is especially relevant for Spanish pre-service teachers, since Spanish educational system has recently included the STEM skill as a mandatory expertise in elementary education (Spanish Government, 2022). Thus, there are new contents in the Primary Education curriculum related with STEM approach that teachers are going to deal with from now on such as the relationship between advances in mathematics, science, engineering and technology for understand the evolution of society in the scientific-technological field, or Science, technology and engineering as human activities: the STEM professions today from a gender perspective. However, most practices labelled as STEM are of deficient educational quality and are scarcely new compared to existing science education approaches (Maass *et al.*, 2019; McComas & Burgin, 2020; Toma & García-Carmona, 2021; White & Delaney, 2021). According to these authors, it is necessary to design and validate meaningful STEM education experiences, with real and proper integrations between the four disciplines of the STEM acronym. Additionally, it is necessary to implement them in initial training of elementary teachers, promoting these students to understand the integrated nature of STEM approach and the connections among the disciplines (Altan & Ercan, 2016; Pimthong & Williams, 2018).

Physics is a hard science to pre-service elementary teachers (Bravo *et al.*, 2022; Brígido *et al.*, 2010; Brígido *et al.*, 2013; Hernández-Barco *et al.*, 2021; Mellado *et al.*, 2014), describing, in relation to its learning and its future teaching, negative emotions (such as nervousness, frustration, worry, uncertainty…) and low levels of self-efficacy. In addition, it is well-known that these students often hold several misconceptions related to Physics contents (Lawrenz, 1986). One of the areas related to misconceptions is simple machines and forces involved in its operation (Diani *et al.*, 2020; Winarno *et al.*, 2019). These are one of the science contents these university students will have to teach in their future professional performance (Spanish Government, 2014). Tacking this into account, it is necessary to improve the training of future teachers in relation to simple machines and its didactic.

Science misconceptions are very resistant to change and hardly modifiable with traditional teaching (Chi, 2005; Roth, 2008). Active methodologies and approaches (as STEM approach) could be a way to improve science knowledge and motivation of future elementary teachers to Physics and to try to overcome these misconceptions and negative attitudes. Practical experiences and engineering approach has been described as adequate to improve the learning of simple machines since Primary Education (Marulcu & Barnett, 2013; Marulcu & Barnett, 2016). In this chapter a STEM education intervention designed to teach simple machines to pre-service elementary teachers is described. The aim of this educational intervention is to train future teachers in: i) simple machines and other physic contents (such as mass, weight, forces…), and ii) STEM education and practical activities to teach science contents in their future professional performance. Finally, a reflection about this experience and, particularly, the difficulties identified in participants is made.

Objectives

The aims of this contribution are to:

- Designing an educational intervention based on the STEM approach to address simple machines with a sample of pre-service Primary teachers.
- Reflecting about the implementation of designed intervention with a sample of pre-service Primary teachers, to identify the main difficulties of these students in order to improve future implementations.

CONTEXT

The STEM experience was carried out by 115 pre-service Primary teachers (in groups around 20 participants), students of the 4th course of the Grade in Primary Education in the context of a subject of didactics of experimental sciences (Faculty of Education and Psychology, University of Extremadura). It is worth noting that this is the last science education subject they attend in the Degree and, before that, they attended other two science subjects related with physics, chemistry, geology, and biology education, and other three related with mathematics education. Moreover, it is significant to consider that, as it mentioned above, these students are not normally keen on science, holding low levels of motivation towards both its learning as university students and its teaching in their future professional performance as Primary teachers (Mellado *et al.*, 2014). This situation is frequently worse for Physics and Technology (Brígido *et al.*, 2010; Brígido *et al.*, 2013; Hernández-Barco *et al.*, 2021), situation that has been related to their previous studies in Upper-Secondary Education (Marcos-Merino *et al.*, 2022).

DESCRIPTION OF THE STEM EXPERIENCE

Previously to the intervention, some basic theoretical contents about simple machines (inclined planes, pulleys, and levers) were imparted in one lecture session (1 hour and 30 minutes long). Moreover, a document about the experience was provided to the students, with a summary of the scientific concepts and a descriptive guide that included the steps (assembly and tests with models of inclined planes, pulleys, and levers) and the activities to carry out. All the materials that were used in the practice served to work on the concept of sustainability, since the vast majority are reused everyday materials (cardboard, wooden sticks, strings, yogurt pots, matchboxes, straws, clips, toilet rolls, expanded polystyrene…), with the aim that our future teachers can achieve this type of STEM experiences in any school. The time of completion of the intervention was approximately 3 hours.

Just before starting the session, the students had to take a short basic knowledge test on simple machines. The test consisted of 5 simple questions about basic concepts such as, "with which of these angles of the inclined plane will it be easier to lift a box? (30, 45 or 60 degrees)" or "which of these objects is a lever of the first class? (tweezers, wheelbarrow or pliers)". In next section the results of the preliminary test will be discussed.

The STEM experience started off the engineering area, that is, the problem to solve: How did the Egyptians transport the stone blocks of the pyramids without having the current machines? This problem is related to the future professional performance of participants, raising the following situation: "Explaining the history of Ancient Egypt in a lesson of Social Sciences in 6th grade of Primary Education, one of your students asked you this question: How did the Egyptians transport the stone blocks of the pyramids without having the current machines? As a teacher, you decide to transfer this question to a science lesson, through a STEM workshop with which to approach simple machines and how they operate."

After focusing the students on the problem, the practice continued with the creation of an inclined plane with a pulley (using cardboard, wooden sticks, strings, yogurt pots, matchboxes, straws, clips, adhesive tape and glue) to lift objects in order to study the necessary forces to elevate an object (figure 1a). Teachers provide a plastic pulley since, although a pulley can be also created with household materials (such as cardboard or straws), it is a complicated process to approach in elementary education. Next, both a simple pulley system and a compound pulley system were assembled. For these models they had to use some materials from the previous assembly. Finally, a simple lever was made (with cardboard, toilet rolls, expanded polystyrene and glue), with the aim to verify the relationship between the distance between the fulcrum and the load (figure 1b). In this model they had to identify two boxes as load (R, derived from "*resistencia*" in Spanish) or effort (P, derived from "*potencia*" in Spanish). Students created these three models in groups of around 4 members, following the descriptive guide where they could find the materials to use and the steps to follow for each model. In the figure 1 are represented some images related to the process of assembly included in this document to guide participants. Moreover, two teachers were in the laboratory to guide students during this process, solving any question.

Figure 1. Images include in the descriptive guide to represent the assembly of the inclined plane with a pulley (a) and the simple lever (b).
Source: Authors.

Once the assembly was over, the students had to carry out some tests with each model (figures 2 and 3), collecting in a sheet the obtained data, analysing them and performing some activities that are described down below.

Figure 2. Pre-service Primary teachers testing the models of an inclined plane with a pulley (on the left) and a simple lever (on the right).
Source: Authors.

Figure 3. Pre-service Primary teachers testing the models of a simple pulley system (on the left) and a compound pulley system (on the right).
Source: Authors.

Model of Inclined Plane With a Pulley

Regarding the model of an inclined plane with a pulley, students had to introduce a weight in the matchbox and then, upload them through the inclined plane introducing several weights of different masses in the yogurt pot. When they achieved this objective, they had to weigh, using a balance, all the components of the system (weights, matchbox, yogurt pot and clip), in order to perform some activities to analyse the implicated forces (table 1). With these activities students can prove how inclined planes reduce the necessary force to upload an object (the weight of the yogurt pot + weights system is lower than the weight of the matchbox + weight + clip system). They can also examine how influence the angle of the inclined plane in the force they should exert to upload the matchbox + weight + clip system.

Table 1. Activities related to the model of an inclined plane with a pulley developed by participants.

Activities Related to the Model of an Inclined Plane With a Pulley
Calculate the weight of the following systems: - Matchbox + weight + clip Weight:__________________ - Yogurt pot+ weights Weight:__________________ Do both weights match? Why?
Measure the angle of the inclined plane with a protractor and calculate the necessary tension force to raise the matchbox + weight + clip system. Does it match the weight of the matchbox + weight + clip system? Does it match the weight of the yogurt pot+ weights system? Why? Angle:___________ Tension force:______________
With the previous data, what would happen if the angle of the inclined plane was 20°? What if it was 70°? Why? Angle: 20° Tension force:______________ Angle: 70° Tension force:______________

Models of Pulleys

In relation with tests and activities related to pulleys, firstly are described those related to the simple pulley system and secondly those related to the compound pulley system. For the simple pulley system participants had to use the yogurt pot joined to the matchbox + weight + clip system thorough a string (used in the model of the inclined plane with a pulley) and place it at the pulley suspended from a stick (figure 4). Then, students had to introduce a weight in the matchbox and upload it introducing several weights of different masses in the yogurt pot. When they achieved this objective, they had to weigh, using a balance, all the components (weights, matchbox, yogurt pot and clip) and perform some activities (table 2). With these activities participants can prove that simple pulleys do not reduce the necessary force to upload an object (the weight of the yogurt pot + weight system is approximately the same than the weight of the matchbox + weight + clip system), it only changes the direction of the force.

Using the yogurt pot, two pulleys, string, a clip and a stick, participants had to build a model of a compound pulley system (figure 4). Afterwards, students join a weight to the mobile pulley (figure 4) and had to introduce several weights of different masses in the yogurt pot to upload the weight joined to the mobile pulley. When they achieved this aim, they had to weigh, using a balance, all the components (weights and yogurt pot) and perform some activities (table 2). With these activities participants can prove how compound pulleys reduce the necessary force to upload an object (the weight of the yogurt pot + weights system is lower than the weight of those joined to the mobile pulley). Additionally, they can test how this force is reduced when more mobile pulleys were incorporated in the system.

Figure 4. Components of the simple pulley system (on the left) and a compound pulley system (on the right) models developed by pre-service Primary teachers.
Source: Authors.

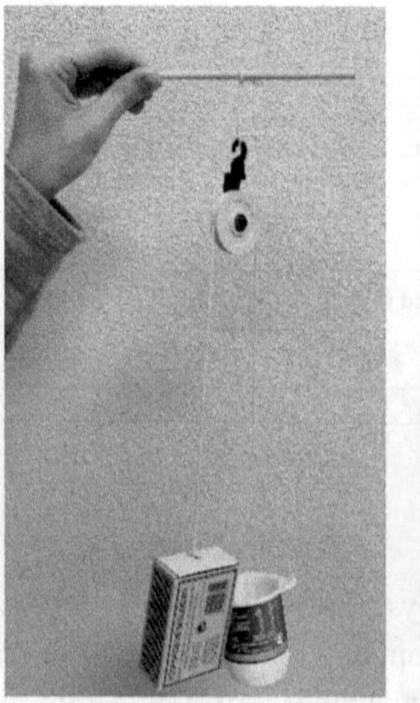

Table 2. Activities related to the models of the simple pulley system and the compound pulley system.

Activities Related to the Model of the Simple Pulley System
Calculate the weight of the following systems: - Matchbox + weight + clip Weight:_____________ - Yogurt pot+ weights Weight:_____________ Do both weights match? Why?
Activities Related to the Model of the Compound Pulley System
Calculate the weight of the following systems: - Weight Weight:_____________ - Yogurt pot+ weights Weight:_____________ Do both weights match? Why?
If you have a compound pulley system with 1 fixed pulley and 3 mobile pulleys, what is the mass of the weights that you should introduce in the yogurt pot to upload the weight? Mass:_____________

Model of Simple Lever

In relation with tests and activities related to simple lever, participants had to introduce 15 coins of 5 cents in the R pot (load) and determine how many coins they need to introduce in the P pot (effort) to upload the R pot. They had to perform this when the fulcrum is at the following positions of the lever: 15 cm (centre of lever), 13 cm (2 cm to centre), 11 cm (4 cm to centre), 9 cm (6 cm to centre) and 7 cm (8 cm to centre). It is important to indicate to participants that every time they change the position of the fulcrum, they have to remove P pot and empty it (this is the only way to test how the lever works to upload the R pot in every position of the fulcrum). With their data, students had to perform some activities (table 3). With these activities participants can prove that more or less number of coins (and, thus, more or less amount of force) is needed depending on the distance from the load to the fulcrum. Also, applying law of the lever, they can calculate the number of coins they would have to introduce in P (effort) pot for any position of the fulcrum.

Table 3. Activities related to the model of simple lever.

Activities Related to the Model of Simple Lever
Based on your data, complete the following table and represent them graphically

Distance from the fulcrum to the centre of the lever (cm)	**Coins difference between pots**

Without using the model of the simple lever, what would happen if you place the fulcrum 5 cm from the R (load) pot?

Once they performed the tests and activities abovementioned, they were asked for their conclusions on how they had approached each of the STEM areas during the practice (table 4)

Table 4. Relation of contents and procedures, related to each STEM area, worked during the intervention.

Science	Technology	Engineering	Mathematics
- Concept of mass - Concept of weight - Difference between mass and weight - Concept of force - Types of forces - Friction force - Simple machines - Scientific method (observation, hypothesis, experimentation, analysing data, reporting conclusions	- Model making - Use of tools and instruments (balance, dynamometer…) - Election and manipulation of materials - Use and interpretation of scales - Use and interpretation of plans and sketches	- Engineering problem as context - Problem solving during design, construction and testing	- Measuring process (angles, distances …) - Units -Performing calculations (formulas, conversion factors…) - Problem solving - Organization of information in tables - Graphical representation - Geometry, shapes,and geometric relationships

Afterwards, using the contents they have learnt related to simple machines, participants had to answer the engineering question from which they started (How did the Egyptians transport the stone blocks of the pyramids without having the current machines?). Eventually, students discuss, under the guidance of teaches, the results from the point of view of the different STEM areas, as well as the didactic aspects of this approach in Primary Education.

Finally, the Table 5 collects the sequence of activities carried out during the intervention and the approximate time for each of them is presented as a summary.

Table 5. Summary of the sequence of activities and their timing.

Activity		Timing
Problem statement		5'
Model building	Inclined plane	25'-35'
	Pulleys	10'-15'
	Simple lever	10'-15'
Model tests	Inclined plane	25'-30'
	Pulleys	10'-15'
	Simple lever	10'-15'
Results discussion	Group discussion	30'-35'
	Teacher-guided discussion	10'-15'
Total timing		180'

DISCUSSION AND REFLEXIONS

This section collects, firstly, the results of the previous knowledge test carried out by the students at the beginning of the intervention and its analysis. Secondly, the reports delivered by the students after the intervention will be discussed and the authors' reflexions on the STEM experience and its suitability for teaching science will be presented.

Analysis of the Preliminary Knowledge Test

As mentioned before, in the first five minutes of the session students carried out a simple test about their knowledge on simple machines. The objective of this test was twofold: to find out if the students had understood the basic concepts explained in the theoretical session, in addition to motivating them to study these concepts before the seminar. Although several other types of machines were explained more extensively in the previous theory session, the test was designed with questions about the simple machines that would be used in the intervention. In this way, the questions to be resolved were related to the interpretation of the law of the lever (relationship between the distance of the fulcrum with the load and the force necessary to lift it), the different types of levers and the classification of various objects depending on the degree of the lever, the relationship between the angle of the inclined plane and the effort required to lift objects and, finally, how many pulleys are needed and what type so that the effort to lift objects is less than their weight. Concepts related to mathematics, technology or engineering were not explicitly asked.

The results obtained after analyzing the 115 tests gave an average score of 7.1 out of 10, with a standard deviation of 2.8 points. The success rate was 77.4%, that is, 89 students achieved a score of 6 out of 10 or higher. Considering that the test was hardly difficult, only 35% (40 students) answered all the questions correctly and obtained the highest score of 10. Figure 5 shows the correct and incorrect answers according to the type of question. The question with the highest success rate is the one related to the inclined plane, practically 80% of the students answered correctly that the lowest angle is the best option. It can be seen that the questions about pulleys, the law of the lever and about first-degree levers have the same level of success, around 71%. The question that entails the greatest difficulty has to do with identifying objects that are levers of the third degree, with 36.5% of the students answering incorrectly or not at all.

Although the question about the angle of the inclined plane offered the best results, it is striking that some students failed in a question that teachers consider to be knowledge of everyday life. Perhaps this is the first mathematical difficulty to face: angles. As will be detailed later, some students have problems in their spatial vision of angles, which makes it difficult for them to reason about physical concepts related to inclined planes. Regarding the pulley question, the students who answered incorrectly confuse mobile pulleys with fixed pulleys, attributing to the latter the property of reducing the effort to be made. Another mathematically related problem can be found in the interpretation of the lever law: the difficulty in understanding and manipulating equations with one unknown variable. This may be the reason why 26% of the students answered that the point of support should be close to the person making the effort instead of close to the object to be lifted. Finally, despite of all the lever objects asked (pliers, tweezers, wheelbarrow, and fishing rod) were explained and classified according to their degree in the theoretical session, the most common error made in this test was to classify the wheelbarrow as a lever of first or third class instead of second class. The interpretation of this mistake may be the awkwardness in identifying the position of the load in the wheelbarrow with respect to the support point.

Figure 5. Correct (in blue), incorrect (in orange) answers and not answered questions (in grey) in the preliminary test according to the different questions.
Source: Authors.

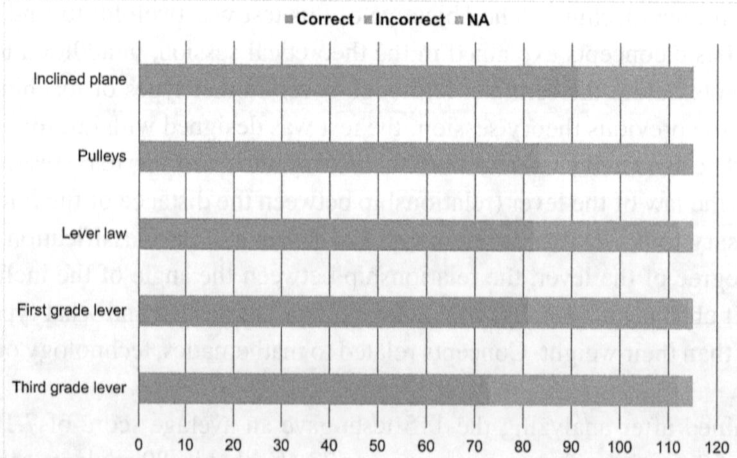

Analysis of the STEM Areas

Throughout the execution of the experience, both the difficulties that arose related to the manipulation and construction of the models and the reasoning of physical and mathematical concepts were solved during the same. From the manipulative point of view, the greatest difficulty was found in the construction of the inclined plane system, despite having a model already built as a sample. In addition, in the activity of the lever, some groups took a while to identify which cups should be the load and effort to move the fulcrum as indicated in the instructions. In general, pulley activity was the easiest to understand and manipulate.

After the completion of the intervention, several hard difficulties with some concepts are detected, especially those related to mathematics. Next, the main problems detected will be broken down according to each STEM area.

Science

Regardless of the type of machine being studied, students had some common mistakes in science area. For example, there was a general tendency not to write the units that correspond to the different physical magnitudes. In many cases, when students do not write down the physical units as they solve a problem, they can mix up magnitudes meaninglessly and give a crazy final answer. In the current intervention, there are no complex calculations that cause unit confusion in the results.

Despite working manipulatively with the inclined plane and the pulleys, students had still confusion between the concepts of mass and force. This error is, unfortunately, more common than it should be in teachers in training and quite persistent. Thus, Métioui and Trudel (2021) in their research on alternative conceptions of force, motion, and gravity research conducted with pupils old 9 to 12 and 85 pre-service teachers of elementary education from Canada, demonstrated that both share many misconceptions, such as the notions of mass and weight are synonymous, despite the difference in age and training. In the following, both concepts correctly learned, and misconceptions are described and analyzed according to each simple machine.

Inclined Plane

An important misconception is to identify all types of forces with weight. In this sense, statements like "It does not match because the inclined plane means that less weight is required to lift the object" were found. However, in this case the error may be due to the way the question was written, that is, students were asked "Do both weights match? Why?" and, since then, they used the term weight to refer to other forces. This will be considered in the elaboration of the questionnaire in future experiences.

Understanding the concept of tension force was extremely problematic for the students, among other reasons, due to their lack of knowledge of basic trigonometry. This topic will be addressed in the analysis of errors in Mathematics. In general, students attributed the differences between the weight of the object and the tension force necessary to lift the object to correct reasons. Thus, they referred to between the surface plane and the box with the object, friction force string of the pulley and only one group also included the air friction. Answers such as "the use of different materials" maybe were a way to relate this force difference with friction force. Some students were aware that their inclined planes were not constructed correctly and indicated this failure as the reason. Despite of experimental results, one group supposed that both values should not be very different.

It is worth noting that the inclined plane involved the greatest cognitive efforts for the students, spending most of the time of the experience in its construction and in the reasoning-explanation of the results obtained.

Pulleys

For the students, the tests with the fixed and mobile pulleys were the easiest to manipulate and understand. In its vast majority, they were able to predict what would happen before carrying out the different tests and to explain why the values of the weights measured experimentally did not coincide with the theoretical values alluding to the friction forces. Only some incomplete answers were found when asked if both weights coincided, for example, "both weights do not coincide, however, we have achieved the balance". Despite understanding the operation of the pulleys, they had difficulties in answering the problem raised about the system of 4 pulleys, 3 mobile and one fixed. The difficulty is associated with mathematical resolution and not with knowledge of physics, so this topic will be dealt with in the area of Mathematics.

First Class Lever

The execution of the tests with the first-class lever hardly caused any problem for the students. In general, the lever law was understood, and most of the students were able to correctly predict the behaviour of the lever as the fulcrum was moved. However, the understanding of the results obtained and their expression in the graphic representation entailed a significant difficulty that was not expected by the teachers. When they are asked about what would happen if you placed the fulcrum 5 cm from the load pot without using the model of the simple lever, it was found that the lever law was not correctly understood. In this way, replies such as "the load pot would fall because it would not have enough weight on the other side to maintain its balance" or "we would not need any force to raise the load pot". In other cases, students were not able to interpret the graph that they had previously represented with the experimental data to answer the question posed. In a first approximation one could think that the reason is that those who already in the initial test did not understand the law of the lever correctly would not have been able to change their knowledge. However, it is quite likely that the most influencing factor is their poor skills in interpreting graphics. This factor will be analyzed in the next section.

Mathematics

Despite the fact that most of the mathematical knowledge needed to solve the project should be learned in the Primary Education stage, most of the teachers in training failed in the mathematical reasoning during the course of the STEM experience. This item has been extensively discussed in the literature. Thus, Rashid *et al.* (2021) claimed that "in physics education, many students memorize equations and formulas, solve problems that require mathematical operations without understanding the subjects and concepts" (pp. 448). For example, one of the difficulties that teachers in training should have resolved before their stage in Secondary Education was the unit conversion factors. However, the reality was very different and in a non-negligible percentage they failed in simple conversion factors.

Other of the most worrying aspects is the lack of knowledge about basic trigonometry. It is worth remembering that in the previous knowledge test, 20% of the students failed the basic question about the best angle on an inclined plane to lift objects. An unfortunate surprise was the detection that some students struggled with tasks as simple as measuring an angle with a traditional angle gauge tool. For example, some students were not able to correctly place the zero value of the tool at the vertex of the angle or they read the complementary angle on the scale instead of the angle actually measured. On the other hand, it is extremely importance that students know basic concepts such as the sine and cosine of an angle to understand inclined planes. Nabie *et al.* (2018) analysed the pre-service teachers' perceptions and knowledge of trigonometric concepts with a sample of 119 second year science/mathematics pre-service teachers from two Colleges of Education in the Northern Region of Ghana. They found that most pre-service teachers lacked the basic knowledge of trigonometric concepts such as the interpretation of the graph of a sine function. Moreover, the few students that knew the concepts did not understand them. Thus, one of the most important obstacles in the workshop discussion was the non-comprehension of the concept of tension force and its difference with weight, since they did not even understand the role of the inclination of the plane, that is, the sine of the angle and that this value, which multiplicate the weight value, varies between 0 and 1 and, therefore, the tension force would be always lower than the weight.

Other disturbing finding was the difficulty of some students to solve equations with only one variable. Van Dooren *et al.* (2003) concluded that Primary teachers did not use algebraic strategies because of their lack of algebra understanding, after examining algebra word problem skills and strategies in Primary pre-service teachers in Belgium. On the other hand, Rule and Hallagan (2007) found that elementary pre-service teachers had great difficulty defining a variable. Both ideas were corroborated by Brown and Berger (2013) in their study with 73 north American pre-service teachers, that is, they do not have an adequate understanding of variables and algebraic expressions and their knowledge base is similar to middle school students.

According to Glazer (2011), graph interpretation competence is affected by many factors among which graph characteristics, the graph's content, and their prior knowledge (misconceptions associated with graphing, prior beliefs, prior knowledge of the content displayed in the graphs and mastery of graphing and explanatory skills). Regarding graph characteristics, students are not used to graphing data. Surprisingly, several groups did not understand what they were supposed to represent and, instead of graphing the obtained experimental data, they drew the lever system. Once the type of representation was explained, they corrected it and made the graph correctly. On the other hand, some of them presented some difficulty in taking advantage of the graph paper, for example, they used very small scales or placed the point (0,0) in the centre of the paper. Moreover, it is usual that they did not adequately indicate the titles of the axes. As mentioned before, the students did not know how to interpret the graph when they were asked about a hypothetical case in which the support point was 5 cm from the load, that is, they were not able to plot on the graph the point corresponding to the number of coins (y axis) that would be needed if the fulcrum was 5 cm away (x axis).

Technology

Of the four STEM areas, technology and engineering are *a priori* the least knowledgeable to our students. In the new Spanish Primary Education curriculum (Spanish Government, 2022), both conceptual and procedural content related to Technology has been largely increased. In fact, Technology and Digitization is one of the three major content groups in the Knowledge of the Natural Environment subject, along with Scientific Culture and Societies and Territories. Until now, part of the contents on technology in this new curriculum used to be taught in Secondary Education. For this reason, they are not contemplated in the current curriculum of the science subjects of the university degree. It is expected that in the medium term the study plan of the degree will be modified so that these contents and the STEM approach are adequately worked by preservice teachers.

Regarding the technology aspects of the intervention, some students had difficulties in the use of some instruments, such as the dynamometer and balance. They did not have to choose material since they were offered all the necessary material, although its handling was not always the most successful. For example, in the construction of the inclined plane, some groups needed more than one carton box since they bended the cardboard a lot by not achieving the desired structure, leaving it unusable for its purpose. In general, they did not have difficulties in the interpretations of scales or plans. The technological difficulties of participants fit with the results of Blanco and Dávila (2017), who found high levels of nervousness in secondary education students with the handling of technological instruments and procedures (especially in girls). These difficulties hinder the development of the technology pedagogical content knowledge, which is necessary to rightly integrate technology in STEM education practices (McDonald, 2016).

Engineering

From the point of view of the engineering aspect, the students achieved the objectives of working as a team and solving the design problems that arose during the intervention. Some groups even proposed alternative designs that improved on the initial design of teachers. Once the seminar was completed, the students had to answer the initial engineering question and object of this study: How did the Egyptians transport the stone blocks of the pyramids? Most of the answers were correct but succinct. Other ones were vague and simplified into "using simple machines". It is curious that some groups disconnected the engineering problem from what was done in the seminar, providing answers inspired by the films of the Egyptian era rather than their recent experience. In this sense, answers such as "the Egyptians transported the stones by rolling them on top of wooden logs" are reminiscent of film fiction. These results suggest that the participants are not prepared to include engineering approaches in their future as teachers. Previous experiences with secondary school students (Brophy *et al.*, 2008) developed effective educational models that connect this approach with science teaching. For this purpose, engineering was addressed through questions such as "What should I make?", "How does it work?," "What factors in the design are critical to my goals?," and "What can I manipulate to achieve my goals? As Simarro and Couso (2022) argues, STEM education has given engineering education an unprecedented role until now, but it is necessary to develop proposals that really address engineering (due to current STEM practices works engineering in a superfluous and deficient way).

Understanding of the STEM Approach

Finally, regarding the understanding of the STEM approach after the execution of the experience students were asked about the implementation of the different areas of the STEM approach. The collected results were mixed: some groups perfectly understood the nature of the STEM approach and the content taught in each of the areas while other groups were not able to correctly define the role of each STEM area. When they commented on the role of mathematics, all mentioned mathematical calculations, although some groups did not indicate more items. The rest of the groups included some but not all these items: measurement, problem solving, unit conversion, tables, graphing, and geometry. Unfortunately, neither group considered trigonometry to be mathematical knowledge. Regarding the science area, and contrary to expectations, the students were not able to relate all the knowledge of physics that was put into play during the realization of the project. Thus, they considered aspects such as physical laws, forces, physical magnitudes, and they forgot to mention, among others, the scientific method, the forces of friction (even though the latter, for example, has been considered in the explanations of the experience). At last, some of them confused technology and engineering, for example, they interchanged the phases of design and construction. In addition, some groups simply answered that the technology area included the three employed simple machines, without going into more detail. These difficulties are consistent with those found by Correia and Baptista (2022) who showed weaknesses in the Pedagogical Content Knowledge (PCK) on STEM teaching in teachers in training. This suggests that the students did not receive basic engineering knowledge either in their secondary education or in their university degree. This shows, as indicated by Di Francesca *et al.* (2014), the need to include these contents in initial teacher training. In fact, the relationship between engineering and multiple science contents would facilitate this inclusion, as Roehring *et al.* (2012) affirm.

FUTURE PROPOSALS TO IMPROVE THE STEM EXPERIENCE

The intention of the authors is to re-implement this experience in successive courses, including a series of improvements with the aim of preparing future teachers for new skills that will be required of them in the performance of their profession.

One of the main weaknesses detected by the authors is that the methodological aspects related to the experience were not addressed. Thus, future implementations will address the relations of methodologies such as design-based process, inquiry- and problem-based learning, collaborative learning, to name a few, with the STEM approach. The inclusion of these aspects would be essential to develop the PCK in STEM education (Correia and Baptista, 2022).

Regarding mathematical misconceptions, it is evident that the learning of mathematical concepts during their Secondary school stage has not been entirely significant and that they easily forget those mathematical tools that they do not use in their daily lives. On the other hand, even though participants have studied three mathematics subjects during the degree, their focus was developed to a greater extent in the didactics of the content and less in the content itself. The first proposal for improvement, therefore, should be a better coordination with teachers of mathematics education subjects so that they reinforce contents such as trigonometry or data representation and its subsequent interpretation. Thus, mathematics teachers should be directly involved in the experience, both in the design of tests or the script of the experience and in person during its execution. This interdisciplinary approach between subjects of different areas could improve the experience and could help students to overcome problems with mathematical tasks. In this sense, Margot and Kettler (2019) and Roehring *et al.* (2012) propose, in their models to integrate STEM areas, the collaboration of teachers from different areas.

So that the students do not lose the objective of the study, that is, to respond to the engineering problem that is initially raised, it should be referred to several times during the elaboration of the experience. Moreover, it could be more motivating if the engineering problem to be solved was more related to the student's daily life. In future implementations, authors could raise other different engineering problems more related to events of everyday life.

With the objective to improve the learning of the STEM approach and until the curriculum of the science subjects of the degree is renewed, more previous sessions should be taught in the classroom that reinforce these contents, not only scientific and mathematical, but also the aspects about the areas of technology and engineering that are quite unknown to students. On the other hand, more examples of the applicability of the STEM approach that are related to the daily life of the student should be introduced, thus that they are aware that interdisciplinary methodologies should be the strategy to follow in solving any type of problem.

Lastly, it could be interesting to ask students about their self-efficacy in teaching sciences employing the STEM approach. In their study, Mateos-Núñez *et al.* (2019) compared levels of self-efficacy in STEM areas at different educational stages (Primary and Secondary and teachers in training). They concluded that the level of self-efficacy decreases as the educational level increases and the differences between Secondary stage and pre-service teachers are significant since the students of Secondary stage presented higher levels of STEM competence than pre-service teachers. Ascertain this could be important to improve initial training of teachers in STEM skills.

CONCLUSION

The new trend in the Spanish educational system towards the STEM approach and its application as a strategy in solving everyday problems makes it necessary for us to reinforce both the didactics and the contents of the different areas involved in the science and mathematics subjects of the degree. For this reason, this type of experiences (combining practice, interdisciplinary and STEM approaches, as well as the use of everyday materials) should be included in the university degree curriculum. However, the analysis of the implementation of this STEM experience indicate that this kind of activities are not easy to execute, finding some problems with some of the STEM areas (mainly with science and mathematics).

These barriers, which has been described in detail in this contribution, can hinder the implementation and the success of these activities. As teacher trainers, science and mathematics educators should consider these difficulties in our subjects of science and mathematics educations, in order to improve the future STEM teaching of our students.

ACKNOWLEDGMENT

This research was supported by the Spanish Ministry of Science and Innovation [Project PID2020-115214RB-I00] and the Government of Extremadura [Project GR21047].

REFERENCES

Altan, E. B., & Ercan, S. (2016). STEM education program for science teachers: Perceptions and competencies. *Journal of Turkish Science Education, 13*, 103–117.

Anilan, B., & Berber, A. (2019). Effect of conversion factor on problem solving and experience of teacher candidates. *International Journal of Research in Education and Science, 5*(1), 118–133.

Blanco, J. M. R., & Dávila, M.A. (2017). Las emociones según el género, en el aprendizaje de la Tecnología del alumnado de primer curso de Educación Secundaria Obligatoria. *Ápice. Revista de Educación Científica, 1*(2), 18–37.

Bravo, E., Brígido, M., Hernández-Barco, M. A., & Mellado, V (2022). Las emociones en ciencias en la formación inicial del profesorado de infantil y primaria. *Revista Interuniversitaria de Formación del Profesorado. Continuación de la antigua Revista de Escuelas Normales, 97*, 57-74.

Brewer, C., & Smith, D. (2011). Vision and change in undergraduate biology education: a call to action. American Association for the Advancement of Science.

Brígido, M., Bermejo, M. L., Conde, M. C., & Mellado, V. (2010). The Emotions in Teaching and Learning Nature Sciences and Physics/Chemistry in Pre-Service Primary Teachers. *US-China Education Review, 7*(12), 25–32.

Brígido, M., Borrachero, A. B., Bermejo, M. L., & Mellado, V. (2013). Prospective primary teachers' self-efficacy and emotions in science teaching. *European Journal of Teacher Education, 36*(2), 200–217. doi:10.1080/02619768.2012.686993

Bybee, R. W. (2010). What is STEM education? *Science, 329*(5995), 996–996. doi:10.1126cience.1194998 PMID:20798284

Chi, M. (2005). Commonsense conceptions of emergent processes: Why some misconceptions are robust. *Journal of the Learning Sciences, 14*(2), 161–199. doi:10.120715327809jls1402_1

Correia, M., & Baptista, M. (2022). Supporting the Development of Pre-Service Primary Teachers PCK and CK through a STEM Program. *Education Sciences, 12*(4), 258. doi:10.3390/educsci12040258

Diani, R., Yuberti, Y., Anggereni, S., Utami, G. N., Iqbal, A., & Kurniawati, I. (2020). ECIRR (Elicit, Confront, Identify, Resolve, Reinforce) learning model with the pictorial riddle method: Is it effective in reducing physics misconceptions? *Journal of Physics*, *1572*(1), 012020.

DiFrancesca, D., Lee, C., & McIntyre, E. (2014). Where is the 'E' in STEM for Young Children? Engineering Design Education in an Elementary Teacher Preparation Program. *Issues in Teacher Education*, *23*(1), 49–64.

Glazer, N. (2011). Challenges with graph interpretation: A review of the literature. *Studies in Science Education*, *47*(2), 183–210. doi:10.1080/03057267.2011.605307

Hernández-Barco, M., Cañada-Cañada, F., Corbacho-Cuello, I., & Sánchez-Martín, J. (2021). An Exploratory Study Interrelating Emotion, Self-Efficacy and Multiple Intelligence of Prospective Science Teachers. *Frontiers in Education*, *6*, 1–12. doi:10.3389/feduc.2021.604791

Labov, J. B., Reid, A. H., & Yamamoto, K. R. (2010). Integrated biology and undergraduate science education: A new biology education for the twenty-first century? *CBE Life Sciences Education*, *9*(1), 10–16. doi:10.1187/cbe.09-12-0092 PMID:20194802

Lawrenz, F. (1986). Misconceptions of Physical Science Concepts Among Elementary School Teachers. *School Science and Mathematics*, *86*(8), 654–660. doi:10.1111/j.1949-8594.1986.tb11669.x

Leszczynski, E. (2014). *The Study of Middle School Mathematics and Science Teachers' Practices, Perceptions, and Attitudes Related to Mathematics and Science Integration* [Unpublished doctoral dissertation]. University of Montclair, Canada.

Maass, K., Geiger, V., Ariza, M. R., & Goos, M. (2019). The role of mathematics in interdisciplinary STEM education. *ZDM*, *51*(6), 869–884. doi:10.100711858-019-01100-5

Marcos-Merino, J. M., Esteban, R., & Ochoa de Alba, J. A. G. (2022). The interplay of prior knowledge, emotions and learning in a science experiment activity. *Enseñanza de las Ciencias*, *40*(1), 107–124.

Margot, K. C., & Kettler, T. (2019). Teachers' perception of STEM integration and education: A systematic literature review. *International Journal of STEM Education*, *6*(1), 1–16. doi:10.118640594-018-0151-2

Marulcu, I., & Barnett, M. (2013). Fifth graders' learning about simple machines through engineering design-based instruction using LEGO™ materials. *Research in Science Education*, *43*(5), 1825–1850. doi:10.100711165-012-9335-9

Marulcu, I., & Barnett, M. (2016). Impact of an engineering design-based curriculum compared to an inquiry-based curriculum on fifth graders' content learning of simple machines. *Research in Science & Technological Education*, *34*(1), 85–104. doi:10.1080/02635143.2015.1077327

Mateos-Núñez, M., Martínez-Borreguero, G., & Naranjo-Correa, F. L. (2019). Comparación de las emociones, actitudes y niveles de autoeficacia ante áreas STEM entre diferentes etapas educativas. *European Journal of Education and Psychology*, *13*(1), 49–64. doi:10.30552/ejep.v13i1.292

McComas, W. F., & Burgin, S. R. (2020). A critique of "STEM" education. *Science & Education*, *29*(4), 805–829. doi:10.100711191-020-00138-2

McCright, A., O'Shea, B., Sweeder, R., Urquhart, G., & Zeleke, A. (2013). Promoting interdisciplinarity through climate change education. *Nature Climate Change*, *3*(8), 713–716. doi:10.1038/nclimate1844

McDonald, C. V. (2016). STEM Education: A review of the contribution of the disciplines of science, technology, engineering and mathematics. *Science Education International*, *27*(4), 530–569.

Mellado, V., Borrachero, A. B., Brígido, M., Melo, L. V., Dávila, M. A., Cañada, F., Conde, M. C., Costillo, E., Cubero, J., Esteban, R., Martínez-Borreguero, G., Ruiz, C., Sánchez, J., Garritz, A., Mellado, L., Vázquez, B., Jiménez, R., & Bermejo, M. L. (2014). Las emociones en la enseñanza de las ciencias. *Enseñanza de las Ciencias*, *32*(3), 11–36. doi:10.5565/rev/ensciencias.1478

Métioui, A., & Trudel, L. (2021). Children and Preservice Teachers' Misconceptions and Scientifically Acceptable Conceptions about Movement, Force, and Gravity. *Modern Perspectives in Language. Literature and Education*, *5*, 26–42.

Nabie, M. J., Akayuure, P., Ibrahim-Bariham, U. A., & Sofo, S. (2018). Trigonometric Concepts: Pre-Service Teachers' Perceptions and Knowledge. *Journal on Mathematics Education*, *9*(2), 169–182. doi:10.22342/jme.9.2.5261.169-182

NGSS Lead States. (2013). *Next Generation Science Standards: For States, By States*. The National Academies Press.

Pimthong, P., & Williams, J. (2018). Preservice teachers' understanding of STEM education. *Kasetsart Journal of Social Sciences*, *1*, 1–7.

Rashid, S., Baha, A., & Abidin, D. (2021). Students' Perception on Learning Unit Conversion Through a Smart Conversion Card Tool. *Advances in Social Science, Education and Humanities Research*, *544*, 448–451. doi:10.2991/assehr.k.210424.087

Roehrig, G. H., Moore, T. J., Wang, H.-H., & Park, M. S. (2012). Is Adding the E Enough? Investigating the Impact of K-12 Engineering Standards on the Implementation of STEM Integration. *School Science and Mathematics*, *112*(1), 31–44. doi:10.1111/j.1949-8594.2011.00112.x

Roth, W. (2008). The nature of scientific conceptions: A discursive psychological perspective. *Educational Research Review*, *3*(1), 30–50. doi:10.1016/j.edurev.2007.10.002

Rule, A., & Hallagan, J. (2007). *Algebra Rules Object Boxes as an Authentic Assessment Task of Preservice Elementary Teacher Learning in a Mathematics Methods Course*. Paper presented at the meeting of Annual Conference of the New York State Association of Teacher Educators, Saratoga Springs, NY, United States.

Simarro, C., & Couso, D. (2022). Didáctica de la ingeniería: tres preguntas con visión de futuro. *Enseñanza de las Ciencias. Revista de investigación y experiencias didácticas*, *40*(3), 147-174.

Spanish Government (2014). *Real Decreto 126/2014, de 28 de febrero, por el que se establece el currículo básico de Educación Primaria.*

Spanish Government (2022). *Real Decreto 157/2022, de 1 de marzo, por el que se establecen la ordenación y las enseñanzas mínimas de la Educación Primaria.*

Tanisli, D., & Kose, N. Y. (2013). Pre-service mathematic teachers' knowledge of students about the algebraic concepts. *The Australian Journal of Teacher Education*, *38*(2), 1–18. doi:10.14221/ajte.2013v38n2.1

Timmis, K., Cavicchioli, R., Garcia, J. L., Nogales, B., Chavarria, M., Stein, L., McGenity, T. J., Webster, N., Singh, B., Handelsman, J., de Lorenzo, V., Pruzzo, C., Timmis, J., Martin, J. L., Verstraete, W., Jetten, M., Danchin, A., Huang, W., Gilbert, J., ... Harper, L. (2019). The urgent need for microbiology literacy in society. *Environmental Microbiology*, *21*(5), 1513–1528. doi:10.1111/1462-2920.14611 PMID:30912268

Toma, R. B., & Garcia-Carmona, A. (2021). "Of STEM we like everything but STEM". A critical analysis of a buzzing educational trend. *Enseñanza de las Ciencias*, *39*(1), 65–80. doi:10.5565/rev/ensciencias.3093

Van Dooren, W., Verschaffel, L., & Onghena, P. (2003). Pre-service teachers' preferred strategies for solving arithmetic and algebra word problems. *Journal of Mathematics Teacher Education*, *6*(1), 27–52. doi:10.1023/A:1022109006658

White, D., & Delaney, S. (2021). Full STEAM Ahead, but Who Has the Map for Integration? A PRISMA Systematic Review on the Incorporation of Interdisciplinary Learning into Schools. *International Journal on Math. Science and Technology Education*, *9*(2), 9–32.

Winarno, N., Widodo, A., Rusdiana, D., Rochintaniawati, D., & Afifah, R. M. A. (2019). Pre-service Science Teachers' Conceptual Understanding of Integrated Science Subject: A Case Study. *Journal of Physics*, *1204*(1), 012104.

ADDITIONAL READING

Couso, D., & Simarro, C. (2020). STEM education through the epistemological lens: Unveiling the challenge of STEM transdisciplinarity. In *Handbook of research on STEM education* (pp. 17–28). Routledge. doi:10.4324/9780429021381-3

Erduran, S. (2020). Nature of "STEM"? *Science & Education*, *29*(4), 781–784. doi:10.100711191-020-00150-6 PMID:32836878

Kayan-Fadlelmula, F., Sellami, A., Abdelkader, N., & Umer, S. (2022). A systematic review of STEM education research in the GCC countries: Trends, gaps and barriers. *International Journal of STEM Education*, *9*(1), 1–24. doi:10.118640594-021-00319-7

Lee, O., & Grapin, S. E. (2022). The role of phenomena and problems in science and STEM education: Traditional, contemporary, and future approaches. *Journal of Research in Science Teaching*, *59*(7), 1301–1309. doi:10.1002/tea.21776

Li, Y., Schoenfeld, A. H., diSessa, A. A., Graesser, A. C., Benson, L. C., English, L. D., & Duschl, R. A. (2019). Design and design thinking in STEM education. *Journal for STEM Education Research*, *2*(2), 93–104. doi:10.100741979-019-00020-z

Martín-Páez, T., Aguilera, D., Perales-Palacios, F. J., & Vílchez-González, J. M. (2019). What are we talking about when we talk about STEM education? A review of literature. *Science Education*, *103*(4), 799–822. doi:10.1002ce.21522

Martínez-Borreguero, G., Mateos-Núñez, M., & Naranjo-Correa, F. L. (2019). Implementation and Didactic Validation of STEM Experiences in Primary Education: Analysis of the Cognitive and Affective Dimension. In Theorizing STEM Education in the 21st Century. IntechOpen.

Ortiz-Revilla, J., Adúriz-Bravo, A., & Greca, I. M. (2020). A framework for epistemological discussion on integrated STEM education. *Science & Education*, *29*(4), 857–880. doi:10.100711191-020-00131-9

KEY TERMS AND DEFINITIONS

First Class Lever: Lever in which the fulcrum is located between the load and the effort (force).

Pulley: A wheel with a grooved rim around which a cord passes, which acts to change the direction of a force applied to the cord and is used to raise heavy weights.

Second Class Lever: Lever in which the load is between the effort (force) and the fulcrum.

Self-Efficacy: People's beliefs about their capabilities to produce designated levels of performance that exercise influence over events that affect their lives.

Simple Machines: Any of the basic mechanical devices for applying a force, such as an inclined plane, wedge, or lever.

STEM: Is an approach to learning and development that integrates the areas of science, technology, engineering, and mathematics.

Third Class Lever: Lever in which the effort is between the load and the fulcrum.

Compilation of References

Abimbola, I. O., & Baba, S. (1996). Misconceptions & Alternative Conceptions in Science Textbooks: The Role of Teachers as Filters. *The American Biology Teacher*, *58*(1), 14–19. doi:10.2307/4450067

Acosta-Pérez, R. M. (2014). *Evidencias de la evolución biológica: aprendiendo evolución desde la metodología científica* [Tesis de licenciatura, Universidad Nacional Autónoma de México]. Repositorio de la Dirección General de Bibliotecas y Servicios Digitales de Información. https://repositorio.unam.mx/contenidos/238472

Adadan, E., & Oner, D. (2014). Exploring the progression in preservice chemistry teachers' pedagogical content knowledge representations: The case of "behavior of gases". *Research in Science Education*, *44*(6), 829–858. doi:10.100711165-014-9401-6

Aguilera, D., & Ortiz-Revilla, J. (2021). STEM vs. STEAM Education and Student Creativity: A Systematic Literature Review. *Education Sciences*, *11*(7), 331. doi:10.3390/educsci11070331

Aguirre-Munoz, Z., Dang, B., & Garro, E. S. (2022). *Impact of integrated science and mathematics instruction on middle school science and mathematics achievement. In Advances in research in STEM education.* IntechOpen.

Aguirre-Muñoz, Z., Dang, B., & Garro, E. S. L. (2022). Impact of Integrated Science and Mathematics Instruction on Middle School Science and Mathematics Achievement. In M. Kalogiannakis & M. Ampartzaki (Eds.), *Advances in Research in STEM Education* (pp. 1–19). IntechOpen. doi:10.5772/intechopen.104082

Aguirre-Muñoz, Z., Yeter, I. H., Garro, E. S. L., & Koca, F. (2021). Building teachers' capacity to integrate science and math content: Implications for professional development and learning. *Journal of Science Teacher Education*, *32*(1), 62–84. doi:10.1080/1046560X.2020.1789814

Ahuja, A. (2021). Backyard evolutionary biology: Investigating local flowers brings learning to life. *Ecology and Evolution*, *11*(8), 3459–3463. doi:10.1002/ece3.7199 PMID:33898002

Ainley, J., Pratt, D., & Hansen, A. (2006). Connecting engagement and focus in pedagogic task design. *British Educational Research Journal*, *32*(1), 23–38. doi:10.1080/01411920500401971

Aivelo, T., & Uitto, A. (2016). Digital gaming for evolutionary biology learning: The case study of parasite race, an augmented reality location-based game. *LUMAT: International Journal on Math, Science and Technology Education*, *4*(1), 1–26. doi:10.31129/LUMAT.4.1.3

Akaike, H. (1987). Factor Analysis and AIC. *Psychometrika*, *52*(3), 317–332. doi:10.1007/BF02294359

Al-Kahtani, N., Abdullah, A., Arun, V. S., Tunny, S., Hend, A., & Duaa, A. (2022). Sustaining Blended and Online Learning during the Normal and New Normal Conditions in a Saudi Higher Education Institution: Health Science Students' Perspectives. *Heliyon*, *8*(10), e10898. doi:10.1016/j.heliyon.2022.e10898 PMID:36247156

Allaby, M., & Gjertsen, D. (Eds.). (2002). *Makers of Science* (Vol. 2). Oxford University Press USA.

Alles, D. L. (2001). Using Evolution as the Framework for Teaching Biology. *The American Biology Teacher*, *63*(1), 20–23. doi:10.2307/4451025

Allison, J. C. (1986). CDT - craft, design and technology. *Physics Education*, *21*(6), 337–339. doi:10.1088/0031-9120/21/6/303

Almeida, A., Valente, B., Silva, M. J., Rodrigues, M., & Manteigas, V. (2019). O projeto GLOCAL – AGIR: Conhecer o território para o valorizar [The Project GLOCAL - AGIR: Know the territory to value it]. In J. Pinhal, C. Cavaco, M. J. Cardona, F. A. Costa, J. Marques, A., & R. Faria (Eds.), Atas do XXV Colóquio da AFIRSE Portugal: A investigação, a formação, as políticas e as práticas em educação – 30 Anos de AFIRSE em Portugal (pp. 92-102). Instituto de Educação da Universidade de Lisboa.

Alonso, J. C. P. (2003). Las matemáticas en la naturaleza [Mathematics in nature]. *Sigma: revista de matemáticas, 22*, 161-171.

Alonso, M., Galagovsky, L. R., Stella, C., & Yankilevich, M. L. (2005). Una simulación con planilla de cálculo para el aprendizaje de mecanismos evolutivos a partir del equilibrio de Hardy-Weinberg. *Enseñanza de las ciencias,* 1-6.

Alrahlah, A. (2016). How Effective the Problem-Based Learning (PBL) in Dental Education. A Critical Review. *The Saudi Dental Journal*, *28*(4), 155–161. doi:10.1016/j.sdentj.2016.08.003 PMID:27872545

Altan, E. B., & Ercan, S. (2016). STEM education program for science teachers: Perceptions and competencies. *Journal of Turkish Science Education, 13*, 103–117.

Álvarez-Pérez, E., & Ruiz, R. (2016). Proposal for Teaching Evolutionary Biology: A bridge between research and educational practice. *Journal of Biological Education*, *50*(2), 123–146. doi:10.1080/00219266.2015.1007887

Amir, N. (2014). Showcasing the creative talents in science of the academically less-inclined students through a values-driven toy story-telling project. In L. C. Lennex & K. F. Nettleton (Eds.), Cases on Instructional Technology in Gifted and Talented Education (pp. 141-179). IGI Global Publishing. doi:10.4018/978-1-4666-6489-0.ch007

Amir, N. (2021a). Fostering joy and creativity amongst students in a Singapore classroom through fun design-and-make STEM projects. In D. Ktoridou, E. Doukanari & N. Eteokleous (Eds.), Fostering Meaningful Learning Experiences through Student Engagement (pp. 51-73). IGI Global Publishing. doi:10.4018/978-1-7998-4658-1.ch003

Amir, N. (2021b). Views from parents of at-risk students in Singapore towards the use of fun design-and-make STEM projects to foster creativity and joy of learning. In C. Gaines & K. Hutson (Eds.), Promoting Positive Learning Experiences in Middle School Education (pp. 106-131). IGI Global Publishing. doi:10.4018/978-1-7998-7057-9.ch006

Amir, N. (2022). Fostering Student Motivation and Engagement Through the Relevant, Appealing, and Personal (RAP) Pedagogical Guideline. In Transformative Teaching Around the World: Stories of Cultural Impact, Technology Integration, and Innovative Pedagogy (pp. 87-91). Routledge.

Amir, N., & Subramaniam, R. (2012). Fostering inquiry in science among kinaesthetic learners through Design & Technology. In L. C. Lennex & K. F. Nettleton (Eds.), Cases on Inquiry Through Instructional Technology in Math and Science: Systemic Approaches (pp. 221-257). IGI Global Publishing. doi:10.4018/978-1-4666-0068-3.ch009

Amir, N. (2020). Strengthening the science-D&T interaction through simple maker-centred projects in a Singapore classroom. *Physics Education*, *55*(4), 045019. doi:10.1088/1361-6552/ab9212

Amir, N., & Subramaniam, R. (2006). Making physics toys fosters creativity. *Physics Education*, *41*(1), 18–20. doi:10.1088/0031-9120/41/1/F05

Amir, N., & Subramaniam, R. (2009). Making a low cost candy floss kit gets students excited about learning physics. *Physics Education*, *44*(4), 420–428. doi:10.1088/0031-9120/44/4/013

Amorós-Hernández, M. (2018). Teoría de la evolución y modelización pedagógica: Una perspectiva ecológica. *Temps d'Educacio*, *55*, 171–190.

Andersen, A. L., & Carin, R. (2021). Continuing Engineering Education (CEE) in Changeable and Reconfigurable Manufacturing Using Problem-Based Learning (PBL). *Procedia CIRP*, *104*, 1035–1040. doi:10.1016/j.procir.2021.11.174

Andersson, B., & Wallin, A. (2006). On developing content-oriented theories taking biological evolution as an example. *International Journal of Science Education, Abingdon*, *28*(6), 673–695. doi:10.1080/09500690500498385

Andrews, S. E., Runyon, C., & Aikens, M. L. (2017). The math-biology values instrument: Development of a tool to measure life science majors' task values of using math in the context of biology. *CBE Life Sciences Education*, *16*(3), 1–12. doi:10.1187/cbe.17-03-0043 PMID:28747355

Anilan, B., & Berber, A. (2019). Effect of conversion factor on problem solving and experience of teacher candidates. *International Journal of Research in Education and Science*, *5*(1), 118–133.

An, S. A. (2017). Preservice teachers' knowledge of interdisciplinary pedagogy: The case of elementary mathematics–science integrated lessons. *ZDM Mathematics Education*, *49*(2), 237–248. doi:10.100711858-016-0821-9

An, S. A., & Tillman, D. A. (2018). Preservice teachers' pedagogical use of "Gerrymandering" to integrate social studies and mathematics. *Journal of Mathematics Education*, *11*(3), 33–53.

Apodaca, M. J., McInerney, J. D., Sala, O. E., Katinas, L., & Crisci, J. V. (2019). A Concept Map of Evolutionary Biology to Promote Meaningful Learning in Biology. *The American Biology Teacher*, *81*(2), 79–87. doi:10.1525/abt.2019.81.2.79

Apple, M. (1993). The politics of official knowledge: Does a national curriculum make sense? *Discourse (Abingdon)*, *14*(1), 1–16. doi:10.1080/0159630930140101

Appleton, L., & Mackie, J. (2019). Using Digital Organism Evolutionary Software in the Classroom. *The American Biology Teacher*, *81*(1), 12–17. doi:10.1525/abt.2019.81.1.12

Araujo, R., & Cantó, J. (2015). El árbol de la vida, un juego para la superación de los obstáculos al aprendizaje de la evolución biológica. *Escritos sobre la Biología y su enseñanza*, *8*(15), 59-71.

Archer, L., Moote, J., Francis, B., DeWitt, J., & Yeomans, L. (2016). Stratifying science: A Bourdieusian analysis of student views and experiences of school selective practices in relation to 'Triple Science' at Key Stage 4 in England. *Research Papers in Education*, 1–21.

Arenson, M., & Coley, J. D. (2018). Anthropocentric by Default? Attribution of Familiar and Novel Properties to Living Things. *Cognitive Science*, *42*(1), 253–285. doi:10.1111/cogs.12501 PMID:28481420

Arias, I. A., & Sánchez, F. M. (2016). *La transversalidad de las temáticas de evolución biológica en el PCLB como posibilidad para la configuración del conocimiento profesional del profesor en Ciencias* [tesis de licenciatura, Universidad Pedagógica Nacional]. Repositorio Institucional UPN: http://hdl.handle.net/20.500.12209/1766

Aristotle. (1910). *Historia Animalium* (W. Ogle, Trans.). Clarendon Press. doi:10.5962/bhl.title.147382

Asghar, A., Hameed, S., & Farahani, N. K. (2014). Evolution in Biology Textbooks: A Comparative Analysis of 5 Muslim Countries. *Religious Education (Chicago, Ill.)*, *41*(1), 1–15. doi:10.1080/15507394.2014.855081

Ashe, A., Colot, V., & Oldroyd, B. P. (2021). How does epigenetics influence the course of evolution?. *Philosophical Transactions of the Royal Society B*, *376*(1826), 20200111.

Aspfors, J., & Eklund, G. (2017). Explicit and implicit perspectives on research-based teacher education: Newly qualified teachers' experiences in Finland. *Journal of Education for Teaching*, *43*(4), 400–413. doi:10.1080/02607476.2017.1297042

Associação Bandeira Azul da Europa (ABAE). (2018). *Programa Eco-Escolas* [Eco-Schools Program]. https://ecoescolas.abae.pt/wp-content/uploads/sites/3/2018/09/Programa-Eco-Escolas-SINTESE.pdf

Association for Science Education. (2016a). *The language of mathematics in science. A guide for teachers of 11-16 science*. Association for Science Education.

Association for Science Education. (2016b). *The language of mathematics in science. Teaching approaches*. Association for Science Education.

Asterhan, C. S. C., & Schwarz, B. B. (2007). The effects of monological and dialogical argumentation on concept learning in evolutionary theory. *Journal of Educational Psychology*, *99*(3), 626–639. doi:10.1037/0022-0663.99.3.626

Athanasiou, K., Katakos, E., & Papadopoulou, P. (2012). Conceptual ecology of evolution acceptance among Greek education students: The contribution of knowledge increase. *Journal of Biological Education*, *46*(4), 234–241. doi:10.1080/00219266.2012.716780

Athanasiou, K., Katakos, E., & Papadopoulou, P. (2016). Acceptance of evolution as one of the factors structuring the conceptual ecology of the evolution theory of Greek secondary school teachers. *Evol Educ Outreach*, *9*(1), 7. doi:10.118612052-016-0058-7

Ausubel, D. (2000). *Adquisición y retención del conocimiento: Una perspectiva cognitiva* (G. Sánchez, Trans.). Paidós.

Ausubel, D. P. (1963). *The psychology of meaningful verbal learning*. Grune & Stratton.

Aydin-Gunbatar, S., Ekiz-Kiran, B., & Oztay, E. S. (2020). Pre-service chemistry teachers' pedagogical content knowledge for integrated STEM development with LESMeR model. *Chemistry Education Research and Practice*, *21*(4), 1063–1082. doi:10.1039/D0RP00074D

Aydin, S., Demirdogen, B., Tarkin, A., Kutucu, S., Ekiz, B., Akin, F. N., Tuysuz, M., & Uzuntiryaki, E. (2013). Providing a set of research-based practices to support preservice teachers' long-term professional development as learners of science teaching. *Science Education*, *97*(6), 903–935. doi:10.1002ce.21080

Ayvacı, H. Ş., & Bebek, G. (2018). A study on evaluation of the problems faced with teaching physics. *Kastamonu Education Journal*, *26*(1), 1–10.

Ayyildiz, Y., & Tarhan, L. (2018). Problem-based learning in teaching chemistry: Enthalpy changes in systems. *Research in Science & Technological Education*, *36*(1), 35–54. doi:10.1080/02635143.2017.1366898

Bagiati, A., Yoon, S. Y., Evangelou, D., & Ngambeki, I. (2010). Engineering curricula in early education: Describing the landscape of open resources. *Early Childhood Research & Practice*, *12*(2), 1–15.

Bahar, M., Yener, D., Yılmaz, M., Emen, H., & Gürer, F. (2018). The changes of standards in the 2018 science curriculum and STEM integration. *Bolu Abant İzzet Baysal University Journal of Faculty of Education*, *18*(2), 702–735.

Ball, S. (1987). *The Micropolitics of the School*. Routledge.

Barak, M., & Assal, M. (2018). Robotics and STEM learning: Students' achievements in assignments according to the P3 Task Taxonomy—practice, problem solving, and projects. *International Journal of Technology and Design Education*, *28*(1), 121–144. doi:10.100710798-016-9385-9

Bardapurkar, A. (2008). Do students see the "selection" in organic evolution?: A critical review of the causal structure of student explanations. *Evolution (New York), 1*(3), 299–305. doi:10.100712052-008-0048-5

Bardin, L. (2013). *L'analyse de contenu* [Content analysis]. PUF. doi:10.3917/puf.bard.2013.01

Barlex, D., & Pitt, J. (2000). *Interaction: The relationship Between Science and Design and Technology in the Secondary School Curriculum*. Engineering Council.

Barlow, N. (Ed.). (1958). *The autobiography of Charles Darwin 1809-1882. With the original omissions restored.* Collins.

Barnes, M. E., Dunlop, H. M., Holt, E. A., Zheng, Y., & Brownell, S. E. (2019). Different evolution acceptance instruments lead to different research findings. *Evolution (New York), 12*(1), 4. doi:10.118612052-019-0096-z

Barrable, A. (2019). Refocusing environmental education in the early years: A brief introduction to a pedagogy for connection. *Education Sciences, 9*(1), 61. doi:10.3390/educsci9010061

Başkan, Z., Alev, N., & Karal, I. S. (2010). Physics and mathematics teachers' ideas about topics that could be related or integrated. *Procedia: Social and Behavioral Sciences, 2*(2), 1558–1562. doi:10.1016/j.sbspro.2010.03.235

Bassachs, M., Cañabate, D., Nogué, L., Serra, T., Bubnys, R., & Colomer, J. (2020). Fostering critical reflection in primary education through STEAM approaches. *Education in Science, 10*(12), 1–14. doi:10.3390/educsci10120384

Basson, I. (2002). Physics and mathematics as interrelated fields of thought development using acceleration as an example. *International Journal of Mathematical Education in Science and Technology, 33*(5), 679–690. doi:10.1080/00207390210146023

Bastos, R. M. B. (2017). The surprising success of the Finnish educational system in a global scenario of commodified education. *Revista Brasileira de Educação, 22*, 802–825. doi:10.15901413-24782017227040

Bateson, W. (1894). *Materials for the study of variation*. McMillan and Co.

Bautista-Perez, J. L., Bustamante-Rosario, M., & de Armas, T. (2021). Desarrollo de razonamiento algebraico elemental a través de patrones y secuencias numéricas y geométricas [Development of elementary algebraic reasoning through numerical and geometric patterns and sequences]. *Educación Matemática, 33*(1), 125–152. doi:10.24844/EM3301.05

Baxter, J. A., Ruzicka, A., Beghetto, R. A., & Livelybrooks, D. (2014). Professional development strategically connecting mathematics and science: The impact on teachers' confidence and practice. *School Science and Mathematics, 114*(3), 102–113. doi:10.1111sm.12060

Beach, P. T., Anderson, R. C., Jacovidis, J. N., & Chadwick, K. L. (2020). *Making the abstract explicit: The role of metacognition in teaching and learning*. Organisation du Baccalauréat International.

Beane, J. A. (2000). O que é um currículo coerente? [What is a coherent curriculum?]. In J. A. Pacheco (Org.), Políticas de integração curricular (pp. 39-58). Porto Editora.

Beane, J. (1992). Creating an integrative curriculum: Making the connections. *NASSP Bulletin, 76*(547), 46–54. doi:10.1177/019263659207654706

Beane, J. A. (1995). Curriculum integration and the disciplines of knowledge. *Phi Delta Kappan, 76*(8), 616–622.

Beane, J. A. (1997). *Curriculum Integration. Designing the Core of Democratic Education.* Teachers College Press.

Beauchamp, G., & Parkinson, J. (2008). Pupils' attitudes towards school science as they transfer from an ICT- rich primary school to a secondary school with fewer ICT resources: Does ICT matter? *Education and Information Technologies, 13*(2), 103–118. doi:10.100710639-007-9053-5

Becker, K., & Park, K. (2011). Effects of integrative approaches among science, technology, engineering, and mathematics (STEM) subjects on students' learning: A preliminary meta-analysis. *Journal of STEM Education: Innovations and Research*, *12*, 23–37.

Bejan, A. (2020). Human evolution is biological & technological evolution. *Bio Systems*, *195*, 104156. doi:10.1016/j.biosystems.2020.104156 PMID:32492467

Belbase, S., Mainali, B. R., Kasemsukpipat, W., Tairab, H., Gochoo, M., & Jarrah, A. (2022). At the dawn of science, technology, engineering, arts, and mathematics (STEAM) education: Prospects, priorities, processes, and problems. *International Journal of Mathematical Education in Science and Technology*, *53*(11), 2919–2955. doi:10.1080/0020739X.2021.1922943

Bell, C., & Denniston, K. J. (Eds.). (2002). *Journeys of transformation II: The impact of the Maryland Collaboration for Teacher Preparation on science and mathematics instruction. Successful strategies and results of a reform-based effort to improve science and mathematics instruction and teacher preparation.* Towson University, Center for Science and Mathematics Education. https://eric.ed.gov/?id=ED467606

Bencze, J. L. (2010). Promoting student-led science and technology projects in elementary teacher education: Entry into core pedagogical practices through technological design. *International Journal of Technology and Design Education*, *20*(1), 43–62. doi:10.100710798-008-9063-7

Beneroso, D., & Robinson, J. (2022). Online Project-Based Learning in Engineering Design: Supporting the Acquisition of Design Skills. *Education for Chemical Engineers*, *38*, 38–47. . doi:10.1016/j.ece.2021.09.002

Beniermann, A. (2019). *Evolution-Von Akzeptanz und Zweifeln, Empirische Studien über Einstellungen zu Evolution und Bewusstsein* [Evolution – of acceptance and doubts, Empirical studies on attitudes towards evolution and the evolution of the human mind]. Springer Fachmedien.

Bequette, J., & Bequette, M. (2012). A Place for Art and Design Education in the STEM Conversation. *Art Education*, *65*(2), 40–47. doi:10.1080/00043125.2012.11519167

Berkman, M. B., Pacheco, J. S., & Plutzer, E. (2008). Evolution and Creationism in America's Classrooms: A National Portrait. *PLoS Biology*, *6*(5), e124. doi:10.1371/journal.pbio.0060124 PMID:18494560

Berlin, D. F., & White, A. L. (1994). The Berlin-White integration science and mathematics model. *School Science and Mathematics*, *94*(1), 12–14. doi:10.1111/j.1949-8594.1994.tb12280.x

Berlin, D. F., & White, A. L. (1995). Connecting school science and mathematics. In P. A. House & A. F. Coxford (Eds.), *Connecting mathematics across the curriculum. 1995 National Council of Teachers of Mathematics Yearbook* (pp. 22–33). National Council of Teachers of Mathematics.

Berlin, D. F., & White, A. L. (2010). Preservice mathematics and science teachers in an integrated teacher preparation program for grades 7–12: A 3-year study of attitudes and perceptions related to integration. *International Journal of Science and Mathematics Education*, *8*(1), 97–115. doi:10.100710763-009-9164-0

Berlin, D., & Lee, H. (2005). Integrating Science and Mathematics Education: Historical Analysis. *School Science and Mathematics*, *105*(1), 15–24. doi:10.1111/j.1949-8594.2005.tb18032.x

Berlin, D., & White, A. L. (2012). A longitudinal look at attitudes and perceptions related to the integration of mathematics, science and technology education. *School Science and Mathematics*, *112*(1), 20–30. doi:10.1111/j.1949-8594.2011.00111.x

Bernard-Cobo, J. (2019). *Errores y Malinterpretaciones De La Teoría De La Evolución* [tesis de maestría, Universidad de Jaén]. Repositorio institucional: https://hdl.handle.net/10953.1/11448

Bernstein, B. (2000). *Pedagogy, symbolic control and identity* (2nd ed.). Rowman & Littlefield.

Besemer, S. P. (1998). Creative product analysis matrix: Testing the model structure and a comparison among products – three novel chairs. *Creativity Research Journal, 11*(4), 333–346. doi:10.120715326934crj1104_7

Bestiantono, Agustina, & Cheng. (2020). Online Learning amid the COVID-19 Pandemic: Students' Perspectives Muhammad. *Journal of Pedagogical Sociology and Psychology, 2*(1), 45–51. doi:10.46627ilet.v1i3.46

Betti, L., Shaw, P., & Behrends, V. (2020). Acceptance of biological evolution by first-year life sciences University Students. *Science & Education, 29*(2), 395–409. doi:10.100711191-020-00110-0

Bezen, S., Başal, C., Aykutlu, I., Seçken, N., & Bayrak, C. (2018). An ınterdisciplinary look at the comparative examination of physics and chemistry coursebooks. *Journal of Qualitative Research in Education, 6*(3), 267–283. doi:10.14689/issn.2148-2624.1.6c3s13m

Bianchi, G., Pisiotis, U., & Cabrera Giraldez, M. (2022). GreenComp – The European sustainability competence framework. In EUR 30955 EN. Publications Office of the European Union.

Bijsmans, P., & Schakel, A. H. (2018). The impact of attendance on first-year study success in problem-based learning. *Higher Education, 76*(5), 865–881. doi:10.100710734-018-0243-4

Bingham, C. B., & Davis, J. P. (2012). Learning Sequences: Their Existence, Effect, and Evolution. *Academy of Management Journal, 55*(3), 611–641. doi:10.5465/amj.2009.0331

Biseth, H., Svenkerud, S. W., Magerøy, S. M., & Rubilar, K. H. (2022). Relevant transformative teacher education for future generations. *Frontiers in Education, 7*(806495). Advance online publication. doi:10.3389/feduc.2022.806495

Bixler, A. (2007). Teaching Evolution with the Aid of Science Fiction. *The American Biology Teacher, 69*(6), 337–340. doi:10.1662/0002-7685(2007)69[337:TEWTAO]2.0.CO;2

Blackstone, N. W. (2016). An Evolutionary Framework for Understanding the Origin of Eukaryotes. *Biology (Basel), 5*(2), 18. doi:10.3390/biology5020018 PMID:27128953

Blanco, J. M. R., & Dávila, M.A. (2017). Las emociones según el género, en el aprendizaje de la Tecnología del alumnado de primer curso de Educación Secundaria Obligatoria. *Ápice. Revista de Educación Científica, 1*(2), 18–37.

Blythe, A. J. C., & Thompson, T. (2022). Virtual Teaching in the COVID Era: Providing Surgical Core Trainee Teaching via Online Webinars and Videoconferencing. *The Surgeon, 20*(6), e405–e409. doi:10.1016/j.surge.2022.03.007 PMID:35431109

Bobonová, I., Ceretková, S., Tirpáková, A., & Markechová, D. (2019). Inclusion of interdisciplinary approach in the mathematics education of biology trainee teachers in Slovakia. In B. Doig, J. Williams, D. Swanson, R. B. Ferri, & P. Drake (Eds.), *Interdisciplinary Mathematics Education. The state of the art and beyond* (pp. 263–279). Springer Open. doi:10.1007/978-3-030-11066-6_17

BOE. (2014). Real Decreto 1105/2014, de 26 de diciembre, por el que se establece el currículo básico de la Educación Secundaria Obligatoria y del Bachillerato. Boletín Oficial del Estado, 3 de enero de 2015, 3, 169-546.

BOE. (2020). Real Decreto 217/2022, de 29 de marzo, por el que se establece la ordenación y las enseñanzas mínimas de la Educación Secundaria Obligatoria. Boletín Oficial del Estado, 30 de marzo de 2022, 76, 1-198.

Borrachero, A. B., Dávila-Acedo, M. A., Brígido, M., Gómez, R., & Mellado, L. (2014). Las emociones que experimentaban los futuros profesores de secundaria en el aprendizaje de las ciencias [The emotions that future secondary school teachers experienced in learning science]. *Revista INFAD de Psicología. International Journal of Developmental and Educational Psychology*, *4*(1), 271–280. doi:10.17060/ijodaep.2014.n1.v4.612

Borrego, M., & Henderson, C. (2014). Increasing the use of evidence-based teaching in STEM higher education: A comparison of eight change strategies. *Journal of Engineering Education*, *103*(2), 220–252. doi:10.1002/jee.20040

Bossé, M., Lee, T., Swinson, M., & Faulconer, J. (2010). The NCTM Process Standards and the Five Es of Science: Connecting Math with Science. *School Science and Mathematics*, *110*(5), 262–276. doi:10.1111/j.1949-8594.2010.00033.x

Brakefield, P. M. (2011). Evo-devo and accounting for Darwin's endless forms. *Philosophical Transactions of the Royal Society of London. Series B, Biological Sciences*, *366*(1574), 2069–2075. doi:10.1098/rstb.2011.0007 PMID:21690125

Brandt, M., Groom, Q., Magro, A., Misevic, D., Narraway, C. L., Bruckermann, T., . . . Jenkins, T. (2022). Promoting scientific literacy in evolution through citizen science. *Proceedings of the Royal Society B*, *289*(1980), 20221077. 10.1098/rspb.2022.1077

Bravo, E., Brígido, M., Hernández-Barco, M. A., & Mellado, V (2022). Las emociones en ciencias en la formación inicial del profesorado de infantil y primaria. *Revista Interuniversitaria de Formación del Profesorado. Continuación de la antigua Revista de Escuelas Normales, 97*, 57-74.

Breiner, J. M., Harkness, S. S., Johnson, C. C., & Koehler, C. M. (2012). What is STEM? A discussion about conceptions of STEM in education and partnerships. *School Science and Mathematics*, *112*(1), 3–11. doi:10.1111/j.1949-8594.2011.00109.x

Brent, R., & Felder, R.M. (2016). *Teaching and Learning in STEM: A Practical Guide*. Academic Press.

Brewer, C., & Smith, D. (2011). Vision and change in undergraduate biology education: a call to action. American Association for the Advancement of Science.

Brewer, C., & Smith, D. (2011). *Vision and change in undergraduate biology education: a call to action*. American Association for the Advancement of Science.

Brewer, M. S., & Gardner, G. E. (2013). Teaching Evolution through the Hardy-Weinberg Principle: A Real-Time, Active-Learning Exercise Using Classroom Response Devices. *The American Biology Teacher*, *75*(7), 476–479. doi:10.1525/abt.2013.75.7.6

Brígido, M., Bermejo, M. L., Conde, M. C., & Mellado, V. (2010). The Emotions in Teaching and Learning Nature Sciences and Physics/Chemistry in Pre-Service Primary Teachers. *US-China Education Review*, *7*(12), 25–32.

Brígido, M., Borrachero, A. B., Bermejo, M. L., & Mellado, V. (2013). Prospective primary teachers' self-efficacy and emotions in science teaching. *European Journal of Teacher Education*, *36*(2), 200–217. doi:10.1080/02619768.2012.686993

Broggy, J., O'Reilly, J., & Erduran, S. (2017). Interdisciplinarity and science education. In *Science Education* (pp. 81–90). Brill Sense. doi:10.1007/978-94-6300-749-8_6

Brookfield, J. F. Y. (2000). Genomic sequencing: The complexity conundrum. *Current Biology*, *10*(14), R514–R515. doi:10.1016/S0960-9822(00)00581-9 PMID:10898997

Brown, D. F. (2016). Curriculum integration. In S. B. Mertens, M. Caskey, & N. Flowers (Eds.), *The Encyclopedia of Middle Grades Education* (2nd ed., pp. 122–125). Information Age Publishing.

Brown, D., & Knowles, T. (2014). *What every middle school teacher should know* (3rd ed.). Heinemann.

Brown, R. E., & Bogiages, C. A. (2019). Professional development through STEM integration: How early career math and science teachers respond to experiencing integrated STEM tasks. *International Journal of Science and Mathematics Education, 17*(1), 111–128. doi:10.100710763-017-9863-x

Brown, S. A., Ronfard, S., & Kelemen, D. (2020). Teaching natural selection in early elementary classrooms: Can a storybook intervention reduce teleological misunderstandings? *Evolution (New York), 13*(1), 12. doi:10.118612052-020-00127-7

Bruckermann, T., Fiedler, D., & Harms, U. (2021). Identifying precursory concepts in evolution during early childhood – a systematic literature review. *Studies in Science Education, 57*(1), 85–127. doi:10.1080/03057267.2020.1792678

Bruff, D. (2009). *Teaching with Classroom Response Systems: Creating Active Learning Environments* (Vol. 33). Jossey-Bass.

Bryan, L. A., Moore, T. J., Johnson, C. C., & Roehrig, G. H. (2015). Integrated STEM education. In C. C. Johnson, E. E. Peters-Burton, & T. J. Moore (Eds.), *STEM road map: A framework for integrated STEM education* (pp. 23–37). Routledge. doi:10.4324/9781315753157-3

Buchan, L. (2019). *Development and evaluation of resources for teaching evolution in primary schools* [tesis de doctorado, University of Batch]. Repositorio institucional https://purehost.bath.ac.uk/ws/portalfiles/portal/194268489/Loredana_Buchan_PhD_thesis_PURE_pdf_Redacted.pdf

Bugallo-Rodríguez, A., & Vázquez-Ben, L. (2020). Estado actual de la teoría de la evolución en la normativa educativa curricular española. *eVOLUCIÓN. Revista de la Sociedad Española de Biología Evolutiva, 14*(11), 47–59.

Buick, J. M. (2007). Investigating the correlation between mathematical pre-knowledge and learning gains in service physics. *European Journal of Physics, 28*(6), 1073–1080. doi:10.1088/0143-0807/28/6/004

Bulger, S. M., Mohr, D. J., & Walls, R. T. (2002). Stack the deck in favor of your students by using the four aces of effective teaching. *The Journal of Effective Teaching, 5*(2).

Burke da Silva, K. (2012). Evolution-centered teaching of biology. *Annual Review of Genomics and Human Genetics, 13*(1), 363–380. doi:10.1146/annurev-genom-090711-163749 PMID:22559325

Burks, R. L., & Boles, L. C. (2007). Evolution of the Chocolate Bar: A Creative Approach to Teaching Phylogenetic Relationships Within Evolutionary Biology. *The American Biology Teacher, 69*(4), 229–237. doi:10.1662/0002-7685(2007)69[229:EOTCBA]2.0.CO;2

Burmeister, A. R., & Smith, J. J. (2016). Evolution across the Curriculum: Microbiology. *Journal of Microbiology & Biology Education, 17*(2), 252–260. doi:10.1128/jmbe.v17i2.988 PMID:27158306

Bursal, M., & Paznokas, L. (2006). Mathematics Anxiety and Pre-service Elementary Teachers' Confidence to Teach Mathematics and Science. *School Science and Mathematics, 106*(4), 173–180. doi:10.1111/j.1949-8594.2006.tb18073.x

Busquets, T., Silva, M. y Larrosa, P. (2016). Reflexiones sobre el aprendizaje de las ciencias naturales: Nuevas aproximaciones y desafíos [Reflections on learning from the natural sciences: New approaches and challenges]. *Estudios pedagógicos, 42*, 117-135. doi:10.4067/S0718-07052016000300010

Buss, R. R. (2010). Efficacy for Teaching Elementary Science and Mathematics Compared to Other Content. *School Science and Mathematics, 110*(6), 290–297. doi:10.1111/j.1949-8594.2010.00037.x

Butcher, F. (2005). Two personal reviews of the European Physics Education Conference in Bonn. *Physics Education, 40*(5), 397–402.

Bybee, R. (2006). Scientific Inquiry and Science Teaching. In L. B. Flick & N. G. Lederman (Eds.), *Scientific inquiry and nature of science: Implications for teaching, learning, and teacher education*. Springer. doi:10.1007/978-1-4020-5814-1_1

Bybee, R. (2013). *The case of STEM education: Challenges and opportunities*. NSTA Press.

Bybee, R. W. (2010). What is STEM education? *Science, 329*(5995), 996–996. doi:10.1126cience.1194998 PMID:20798284

Bybee, R. W. (2013). *The Case for Education Challenges and Opportunities*. National Science Teachers Association.

Bybee, R., Taylor, J., Gardner, A., Van Scotter, P., Powell, J., Westbrook, A., & Landes, N. (2006). *The BSCS 5E Instructional Model: Origins, Effectiveness and Applications*. BSCS.

Cabello-Garrido, A., España-Ramos, E., & Blanco-López, Á. (2018). Developing a Human Nutrition Learning Progression. *International Journal of Science and Mathematics Education, 16*(7), 1269–1289. doi:10.100710763-017-9838-y

Cady, J. A., & Rearden, K. (2007). Pre-service teachers' beliefs about knowledge, mathematics, and science. *School Science and Mathematics, 107*(6), 237–245. doi:10.1111/j.1949-8594.2007.tb18285.x

Caetano-Anollés, G., & Janko, R. (2022). The rise of hierarchy and modularity in biological networks explained by empedocles' double tale ~2,400 years before darwin and systems biology. *Frontiers in Genetics, 13*, 973233. doi:10.3389/fgene.2022.973233 PMID:36061206

Caine, R. N., & Caine, G. (1991). *Making connections: teaching and the human brain*. Association for Supervision and Curriculum Development.

Cajas, F. (2001). Alfabetización científica y tecnológica: la transposición didáctica del conocimiento tecnológico [Scientific and technological literacy: the didactic transposition of technological knowledge]. *Enseñanza de las ciencias: Revista de investigación y experiencias didácticas, 19*(2), 243-254.

Camacho, G., Martins, M. C., & Correia, M. (2022). *Os meus itinerários* [My itineraries]. Casa das Ciências. https://www.casadasciencias.org/recurso/8998

Campos, R., & Sá-Pinto, X. (2013). Early evolution of evolutionary thinking: Teaching biological evolution in elementary schools. *Evolution (New York), 6*(25), 1–13. doi:10.1186/1936-6434-6-25

Cañal, P. (2009). Acerca de la enseñanza sobre la evolución biológica en la escuela infantil y primaria. *Alambique, 62*, 75–91.

Cano-Espinosa, D. (2015). La Evolución biológica en la Biofilosofía de F. J. Ayala. *Pensamiento. Revista De Investigación E Información Filosófica, 71*(269), 1015-1054. . doi:10.14422/pen.v71.i269.y2015.001

Capone, R. (2022). Blended Learning and Student-Centered Active Learning Environment: A Case Study with STEM Undergraduate Students. *Canadian Journal of Science, Mathematics and Technology Education, 22*(1), 210–236. doi:10.100742330-022-00195-5

Caprara, G. V., & Cervone, D. (2000). *Personality: determinants, dynamics, and potentials*. Cambridge University Press. doi:10.1017/CBO9780511812767

Carchon, R., & Segers, D. (2021). The Use of the Museum Collection for Educational Purposes. In *Historical Scientific Instruments in Contemporary Education* (pp. 139–157). Brill.

Carr, R. L., & Strobel, J. (2011). Integrating Engineering Design Challenges into Secondary STEM Education. National Center for Engineering and Technology Education.

Carroll, S. P., Jørgensen, P. S., Kinnison, M. T., Bergstrom, C. T., Denison, R. F., Gluckman, P., Smith, T. B., Strauss, S. Y., & Tabashnik, B. E. (2014). Applying evolutionary biology to address global challenges. *Science*, *346*(6207), 1245993. Advance online publication. doi:10.1126cience.1245993 PMID:25213376

Carvalho, V., Rodrigues, C., Minas, G., Lima, R., Teixeira, J. C., & Teixeira, S. (2021). I9Masks: From a Multidisciplinary Summer Project to a Non-Accredited Short Course. In *Engineering Education* (Vol. 9). American Society of Mechanical Engineers.

Castro, L. (2007). Docencia y evolución: La biología evolutiva de la enseñanza secundaria. *eVOLUCIÓN. Revista de la Sociedad Española de Biología Evolutiva*, *2*(1), 63–66.

Castro-Moreno, J.A. (2012). La Biología como ciencia histórica: el caso de la evolución biológica. *Bio-grafía: Escritos sobre la Biología y su Enseñanza*, *5*(9), 19-37.

Catley, K. M., & Novick, L. R. (2008). Seeing the Wood for the Trees: An Analysis of Evolutionary Diagrams in Biology Textbooks. *Bioscience*, *58*(10), 976–987. doi:10.1641/B581011

Catley, K. M., Novick, L. R., & Shade, C. K. (2010). Interpreting evolutionary diagrams: When topology and process conflict. *Journal of Research in Science Teaching*, *47*(7), 861–882. doi:10.1002/tea.20384

Cavadas, B., & Branco, N. (2020). *Abelhas STEM (STEM Bees)*. GO-LAB. https://www.golabz.eu/ils/abelhas-stem-rp

Cavadas, B., Rezio, S., Nogueira, J. R., & Branco, N. (2022). A framework and a research design proposal to identify preservice teachers' integration performance of science and mathematics. *Canadian Journal of Science. Mathematics and Technology Education*, *22*(1), 101–129. doi:10.100742330-022-00198-2

Ceballos, M., Vílchez, J. E., & Escobar, T. (2017). La enseñanza de la Evolución en Primaria. Opinión del profesorado y exploración de ideas inadecuadas en los niños. *Boletín de la Real Sociedad Española de Historia Natural Sección Aula. Museos y Colecciones*, *4*, 55–68.

Ceuppens, S. (2019). *Learning of physics and mathematics concepts in an integrated STEM curriculum*. KU Leuven Faculty of Science.

Chabay, R., & Sherwood, B. (2008). Computational physics in the introductory calculus-based course. *American Journal of Physics*, *76*(4), 307–313. doi:10.1119/1.2835054

Chaovanapricha, K., & Chaturongakul, P. (2020). Interdisciplinary teacher collaboration in English for specific purposes subjects in a Thai University. *English Language Teaching*, *13*(5), 139–148. doi:10.5539/elt.v13n5p139

Chaves-Mejía, G. A. (2017). La enseñanza de la evolución biológica desde la historia y la filosofía de la biología: aportes a la formación continua del profesorado. *Tecné, Episteme y Didaxis: TED. Número Extraordinario*. Memorias, Séptimo Congreso Internacional sobre Formación de Profesores de Ciencias, 12 al 14 de octubre de 2016, Bogotá.

Chaves-Mejía, G. A. (2010). ¿Enseñamos Evo-Devo? Implicaciones de las nuevas hipótesis sobre evolución biológica y su incidencia en la escuela. *Bio-grafía*, *3*(4), 155–166. doi:10.17227/20271034.vol.3num.4bio-grafia155.166

Chaves-Mejía, G. A. (2016, July-December). La enseñanza y el aprendizaje de la evolución biológica (eb) con la perspectiva teórica del perfil conceptual: Implicaciones en la formación continua del profesorado. *Bio-grafía*, *9*(17). Advance online publication. doi:10.17227/20271034.vol.9num.17bio-grafia109.117

Chavez-Mejía, G. A. (2012). ¿Es la endosimbiosis seriada una alternativa a la teoría sintética de la evolución? Implicaciones epistemológicas y didácticas. *Bio-grafía*, *5*(9), 67–79. doi:10.17227/20271034.9biografia67.79

Chawla, L. (2020). Childhood nature connection and constructive hope: A review of research on connecting with nature and coping with environmental loss. *People and Nature*, *2*(3), 619–642. doi:10.1002/pan3.10128

Cheung, G. W., & Rensvold, R. B. (2002). Evaluating goodness-of-fit indexes for testing measurement invariance. *Structural Equation Modeling*, *9*(2), 233–255. doi:10.1207/S15328007SEM0902_5

Chi, M. (2005). Commonsense conceptions of emergent processes: Why some misconceptions are robust. *Journal of the Learning Sciences*, *14*(2), 161–199. doi:10.120715327809jls1402_1

Chiu, M., & Duit, R. (2011). Globalization: Science education from an international perspective. *Journal of Research in Science Teaching*, *48*(6), 553–566. doi:10.1002/tea.20427

Chowdhary, B., Liu, X., Yerrick, R., Smith, E., & Grant, B. (2014). Examining science teachers' development of interdisciplinary science inquiry pedagogical knowledge and practices. *Journal of Science Teacher Education*, *25*(8), 865–884. doi:10.100710972-014-9405-0

Chuang-Tzu. (2005). *Obra completa*. Traducción de Cristóbal Serra Simó. Edicions Cort.

Cleaves, A. (2005). The formation of science choices in secondary school. *International Journal of Science Education*, *27*(4), 471–486. doi:10.1080/0950069042000323746

Cofré, H. L., Santibáñez, D. P., Jiménez, J. P., Spotorno, A., Carmona, F., Navarrete, K., & Vergara, C. A. (2018). The effect of teaching the nature of science on students' acceptance and understanding of evolution: Myth or reality? *Journal of Biological Education*, *52*(3), 248–261. doi:10.1080/00219266.2017.1326968

Cohen, L., Manion, L., & Morrison, K. (2017). *Research methods in education* (6th ed.). Routledge.

Colucci-Gray, L., Burnard, P., Cooke, C., Davies, R., Gray, D., & Trowsdale, J. (2017). *BERA Research Commission: Reviewing the potential and challenges of developing STEAM education through creative pedagogies for 21st learning: how can school curricula be broadened towards a more responsive, dynamic, and inclusive form of education?* BERA.

Cook, K. (2009). A Suggested Project-Based Evolution Unit for High School: Teaching Content through Application. *The American Biology Teacher*, *71*(2), 95–98. doi:10.1662/005.071.0209

Cooper, V. S., Warren, T. M., Matela, A. M., Handwork, M., & Scarponi, S. (2019). EvolvingSTEM: A microbial evolution-in-action curriculum that enhances learning of evolutionary biology and biotechnology. *Evolution (New York)*, *12*(1), 12. doi:10.118612052-019-0103-4 PMID:32647555

Corona-Cruz, A. (2008). ¿Qué hace al buen maestro?: La visión del estudiante de ciencias físico-matemáticas [What makes the good teacher?: The vision of the physical-mathematical science student]. *Latin-American Journal of Physics Education*, *2*(2), 148–151.

Correia, M., & Martins, M. C. (2020). *CreativeLab_SciMath A Medida do Sistema Solar* [CreativeLab_SciMath The Measurement of the Solar System]. Casa das Ciências. https://www.casadasciencias.org/recurso/8867

Correia, M., & Baptista, M. (2022). Supporting the development of pre-service primary teachers PCK and CK through a STEM Program. *Education Sciences*, *12*(4), 258. doi:10.3390/educsci12040258

Couso, D. (2017). Per a què estem a STEM? Un intent de definir l'alfabetització STEM per a tothom i amb valors [What are we for in STEM? An attempt to define STEM literacy for all and with values]. *Ciències*, *34*(34), 22–30. doi:10.5565/rev/ciencies.403

Coutinho, C. P. (2019). Metodologia de investigação em ciências sociais e humanas: Teoria e prática [Research methodology in social and human sciences: Theory and practice] (2nd ed.). Almedina.

Creswell, J. W. (2008). *Educational research: Planning, conducting, and evaluating quantitative and qualitative research* (3rd ed.). Merrill Prentice Hall.

Creswell, J. W. (2014). *Research design: Qualitative, quantitative and mixed methods approaches* (4th ed.). Sage.

Crombie, A. C. (1994). *Styles of scientific thinking in the European tradition*. Duckworth.

Cronbach, L. J. (1970). *Essentials of psychological testing* (3rd ed.). Harper and Row.

Cunningham, D. L., & Wescott, D. J. (2009). Still More "Fancy" and "Myth" than "Fact" in Students' Conceptions of Evolution. *Evolution (New York)*, *2*(3), 505–517. doi:10.100712052-009-0123-6

Cura, G., & Ercan Yalman, F. (2019). Investigating science teacher candidates' use of interdisciplinary instructional approach. *Online Science Education Journal*, *4*(2), 131–153.

Czerniak, C. M., & Johnson, C. C. (2014). Interdisciplinary science teaching. In S. K. Abell & N. G. Lederman (Eds.), *Handbook of Research on Science Education* (2nd ed., pp. 395–411). Routledge.

Czerniak, C., & Johnson, C. (2014). Interdisciplinary science teaching. In N. Lederman & S. Abell (Eds.), *Handbook of research on science education* (Vol. 2, pp. 395–411). Routledge.

Dare, E. A., Ellis, J. A., & Roehrig, G. H. (2014). Driven by beliefs: Understanding challenges physical science teachers face when integrating engineering and physics. *Journal of Pre-College Engineering Education Research*, *4*(2), 47–61. doi:10.7771/2157-9288.1098

Darwin, C. (1859). *On the Origin of Species by Means of Natural Selection, or the Preservation of Favoured Races in the Struggle for Life*. John Murray. doi:10.5962/bhl.title.82303

Darwin, C., & Wallace, A. R. (1858). On the tendency of species to form varieties; and on the perpetuation of varieties and species by natural means of selection. *Journal of the Proceedings of the Linnean Society Zoology*, *20*, 45–62.

Daugherty, M., Carter, V., & Swagerty, L. (2014). Elementary STEM education: The future for technology and engineering education? *The Journal of STEM Teacher Education*, *49*(1), 45–56. doi:10.30707/JSTE49.1Daugherty

Davidson, M., Evens, H., & McCormick, R. (1998). Bridging the gap: the use of concepts from science and mathematics in design and technology at KS 3. *International Design and Technology Education Research and Curriculum Development (IDATER), Loughborough University of Technology*, 48-53. Retrieved July 1, 2022, from https://dspace.lboro.ac.uk/dspace-jspui/handle/2134/1419

Dávila Acedo, M. A. (2017). Las emociones y sus causas en el aprendizaje de Física y Química, en el alumnado de Educación Secundaria [Emotions and their causes in the learning of Physics and Chemistry, in the students of Secondary Education]. *Revista Eureka sobre Enseñanza y Divulgación de las Ciencias*, *14*(3), 570–586.

Dávila, M. A., Cañada, F., Sánchez-Martín, J., & Mellado, V. (2016). Las emociones en el aprendizaje de física y química en educación secundaria. Causas relacionadas con el estudiante [Emotions in the learning of physics and chemistry in secondary education. Student-related causes]. *Educación en la Química*, *27*(3), 217–225. doi:10.1016/j.eq.2016.04.001

Davis, K. S. (2003). "Change is hard": What science teachers are telling us about reform and teacher learning of innovative practices. *Science Education*, *87*(1), 3–30. doi:10.1002ce.10037

Davison, D. M., Miller, K. W., & Metheny, D. L. (1995). What does integration of science and mathematics really mean? *School Science and Mathematics*, *95*(5), 226–230. doi:10.1111/j.1949-8594.1995.tb15771.x

De Ataíde, A. R. P., & Greca, I. M. (2013). Epistemic views of the relationship between physics and mathematics: Its influence on the approach of undergraduate students to problem solving. *Science & Education*, *22*(6), 1405–1421. doi:10.100711191-012-9492-2

De Fraga, F. (2018). Towards an Evolutionary Perspective in Teaching and Popularizing Microbiology. *Journal of Microbiology & Biology Education, 19*(1), 19.1.45. . doi:10.1128/jmbe.v19i1.1531

de Greef, L., Post, G., Vink, C., & Wenting, L. (2017). *Design Interdisciplinary Education. A practical handbook for university teachers.* Amsterdam University Press. doi:10.5117/9789462984769

de Jong, T. (2006). Technological advances in inquiry learning. *Science*, *312*(5773), 532–533. doi:10.1126cience.1127750 PMID:16645080

De Souza, R. F., De Carvalho, M., Matsuo, T., & Zaia, D. A. M. (2010). Study on the opinion of university students about the themes of the origin of Universe and evolution of life. *International Journal of Astrobiology*, *9*(2), 109–117. doi:10.1017/S1473550410000054

Debortoli, N., Li, X., Eyres, I., Fontaneto, D., Hespeels, B., Tang, C. Q., Flot, J. F., & Van Doninck, K. (2016). Genetic Exchange among Bdelloid Rotifers Is More Likely Due to Horizontal Gene Transfer Than to Meiotic Sex. *Current Biology*, *26*(6), 723–732. doi:10.1016/j.cub.2016.01.031 PMID:26948882

Decker, T., Summers, G., & Barrow, L. (2007). The treatment of geologic time & the history of life on Earth in high school biology textbooks. *The American Biology Teacher*, *69*(7), 401–405. doi:10.1662/0002-7685(2007)69[401:TTOGTT]2.0.CO;2

Dehipawala, S., Shekoyan, V., & Yao, H. (2014). Using mathematics review to enhance problem solving skills in general physics classes. In *Proceedings of the 2014 Zone 1 Conference of the American Society for Engineering Education* (pp. 1- 4). IEEE. 10.1109/ASEEZone1.2014.6820631

Dejarnette, N. (2012). America's children: Providing early exposure to STEM (science, technology, engineering, and math) initiatives. *Education*, *133*(1), 77–84.

DeLuca, C., Ogden, H., & Pero, E. (2015). Reconceptualising Elementary Preservice Teacher Education: Examining an Integrated-Curriculum Approach. *New Educator*, *11*(3), 227–250. doi:10.1080/1547688X.2014.960986

Deniz, H., & Sahin, E. A. (2016). Exploring the factors related to acceptance of evolutionary theory among Turkish preservice biology teachers and the relationship between acceptance and teaching preference. *The Electronic Journal of Science Education*, *20*(4), 21–42.

Department for Education and Skills. (2006). *The STEM programme report*. HM Government.

Derman, İ. (2019). *The level of association of science course with daily life* [Unpublished doctoral dissertation]. Hacettepe University, Ankara, Turkey.

Desantis, L. R. G. (2009). Teaching Evolution through Inquiry-Based Lessons of Uncontroversial Science. *The American Biology Teacher*, *71*(2), 106–111. doi:10.2307/27669381

Diani, R., Yuberti, Y., Anggereni, S., Utami, G. N., Iqbal, A., & Kurniawati, I. (2020). ECIRR (Elicit, Confront, Identify, Resolve, Reinforce) learning model with the pictorial riddle method: Is it effective in reducing physics misconceptions? *Journal of Physics*, *1572*(1), 012020.

Dias, D., Correia, M., & Martins, C. (2017). *As ciências em tempo de Natal* [The sciences in Christmas time]. Uma proposta didática interdisciplinar para o 1.º Ciclo. Casa das Ciências. https://www.casadasciencias.org/recurso/8629

DiFrancesca, D., Lee, C., & McIntyre, E. (2014). Where is the 'E' in STEM for Young Children? Engineering Design Education in an Elementary Teacher Preparation Program. *Issues in Teacher Education*, *23*(1), 49–64.

Direção Geral de Educação (DGE). (2021). *Novas Aprendizagens Essenciais de Matemática - Ensino Básico* [New Essential Learning of Mathematics - Basic Education]. Ministério da Educação. https://www.dge.mec.pt/aprendizagens-essenciais-ensino-basico

Direção Geral de Educação. (2018). *Aprendizagens Essenciais - Ensino Básico* [Essential Learning - Basic Education]. Ministério da Educação. https://www.dge.mec.pt/aprendizagens-essenciais-ensino-basico

diSessa, A. A. (2018). Computational Literacy and "The Big Picture" Concerning Computers in Mathematics Education. *Mathematical Thinking and Learning*, *20*(1), 3–31. doi:10.1080/10986065.2018.1403544

Dissanayake, E. (2017). Ethology, Interpersonal Neurobiology, and Play: Insights into the Evolutionary Origin of the Arts. *American Journal of Play*, *9*(2), 143–168.

Doppelt, Y., Mehalik, M., Schunn, C. D., Silk, E., & Krysinski, D. (2008). Engagement and achievements: A case study of design-based learning in a science context. *Journal of Technology Education*, *19*(2), 22–39.

Downie, J. R. (2004). Evolution in Health and Disease: The Role of Evolutionary Biology in the Medical Curriculum. *Bioscience Education*, *4*(1), 1–18. doi:10.3108/beej.2004.04000004

Drake, S. M., & Reid, J. L. (2020). 21st Century competencies in light of the history of integrated curriculum. *Frontiers in Education*, *5*(122), 122. Advance online publication. doi:10.3389/feduc.2020.00122

Duarte, J., & Nogueira, J. R. (2019). From the Hidden Protest of Students towards Problem-Based Learning. *Education Sciences*, *9*(3), 215. doi:10.3390/educsci9030215

Duit, R., Schecker, H., Höttecke, D., & Niedderer, H. (2007). *Teaching physics handbook of research on science education*. Routledge.

Eccles, J. S., & Wigfield, A. (2020). From expectancy-value theory to situated expectancy-value theory: A developmental, social cognitive, and sociocultural perspective on motivation. *Contemporary Educational Psychology*, *61*, 101859. doi:10.1016/j.cedpsych.2020.101859

Eco-Sensors4Health Project. (2019). *Eco-Sensors4Health Toolkit: Guia Eco-sensores para a saúde* [Eco-Sensors4Health Toolkit: Eco-Sensors Guide to Health]. Instituto Politécnico de Lisboa, Instituto Politécnico de Viseu, Ciência Viva – ANCCT, Município de Viseu. https://ecosensors4health.files.wordpress.com/2019/03/toolkit-eco-sensors4health_cv_vf.pdf

Eder, E., Turic, K., Milasowszky, N., Van Adzin, K., & Hergovich, A. (2011). The relationships between paranormal belief, creationism, intelligent design and evolution at secondary schools in Vienna (Austria). *Science & Education*, *20*(5-6), 517–534. doi:10.100711191-010-9327-y

Ekmen, C., & Bakar, E. (2019). Digital competencies in the curriculum and texsbooks of primary education. *Milli Eğitim Dergisi*, *48*(221), 5–35.

Eldredge, N., & Gould, S. J. (1972). Punctuated equilibria: an alternative to phyletic gradualism. *Models in Paleobiology*, 82-115.

El-Soussi, A. (2022). The Shift from Face-to-Face to Online Teaching Due to COVID-19: Its Impact on Higher Education Faculty's Professional Identity. *International Journal of Educational Research Open, 3*(January), 100139. doi:10.1016/j.ijedro.2022.100139 PMID:35233543

Emmons, N., Lees, K., & Kelemen, D. (2017). Young children's near and far transfer of the basic theory of natural selection: An analogical storybook intervention. *Journal of Research in Science Teaching, 55*(3), 321–347. doi:10.1002/tea.21421

English, L. (2016). STEM education K-12: Perspectives on integration. *International Journal of STEM Education, 3*(3), 1–8. doi:10.118640594-016-0036-1

English, L. D. (2009). Promoting interdisciplinarity through mathematical modelling. *ZDM, 41*(1), 161–181. doi:10.100711858-008-0106-z

English, L. D., King, D., & Smeed, J. (2017). Advancing integrated STEM learning through engineering design: Sixth-grade students' design and construction of earthquake resistant buildings. *The Journal of Educational Research, 110*(3), 255–271. doi:10.1080/00220671.2016.1264053

Eriksen, D., Stasiuk, J., & Frank, M. (1995). Bringing Pythagoras to life. *Mathematics Teacher, 88*(9), 744–747. doi:10.5951/MT.88.9.0744

Ersoy, A. (2015). Investigation of Phd students' initial qualitative research experiences via their diaries. *Pegem Journal of Education and Instruction, 5*(5), 549–568.

Escobar-Pérez, J., & Cuervo-Martínez, A. (2008). Validez de contenido y juicio de expertos: Una aproximación a su utilización [Content validity and expert judgment: an approximation to its use]. *Avances en Medición, 6*, 27–36.

Escola Superior de Educação de Lisboa (ESELx). (2021). *Projeto Plant@Eselx* [Project Plant@Eselx]. https://www.eselx.ipl.pt/plantas-eselx

Estapa, A., & Tank, K. (2017). Supporting integrated STEM in the elementary classroom: A professional development approach centered on an engineering design challenge. *International Journal of STEM Education, 4*(6), 6. Advance online publication. doi:10.118640594-017-0058-3

Etkin, A., Büchel, C., & Gross, J. J. (2015). The neural bases of emotion regulation. In Nature Reviews Neuroscience (Vol. 16, Issue 11, pp. 693–700). doi:10.1038/nrn4044

European Comission. (2010). *Science and Technology. Special Eurobarometer340.* European Commission.

European Commission. (2015). *Shaping career-long perspectives on teaching. A guide on policies to improve initial teacher education.* Publications Office of the European Union.

Evans, M. (2010). Conceptual change and evolutionary biology: a developmental analysis. In S. Vosniadou (Ed.), *International handbook of research in conceptual change* (pp. 263–294). Routledge.

Fan, S. C., & Yu, K. C. (2019). Teaching engineering-focused STEM curriculum: PCK needed for teachers. In Y. S. Hsu & Y. F. Yeh (Eds.), *Asia-Pacific STEM Teaching Practices* (pp. 103–116). Springer. doi:10.1007/978-981-15-0768-7_7

Fensham, P. (2009). The link between policy and practice in science education: The role of research. *Science Education, 93*(6), 1076–1095. doi:10.1002ce.20349

Fernandes, H. V. J. (2021). From Student to Tutor: A Journey in Problem-Based Learning. *Currents in Pharmacy Teaching & Learning, 13*(12), 1706–1709. doi:10.1016/j.cptl.2021.09.037 PMID:34895682

Fernández, J. J., & Sanjosé, V. (2007). Permanencia de ideas alternativas sobre Evolución de las Especies en la población culta no especializada. *Didáctica de las Ciencias Experimentales y Sociales, 21*, 129–149.

Fernández, N. (2006). Estrategias de enseñanza para favorecer el aprendizaje significativo [Teaching strategies to promote meaningful learning]. *Revista Cognición, 1*(6), 12–28.

Ferrada, C., Díaz-Levicoy, D., & Salgado-Orellana, N. (2018). Análisis de actividades STEM en libros de texto chilenos y españoles de ciencias [Analysis of STEM activities in Chilean and Spanish science textbooks]. *Revista de Pedagogía, 39*(105), 111–130.

Ferri, R. B. (2019). Teacher education and teacher development. In B. Doig, J. Williams, D. Swanson, R. B. Ferri, & P. Drake (Eds.), *Interdisciplinary Mathematics Education. The state of the art and beyond* (pp. 259–261). Springer Open. doi:10.1007/978-3-030-11066-6_16

Finbråten, H. S., Grønlien, H. K., Pettersen, K. S., Foss, C., & Guttersrud, Ø. (2022). Nursing Students' Experiences with Concept Cartoons as an Active Learning Strategy for Developing Conceptual Understanding in Anatomy and Physiology: A Mixed-Method Study. *Nurse Education in Practice, 65*(November), 103493. doi:10.1016/j.nepr.2022.103493 PMID:36401995

Finch, W. H. (2020). Using fit statistic differences to determine the optimal number of factors to retain in an exploratory factor analysis. *Educational and Psychological Measurement, 80*(2), 217–241. doi:10.1177/0013164419865769 PMID:32158020

Fink, D.L. (2003). *Creating Significant Learning Experiences: An Integrated Approach to Designing College Courses.* Jossey Bass.

Fllis, A., & Fouts, J. (2001). Interdisciplinary curriculum: The research base: The decision to approach music curriculum from an interdisciplinary perspective should include a consideration of all the possible benefits and drawbacks. *Music Educators Journal, 87*(22), 22–26, 68. doi:10.2307/3399704

Fogarty, R. (1991). Ten ways to integrate curriculum. *Educational Leadership, 49*(2), 61–65.

Folguera, G., & Gonzalez-Galli L. (2012). La extensión de la síntesis evolutiva y los alcances sobre la enseñanza de la teoría de la evolución. *Biografia: Escritos sobre biología y su enseñanza, 5*(9), 4-18.

Forsblom, L., Pekrun, R., Loderer, K., & Peixoto, F. (2022). Cognitive appraisals, achievement emotions, and students' math achievement: A longitudinal analysis. *Journal of Educational Psychology, 114*(2), 346–367. doi:10.1037/edu0000671

Fortus, D., Krajcikb, J., Dershimerb, R. C., Marxc, R. W., & Mamlok-Naamand, R. (2004). Design-based science and student learning. *Journal of Research in Science Teaching, 41*(10), 1081–1110. doi:10.1002/tea.20040

Fortus, D., Krajcikb, J., Dershimerb, R. C., Marxc, R. W., & Mamlok-Naamand, R. (2005). Design-based science and real-world problem-solving. *International Journal of Science Education, 27*(7), 855–879. doi:10.1080/09500690500038165

Foundation for Environmental Education (FEE). (2022). *About Eco-Schools.* https://www.ecoschools.global/how-does-it-work

Frade, C., Winbourne, P., & Braga, S. M. (2009). A mathematics-science community of practice: Reconceptualising transfer in terms of crossing boundaries. *For the Learning of Mathematics, 29*, 14–22.

Franco-Goyena, J. (2019). *La evolución biológica en secundaria: contribución a la incorporación de nuevos contenidos y metodologías* [tesis de máster, Universidad Pública de Navarra]. Recuperado de: https://hdl.handle.net/2454/33655

Fredericks, J., Blumenfeld, P., & Paris, A. (2004). School Engagement: Potential of the Concept, State of the Evidence. *Review of Educational Research, 74*(1), 59–109. doi:10.3102/00346543074001059

Freeman, S., Eddy, S. L., McDonough, M., Smith, M. K., Okoroafor, N., Jordt, H., & Wenderoth, M. P. (2014). Active learning increases student performance in science, engineering, and mathematics. *Proceedings of the National Academy of Sciences of the United States of America*, *111*(23), 8410–8415. doi:10.1073/pnas.1319030111 PMID:24821756

Friedman, T. L. (2005). *The world is flat. A brief history of the twenty-first century*. Farrar, Straus and Giroux.

Friedrichsen, P. J., Linke, N., & Barnett, E. (2016). Biology teachers' professional development needs for teaching evolution. *Science Educator*, *25*(1), 51–61.

Friesen, S., & Scott, D. (2013). *Inquiry-based learning: A review of the research literature*. Paper prepared for the Alberta Ministry of Education.

Frykholm, J., & Glassom, G. (2005). Connecting Science and Mathematics Instruction: Pedagogical context knowledge for teachers. *School Science and Mathematics*, *105*(3), 127–141. doi:10.1111/j.1949-8594.2005.tb18047.x

Furner, J. M., & Kumar, D. D. (2007). The mathematics and science integration argument: A stand for teacher education. *Eurasia Journal of Mathematics, Science and Technology Education*, *3*(3), 185–189. doi:10.12973/ejmste/75397

Furrow, R. E., Kim, H. G., Abdelrazek, S. M. R., Dahlhausen, K., Yao, A. I., Eisen, J. A., Goldman, M. S., Albeck, J. G., & Facciotti, M. T. (2020). Combining Microbial Culturing With Mathematical Modeling in an Introductory Course-Based Undergraduate Research Experience. *Frontiers in Microbiology*, *11*, 581903. doi:10.3389/fmicb.2020.581903 PMID:33250873

Futuyma, D. J. (1998). *Biología Evolutiva* (3rd ed.). Sunderland Associates.

Gallego, A., & Muñoz, A. (2015). Análisis de las hipótesis evolutivas en alumnos de Educación Secundaria y Bachillerato. *Revista Electrónica de Enseñanza de las Ciencias*, *14*(1), 35–54.

Gao, X., Li, P., Shen, J., & Sun, H. (2020). Reviewing assessment of student learning in interdisciplinary STEM education. *International Journal of STEM Education*, *7*(4), 24. Advance online publication. doi:10.118640594-020-00225-4

García de la Torre, E., Sequeiros, L., & Pedrinaci, E. (1995). Tectónica de placas y evolución biológica: Construcción de un paradigma e implicaciones didácticas. *Enseñanza de las Ciencias de la Tierra*, *3*(1), 14–22.

García-Carmona, A., & Toma, R. (2021, Fevereiro). *STEM integrado: significados, dificultades y perspectivas en el contexto educativo español* [Integrated STEM: meanings, difficulties and perspectives in the Spanish educational context]. Comunicação apresentada em 29 Encuentros de Didáctica de las Ciencias Experimentales y 5ª Escuelas de Doctorado. Universidad de Córdoba y APICE (Asociación Española de Profesores e Investigadores en Didáctica de las Ciencias Experimentales).

García-Penas, A. (2019). *"Simios contra el reloj": mecánicas de juego y role-playing como herramienta en la enseñanza de la Evolución Biológica* [tesis de maestría, Universidad de Zaragoza]. Repositorio institucional: https://zaguan.unizar.es/record/86683

Gardner, P. (1994). Representations of the relationship between science and technology in the curriculum. *Studies in Science Education*, *24*(1), 29–47. doi:10.1080/03057269408560037

Gefaell, J., Prieto, T., Abdelaziz, M., Álvarez, I., Antón, J., Arroyo, J., Bella, J. L., Botella, M., Bugallo, A., Claramonte, V., Gijón, J., Lizarte, E., Maroto, R. M., Megías, M., Milá, B., Ramón, C., Vila, M., & Rolán-Alvarez, E. (2020). Acceptance and knowledge of evolutionary theory among third-year university students in Spain. *PLoS One*, *15*(9), e0238345. doi:10.1371/journal.pone.0238345 PMID:32881957

Geiger, V., Mulligan, J., Date-Huxtable, L., Ahlip, R., Jones, D. H., May, E. J., Rylands, L., & Wright, I. (2018). An interdisciplinary approach to designing online learning: Fostering pre-service mathematics teachers' capabilities in mathematical modelling. *ZDM Mathematics Education, 50*(1-2), 217–232. doi:10.100711858-018-0920-x

Gelman, S. A., & Rhodes, M. (2012). "Two-Thousand Years of Stasis": How Psychological Essentialism Impedes Evolutionary Understa. In *Evolution Challenges: Integrating Research and Practice in Teaching and Learning about Evolution.* Oxford University Press. doi:10.1093/acprof:oso/9780199730421.003.0001

Gendron, R. P. (2000). The Classification & Evolution of Caminalcules. *The American Biology Teacher, 62*(8), 570–576. doi:10.2307/4450980

Gifford, J. D., & Finkelstein, N. D. (2021). Applying a mathematical sense-making framework to student work and its potential for curriculum design. *Physical Review. Physics Education Research, 17*(1), 010138-1–27. doi:10.1103/PhysRevPhysEducRes.17.010138

Gilbert, S. F. (2020). Evolutionary developmental biology and sustainability: A biology of resilience. *Evolution & Development, 12366.* Advance online publication. doi:10.1111/ede.12366 PMID:33400344

Giménez, J., Abdounur, O. J., Badillo, E., Balbás, S., Corbalán, F., Dos Santos, J. M., & Spinadel, V. W. (2009). *La proporción: arte y matemáticas* [The proportion: Art and mathematics]. Graó.

Gingras, Y. (2001). What did mathematics do to physics? *History of Science, 39*(4), 383–416. doi:10.1177/007327530103900401

Gkouskou, E., & Koliopoulos, D. (2021). Describing the educational role of Natural History Museums: An analysis tool for pre-service and in-service teachers. *European Journal of Education Studies, 8*(1). Advance online publication. doi:10.46827/ejes.v8i1.3526

Glaze, A. L., & Goldston, M. J. (2015). U.S. science teaching and learning of evolution: A critical review of the literature 2000-2014. *Science Education, 99*(3), 500–518. doi:10.1002ce.21158

Glazer, N. (2011). Challenges with graph interpretation: A review of the literature. *Studies in Science Education, 47*(2), 183–210. doi:10.1080/03057267.2011.605307

Glendon, K., & Ulrich, D. (2005). Using Games as a Teaching Strategy. *The Journal of Nursing Education, 44*(7), 338–339. doi:10.3928/01484834-20050701-11 PMID:16094796

Goddard, Y. L., Goddard, R. D., & Tschannen-Moran, M. (2007). A theoretical and empirical investigation of teacher collaboration for school improvement and student achievement in public elementary schools. *Teachers College Record, 109*(4), 877–896. doi:10.1177/016146810710900401

Godec, S., King, H., Archer, L., Dawson, E., & Seakins, A. (2018). Examining Student Engagement with Science Through a Bourdieusian Notion of Field. *Science and education, 27*(5-6), 501–521. doi:10.100711191-018-9988-5

Goldman, S. R., Radinsky, J., Tozer, S., & Wink, D. (2010). Learning as Inquiry. In P. Peterson, E. Baker, & B. McGaw (Eds.), *International encyclopedia of education* (3rd ed., pp. 297–302). Elsevier., doi:10.1016/B978-0-08-044894-7.00495-4

Goldsworthy, A., Watson, R., & Wood-Robinson, V. (1999). *Getting to grips with graphs.* ASE.

Gómez Puente, S. M., van Eijck, M., & Jochems, W. (2013). A sampled literature review of design-based learning approaches: A search for key characteristics. *International Journal of Technology and Design Education, 23*(3), 717–732. doi:10.100710798-012-9212-x

Gómez-Moreno, D. K. (2020). *Enseñanza de la evolución biológica a través de historietas conceptuales contextualizadas* [tesis de maestría, Universidad Distrital Francisco José de Caldas]. Repositorio institucional: http://hdl.handle.net/11349/24454

González-Galli, L. (2017). Darwin y los naturalistas del Plata: una propuesta para la utilización de la historia y la sociología de la ciencia en la enseñanza de la teoría de la evolución. Historia de la ciencia e investigación en formación de profesores de ciencia, 34-59.

González-Galli, L., & Meinardi, S. (2017). Obstáculos para el aprendizaje del modelo de evolución por selección natural, en estudiantes universitarios de biología. *Revista Eureka sobre Enseñanza y Divulgación de las Ciencias, 14*(3), 435-449. . doi:10.25267/Rev_Eureka_ensen_divulg_cienc.2017.v14.i3.03

González-Galli, L., & Meinardi, E. (2013). ¿Está en crisis el darwinismo? Los nuevos modelos de la biología evolutiva y sus implicaciones didácticas. *Didáctica de las Ciencias Experimentales y Sociales*, *27*(27), 219–234. doi:10.7203/dces.27.2458

González-Galli, L., & Meinardi, S. (2015). Obstáculos para el aprendizaje del modelo de evolución por selección natural, en estudiantes de escuela secundaria de la Argentina. *Ciência & Educação (Bauru)*, *21*(1), 101–122. doi:10.1590/1516-731320150010007

González-Gómez, D., & Jeong, J. S. (2022). Approaches and Methods of Science Teaching and Sustainable Development. *Sustainability (Switzerland)*, *14*(3), 1546. Advance online publication. doi:10.3390u14031546

Goodson, I. (1995). Becoming a School Subject. In I. Goodson (Ed.), *The making of the curriculum* (pp. 155–177). The Falmer Press.

Gould, S. (2004). *La Estructura de la Teoría Evolutiva*. Tusquets.

Goyal, M., Gupta, C., & Gupta, V. (2022). A Meta-Analysis Approach to Measure the Impact of Project-Based Learning Outcome with Program Attainment on Student Learning Using Fuzzy Inference Systems. *Heliyon*, *8*(8), e10248. doi:10.1016/j.heliyon.2022.e10248 PMID:36042720

Greca, I., Ortiz-Revilla, J., & Arriassecq, I. (2021). Diseño y evaluación de una secuencia de enseñanza aprendizaje STEAM para educación primaria [Design and evaluation of a STEAM teaching-learning sequence for primary education]. *Revista Eureka sobre Enseñanza y Divulgación de las Ciencias, 18*(1), 1802–1819. doi:10.25267/Rev_Eureka_ensen_divulg_cienc.2021.v18.i1.1802

Greenwood, R., Austin, S., Bacon, K., & Pike, S. (2022). Enquiry-based learning in the primary classroom: student teachers' perceptions. *Education 3-13*, *50*(3), 404-418. doi:10.1080/03004279.2020.1853788

Gregory, T. R. (2008). Understanding Evolutionary Trees. *Evolution (New York)*, *1*(2), 121–137. doi:10.100712052-008-0035-x

Gregory, T. R., & Ellis, C. A. J. (2009). Conceptions of Evolution among Science Graduate Students. *Bioscience*, *59*(9), 792–799. doi:10.1525/bio.2009.59.9.11

Gresnigt, R., Taconis, R., van Keulen, H., Gravemeijer, K., & Baartman, L. (2014). Promoting science and technology in primary education: A review of integrated curricula. *Studies in Science Education*, *50*(1), 47–84. doi:10.1080/03057267.2013.877694

Griffith, J. A., & Brem, S. K. (2004). Teaching evolutionary biology: Pressures, stress, and coping. *Journal of Research in Science Teaching*, *41*(8), 791–809. doi:10.1002/tea.20027

Grimley, M., Green, R., Nilsen, T., Thompson, D., & Tomes, T. (2011). Using Computer Games for Instruction: The Student Experience. *Active Learning in Higher Education, 12*(1), 45–56. doi:10.1177/1469787410387733

Grossman, P., & Stodolsky, S. (1995). Content as Context: The Role of School Subjects in Secondary School Teaching. *Educational Researcher, 14*(5), 5–23. doi:10.3102/0013189X024008005

Guilford, J. P. (1959). Traits of Creativity. In H. H. Anderson (Ed.), *Creativity and Its Cultivation* (pp. 142–161). Harper & Row.

Guyotte, K. W., Sochacka, N. W., Costantino, T. E., Kellam, N., Kellam, N. N., & Walther, J. (2015). Collaborative creativity in STEAM: Narratives of art education students' experiences in transdisciplinary spaces. *International Journal of Education & the Arts, 16*(15). http://www.ijea.org/v16n15/

Guzey, S. S., Harwell, M., Moreno, M., Peralta, Y., & Moore, T. J. (2017). The impact of design-based STEM integration curricula on student achievement in engineering, science, and mathematics. *Journal of Science Education and Technology, 26*(2), 207–222. doi:10.100710956-016-9673-x

Guzey, S. S., Tank, K., Wang, H. H., Roehrig, G., & Moore, T. (2014). A high-quality professional development for teachers of grades 3–6 for implementing engineering into classrooms. *School Science and Mathematics, 114*(3), 139–149. doi:10.1111sm.12061

Guzey, S., Moore, T., & Harwell, M. (2016). Building up STEM: An analysis of teacher-developed engineering design-based STEM integration curricular materials. *Journal of Pre-College Engineering Education Research, 6*(1), 11–29. doi:10.7771/2157-9288.1129

Haatainen, O., Turkka, J., & Aksela, M. (2021). Science teachers' perceptions and self-efficacy beliefs related to integrated science education. *Education Sciences, 11*(272), 1–20. doi:10.3390/educsci11060272

Hadjichambis, A. C., & Paraskeva-Hadjichambi, D. (2020). Education for Environmental Citizenship: The Pedagogical Approach. *Environmental Discourses in Science Education, 237*, 237–261. Advance online publication. doi:10.1007/978-3-030-20249-1_15

Hagman, M., Olander, C., & Wallin, A. (2003). Research-based teaching about biological evolution. In J. Lewis, A. Margo & L. Simonneaux (Eds.), *Biology education for the real world. Student - Teacher – Citizen. Proceedings of the IV Conference of European Researchers in Didactic of Biology (ERIDOB)* (pp. 105-119). Ecole national de formation agronomique.

Hagolani, P. F., Zimm, R., Vroomans, R., & Salazar-Ciudad, I. (2021). On the evolution and development of morphological complexity: A view from gene regulatory networks. *PLoS Computational Biology, 17*(2), e1008570. doi:10.1371/journal.pcbi.1008570 PMID:33626036

Haig, D. (2007). Weismann Rules! OK? Epigenetics and the Lamarckian temptation. *Biology & Philosophy, 22*(3), 415–428. doi:10.100710539-006-9033-y

Hall, B. K. (2012). Evolutionary Developmental Biology (Evo-Devo): Past, Present, and Future. *Evolution (New York), 5*(2), 184–193. doi:10.100712052-012-0418-x

Ha, M., & Nehm, R. H. (2014). Darwin's difficulties and student's struggles with trait loss: Cognitive-historical parallelisms in evolutionary explanation. *Science & Education, 23*(5), 1051–1074. doi:10.100711191-013-9626-1

Hammad, S., Graham, T., Dimitriadis, C., & Taylor, A. (2022). Effects of a Successful Mathematics Classroom Framework on Students' Mathematics Self-Efficacy, Motivation, and Achievement: A Case Study with Freshmen Students at a University Foundation Programme in Kuwait. *International Journal of Mathematical Education in Science and Technology*, *53*(6), 1502–1527. doi:10.1080/0020739X.2020.1831091

Hancock, Z. B., Lehmberg, E. S., & Bradburd, G. S. (2021). Neo-darwinism still haunts evolutionary theory: A modern perspective on Charlesworth, Lande, and Slatkin (1982). *Evolution*, *75*(6), 1244–1255. doi:10.1111/evo.14268 PMID:33999415

Hanisch, S., & Eirdosh, D. (2020). Educational potential of teaching evolution as an interdisciplinary science. *Evolution (New York)*, *13*(1), 25. doi:10.118612052-020-00138-4

Hansen, P. J. K. (2009). Analysing cases in technology and design education: How could designing and making technological products be a vehicle for enhancing understanding of natural science principles? *Design and Technology Education: An International Journal*, *14*(2), 45–52.

Hansson, L., Hansson, Ö., Juter, K., & Redfors, A. (2021). Curriculum emphases, mathematics and teaching practices: Swedish upper-secondary physics teachers' views. *International Journal of Science and Mathematics Education*, *19*(3), 499–515. doi:10.100710763-020-10078-6

Hargreaves, A., & Macmillan, R. (1995). The Balkanization of Secondary School Teaching. In L. Siskin & J. W. Little (Eds.), *The Subjects in Question* (pp. 141–171). Teachers College Press.

Harmin, M., & Toth, M. (2006). *Inspiring Active Learning: A Complete Handbook for Today's Teachers*. Association for Supervision and Curriculum Developmen.

Hartelt, T., Martens, H., & Minkley, N. (2022). Teachers' ability to diagnose and deal with alternative student conceptions of evolution. *Science Education*, *106*(3), 706–738. doi:10.1002ce.21705

Hartikainen, S., Rintala, H., Pylväs, L., & Nokelainen, P. (2019). The Concept of Active Learning and the Measurement of Learning Outcomes: A Review of Research in Engineering Higher Education. *Education Sciences*, *9*(4), 9–12. doi:10.3390/educsci9040276

Hartman, S., & Hines-Bergmeier, J. (2015). Building connections: Strategies to address rurality and accessibility challenges. *Journal of Museum Education*, *40*(3), 288–303. doi:10.1179/1059865015Z.000000000105

Hasnine, M. N., Ueda, H., & Ahmed, M. M. H. (2022). Adaptation of AL-TST Active Learning Model in Hybrid Classroom: Findings from Teaching during COVID-19 Pandemic in Egypt. *Procedia Computer Science*, *207*, 3226–3233. doi:10.1016/j.procs.2022.09.380 PMID:36275371

Hattie, J. (2015). *What Works Best in Education: The Politics of Collaborative Expertise*. Pearson.

Hawley, P. H., & Sinatra, G. M. (2019). Declawing the dinosaurs in the science classroom: Reducing Christian teachers' anxiety and increasing their efficacy for teaching evolution. *Journal of Research in Science Teaching*, *56*(4), 375–401. doi:10.1002/tea.21479

Henriksen, D. (2014). Full STEAM ahead: Creativity in excellent STEM teaching practices. *The STEAM Journal*, *1*(2), 1–9. doi:10.5642team.20140102.15

Henze, I., & Barendsen, E. (2019). Unravelling student science teachers' pPCK development and the influence of personal factors using authentic data sources. In A. Hume, R. Cooper, & A. Borowski (Eds.), *Repositioning pedagogical content knowledge in teachers' knowledge for teaching science* (pp. 201–221). Springer. doi:10.1007/978-981-13-5898-2_9

Hermann, R. S., Shane, J. W., Meadows, L., & Binns, I. C. (2020). Understanding of Evolution Law among K–12 Public School Teachers. *The American Biology Teacher, 82*(2), 86–92. doi:10.1525/abt.2020.82.2.86

Hernández-Barco, M., Cañada-Cañada, F., Corbacho-Cuello, I., & Sánchez-Martín, J. (2021). An Exploratory Study Interrelating Emotion, Self-Efficacy and Multiple Intelligence of Prospective Science Teachers. *Frontiers in Education, 6*, 1–12. doi:10.3389/feduc.2021.604791

Hernández, R. M. C., Álvarez Pérez, E., & Ruíz Gutiérrez, R. (2009). La selección natural: Aprendizaje de un paradigma. *Teorema, 28*(2), 107–121.

Hill, M. J. (2013). *The Public Policy Process* (5th ed.). Pearson.

Hobbs, L., Doig, B., & Plant, B. (2019). The successful students STEM project: A medium scale case study. In B. Doig, J. Williams, D. Swanson, R. B. Ferri, & P. Drake (Eds.), *Interdisciplinary Mathematics Education. The state of the art and beyond* (pp. 209–227). Springer Open. doi:10.1007/978-3-030-11066-6_13

Hodgen, J., Pepper, D. S., & Ruddock, G. (2010). *Is the UK an Outlier? An international comparison of upper secondary mathematics education*. Nuffield Foundation.

Holmlund, T., Lesseig, K., & Slavit, D. (2018). Making sense of "STEM education" in K-12 contexts. *International Journal of STEM Education, 5*(32), 32. Advance online publication. doi:10.118640594-018-0127-2 PMID:30631722

Honey, M., Pearson, G., & Schweingruber, A. (2014). *STEM integration in K-12 education: status, prospects, and an agenda for research*. National Academies Press.

Honey, M., Pearson, G., & Schweingruber, A. (2014). *STEM integration in K-12. Status, Prospects, and an agenda for research*. National Academies Press.

Honey, M., Pearson, G., & Schweingruber, H. (2014). *STEM Integration in K-12 Education: Status, Prospects, and an Agenda for Research*. National Academies Press. Retrieved from https://www.nap.edu/catalog.php?record_id=18612

Hossein-Mohand, H., Trujillo-Torres, J. M., Gómez-García, M., Hossein-Mohand, H., & Campos-Soto, A. (2021). Analysis of the Use and Integration of the Flipped Learning Model, Project-Based Learning, and Gamification Methodologies by Secondary School Mathematics Teachers. *Sustainability (Switzerland), 13*(5), 1–18. doi:10.3390u13052606

Hourigan, M., O'Dwyer, A., Leavy, A., & Corry, E. (2021). Integrated STEM – a step too far in primary education contexts? *Irish Educational Studies*. Advance online publication. doi:10.1080/03323315.2021.1899027

Howe, C., Tolmie, A., Thurston, A., Topping, K., Christie, D., Livingston, K., Jessiman, E., & Donaldson, C. (2007). Group work in elementary science: Organisational principles for supporting pupil learning. *Learning and Instruction, 17*(5), 549–563. doi:10.1016/j.learninstruc.2007.09.004

Howell, R. A. (2021). Engaging Students in Education for Sustainable Development: The Benefits of Active Learning, Reflective Practices and Flipped Classroom Pedagogies. *Journal of Cleaner Production, 325*(October), 129318. doi:10.1016/j.jclepro.2021.129318

Hsu, Y.-S., & Fang, S.-C. (2019). Opportunities and Challenges of STEM Education. In Y.-S. Hsu & S.-C. Fang (Eds.), Asia-Pacific STEM Teaching Practices (pp. 1-16). Springer International Publishing. doi:10.1007/978-981-15-0768-7_1

Hué, C. (2012). Bienestar docente y pensamiento emocional. *Revista Fuentes, 12*, 47–68.

Huntley, M. A. (1998). Design and implementation of framework for defining integrated mathematics and science education. *School Science and Mathematics, 98*(1), 320–327. doi:10.1111/j.1949-8594.1998.tb17427.x

Hurd, P. D. (1991). Why we must transform science education. *Educational Leadership, 49*(2), 33–35.

Hurley, J., & Hallissy, M. (2017). *STEM Education Policy Statement 2017–2026*. Department of Education and Skills.

Hussain, M., Sahudin, S., Samah, N. H. A., & Anuar, N. K. (2019). Students Perception of an Industry Based Approach Problem Based Learning (PBL) and Their Performance in Drug Delivery Courses. *Saudi Pharmaceutical Journal*, *27*(2), 274–282. doi:10.1016/j.jsps.2018.11.009 PMID:30766440

Institute of Physics. (2017). *Improving gender balance*. Institute of Physics. Retrieved from https://www.iop.org/sites/default/files/2019-07/IGB-reflections-intervention.pdf

International Test Commission. (2005). International guidelines on computer-based and internet-delivered testing. *International Journal of Testing*, *6*(2), 143–171.

International Test Commission. (2017). *The ITC Guidelines for Translating and Adapting Tests*. https://www.intestcom.org

James, P., & Clark, I. (2006). Overcoming geological misconceptions. *Planet*, *17*(1), 10–13. doi:10.11120/plan.2006.00170010

Janes, G., Ekpenyong, M. S., Mbeah-Bankas, H., & Serrant, L. (2022). The Role of Blended Learning in Preparing Nursing and Midwifery Students for Initial Professional Registration: An International, Cross-Sectional, Sequential Mixed Methods Study. *Nurse Education in Practice*, *103514*. Advance online publication. doi:10.1016/j.nepr.2022.103514

Jensen, M. S., & Finley, F. N. (1996). Changes in students' understanding of evolution resulting from different curricular and instructional strategies. *Journal of Research in Science Teaching*, *33*(8), 879–900. doi:10.1002/(SICI)1098-2736(199610)33:8<879::AID-TEA4>3.0.CO;2-T

Jiménez-Aleixandre, M. P. (1994). Teaching evolution and natural selection: A look at textbooks and teachers. *Journal of Research in Science Teaching*, *31*(5), 519–535. doi:10.1002/tea.3660310507

Johnston, J. (2005). *Early explorations in science*. Mcgraw-Hill Education.

Johnston, J., Walshe, G., & Riordain, M. (2020). Supporting key aspect of practice in making mathematics explicit in science lessons. *International Journal of Science and Mathematics Education*, *18*(7), 1399–1417. doi:10.100710763-019-10016-1

Jonassen, D. H. (2010). *Learning to Solve Problems: A Handbook for Designing Problem-Solving Learning Environments* (1st ed.). Routledge. doi:10.4324/9780203847527

Jones, I., Lake, V. E., & Dagli, U. (2005). Integration of science and mathematics methods and preservice teachers' understanding of constructivism. *Journal of Early Childhood Teacher Education*, *25*(2), 165–172. doi:10.1080/1090102050250209

Judson, E. (2013). Development of an Instrument to Assess and Deliberate on the Integration of Mathematics into Student-Centered Science Learning. *School Science and Mathematics*, *113*(2), 56–68. doi:10.1111sm.12004

Juhler, M. V. (2016). The use of lesson study combined with content representation in the planning of physics lessons during field practice to develop pedagogical content knowledge. *Journal of Science Teacher Education*, *27*(5), 533–553. doi:10.100710972-016-9473-4

Julius, M. L., & Schoenfuss, H. L. (2006). Phylogenetic reconstruction as a broadly applicable teaching tool in the biology classroom: The value of data in estimating likely answers. *Journal of College Science Teaching*, *35*(7), 40–45.

Jungck, J. R., & Weisstein, A. E. (2013). Mathematics and evolutionary biology make bioinformatics education comprehensible. *Briefings in Bioinformatics*, *14*(5), 599–609. doi:10.1093/bib/bbt046 PMID:23821621

Jung, J. Y., Lim, Y., & Huh, M. K. (2017). Flipped classroom pedagogy enhances student satisfaction and validated motivated strategies in Evolution Biology classrooms. *International Journal of Academic Research and Reflection*, *5*(6), 1–7.

Kahn, P. H. Jr, Weiss, T., & Harrington, K. (2018). Modeling child–nature interaction in a nature preschool: A proof of concept. *Frontiers in Psychology*, *835*, 835. Advance online publication. doi:10.3389/fpsyg.2018.00835 PMID:29896143

Kampourakis, K. (2013). Teaching About Adaptation: Why Evolutionary History Matters. *Science & Education*, *22*(2), 173–188. doi:10.100711191-011-9363-2

Kampourakis, K., & Zogza, V. (2008). Students' intuitive explanations of the causes of homologies and adaptations. *Science & Education*, *17*(1), 27–47. doi:10.100711191-007-9075-9

Kang, Y., McNeish, D. M., & Hancock, G. R. (2016). The role of measurement quality on practical guidelines for assessing measurement and structural invariance. *Educational and Psychological Measurement*, *76*(4), 533–561. doi:10.1177/0013164415603764 PMID:29795877

Karataş, A. (2020). Preservice Science Teachers' Misconceptions About Evolution. *Journal of Education and Training Studies*, *8*(2), 38–46. doi:10.11114/jets.v8i2.4690

Kelemen, D., Emmons, N. A., Schillaci, R. S., & Ganea, P. A. (2014). Young children can be taught basic natural selection using a picture storybook intervention. *Psychological Science*, *25*(4), 893–902. doi:10.1177/0956797613516009 PMID:24503874

Kelemen, D., & Rosset, E. (2009). The human function compunction: Teleological explanation in adults. *Cognition*, *111*(1), 138–143. doi:10.1016/j.cognition.2009.01.001 PMID:19200537

Kelley, T., & Knowles, J. (2016). A conceptual framework for integrated STEM education. *International Journal of STEM Education*, *3*(1), 11. doi:10.118640594-016-0046-z

Kennedy, T. J., & Odell, M. R. L. (2014). Engaging students in STEM education. *Science Education International*, *25*(3), 246–258.

Kerfeld, C. A., & Scott, K. M. (2011). Using BLAST to teach "E-value-tionary" concepts. *PLoS Biology*, *9*(2), e1001014. doi:10.1371/journal.pbio.1001014 PMID:21304918

Kim, D., & Bolger, M. (2017). Analysis of Korean elementary pre-service teachers' changing attitudes about integrated STEAM pedagogy through developing lesson plans. *International Journal of Science and Mathematics Education*, *15*(4), 587–605. doi:10.100710763-015-9709-3

Kim, M. K., & Cho, M. K. (2015). Design and implementation of integrated instruction of mathematics and science in Korea. *Eurasia Journal of Mathematics, Science and Technology Education*, *11*(1), 3–15. doi:10.12973/eurasia.2015.1301a

Kim, M. M., Andrews, R. L., & Carr, D. L. (2004). Traditional versus integrated preservice teacher education curriculum: A case study. *Journal of Teacher Education*, *55*(4), 341–356. doi:10.1177/0022487104266778

Kimura, M. (1968). Evolutionary rate at the molecular level. *Nature*, *217*(5129), 624–626. doi:10.1038/217624a0 PMID:5637732

King, C. (1877). Catastrophism and evolution. *American Naturalist*, *11*(8), 449–470. doi:10.1086/271929

King, D., Ritchie, S., Sandhu, M., & Henderson, S. (2015). Emotionally intense science activities. *International Journal of Science Education*, *37*(12), 1886–1914. doi:10.1080/09500693.2015.1055850

King, K., & Wiseman, D. (2001). Comparing science efficacy beliefs of elementary education majors in integrated and non-integrated teacher education coursework. *Journal of Science Teacher Education, 12*(2), 143–153. doi:10.1023/A:1016681823643

Kiray, S. A., & Kaptan, F. (2012). A new model for the integration of science and mathematics: The balance model. *Energy Education Science and Technology, Part B. Social and Educational Studies, 4*(3), 1181–1196.

Kline, R. B. (2011). *Principles and practice of structural equation modeling*. Guilford.

Klusmann, B., Trippenzee, M., Fokkens-Bruinsma, M., Sanderman, R., & Schroevers, M. J. (2022). Providing Emergency Remote Teaching: What Are Teachers' Needs and What Could Have Helped Them to Deal with the Impact of the COVID-19 Pandemic? *Teaching and Teacher Education, 118*, 103815. doi:10.1016/j.tate.2022.103815 PMID:35874644

Koirala, H. P., & Bowman, J. K. (2003). Preparing middle level preservice teachers to integrate mathematics and science: Problems and possibilities. *School Science and Mathematics, 103*(3), 145–154. doi:10.1111/j.1949-8594.2003.tb18231.x

Kong, Y., Thawani, A., Anderson, T., & Pelaez, N. (2017). A Model of the Use of Evolutionary Trees (MUET) to Inform K-14 Biology Education. *The American Biology Teacher, 79*(2), 81–90. doi:10.1525/abt.2017.79.2.81

Koonin, E. V. (2009). Towards a postmodern synthesis of evolutionary biology. *Cell Cycle (Georgetown, Tex.), 8*(6), 799–800. doi:10.4161/cc.8.6.8187 PMID:19242109

Koonin, E. V., & Wolf, Y. I. (2009). Is evolution Darwinian or/and Lamarckian? *Biology Direct, 4*(1), 42. doi:10.1186/1745-6150-4-42 PMID:19906303

Krajcik, J., & Czerniak, C. M. (2014). *Teaching science to children: A project-based science approach*. Routledge. doi:10.4324/9780203113660-7

Kraska-Miller, M. (2014). *Nonparametric statistics for social and behavioral sciences*. CRC Press.

Kuschmierz, P., Meneganzin, A., Pinxten, R., Pievani, T., Cvetković, D., Mavrikaki, E., Graf, D., & Beniermann, A. (2020). Towards common ground in measuring acceptance of evolution and knowledge about evolution across Europe: A systematic review of the state of research. *Evolution (New York), 13*(1), 18. doi:10.118612052-020-00132-w

Kutschera, U. (2011). From the scala naturae to the symbiogenetic and dynamic tree of life. *Biology Direct, 6*(1), 33. doi:10.1186/1745-6150-6-33 PMID:21714937

Labov, J. B., Reid, A. H., & Yamamoto, K. R. (2010). Integrated biology and undergraduate science education: A new biology education for the twenty-first century? *CBE Life Sciences Education, 9*(1), 10–16. doi:10.1187/cbe.09-12-0092 PMID:20194802

Lage-Gómez, C., & Ros, G. (2021). Transdisciplinary integration and its implementation in primary education through two STEAM projects. *Journal for the Study of Education and Development*. doi:10.1080/02103702.2021.1925474

Lahm, K. (2007). Validity, quantitative. In G. Ritzer (Ed.), *The blackwell encyclopedia of sociology* (pp. 5173–5175). Blackwell Publishing. doi:10.1002/9781405165518.wbeosv002

Lamarck, J. B. (1809). Philosophie Zoologique, ou Exposition des Considérations relatives à l'Histoire naturelle des Animaux. Dentu et 'l Auteur.

Land, M. (2013). Full STEAM ahead: The benefits of integrating the arts into STEM. *Procedia Computer Science, 20*, 547–552. doi:10.1016/j.procs.2013.09.317

Latham, L. G. II, & Scully, E. P. (2008). Critters! A Realistic Simulation for Teaching Evolutionary Biology. *The American Biology Teacher, 70*(1), 30–33. doi:10.1662/0002-7685(2008)70[30:CARSFT]2.0.CO;2

Lau, M., & Multani, S. (2018). Engineering STEM teacher learning: Using a museum-based field experience to foster STEM teachers' pedagogical content knowledge for engineering. In S. M. Uzzo, S. B. Graves, E. Shay, M. Harford, & R. Thompson (Eds.), *Pedagogical content knowledge in STEM* (pp. 195–213). Springer. doi:10.1007/978-3-319-97475-0_11

Lawrenz, F. (1986). Misconceptions of Physical Science Concepts Among Elementary School Teachers. *School Science and Mathematics*, *86*(8), 654–660. doi:10.1111/j.1949-8594.1986.tb11669.x

Lawson, A. E., & Thompson, L. D. (1988). Formal reasoning ability and misconceptions concerning genetics and natural selection. *Journal of Research in Science Teaching*, *25*(9), 733–746. doi:10.1002/tea.3660250904

Lebreton, M., Abitbol, R., Daunizeau, J., & Pessiglione, M. (2015). Automatic integration of confidence in the brain valuation signal. *Nature Neuroscience*, *18*(8), 1159–1167. doi:10.1038/nn.4064 PMID:26192748

Lederman, N. G., & Lederman, J. S. (2013). Is it STEM or "S & M" that we truly love? *Journal of Science Teacher Education*, *24*(8), 1237–1240. doi:10.100710972-013-9370-z

Lederman, N., & Niess, M. (1998). 5 Apples + 4 oranges =? *School Science and Mathematics*, *98*(6), 281–284. doi:10.1111/j.1949-8594.1998.tb17422.x

Lee, M. (2007). Spark up the American revolution with math, science, and more: An example of an integrative curriculum unit. *Social Studies*, *98*(4), 159–164. doi:10.3200/TSSS.98.4.159-164

Lee, M. M., Chauvot, J. B., Vowell, J., Culpepper, S. M., & Plankis, B. J. (2013). Stepping into iSMART: Understanding science–mathematics integration for middle school science and mathematics teachers. *School Science and Mathematics*, *113*(4), 159–169. doi:10.1111sm.12015

Lehman, J. D., Kim, W., & Harris, C. (2014). Collaborations in a community of practice working to integrate engineering design in elementary science education. *Journal of STEM Education: Innovations and Research*, *15*(3), 21–28.

Lehman, J. R. (1994). Integrating science and mathematics: Perceptions of preservice and practicing elementary teachers. *School Science and Mathematics*, *94*(2), 58–64. doi:10.1111/j.1949-8594.1994.tb12293.x

Lesseig, K., Nelson, T. H., Slavit, D., & Seidel, R. A. (2016). Supporting middle school teachers' implementation of STEM design challenges. *School Science and Mathematics*, *116*(4), 177–188. doi:10.1111sm.12172

Leszczynski, E. (2014). *The Study of Middle School Mathematics and Science Teachers' Practices, Perceptions, and Attitudes Related to Mathematics and Science Integration* [Unpublished doctoral dissertation]. University of Montclair, Canada.

Levine, T. H., & Marcus, A. S. (2010). How the structure and focus of teachers' collaborative activities facilitate and constrain teacher learning. *Teaching and Teacher Education*, *26*(3), 389–398. doi:10.1016/j.tate.2009.03.001

Lewontin, R. C. (1970). The Units of Selection. *Annual Review of Ecology and Systematics*, *1*(1), 1–18. doi:10.1146/annurev.es.01.110170.000245

Libarkin, J. C., Kurdziel, J. P., & Anderson, S. W. (2007). College student conceptions of geological time and the disconnect between ordering and scale. *Journal of Geoscience Education*, *55*(5), 413–422. doi:10.5408/1089-9995-55.5.413

Lichtenberger, A., Wagner, C., Hofer, S. I., Stern, E., & Vaterlaus, A. (2017). Validation and structural analysis of the kinematics concept test. *Physical Review. Physics Education Research*, *13*(1), 010115. doi:10.1103/PhysRevPhysEducRes.13.010115

Li, H., Xiong, Y., Hunter, C. V., Guo, X., & Tywoniw, R. (2020). Does peer assessment promote student learning? A meta-analysis. *Assessment & Evaluation in Higher Education*, *45*(2), 193–211. doi:10.1080/02602938.2019.1620679

Li, J., Luo, H., Zhao, L., Zhu, M., Ma, L., & Liao, X. (2022). Promoting STEAM education in primary school through cooperative teaching: A design-based research study. *Sustainability, 14*(16), 10333. doi:10.3390u141610333

Linnæus, C. (1758). Systema naturæ per regna tria naturæ, secundum classes, ordines, genera, species, cum characteribus, differentiis, synonymis, locis. Tomus I. Editio decima, reformata.

Lino-Neto, T., Ribeiro, E., Rocha, M., & Costa, M. J. (2022). Going virtual and going wide: Comparing Team-Based Learning in-class versus online and across disciplines. *Education and Information Technologies, 27*(2), 2311–2329. doi:10.100710639-021-10683-0 PMID:34421327

Liu, Y., and Pásztor, A. (2022). Effects of Problem-Based Learning Instructional Intervention on Critical Thinking in Higher Education: A Meta-Analysis. *Thinking Skills and Creativity, 45.* . doi:10.1016/j.tsc.2022.101069

Livio, M. (2008). *The golden ratio: The story of phi, the world's most astonishing number.* Crown.

Lohbeck, A., & Retelsdorf, J. (2021). Assessing value beliefs among university students: Validation of the Value Beliefs Questionnaire for University Students (VBQU). *Studies in Educational Evaluation, 70,* 101052. doi:10.1016/j.stueduc.2021.101052

Loison, L. (2018). Lamarckism and epigenetic inheritance: A clarification. *Biology & Philosophy, 33*(3), 1–17.

LOMLOE. (2020). Ley Orgánica 3/2020, de 29 de diciembre de 2020, por la que se modifica la Ley Orgánica 2/2006, de 3 de mayo, de Educación. BOE núm. 340, de 30 de diciembre de 2020, páginas 122868 a 122953.

Lonergan, R., Cumming, T.M., and O'Neill, S.C. (2022). Exploring the Efficacy of Problem-Based Learning in Diverse Secondary School Classrooms: Characteristics and Goals of Problem-Based Learning. *International Journal of Educational Research, 112.* doi:10.1016/j.ijer.2022.101945

Lonning, R. A., & DeFranco, T. C. (1997). Integration of science and mathematics: A theoretical model. *School Science and Mathematics, 97*(4), 212–215. doi:10.1111/j.1949-8594.1997.tb17369.x

López-Fernández, D., Ezquerro, J. M., Rodríguez, J., Porter, J., & Lapuerta, V. (2019). Motivational Impact of Active Learning Methods in Aerospace Engineering Students. *Acta Astronautica, 165,* 344–354. doi:10.1016/j.actaastro.2019.09.026

Loughran, J., Mulhall, P., & Berry, A. (2004). In search of pedagogical content knowledge in science: Developing ways of articulating and documenting professional practice. *Journal of Research in Science Teaching, 41*(4), 370–391. doi:10.1002/tea.20007

Love, A. C. (2013). Interdisciplinary Lessons for the Teaching of Biology from the Practice of Evo-Devo. *Science & Education, 22*(2), 255–278. doi:10.100711191-011-9351-6

Lozada-Guzmán, M. M. (2016). *Implementación De Una Secuencia De Enseñanza-Aprendizaje (SEA) En Teoría De Evolución Biológica: Aproximación A La Transformación De Las Concepciones De Naturaleza De La Ciencia De Profesores En Formación* [tesis de maestría, Universidad de Tolima]. Repositorio institucional: http://repository.ut.edu.co/handle/001/1737/statistics

Lucas da Rocha Cunha, M., Amendola, F., Fernandez Samperiz, M. M., & Gomes da Costa Mohallem, A. (2018). Evaluation of Student Perception of the Team-Based Learning Method (APA-TBL): Instrument Construction and Validation. *Nurse Education in Practice, 33*(April), 141–147. doi:10.1016/j.nepr.2018.09.008 PMID:30296726

Luo, Y. (2019). The influence of problem-based learning on learning effectiveness in students of varying learning abilities within physical education. *Innovations in Education and Teaching International, 56*(1), 3–13. doi:10.1080/14703297.2017.1389288

Luttrell, V. R., Callen, B. W., Allen, C. S., Wood, M. D., Deeds, D. G., & Richard, D. C. S. (2010). The Mathematics Value Inventory for General Education Students: Development and Initial Validation. *Educational and Psychological Measurement, 70*(1), 142–160. doi:10.1177/0013164409344526

Lyons, T. (2006). Different countries, same science classes: Students' experiences of school science in their own words. *International Journal of Science Education, 28*(6), 591–613. doi:10.1080/09500690500339621

Lytras, M. D., Serban, A. C., Torres Ruiz, M. J., Ntanos, S., & Sarirete, A. (2022). Translating Knowledge into Innovation Capability: An Exploratory Study Investigating the Perceptions on Distance Learning in Higher Education during the COVID-19 Pandemic - the Case of Mexico. *Journal of Innovation and Knowledge, 7*(4), 100258. doi:10.1016/j.jik.2022.100258

Maass, K., Geiger, V., Ariza, M. R., & Goos, M. (2019). The role of mathematics in interdisciplinary STEM education. *ZDM, 51*(6), 869–884. doi:10.100711858-019-01100-5

MacKinnon, P. J., Hine, D., & Barnard, R. T. (2013). Interdisciplinary science research and education. *Higher Education Research & Development, 32*(3), 407–419. doi:10.1080/07294360.2012.686482

MacMath, S., Wallace, J., & Chi, X. (2009). Curriculum integration: Opportunities to maximize assessment as, of, and for learning. *McGill Journal of Education, 44*(3), 451–466. doi:10.7202/039949ar

Madden, L., Beyers, J., & O'Brien, S. (2016). The Importance of STEM Education in the Elementary Grades: Learning from Pre-service and Novice Teachers' Perspectives. *The Electronic Journal of Science Education, 20*, 1–18.

Magalhães, S., Blanchet, E., Egas, M., & Olivieri, I. (2009). Are adaptation costs necessary to build up a local adaptation pattern? *BMC Evolutionary Biology, 9*(1), 182. doi:10.1186/1471-2148-9-182 PMID:19650899

Magnusson, S., Krajcik, J., & Borko, H. (1999). Nature, sources and development of pedagogical content knowledge for science teaching. In J. Gess-Newsome & N. G. Lederman (Eds.), *Examining pedagogical content knowledge: The construct and its implications for science education* (pp. 95–132). Kluwer Academic Publishers.

Maguire, M., Ball, S., & Braun, A. (2012). *How schools do policy: policy enactments in secondary schools*. Routledge.

Mahoney, M. (2010). Student attitudes toward STEM: Development of an instrument for high school STEM-based programs. *The Journal of Technology Studies, 36*(1), 24–34. doi:10.21061/jots.v36i1.a.4

Maia, D., & Dos Santos, S. C. (2022). Monitoring Students' Professional Competencies in PBL: A Proposal Founded on Constructive Alignment and Supported by AI Technologies. *Proceedings - Frontiers in Education Conference*. 10.1109/FIE56618.2022.9962459

Major, T., & Mulvihill, T. M. (2018). Problem-Based Learning Pedagogies in Teacher Education: The Case of Botswana. *The Interdisciplinary Journal of Problem-Based Learning, 12*(1). Advance online publication. doi:10.7771/1541-5015.1543

Maknun, J. (2021). Physics Problem Solving Skills of Civil Engineering Students by Implementing Problem-Based Learning. *Review of International Geographical Education, 11*(3), 594–603.

Malgieri, M., & Onorato, P. (2020). Teaching the heat transfer law using a stochastic toy model. *European Journal of Physics, 41*, 015103.

Malik, A. H., Ziermann, J. M., & Diogo, R. (2018). An untold story in biology: The historical continuity of evolutionary ideas of Muslim scholars from the 8th century to Darwin's time. *Journal of Biological Education, 52*(1), 3–17. doi:10.1080/00219266.2016.1268190

Malthus, T. R. (1798). *An essay on the principle of population*. Oxford World's Classics reprint.

Malthus, T. (1798). *An essay on the principle of population, as it affects the future improvement of society with remarks on the speculations of Mr. Godwin, M. Condorcet and other writers*. Eletronic Scholarly Publishing Project.

Manning, M. L., & Bucher, K. T. (2012). *Teaching in the middle school* (4th ed.). Pearson.

Marbá, A., & Márquez, C. (2010). ¿Qué opinan los estudiantes de las clases de ciencias? Un estudio transversal de sexto de Primaria a cuarto de ESO [What do students think of science classes? A cross-sectional study from sixth grade of Primary to fourth year of ESO]. *Enseñanza de las Ciencias*, *28*(1), 19–30. doi:10.5565/rev/ensciencias.3618

Marble, S. (2007). Inquiring into teaching: Lesson study in elementary science methods. *Journal of Science Teacher Education*, *18*(6), 935–953. doi:10.100710972-007-9071-6

Marcelos, M. F., & Nagem, R. (2012). Use of the "Tree" Analogy in Evolution Teaching by Biology Teachers. *Science & Education*, *21*(4), 507–541. doi:10.100711191-011-9370-3

Marcos-Merino, J. M., Esteban, R., & Ochoa de Alba, J. A. G. (2022). The interplay of prior knowledge, emotions and learning in a science experiment activity. *Enseñanza de las Ciencias*, *40*(1), 107–124.

Marcos-Merino, J. M., Gallego, R. E., & Ochoa de Alda, J. A. G. (2020). Valor subjetivo y emociones hacia el uso de Química en una práctica activa interdisciplinar. *Educación en la Química*, *31*(4), 101–111. doi:10.22201/fq.18708404e.2020.4.76221

Margot, K. C., & Kettler, T. (2019). Teachers' perception of STEM integration and education: A systematic literature review. *International Journal of STEM Education*, *6*(1), 1–16. doi:10.118640594-018-0151-2

Margulis, L. (1981). *Symbiosis in Cell evolution*. Freeman.

Mariano-Pérez, G., Gómez Galindo, A. A., & González Galli, L. (2018). Enseñanza de la evolución: fundamentos para el diseño de una propuesta didáctica basada en la modelización y la metacognición sobre los obstáculos epistemológicos. *Revista Eureka sobre Enseñanza y Divulgación de las Ciencias, 15*(2), 210101-210113.

Martin, S., & Bradbeer, C. (2016). Creating collaborative effectiveness. One school's approach. *Teachers as Communities of Learning professionals, 2*, 48-52. doi:10.18296/set.0046

Martínez-Borreguero, G., Mateos-Núñez, M., & Naranjo-Correa, F. L. (2019). Implementation and Didactic Validation of STEM Experiences in Primary Education: Analysis of the Cognitive and Affective Dimension. In Theorizing STEM Education in the 21st Century. IntechOpen.

Martínez-Borreguero, G., Mateos-Núñez, M., & Naranjo-Correa, F. L. (2020). Emotions of elementary school students: comparative analysis between different school subjects. In E. J. Byker & A. Horton (Eds.), *Elementary Education. Global Perspectives Challenges and Issues Of The 21st Century*. Nova Science Publishers, Inc.

Martínez-Borreguero, G., Naranjo-Correa, F., & Mateos-Núñez, M. (2022). Cognitive and Emotional Development of STEM Skills in Primary School Teacher Training through Practical Work. *Education Sciences*, *12*(7), 470. doi:10.3390/educsci12070470

Martínez-Hernández, F., Mendoza-Fernández, A. J., Salmerón-Sánchez, E., Mota-Poveda, J. F., & Garzón, A. (2017). Sobre el grado de aceptación de la teoría evolutiva de los alumnos de Enseñanza Secundaria Obligatoria. *Enseñanza de las ciencias: revista de investigación y experiencias didácticas,* 1281-1288. https://www.raco.cat/index.php/Ensenanza/article/view/335260

Martín-Garin, A., Millán-García, J. M., Leon, I., Oregi, X., Estevez, J., & Marieta, C. (2021). Pedagogical Approaches for Sustainable Development in Higher Studies in Building. *INTED2021 Proceedings, 1*, 6777–83. 10.21125/inted.2021.1348

Martin-Paez, T., Aguilera, D., Perales-Palacios, F. J., & Vílchez-González, J. M. (2019). What are we talking about when we talk about STEM education? A review of literature. *Science Education*, *103*(4), 799–822. doi:10.1002ce.21522

Martín-Peciña, M., & Rams, S. (2022). Didáctica de la Teoría de la Evolución Biológica desde la perspectiva del profesorado de educación secundaria en activo y en formación inicial. In Investigación educativa ante los actuales retos migratorios (pp. 1173–1185). Dykinson.

Martín-PeciñaM. (2022). SHAPE MATTERS! Una práctica de filogenias moleculares con datos reales orientada a superar concepciones alternativas sobre la Evolución Biológica en 4° de ESO. Zenodo. doi:10.5281/zenodo.5946898

Martín-PeciñaM.RamsS. (2021). Instrumento de toma de datos del Trabajo Fin de Máster «Secuencia de Actividades para Biología Evolutiva a partir de los Resultados de una Investigación Educativa sobre la percepción didáctica de la materia en el profesorado (SABE y RIE)» del Máster Universitario en Profesorado de Enseñanza Secundaria Obligatoria y Bachillerato, Formación Profesional y Enseñanzas de Idiomas de la Universidad de Granada (España). Zenodo. doi:10.5281/zenodo.5649365

Martins, G., Carvalho, V., Afect, C., Teixeira, S. F. C. F., & Rodrigues, C. (2022). Getting Contact to Elderly Associates through the ICT: An Exploratory Study. *EAI MobiHealth 2022 - 11th EAI International Conference on Wireless Mobile Communication and Healthcare.*

Martins, G., Gomes, C., Brocardo, J., Pedroso, J., Carrillo, J., Silva, L., ... Rodrigues, S. (2017). *Perfil dos alunos à saída da Escolaridade Obrigatória* [Profile of students leaving compulsory schooling]. Ministério da Educação/Direção-Geral da Educação (DGE).

Martins, G., Carvalho, V., Afect, C., Teixeira, S. F. C. F., & Rodrigues, C. (2022). *Um Estudo Colaborativo Das Escolas Industriais e Do Seu Impacto Na Região de Braga. 9º Encontro Nacional de Engenharia e Gestão Industrial.*

Marulcu, I., & Barnett, M. (2013). Fifth graders' learning about simple machines through engineering design-based instruction using LEGO™ materials. *Research in Science Education*, *43*(5), 1825–1850. doi:10.100711165-012-9335-9

Marulcu, I., & Barnett, M. (2016). Impact of an engineering design-based curriculum compared to an inquiry-based curriculum on fifth graders' content learning of simple machines. *Research in Science & Technological Education*, *34*(1), 85–104. doi:10.1080/02635143.2015.1077327

Maskiewicz, A. C., & Lineback, J. E. (2013). Misconceptions are "so yesterday!". *CBE Life Sciences Education*, *12*(3), 352–356. doi:10.1187/cbe.13-01-0014 PMID:24006383

Mason, T. C. (1996). Integrated curricula: Potential and problems. *Journal of Teacher Education*, *47*(4), 263–270. doi:10.1177/0022487196474004

Mateos-Núñez, M., Martínez-Borreguero, G., & Naranjo-Correa, F. L. (2019a). Comparación de las emociones, actitudes y niveles de autoeficacia ante áreas STEM entre diferentes etapas educativas [Comparison of emotions, attitudes and levels of self-efficacy in STEM areas between different educational stages]. *European Journal of Education and Psychology*, *13*(1), 251–267. doi:10.30552/ejep.v13i1.292

Mateos-Núñez, M., Martínez-Borreguero, G., & Naranjo-Correa, F. L. (2019b). Análisis cognitivo y emocional hacia competencias científico-tecnológicas en el alumnado de 6º de Educación Primaria [Cognitive and emotional analysis towards scientific-technological competences in the students of 6th year of Primary Education]. In A. B. Corsini (Ed.), *Conectando pilares para la educación* (pp. 113–124). Global Knowledge Academics.

Mateos-Núñez, M., Martínez-Borreguero, G., & Naranjo-Correa, F. L. (2020). Learning science in primary education with STEM workshops: Analysis of teaching effectiveness from a cognitive and emotional perspective. *Sustainability*, *12*(8), 3095. doi:10.3390u12083095

Matthews, B., Jokela, J., Narwani, A., Räsänen, K., Pomati, F., Altermatt, F., Spaak, P., Robinson, C. T., & Vorburger, C. (2020). On biological evolution and environmental solutions. *The Science of the Total Environment, 724*, 138194. doi:10.1016/j.scitotenv.2020.138194 PMID:32251887

Mayes, R. (2019). Quantitative reasoning and its role in interdisciplinarity. In B. Doig, J. Williams, D. Swanson, R. B. Ferri, & P. Drake (Eds.), *Interdisciplinary Mathematics Education. The state of the art and beyond* (pp. 113–133). Springer Open. doi:10.1007/978-3-030-11066-6_8

Mayr, E. (1984). What is Darwinism Today? *PSA: Proceedings of the Biennial Meeting of the Philosophy of Science Association, 1984*(2), 145–156. https://www.jstor.org/stable/192502

May, R. (2004). Uses and abuses of mathematics in biology. *Science, 303*(5659), 790–793. doi:10.1126cience.1094442 PMID:14764866

McCain, K., & Kampourakis, K. (2018). Which question do polls about evolution and belief really ask, and why does it matter? *Public Understanding of Science (Bristol, England), 27*(1), 2–10. doi:10.1177/0963662516642726 PMID:27117489

McComas, W. F. (1991). Resources for Teaching Evolutionary Biology Labs: An Analysis. *The American Biology Teacher, 53*(4), 205–209. doi:10.2307/4449270

McComas, W., & Burgin, S. (2020). A Critique of "STEM" education. *Science and Education, 29*(4), 805–829. doi:10.100711191-020-00138-2

McCright, A., O'Shea, B., Sweeder, R., Urquhart, G., & Zeleke, A. (2013). Promoting interdisciplinarity through climate change education. *Nature Climate Change, 3*(8), 713–716. doi:10.1038/nclimate1844

McDonald, C. V. (2016). STEM Education: A review of the contribution of the disciplines of science, technology, engineering and mathematics. *Science Education International, 27*(4), 530–569.

McGuire, R. M., Hayashi, K. T., Yan, X., Caritá Vaz, M., Cinoğlu, D., Cowen, M. C., Martínez-Blancas, A., Sullivan, L. L., Vazquez-Morales, S., & Kandlikar, G. S. (2022). EcoEvoApps: Interactive apps for theoretical models in ecology and evolutionary biology. *Ecology and Evolution, 12*(12), e9556. doi:10.1002/ece3.9556 PMID:36479028

McHugh, L., Kelly, A. M., & Burghardt, M. D. (2018). Professional development for a middle school mathematics-infused science curriculum. *Journal of Science Teacher Education, 29*(8), 804–828. doi:10.1080/1046560X.2018.1514825

McKay, M.A., Pariseault, C.A., Whitehouse, C.R., Smith, T., and Gunberg Ross, J. (2022). The Experience of Baccalaureate Clinical Nursing Faculty Transitioning to Emergency Remote Clinical Teaching during the COVID-19 Pandemic: Lessons for the Future. *Nurse Education Today, 111*. doi:10.1016/j.nedt.2022.105309

McMillan, J. H. (2004). *Educational research: Fundamentals for the consumer* (4th ed.). Pearson Education, Inc.

Mead, R., Hejmadi, M., & Hurst, L. D. (2017). Teaching genetics prior to teaching evolution improves evolution understanding but not acceptance. *PLoS Biology, 15*(5), e2002255. doi:10.1371/journal.pbio.2002255 PMID:28542179

Medina-Talero, N. A. (2017). La enseñanza de la evolución biológica y otros conceptos asociados por medio de la estrategia didáctica "The Caminalcules". *Tecné, Episteme y Didaxis: TED. Número Extraordinario.* Memorias, Séptimo Congreso Internacional sobre Formación de Profesores de Ciencias, 12 al 14 de octubre de 2016, Bogotá.

Meisel, R. P. (2010). Teaching Tree-Thinking to Undergraduate Biology Students. *Evolution, 3*(4), 621–628. doi:10.100712052-010-0254-9 PMID:21572571

Mellado, V., Borrachero, A. B., Brígido, M., Melo, L. V., Dávila, M. A., Cañada, F., Conde, M. C., Costillo, E., Cubero, J., Esteban, R., Martínez, G., Ruiz, C., Sánchez, J., Garritz, A., Mellado, L., Vázquez, B., Jiménez, R., & Bermejo, M. L. (2014). Las Emociones en la Enseñanza de las Ciencias [Emotions in Science Education]. *Enseñanza de las Ciencias, 32*(3), 11–36. doi:10.5565/rev/ensciencias.1478

Mellizo-Soto, M., & Romero, M. (2015). Alfabetización científica. ¿Qué ha cambiado en la última década? In *Percepción Social de la Ciencia y la Tecnología* (pp. 71–98). FECYT, Fundación Española para la Ciencia y la Tecnología.

Melo, N., Telo, C., & Sarreira, P. (2022). O Projeto Plant@ESELx e a participação dos estudantes na caracterização do património vegetal da ESELx [The Project Plant@ESELx and the participation of students in the characterization of eselx's plant heritage]. *Da Investigação às Práticas. Estudos de Natureza Educacional, 12*(1), 188–208. doi:10.25757/invep.v12i1.321

Meltzer, D. E. (2002). The relationship between mathematics preparation and conceptual learning gains in physics: A possible "hidden variable" in diagnostic pretest scores. *American Journal of Physics, 70*(12), 1259–1268. doi:10.1119/1.1514215

Mengascini, A., & Menegaz, A. (2005). "El juego de las mariposas". Propuesta didáctica para el tratamiento del cambio biológico. *Revista Eureka sobre Enseñanza y Divulgación de las Ciencias, 2*(3), 403-415.

Merritt, J., Lee, M., Rillero, P., & Kinach, B. M. (2017). Problem-Based Learning in K–8 Mathematics and Science Education: A Literature Review. *The Interdisciplinary Journal of Problem-Based Learning, 11*(2). Advance online publication. doi:10.7771/1541-5015.1674

Mestrinho, N., & Cavadas, B. (2018). Innovation in teacher education: An integrative approach to teaching and learning science and mathematics. *Proceedings, 2*(21), 1343.

Métioui, A., & Trudel, L. (2021). Children and Preservice Teachers' Misconceptions and Scientifically Acceptable Conceptions about Movement, Force, and Gravity. *Modern Perspectives in Language. Literature and Education, 5*, 26–42.

Microsoft Makecode. (2022). *Light Sensor Tone control. Version 5.0.12.* https://makecode.microbit.org/v0/projects/guitar/lightsensor

Miglino, O., Rubinacci, F., Pagliarini, L., & Lund, H. H. (2004). Using artificial life to teach evolutionary biology. *Cognitive Processing, 5*(2), 123–129. doi:10.100710339-004-0009-z

Millar, V. (2020). Trends, Issues and Possibilities for an Interdisciplinary STEM Curriculum. *Science and Education, 29*(4), 929–948. doi:10.100711191-020-00144-4

Miller, J., & Knezek, G. (2013). STEAM for Student Engagement. In R. McBride, & M. Searson (Eds.), *Proceedings of SITE 2013--Society for Information Technology & Teacher Education International Conference* (pp. 3288-3298). Association for the Advancement of Computing in Education (AACE).

Miller, D. M., Khalil, K., Iskaros, O., & Van Amburgh, J. A. (2017). Professional and Pre-Professional Pharmacy Students' Perceptions of Team Based Learning (TBL) at a Private Research-Intensive University. *Currents in Pharmacy Teaching & Learning, 9*(4), 666–670. doi:10.1016/j.cptl.2017.03.001 PMID:29233441

Ministério da Educação (ME). (2018a). *Aprendizagens Essenciais. Terceiro ano: Educação Básica. Primeiro Ciclo* [Essential Learning. Third year: Basic Education. First Cycle]. Estudo do Meio. https://www.dge.mec.pt/sites/default/files/Curriculo/Aprendizagens_Essenciais/1_ciclo/3_estudo_do_meio.pdf

Ministério da Educação (ME). (2018b). *Aprendizagens Essenciais. Quarto ano: Educação Básica. Primeiro Ciclo* [Essential Learning. Fourth year: Basic Education. First Cycle]. Estudo do Meio. https://www.dge.mec.pt/sites/default/files/Curriculo/Aprendizagens_Essenciais/1_ciclo/3_estudo_do_meio.pdf

Ministério da Educação (ME). (2021). *Aprendizagens Essenciais. Quarto ano: Educação Básica. Primeiro Ciclo* [Essential Learning. Fourth year: Basic Education. First Cycle]. Matemática. https://www.dge.mec.pt/sites/default/files/Curriculo/Aprendizagens_Essenciais/1_ciclo/ae_mat_4.o_ano.pdf

Ministry of National Education (MEB). (2018a). *Ortaöğretim fizik dersi öğretim programı* [Secondary education physics curriculum]. Talim ve Terbiye Kurulu Başkanlığı. http://mufredat.meb.gov.tr/Dosyalar/201812103112910-ortaöğretim_fizik_son.pdf

Ministry of National Education (MEB). (2018b). *Ortaöğretim matematik dersi öğretim programı* [Secondary education mathematics curriculum]. Talim ve Terbiye Kurulu Başkanlığı. http://mufredat.meb.gov.tr/Dosyalar/20182110272710 1-OGM%20MATEMATİK%20PRG%2020.01.2018.pdf

Ministry of National Education. (2018a). *Fen bilimleri dersi öğretim programı (ilkokul ve ortaokul 3, 4, 5, 6, 7, ve 8. sınıflar)* [Science course curriculum (Primary and middle school 3, 4, 5, 6, 7, and 8th grades)]. Ministry of National Education.

Ministry of National Education. (2018b). *Matematik dersi öğretim programı (ilkokul ve ortaokul 3, 4, 5, 6, 7, ve 8. sınıflar)* [Mathematics course curriculum (Primary and middle school 3, 4, 5, 6, 7, and 8th grades)]. Ministry of National Education.

Miron, C., & Staicu, I. (2010). The impact of interdisciplinarity on the physics-mathematics scientific education in high schools. *Romanian Reports in Physics*, *62*(4), 907–917.

Mohr-Schroeder, M., Cavalcanti, M., & Blyman, K. (2015). STEM education: understanding the changing landscape. In A. Sahin (Ed.), *A practice-based model of STEM teaching* (pp. 3–14). Sense. doi:10.1007/978-94-6300-019-2_1

Molefe, L. (2013). *The pedagogical content knowledge of Life Sciences teachers teaching evolution for the first time* [tesis de maestría, University of the Witwatersrand]. Repositorio institucional: http://hdl.handle.net10539/13385.

Molina-Martínez, A. L. (2015). Propuesta didáctica para la enseñanza-aprendizaje de "la historia y evolución de la vida en la tierra" dirigida a estudiantes de grado octavo utilizando como estrategia el juego [tesis de mestría, Universidad Nacional de Colombia]. Repositorio institucional: https://repositorio.unal.edu.co/handle/unal/78048

Moore, J. A. (1984). Science as a way of knowing—Evolutionary biology. *American Zoologist*, *24*(2), 467–534. doi:10.1093/icb/24.2.467

Moore, R., Cotner, S., & Bates, A. (2009). The influence of religion and high school biology courses on students' knowledge of evolution when they enter college. *The Journal of Effective Teaching*, *9*(2), 4–12.

Morais, P., Ferreira, M. J., & Veloso, B. (2021). Improving Student Engagement with Project-Based Learning: A Case Study in Software Engineering. *Revista Iberoamericana de Tecnologias Del Aprendizaje*, *16*(1), 21–28. doi:10.1109/RITA.2021.3052677

Moran, L. A. (2007). *The deflated ego problem.* https://sandwalk.blogspot.com/2007/05/deflated-ego-problem.html

Moreno, V., Ferreira, M., DePasquale, J., Malicki, P., Schier, L., Morris, S., & Smith, H. (2017). Learning Experiences from an Industrially Sponsored Senior Design Capstone Program - A Student Perspective. *ASME International Mechanical Engineering Congress and Exposition Proceedings*, *5*, 1–6. doi:10.1115/IMECE2017-70916

Morin, E. (2001). L' enseignement des connaissances [The Project Plant@ESELx and the participation of students in the characterization of eselx's plant heritage]. In F. C. Gulbenkian (Ed.), *Novo conhecimento, nova aprendizagem* (pp. 25–33). Fundação Calouste Gulbenkian.

Morón-Monge, H., Morón-Monge, M. C., Wamba-Aguado, A. M., & Jiménez-Pérez, R. (2012). Una propuesta metodológica para la enseñanza de la biología y geología en la educación secundaria. *Revista de Educación en Biología, 2*, 58–68.

Morrison, J. (2006). *TIES STEM education monograph series, Attributes of STEM education.* TIES.

Müller, F. A., & Wulf, T. (2022). Blended Learning Environments and Learning Outcomes: The Mediating Role of Flow Experience. *International Journal of Management Education, 20*(3), 100694. doi:10.1016/j.ijme.2022.100694

Mumthas, N. S., & Abdulla, S. U. (2019). Substandard Performance in Mathematical Problem Solving in Physics among Higher Secondary School Students in Kerala—An Investigation on Teacher Perceptions and Student Difficulties. *Online Submission, 7*(1), 31–39. doi:10.15415/iie.2019.71005

Murphy, C., & Beggs, J. (2003). Children's perceptions of school science. *The School Science Review, 84*(308), 109–116.

Nabie, M. J., Akayuure, P., Ibrahim-Bariham, U. A., & Sofo, S. (2018). Trigonometric Concepts: Pre-Service Teachers' Perceptions and Knowledge. *Journal on Mathematics Education, 9*(2), 169–182. doi:10.22342/jme.9.2.5261.169-182

Nadelson, L. S., Callahan, J., Pyke, P., Hay, A., Dance, M., & Pfiester, J. (2013). Teacher STEM Perception and Preparation: Inquiry-Based STEM Professional Development for Elementary Teachers. *The Journal of Educational Research, 106*(2), 157–168. doi:10.1080/00220671.2012.667014

Nadelson, L., Culp, R., Bunn, S., Burkhart, R., Shetlar, R., Nixon, K., & Waldron, J. (2009). Teaching evolution concepts to early elementary school students. *Evolution (New York), 4*(3), 267–274. doi:10.100712052-009-0148-x

Naraine, B., Dotseth, K. A., Kim, G., & Moody, M. (1993). If Pythagoras had a geoboard. *Mathematics Teacher, 86*(2), 137–139. doi:10.5951/MT.86.2.0137

Nasr El-Din, W.A., Atwa, H., Potu, B.K., Deifalla, A., and Fadel, R.A. (2022). Checklist-Based Active Learning in Anatomy Demonstration Sessions during the COVID-19 Pandemic: Perception of Medical Students. *Morphologie : Bulletin de l'Association Des Anatomistes*. doi:10.1016/j.morpho.2022.10.001

National Academy of Sciences. (2014). *STEM integration in K-12 education: Status, prospects, and an agenda for research.* National Academies Press.

National Research Council. (2012). *A framework for K-12 science education: Practices, crosscutting concepts, and core ideas.* The National Academies Press.

National Research Council. (2012). *A Framework for K-12 Science Education: Practices, crosscutting concepts, and core ideas.* The National Academies Press. https://www.nap.edu/catalog/13165/a-framework-for-k-12-science-education-practices-crosscutting-concepts

National Science Board. (2004). *Science and Engineering Indicators 2004 (NSB 04-01).* National Science Foundation.

Nehm, R. H., & Schonfeld, I. S. (2007). Does Increasing Biology Teacher Knowledge of Evolution and the Nature of Science Lead to Greater Preference for the Teaching of Evolution in Schools? *Journal of Science Teacher Education, 18*(5), 699–723. doi:10.100710972-007-9062-7

Neira, J. C. R. (2021). La experimentación en ciencias naturales como estrategia de alfabetización científica [Experimentation in natural sciences as a strategy for scientific literacy]. *UCMaule*, (60), 102–116. doi:10.29035/ucmaule.60.102

Nelson, C. E., Scharmann, L. C., Beard, J., & Flammer, L. I. (2019). The nature of science as a foundation for fostering a better understanding of evolution. *Evolution (New York), 12*(1), 6. doi:10.118612052-019-0100-7

Nesse, R. M., & Natterson-Horowitz, B. (2019). Evolutionary Medicine – A Great Way to Teach Biology. *The American Biology Teacher, 81*(8), 533. doi:10.1525/abt.2019.81.8.533

Newman, D. (2003). On borders and power: A theoretical framework. *Journal of Borderlands Studies, 18*(1), 13–25. doi:10.1080/08865655.2003.9695598

Next Generation Science Standards (NGSS). (2013). http://www.nextgenscience.org

Next Generation Science Standards Lead States. (2013). *Next generation science standards: For states, by states*. National Academies Press.

Ng, S. B. (2019). Exploring STEM competences for the 21st century. *Current and Critical Issues in Curriculum, Learning and Assessment, 30*. https://unesdoc.unesco.org/ark:/48223/pf0000368485

Ng, B. S. (2019). *Exploring STEM competences for the 21st century. Current and Critical Issues in Curriculum, Learning and Assessment, 30.*

NGSS Lead States. (2013). *Next Generation Science Standards: For States, By States*. The National Academies Press.

Nilsson, P. (2008). Teaching for understanding: The complex nature of pedagogical content knowledge in preservice education. *International Journal of Science Education, 30*(10), 1281–1299. doi:10.1080/09500690802186993

Nogueira, J. R., Alves, R., & Marques, P. C. (2019). Computational Programming as a Tool in the Teaching of Electromagnetism in Engineering Courses: Improving the Notion of Field. *Education Sciences, 9*(1), 64. doi:10.3390/educsci9010064

Norman, E. (1993). Science for design. *Physics Education, 28*(5), 301–306. doi:10.1088/0031-9120/28/5/010

Norman, E. (1998). The Nature of Technology for Design. *International Journal of Technology and Design Education, 8*(1), 67–87. doi:10.1023/A:1008827820764

Norton, S. J. (2007). The use of design practice to teach mathematics and science. *International Journal of Technology and Design Education, 18*(1), 19–44. doi:10.100710798-006-9019-8

Nurse-Clarke, N., & Joseph, M. (2022). An Exploration of Technology Acceptance among Nursing Faculty Teaching Online for the First Time at the Onset of the COVID-19 Pandemic. *Journal of Professional Nursing, 41*, 8–18. . doi:10.1016/j.profnurs.2022.04.002

O'Malley, M. A. (2015). Endosymbiosis and its implications for evolutionary theory. *Proceedings of the National Academy of Sciences of the United States of America, 112*(33), 10270–10277. doi:10.1073/pnas.1421389112 PMID:25883268

Ocaña, G., Romero, I.M. & Gil, F. (2017). Educación STEM para integrar conocimientos científicos en la asignatura "tecnología industrial" de bachillerato [STEM education to integrate scientific knowledge in the subject "industrial technology" of baccalaureate]. *Enseñanza de las ciencias: revista de investigación y experiencias didácticas*, (Extra), 5327-5336.

OECD. (2018). *Curriculum Flexibility and Autonomy in Portugal: An OECD Review*. http://www.oecd.org/ education/2030/Curriculum-Flexibility-and-Autonomy-in-Portugal-an-OECD-Review.pdf

OECD. (2018). The Future of Education and Skills. *Education*.

Offer, J., & Vasquez-Mireles, S. (2009). Mix it up: Teachers' beliefs on mixing mathematics and science. *School Science and Mathematics*, *109*(3), 146–152. doi:10.1111/j.1949-8594.2009.tb17950.x

Ogborn, J. (2010). Curriculum development as practical activity. En K. Kortland & K. Klaassen (Eds.), Designing Theory-Based Teaching-Learning Sequences for Science Education (pp. 69-78). Utrecht: CDBeta Press – Freudenthal Institute for science and mathematics education (FIsme), Utrecht University – FIsme series on Research in Science Education.

Oltvai, Z., & Barabási, A. (2002). Life's complexity pyramid. *Science*, *298*(5594), 763–764. doi:10.1126cience.1078563 PMID:12399572

Opitz, S. T. (2016). *Students' Progressing Understanding Of The Energy Concept. An Analysis Of Learning In Biological And Cross- Disciplinary Contexts* [tesis de doctorado, Christian-Albrechts University At Kiel]. Repositorio insitucional https://nbn-resolving.org/urn:nbn:de:gbv:8-diss-190054

Orton, T., & Roper, T. (2000). Science and Mathematics: A relationship in need of counselling? *Studies in Science Education*, *35*(1), 123–153. doi:10.1080/03057260008560157

Osborne, J. (2014). Teaching scientific practices: Meeting the challenge of change. *Journal of Science Teacher Education*, *25*(2), 177–196. doi:10.100710972-014-9384-1

Osborne, J., Simon, S., & Collins, S. (2003). Attitudes towards science: A review of the literature and its implications. *International Journal of Science Education*, *25*(9), 1049–1079. doi:10.1080/0950069032000032199

Othman, O., Iksan, Z. H., & Yasin, R. M. (2022). Creative Teaching STEM Module: High School Students' Perception. *Creative Teaching STEM Module: High School Students' Perception*, *11*(4), 2127–2137.

Ottevanger, W. (2001). *Teacher support materials as a catalyst for science curriculum implementation Namibia* [tesis de doctorado, University of Twente]. Repositorio institucional https://research.utwente.nl/en/publications/teacher-support-materials-as-a-catalyst-for-science-curriculum-im

Pang, J., & Good, R. (2000). A Review of the Integration of Science and Mathematics: Implications for Further Research. *School Science and Mathematics*, *100*(2), 73–82. doi:10.1111/j.1949-8594.2000.tb17239.x

Park, M., Dimitrov, D. M., Patterson, L. G., & Park, D. (2017). Early childhood teachers' beliefs about readiness for teaching science, technology, engineering, and mathematics. *Journal of Early Childhood Research*, *15*(3), 275–291. doi:10.1177/1476718X15614040

Park, W., Kim, D., & Kang, D. Y. (2021). Research trends in science and mathematics education in South Korea 2014-2018: A cross-disciplinary analysis of publications in selected local journals. *Asia-Pasific Science Education*, *7*(2), 280–308. doi:10.1163/23641177-bja10029

Paul, A. (2020). Memes as means for understanding interpersonal communication: A formative assignment. *Communication Teacher*, *34*(4), 346–354. doi:10.1080/17404622.2019.1681484

Pawlicka, A., Tomaszewska, R., Krause, E., Jaroszewska-Choraś, D., Pawlicki, M., & Choraś, M. (2022). Has the pandemic made us more digitally literate? *Journal of Ambient Intelligence and Humanized Computing*, 1–11. doi:10.100712652-022-04371-1 PMID:35971559

Pazza, R., Penteado, P. R., & Kavalco, K. F. (2010). Misconceptions About Evolution in Brazilian Freshmen Students. *Evolution (New York)*, *3*(1), 107–113. doi:10.100712052-009-0187-3

Pedaste, M., Mäeots, M., Siiman, L. A., de Jong, T., van Riesen, S. A. N., Kamp, E. T., Manoli, C. C., Zacharia, Z. C., & Tsourlidaki, E. (2015). Phases of inquiry-based learning: Definitions and the inquiry cycle. *Educational Research Review*, *14*, 47–61. doi:10.1016/j.edurev.2015.02.003

Pekrun, R. (2006). The control-value theory of achievement emotions: Assumptions, corollaries, and implications for educational research and practice. *Educational Psychology Review*, *18*(4), 315–341. doi:10.100710648-006-9029-9

Pektaş, M., Sadak, M., İncikabı, L., & Kayabaşı, B. (2022). Examining the mathematics content of science textbooks: Connected textbooks. *Mehmet Akif Ersoy University Journal of Education Faculty*, *61*, 98–121.

Peñaloza, G. (2016). La Importancia De La Demarcación: Una Revisión Del Concepto De "Aceptación De La Evolución". *Bio-grafía Escritos sobre la biología y su enseñanza, 9*(17), 17-28. . doi:10.17227/20271034.vol.9num.17bio-grafia17.28

Pennell, M. W., Harmon, L. J., & Uyeda, J. C. (2014). Is there room for punctuated equilibrium in macroevolution? *Trends in Ecology & Evolution*, *29*(1), 23–32. doi:10.1016/j.tree.2013.07.004 PMID:23978567

Perales-Palacios, F., & Aguilera, D. (2020). Ciencia-Tecnología-Sociedad vs. STEM: ¿evolución, revolución o disyunción? [Science-Technology-Society vs. STEM: evolution, revolution or disjunction?]. *Ápice. Revista de Educación Científica*, *4*(1), 1–15. doi:10.17979/arec.2020.4.1.5826

Pérez, A., & de Pro, A. (2013). Estudio demoscópico de lo que sienten y piensan los niños y adolescentes sobre la enseñanza formal de las ciencias [A demographic study of what children and adolescents feel and think about formal science education]. In V. Mellado, L. J. Blanco, A. B. Borrachero, & J. A. Cárdenas (Eds.), *Las Emociones en la Enseñanza y el Aprendizaje de las Ciencias y las Matemáticas* (pp. 495–520). DEPROFE.

Perignat, E., & Katz-Buonincontro, J. (2019). STEAM in practice and research: An integrative literature review. *Thinking Skills and Creativity*, *31*, 31–43. doi:10.1016/j.tsc.2018.10.002

Pessoa da Silva, K. A., & Dalto, J. O. (2017). Uma estratégia de Avaliação de Atividades de Modelagem Matemática [A strategy of Evaluating Mathematical Modeling Activities]. *Revista Electrónica de Investigación en Educación en Ciencias*, *12*(2), 1–14. doi:10.54343/reiec.v12i2.228

Pierce, A. A., & de Man, T. (2019). Antibiotic resistant pathogen outbreak investigation: An interdisciplinary module to teach fundamentals of evolutionary biology. *Journal of Biological Education*, *53*(2), 150–156. doi:10.1080/00219266.2018.1447003 PMID:31073246

Pietrocola, M. (2008). Mathematics as structural language of physical thought. *Connecting Research in Physics Education with Teacher Education*, *18*(2), 193–205.

Pigliucci, M., & Muller, G. B. (Eds.). (2010). *Evolution–the extended synthesis*. MIT Press. doi:10.7551/mitpress/9780262513678.001.0001

Pimthong, P., & Williams, J. (2018). Preservice teachers' understanding of STEM education. *Kasetsart Journal of Social Sciences*, *1*, 1–7.

Pinxten, R., Vandervieren, E., & Janssenswillen, P. (2020). Does integrating natural selection throughout upper secondary biology education result in a better understanding? A cross-national comparison between Flanders, Belgium and the Netherlands. *International Journal of Science Education*, *42*(10), 1609–1634. doi:10.1080/09500693.2020.1773005

Plomp, T. (2013). Educational design research: An introduction. In T. Plomp & N. Nieveen (Eds.), Educational design research - part a: an introduction (pp. 10 – 51). Netherlands Institute for Curriculum Development (SLO).

Pobiner, B. (2012). Use human examples to teach evolution. *The American Biology Teacher*, *74*(2), 71–72. doi:10.1525/abt.2012.74.2.2

Pobiner, B., Beardsley, P. M., Bertka, C. M., & Watson, W. A. (2018). Using human case studies to teach evolution in high school A.P. biology classrooms. *Evolution (New York)*, *11*(1), 3. doi:10.118612052-018-0077-7

Polverini, G. (2021). *Exploring interdisciplinarity between physics and mathematics: the design of a linguistic and an epistemological tool for analysing texts about the parabolic motion* [Unpublished master's thesis]. Alma Mater Studiorum-University of Bologna, Italy.

Ponte. (2020). A didática da matemática e o trabalho do professor [The didactics of mathematics and the work of the teacher]. *RBECM, Passo Fundo, 3*(3), 809-826.

Ponte, J. P., Boavida, A. M., Graça, M., & Abrantes, P. (1997). *Didáctica da Matemática* [Mathematics didactics]. Editorial do Ministério da Educação.

Ponte, J. P., Branco, N., & Matos, A. (2009). *Álgebra no Ensino Básico* [Algebra in elementary school]. Direção Geral da Inovação e do Desenvolvimento Curricular.

Pospiech, G. (2015). Interplay of mathematics and physics in physics education. *International Symposium of Mathematics and its Connections to the Arts and Sciences* (s. 36-43). University of Education Schwäbisch Gmünd.

Pozo, J. I., & Gómez- Crespo, M.A. (1998) *Aprender y enseñar ciencia. Del conocimiento cotidiano al conocimiento científico* [Learn and teach science. From everyday knowledge to scientific knowledge]. Ediciones Morata.

Presmeg, N. C. (1986). Visualisation and mathematical giftedness. *Educational Studies in Mathematics, 17*(3), 297–311. doi:10.1007/BF00305075

Priestley, M., Philippou, S., Alvunger, D., & Soini, T. (2021). Curriculum making: a conceptual framing. In M. Priestley, D. Alvunger, S. Philippou, & T. Soini (Eds.), *Curriculum making in Europe: policy and practice within and across diverse contexts* (pp. 1–28). Emerald Publishing Limited.

Prince, M. (2004). Does Active Learning Work? A Review of the Research. *Journal of Engineering Education, 93*(3), 223–231. doi:10.1002/j.2168-9830.2004.tb00809.x

Prinou, L., Halkia, L., & Skordoulis, C. (2011). The inability of Primary School to introduce children to the Theory of Biological Evolution. *Evolution (New York), 4*(2), 275–285. doi:10.100712052-011-0323-8

Prinou, L., Halkia, L., & Skordoulis, G. (2008). What conceptions do Greek school students form about biological evolution? *Evolution (New York), 1*(3), 312–317. doi:10.100712052-008-0051-x

Putnick, D. L., & Bornstein, M. H. (2016). Measurement invariance conventions and reporting: The state of the art and future directions for psychological research. *Developmental Review, 41*, 71–90. doi:10.1016/j.dr.2016.06.004 PMID:27942093

Pyke, C. L., & Lynch, S. (2005). Mathematics and science teachers' preparation for national board of teaching standards certification. *School Science and Mathematics, 105*(5), 25–35. doi:10.1111/j.1949-8594.2005.tb18033.x

Quessada, M. P., Clément, P., Oerke, B., & Valente, A. (2008). Human evolution in science textbooks: A survey in eighteen countries. *Science Education International, 19*(2), 147–162.

Quigley, C., & Herro, D. (2016). "Finding the Joy in the Unknown": Implementation of STEAM Teaching Practices in Middle School Science and Math Classrooms. *Journal of Science Education and Technology, 25*(3), 410–426. doi:10.100710956-016-9602-z

Quigley, C., Herro, D., & Jamil, F. (2017). Developing a conceptual model of STEAM teaching practices. *School Science and Mathematics, 117*(1–2), 1–12. doi:10.1111sm.12201

Quintana-Pinto, A. L. (2018). Modelo de argumentación de toulmin: Una propuesta didáctica para el aprendizaje de la herencia biológica. *Paradigma, 39*(1), 11-22. http://revistaparadigma.online/ojs/index.php/paradigma/article/view/639/636

Radloff, J., & Guzey, S. (2017). Investigating changes in preservice teachers' conceptions of STEM education following video analysis and reflection. *School Science and Mathematics*, *117*(3-4), 158–167. doi:10.1111sm.12218

Ramírez-Olaya, L. C. J. (2014). Juego de Simulación para enseñar evolución por selección natural a estudiantes de noveno grado. *Revista Perspectivas Educativas*, *7*(1), 115–133.

Ramírez-Olaya, L. C. J. (2017). Juego De Simulación Como Propuesta Didáctica Para Abordar El Concepto Evolución Biológica. *Bio-grafía*, *83*, 92. doi:10.17227/20271034.vol.0num.0bio-grafia83.92

Rashid, S., Baha, A., & Abidin, D. (2021). Students' Perception on Learning Unit Conversion Through a Smart Conversion Card Tool. *Advances in Social Science, Education and Humanities Research*, *544*, 448–451. doi:10.2991/assehr.k.210424.087

Ratan, R., Ucha, C., Lei, Y., Lim, C., Triwibowo, W., Yelon, S., ... Chen, V. H. H. (2022). How do social presence and active learning in synchronous and asynchronous online classes relate to students' perceived course gains? *Computers & Education*, *191*, 104621. doi:10.1016/j.compedu.2022.104621

Ratcliff, W. C., Raney, A., Westreich, S., & Cotner, S. (2014). A Novel Laboratory Activity for Teaching about the Evolution of Multicellularity. *The American Biology Teacher*, *76*(2), 81–87. doi:10.1525/abt.2014.76.2.3

Raynaut, C. (2018). Paradoxos e ambiguidades na ideia de interdisciplinaridade. *Desenvolvimento e Meio Ambiente*, *47*, 13–48. doi:10.5380/dma.v47i0.62428

Redish, E. F. (2017). Analysing the competency of mathematical modelling in physics. In Key competences in physics teaching and learning (pp. 25-40). Springer. doi:10.1007/978-3-319-44887-9_3

Redish, E. F., & Kuo, E. (2015). Language of Physics, Language of Math: Disciplinary Culture and Dynamic Epistemology. *Science and education*, *24*(5-6), 561–590. doi:10.100711191-015-9749-7

Rennie, L., Venville, G., & Wallace, J. (2012). *Integrating science, technology, engineering, and mathematics: Issues, reflections, and ways forward*. Routledge. doi:10.4324/9780203803899

Retnawati, H., Arlinwibowo, J., Wulandari, N. F., & Pradani, R. G. (2018). Teachers' difficulties and strategies in physics teaching and learning that applying mathematics. *Journal of Baltic Science Education*, *17*(1), 120–135. doi:10.33225/jbse/18.17.120

Reynante, B., Selbach-Allen, M., & Pimentel, D. (2020). Exploring the promises and perils of integrated STEM through disciplinary practices and epistemologies. *Science and Education*, *29*(4), 785–903. doi:10.100711191-020-00121-x

Rézio, S., Cavadas, B., Nogueira, J. R., & Branco, N. (2020). *Bola Saltitona*. GO-LAB. https://www.golabz.eu/ils/bola-saltitona

Ricaurte, M., & Viloria, A. (2020). Project-based learning as a strategy for multi-level training applied to undergraduate engineering students. *Education for Chemical Engineers*, *33*, 102–111. doi:10.1016/j.ece.2020.09.001

Richard, M., Coley, J. D., & Tanner, K. D. (2017). Investigating Undergraduate Students' Use of Intuitive Reasoning and Evolutionary Knowledge in Explanations of Antibiotic Resistance. *CBE Life Sciences Education*, *16*(3), ar55. doi:10.1187/cbe.16-11-0317 PMID:28821540

Riegle-Crumb, C., Morton, K., Moore, C., Chimonidou, A., Labrake, C., & Kopp, S. (2015). Do Inquiring Minds Have Positive Attitudes? The Science Education of Pre-service Elementary Teachers. *Science Education*, *99*(5), 819–836. doi:10.1002ce.21177 PMID:27667862

Ring, E. A., Dare, E. A., Crotty, E. A., & Roehrig, G. H. (2017). The evolution of teacher conceptions of STEM education throughout an intensive professional development experience. *Journal of Science Teacher Education, 28*(5), 444–467. doi:10.1080/1046560X.2017.1356671

Ríordáin, M. N., Johnston, J., & Walshe, G. (2016). Making mathematics and science integration happen: Key aspects of practice. *International Journal of Mathematical Education in Science and Technology, 47*(2), 233–255. doi:10.1080/0020739X.2015.1078001

Robles-Garrote, P., & Rojas, M. C. (2015). La validación por juicio de expertos: Dos investigaciones cualitativas en Lingüística aplicada [Validation by expert judgment: two qualitative investigations in applied linguistics]. *Revista Nebrija de Lingüística Aplicada a La Enseñanza de Las Lenguas, 18*, 1–16.

Rocha, J., Pessoa, P., Gomes, J., Sá-Pinto, X., & Lopes, J. (2020). BiblioLab Project: Teachers, Parents and Students' Perspectives About the Usability and Usefulness of an Educational Distance Learning Platform. In A. Reis, J. Barroso, J. B. Lopes, T. Mikropoulos, & C. W. Fan (Eds.), *Technology and Innovation in Learning, Teaching and Education. TECH-EDU 2020. Communications in Computer and Information Science* (Vol. 1384, pp. 90–110). Springer.

Rodrigues, M., Caseiro, A., Silva, M. J., Loução, A., & Monteiro, S. (2019). Integração curricular de matemática e ciências naturais na abordagem de problemas de saúde ambiental na formação para a docência [Curricular integration of mathematics and natural sciences in the approach of environmental health problems in teaching training]. In M. J. Silva & R. Brito (Coords.), *Utilização pedagógica de sensores eletrónicos para a participação na saúde ambiental das escolas* (pp. 30-50). Centro Interdisciplinar de Estudos Educacionais.

Rodrigues, N., Teixeira, I., Carvalho, V., Abreu, I., Gonçalves, I., Graçoeiro, D., ... Teixeira, S. (2022). Summer Grants 'Verão Com Ciência', from Foundation for Science and Technology, in Portugal: Experience and Achievements. *Proceedings of the ASME 2022 - International Mechanical Engineering Congress and Exposition.*

Roehrig, G., Moore, T., Wang, H.-H., & Park, M. (2012). Is Adding the E Enough? In-vestigating the Impact of K-12 Engineering Standards on the Implementation of STEM Integration. *School Science and Mathematics, 112*(1), 31–44. doi:10.1111/j.1949-8594.2011.00112.x

Roldão, M. C. (2021). Investigação como instrumento da formação profissional de docentes. [Research as an instrument of professional training of teachers]. *Revista Formação Docente – Revista Brasileira de Pesquisa sobre Formação de Professores, 13*(28), 79-90. doi:10.31639/rbpfp.v13i28.545

Roldão, M. C. (2017). Conhecimento, didáctica e compromisso: O triângulo virtuoso de uma profissionalidade em risco [Knowledge, didactics and commitment: the virtuous triangle of a professionality at risk]. *Cadernos de Pesquisa, 47*(166), 1134–1149. doi:10.1590/198053144367

Romero Ariza, M., Armenteros, A. Q., & Castro, A. E. (2021). Promoting critical thinking through mathematics and science teacher education: The case of argumentation and graphs interpretation about climate change. *European Journal of Teacher Education*, 1–19. doi:10.1080/02619768.2021.1961736

Romero-Ariza, M., Gallego, A. M. A., Quesada, A., & Martín-Peciña, M. (2022). MOST: alfabetización científica y educación para la sostenibilidad a través de Proyectos de Escuela Abierta. In A. Benarroch (Ed.), 30 Encuentros Internacionales de Didáctica de las Ciencias Experimentales. La enseñanza de las ciencias en un entorno intercultural (pp. 311-316). Universidad de Granada.

Ronfeldt, M., Farmer, S., McQueen, K., & Grissom, J. (2015). Teacher collaboration in instructional teams and student achievement. *American Educational Research Journal, 52*(3), 475–514. doi:10.3102/0002831215585562

Rosa, S. M., & Tricarico, H. (2016). Uso de árboles evolutivos para contextualizar científicamente la enseñanza de la biodiversidad vegetal. *Revista Eureka sobre Enseñanza y Divulgación de las Ciencias, 13*(2), 384-394. . doi:10.25267/Rev_Eureka_ensen_divulg_cienc.2016.v13.i2.11

Rosales-López, C. (2015). Evolución y desarrollo actual de los Temas Transversales: Posibilidades y límites. *Foro de Educación, 13*(18), 143–160. doi:10.14516/fde.2015.013.018.008

Rosa, S. S., Coutinho, C. P., & Flores, M. A. (2016). Online peer assessment: Method and digital technologies. *Procedia: Social and Behavioral Sciences, 228*, 418–423.

Rosenthal, D. B. (1985). Evolution in high school biology textbooks: 1963-1983. *Science Education, 69*(5), 637–648. doi:10.1002ce.3730690506

Rossi-Linnemann, C., & De Martini, G. (Eds.). (2019). *Art in Science Museums: Towards a Post-Disciplinary Approach*. Routledge. doi:10.4324/9780429491597

Roth, W. (2008). The nature of scientific conceptions: A discursive psychological perspective. *Educational Research Review, 3*(1), 30–50. doi:10.1016/j.edurev.2007.10.002

Rubinoff, D., & Holland, B. S. (2005). Between two extremes: Mitochondrial DNA is neither the panacea nor the nemesis of phylogenetic and taxonomic inference. *Systematic Biology, 54*(6), 952–961. doi:10.1080/10635150500234674 PMID:16385775

Rule, A., & Hallagan, J. (2007). *Algebra Rules Object Boxes as an Authentic Assessment Task of Preservice Elementary Teacher Learning in a Mathematics Methods Course*. Paper presented at the meeting of Annual Conference of the New York State Association of Teacher Educators, Saratoga Springs, NY, United States.

Russell, J., & Zembylas, M. (2007). Arts integration in the curriculum: A review of research and implications for teaching and learning. In L. Bresler (Ed.), *International Handbook of Research in Arts Education* (pp. 287–312). Springer Netherlands. doi:10.1007/978-1-4020-3052-9_18

Rutledge, M. L., & Warden, M. A. (1999). The Development and Validation of the Measure of Acceptance of the Theory of Evolution Instrument. *School Science and Mathematics, 99*(1), 13–18. doi:10.1111/j.1949-8594.1999.tb17441.x

Sackes, M., Flevares, L. M., Gonya, J., & Trundle, K. C. (2012). Preservice early childhood teachers' sense of efficacy for integrating mathematics and science: Impact of a methods course. *Journal of Early Childhood Teacher Education, 33*(4), 349–364. doi:10.1080/10901027.2012.732666

Sacristán, J. (2000). O currículo: Uma reflexão sobre a prática [The curriculum: A reflection on the practice] (3rd ed.). Artmed. (Original work in Spanish published in 1991)

Sage, J., & Steeg, T. (1993). Linking the learning of mathematics, science and technology within key stage 4 of the national curriculum. *International Conference on Design and Technology Educational Research and Curriculum Development, Loughborough University of Technology*, 58–64. Retrieved July 1, 2022, from: https://dspace.lboro.ac.uk/dspace-jspui/bitstream/2134/1579/3/sage93.pdf

Samson, G. (2014). From writing to doing: The challenges of implementing integration (and interdisciplinarity) in the teaching of mathematics, sciences, and technology. *Canadian Journal of Science. Mathematics and Technology Education, 14*(4), 346–358. doi:10.1080/14926156.2014.964883

Sanabria-Totaitive, I. A., & Arango-Martínez, A. V. (2021). La analogía como estrategia en la enseñanza de la evolución biológica. *Praxis (Bern), 17*(1), 11–26. doi:10.21676/23897856.3312

Sánchez, J. Conde, M., & Zapata, V. (2017). Concepciones alternativas sobre evolución. Un estudio en futuros maestros. *Enseñanza de las Ciencias, Número Extraordinario*, 2219-2224.

Sanders, M. (2009). STEM, STEM education, STEM mania. *Technology Teacher*, *68*(4), 20–26.

Sanders, M. (2009). STEM, STEM Education, STEMAnia. *Technology Teacher*, *68*(4), 20–27.

Sanders, M., & Makotsa, D. (2016). The possible influence of curriculum statements and textbooks on misconceptions: The case of evolution. *Education as Change*, *20*(1), 1–23. doi:10.17159/1947-9417/2015/555

Sanders, M., & Ngxola, N. (2009). Addressing teachers' concerns about teaching evolution. *Journal of Biological Education*, *43*(3), 121–128. doi:10.1080/00219266.2009.9656166

Sandvik, H. (2008). Tree thinking cannot taken for granted: challenges for teaching phylogenetics. *Theory in biosciences = Theorie in den Biowissenschaften, 127*(1), 45–51. . doi:10.1007/s12064-008-0022-3

Santaolalla, E., Urosa, B., Martín, O., Verde, A., & Díaz, T. (2020). Interdisciplinarity in teacher education: Evaluation of the effectiveness of an educational innovation project. *Sustainability*, *12*(17), 6748. doi:10.3390u12176748

Sá-Pinto, X., Pinto, A., Ribeiro, J., Sarmento, I., Pessoa, P., Rodrigues, L. R., Vázquez-Ben, L., Mavrikaki, E., & Bernardino Lopes, J. (2021). Following Darwin's footsteps: Evaluating the impact of an activity designed for elementary school students to link historically important evolution key concepts on their understanding of natural selection. *Ecology and Evolution*, *11*(18), 12236–12250. doi:10.1002/ece3.7849 PMID:34594496

Satchwell, R., & Loepp, F. (2002). Designing and Implementing an Integrated Mathematics, Science, and Technology Curriculum for the Middle School. *Journal of Industrial Teacher Education*, *39*(3), 41–66.

Sattarova, U., Groot, W., & Arsenijevic, J. (2021). Student and Tutor Satisfaction with Problem-Based Learning in Azerbaijan. *Education in Science*, *11*, 288.

Schoenfeld, A. H. (2004). Multiple learning communities: Students, teachers, instructional designers, and researchers. *Journal of Curriculum Studies*, *36*(2), 237–255. doi:10.1080/0022027032000145561

Schramm, T., Jose, A., & Schmiemann, P. (2021). Seeing the Woods for the Trees Again: Analyzing Evolutionary Diagrams in German and US University-Level Textbooks. *Education in Science*, *11*(8), 367. doi:10.3390/educsci11080367

Scientific Advisory Group for Emergencies. (2022). *Coronavirus (COVID-19)*. Retrieved from gov.uk: https://www.gov.uk/government/organisations/scientific-advisory-group-for-emergencies

Seligmann, T. (2014). Learning museum. *Journal of Museum Education*, *39*(1), 42–53. doi:10.1080/10598650.2014.11510794

Shahali, E., Halim, L., Rasul, M., Osman, K., & Zulkifeli, M. (2017). STEM learning through engineering design: Impact on middle secondary students' interest towards STEM. *Eurasia Journal of Mathematics, Science and Technology Education*, *13*(5), 1189–1211. doi:10.12973/eurasia.2017.00667a

Shakenova, L. (2017). The theoretical framework of teacher collaboration. *Khazar Journal of Humanities and Social Sciences*, *20*(2), 34-48. http://jhss-khazar.org/wp-content/uploads/2016/11/2.Lyaila.20.02.2017.2.pdf

Sharkawy, A., Barlex, D., Welch, M., McDuff, J., & Craig, N. (2009). Adapting a Curriculum Unit to Facilitate Interaction Between Technology, Mathematics and Science in the Elementary Classroom: Identifying Relevant Criteria. *Design and Technology Education: An International Journal*, *14*(1), 7–20.

Sharma, A., Janke, K. K., Larson, A., & Peter, W. S. (2017). Understanding the early effects of team-based learning on student accountability and engagement using a three session TBL pilot. *Currents in Pharmacy Teaching & Learning, 9*(5), 802–807. doi:10.1016/j.cptl.2017.05.024 PMID:29233307

Sharot, T., & Sunstein, C. R. (2020). How people decide what they want to know. *Nature Human Behaviour, 4*(1), 14–19. doi:10.103841562-019-0793-1 PMID:31932690

Shchukina, I., Kudinova, V., Kudinova, O., Kondratenko, N., & Shashkova, S. (2022). Using internet memes to develop students' motivation and cross-cultural competence at the lessons of English. *INTED2022 Proceedings*, 7364-7369.

Shifman, L. (2014). *Memes in digital culture*. MIT Press.

Shogren, E. H., Anciães, M., Barske, J., Cestari, C., DuVal, E. H., Gaiotti, M. G., Johnson, E. I., Kimball, R. T., Marini, M. A., Ryder, T. B., Scholer, M. N., Ungvári, J., White, S. A., & Boyle, W. A. (2022). Dancing drives evolution of sexual size dimorphism in manakins. *Proceedings of the Royal Society B, 289*(1974), 20212540. 10.1098/rspb.2021.2540

Shulman, L. S. (1987). Knowledge and training: Foundations of the new reform. *Harvard Educational Review, 57*(1), 1–22. doi:10.17763/haer.57.1.j463w79r56455411

Siagian, M. V., Saragih, S., & Sinaga, B. (2019). Development of Learning Materials Oriented on Problem-Based Learning Model to Improve Students' Mathematical Problem Solving Ability and Metacognition Ability. *International Electronic Journal of Mathematics Education, 14*(2), 331–340.

Sickel, A. J., & Friedrichsen, P. (2013). Examining the evolution education literature with a focus on teachers: Major findings, goals for teacher preparation, and directions for future research. *Evolution (New York), 6*(1), 23. doi:10.1186/1936-6434-6-23

Sidawi, M. (2009). Teaching science through designing technology. *International Journal of Technology and Design Education, 19*(3), 269–287. doi:10.100710798-007-9045-1

Siew, N. M., Amir, N., & Chong, C. L. (2015). The perceptions of pre-service and in-service teachers regarding a project-based STEM approach to teaching science. *SpringerPlus, 4*(1), 8. doi:10.1186/2193-1801-4-8 PMID:25674494

Silva, M. J., Almeida, A., Valente, B., Rodrigues, M., & Manteigas, V. (2017). Sensing locally in the global environment: Using sensors in teachers' education. In *Proceedings of IADIS International Conference Educational Technologies 2017 - ICEduTech: IADIS International Conference Educational Technologies 2017 - ICEduTech* (pp. 166-170). IADIS.

Silva, M. J., Ferreira, E., Souza, A., & Alves, A. R. (2018). Eco-Sensors4Health Toolkit: Scaffolding children participation in schools' environmental health. In *Proceedings of Global Health 2018, The Seventh International Conference on Global Health Challenges.: The Seventh International Conference on Global Health Challenges* (pp. 47-52). https://thinkmind.org/index.php?view=article&articleid=global_health_2018_4_10_70026

Silva, M. J. (2020). Children using electronic sensors to create and use knowledge on environmental health. *First Monday, 25*(3). Advance online publication. doi:10.5210/fm.v25i3.9646

Simarro, C., & Couso, D. (2022). Didáctica de la ingeniería: tres preguntas con visión de futuro. *Enseñanza de las Ciencias. Revista de investigación y experiencias didácticas, 40*(3), 147-174.

Siskin, L. S. (1994). *Realms of Knowledge: Academic Departments in Secondary Schools*. The Falmer Press.

Skoog, G. (2005). The coverage of human evolution in high school biology textbooks in the 20th century and in current state science standards. *Science & Education, 14*(3-5), 395–422. doi:10.100711191-004-5611-z

Slater, L. (2004). Collaboration: A framework for school improvement. *International Electronic Journal for Leadership in Learning, 8*(5), 1-19. https://files.eric.ed.gov/fulltext/EJ984546.pdf

Smith, E., & White, P. (2019). Where do all the STEM Graduates Go? Higher Education, the Labour Market and Career Trajectories in the UK. *Journal of Science Education and Technology, 28*(1), 26–40. doi:10.100710956-018-9741-5

Smith, J. J., Johnson, W. R., Lark, A. M., Mead, L. S., Wiser, M. J., & Pennock, R. T. (2016). An Avida-ED digital evolution curriculum for undergraduate biology. *Evolution (New York), 9*(1), 9. doi:10.118612052-016-0060-0

Smith, M. U. (2010). Current status of research in teaching and learning evolution: I. philosophical/epistemological issues. *Science & Education, 19*(6-8), 523–538. doi:10.100711191-009-9215-5

Soderberg, P., & Price, F. (2003). An examination of problem-based teaching and learning in population genetics and evolution using EVOLVE, a computer simulation. *International Journal of Science Education, 25*(1), 35–55. doi:10.1080/09500690110095285

Soler, M. (2002). *Evolución: la base de la biología.* Proyecto Sur.

Solinas, M. (2015). *From Aristotle's Teleology to Darwin's Genealogy: The stamp of inutility.* Springer.

Šorgo, A. (2010). Connecting biology and mathematics: First prepare the teachers. *CBE Life Sciences Education, 9*(3), 196–200. doi:10.1187/cbe.10-03-0014 PMID:20810951

Spanish Government (2014). *Real Decreto 126/2014, de 28 de febrero, por el que se establece el currículo básico de Educación Primaria.*

Spanish Government (2022). *Real Decreto 157/2022, de 1 de marzo, por el que se establecen la ordenación y las enseñanzas mínimas de la Educación Primaria.*

Spanish Government. (2022). *Real Decreto 157/2022, de 1 de marzo, por el que se establecen la ordenación y las enseñanzas mínimas de la Educación Primaria* [Royal Decree 157/2022, of March 1, establishing the organization and minimum standards for primary education].

Spendlove, D., & Hopper, M. (2005). D&T and ICT. In *Creativity in the Primary Curriculum* (pp. 75–90). Routledge.

Spielvogel, J. (2015). *Western Civilization* (9th ed., Vol. 2). Cengage Learning.

Springer, M. (2013). Charting the course of curriculum integration. In P. G. Andrews (Ed.), *Research to guide practice in middle grades education* (pp. 187–216). Association for Middle Level Education.

St. Clair, B., & Hough, D. L. (1992). *Interdisciplinary teaching: A review of the literature.* Southwest Missouri State University Department of Curriculum and Instruction. https://files.eric.ed.gov/fulltext/ED373056.pdf

Stake, R. E. (1995). *The art of case study research.* Sage Publications.

Stanchev, P. (2015). Teaching the Entrepreneurial Mindset in Database Class. *11th Annual International Conference on Computer Science and Education in Computer Science (CSECS).*

Steflitsch, D. (2022). Math teachers' visions of an ideal math class: What do they tell about bringing innovations into the classroom? *In Twelfth Congress of the European Society for Research in Mathematics Education (CERME12).*

Stentoft, D. (2017). From saying to doing interdisciplinary learning: Is problem-based learning the answer? *Active Learning in Higher Education, 18*(1), 51–61.

Stern, L. (2004). Effective assessment: Probing students' understanding of natural selection. *Journal of Biological Education, 39*(1), 12–17. doi:10.1080/00219266.2004.9655948

Stewart, J. (2010). Modeling for Understanding in Science Education (Project MUSE): Involving High School Students in Evolutionary Biology with Realistic Problems and Causal Models. *Biology International: Darwin 200 – Evolution in Action. 47,* 78-90.

Stewart, A. F., Williams, A. L., Lofgreen, J. E., Edgar, L. J., Hoch, L. B., & Dicks, A. P. (2016). Chemistry writing instruction and training: Implementing a comprehensive approach to improving student communication skills. *Journal of Chemical Education, 93*(1), 86–92. doi:10.1021/acs.jchemed.5b00373

Stinson, K., Harkness, S. S., Meyer, H., & Stallworth, J. (2009). Mathematics and science integration: Models and characterizations. *School Science and Mathematics, 109*(3), 153–161. doi:10.1111/j.1949-8594.2009.tb17951.x

Stohlmann, M., Moore, T. J., & Roehrig, G. H. (2012). Considerations for teaching integrated STEM education. *Journal of Pre-College Engineering Education Research, 2*(1), 28–34. Advance online publication. doi:10.5703/1288284314653

Stone, E. M. (2014). Guiding students to develop an understanding of scientific inquiry: A science skills approach to instruction and assessment. *CBE Life Sciences Education, 13*(1), 90–101. doi:10.1187/cbe-12-11-0198 PMID:24591508

Struyf, A., De Loof, H., Boeve-de Pauw, J., & Van Petegem, P. (2019). Students' engagement in different STEM learning environments: Integrated STEM education as promising practice? *International Journal of Science Education, 41*(10), 1387–1407. doi:10.1080/09500693.2019.1607983

Succar, T., Beaver, H. A., & Lee, A. G. (2022). Impact of COVID-19 pandemic on ophthalmology medical student teaching: Educational innovations, challenges, and future directions. *Survey of Ophthalmology, 67*(1), 217–225. doi:10.1016/j.survophthal.2021.03.011 PMID:33838164

Sudarshan, A., & Khin, M. A. (2008). Teacher Practices in Mathematics Classrooms with At-risk Students. *Proceedings of the Meeting of the Australian Association for Research in Education.*

Sukackė, V., Guerra, A. O. P. D. C., Ellinger, D., Carlos, V., Petronienė, S., Gaižiūnienė, L., Blanch, S., Marbà-Tallada, A., & Brose, A. (2022). Towards active evidence-based learning in engineering education: A systematic literature review of PBL, PjBL, and CBL. *Sustainability, 14*(21), 13955. doi:10.3390u142113955

Sun, J. (2005). *Assessing goodness of fit in confirmatory factor analysis. Measurement and Evaluation in Counseling and Development.* SAGE Publications Inc.

Surma, K. (2020). *Does the mathematics level affect student success in high school physics?* School of Education Student Capstone Theses and Dissertations. https://digitalcommons.hamline.edu/hse_all/4510

Taboada, N. J., & Santiago, V. A. O. (2021). Evaluación de talleristas científicos: Propuesta desde la educación y la comunicación científica [Evaluation of scientific workshops: Proposal from education and scientific communication]. *Emerging Trends in Education, 4*(7). Advance online publication. doi:10.19136/etie.a4n7.4458

Takeuchi, M. A., Sengupta, P., Shanahan, M.-C., Adams, J. D., & Hachem, M. (2020). Transdisciplinarity in STEM education: A critical review. *Studies in Science Education, 56*(2), 213–253. doi:10.1080/03057267.2020.1755802

Tamayo, M. (2010). Dificultades en la enseñanza de la evolución biológica. *eVOLUCIÓN. Revista de la Sociedad Española de Biología Evolutiva, 5*(2), 23–27.

Tanik-Onal, N., & Saylan-Kirmizigul, A. (2021). A Makey-Makey based STEM activity for children. *Science Activities, 58*(4), 166–182. doi:10.1080/00368121.2021.2011086

Tanisli, D., & Kose, N. Y. (2013). Pre-service mathematic teachers' knowledge of students about the algebraic concepts. *The Australian Journal of Teacher Education, 38*(2), 1–18. doi:10.14221/ajte.2013v38n2.1

Tehrani, J. J. (2013). The phylogeny of little red riding hood. *PLoS One*, *8*(11), e78871. doi:10.1371/journal.pone.0078871 PMID:24236061

Teles, N., & Fonseca, M. J. (2019). A importância do microscópio ótico na Revolução Científica - Das práticas educacionais à Representação Museológica [The importance of the optical microscope in the Scientific Revolution - From educational practices to Museological Representation]. *História Da Ciência e Ensino: Construindo Interfaces*, *20*, 126–140. doi:10.23925/2178-2911.2019v20espp126-140

Theobald, E. J., Hill, M. J., Tran, E., Agrawal, S., Arroyo, E. N., Behling, S., Chambwe, N., Cintrón, D. L., Cooper, J. D., Dunster, G., Grummer, J. A., Hennessey, K., Hsiao, J., Iranon, N., Jones, L. II, Jordt, H., Keller, M., Lacey, M. E., Littlefield, C. E., ... Freeman, S. (2020). Active learning narrows achievement gaps for underrepresented students in undergraduate science, technology, engineering, and math. *Proceedings of the National Academy of Sciences of the United States of America*, *117*(12), 6476–6483. doi:10.1073/pnas.1916903117 PMID:32152114

Thibaut, L., Ceuppens, S., De Loof, H., De Meester, J., Goovaerts, L., Struyf, A., Boeve-de Pauw, J., Dehaene, W., Deprez, J., De Cock, M., Hellinckx, L., Knipprath, H., Langie, G., Struyven, K., Van de Velde, D., Van Petegem, P., & Depaepe, F. (2018). Integrated STEM Education: A Systematic Review of Instructional Practices in Secondary Education. *European Journal of STEM Education*, *3*(1), 2. doi:10.20897/ejsteme/85525

Thomas, B., & Watters, J. J. (2015). Perspectives on Australian, Indian and Malaysian approaches to STEM education. *International Journal of Educational Development*, *45*, 42–53. doi:10.1016/j.ijedudev.2015.08.002

Thompson, K. V., Cooke, T. J., Fagan, W. F., Gulick, D., Levy, D., Nelson, K. C., Redish, E. F., Smith, R. F., & Presson, J. (2013). Infusing quantitative approaches throughout the biological sciences curriculum. *International Journal of Mathematical Education in Science and Technology*, *44*(6), 817–833. doi:10.1080/0020739X.2013.812754

Tidon, R., & Lewontin, R. C. (2004). Teaching evolutionary biology. *Genetics and Molecular Biology*, *27*(1), 124–131. doi:10.1590/S1415-47572004000100021

Timmis, K., Cavicchioli, R., Garcia, J. L., Nogales, B., Chavarria, M., Stein, L., McGenity, T. J., Webster, N., Singh, B., Handelsman, J., de Lorenzo, V., Pruzzo, C., Timmis, J., Martin, J. L., Verstraete, W., Jetten, M., Danchin, A., Huang, W., Gilbert, J., ... Harper, L. (2019). The urgent need for microbiology literacy in society. *Environmental Microbiology*, *21*(5), 1513–1528. doi:10.1111/1462-2920.14611 PMID:30912268

To, C., Tenenbaum, H. R., & Hogh, H. (2017). Secondary school students' reasoning about evolution. *Journal of Research in Science Teaching*, *54*(2), 247–273. doi:10.1002/tea.21347

Toma, R. B., & Garcia-Carmona, A. (2021). "Of STEM we like everything but STEM". A critical analysis of a buzzing educational trend. *Enseñanza de las Ciencias*, *39*(1), 65–80. doi:10.5565/rev/ensciencias.3093

Torkar, G., & Šorgo, A. (2020). Evolutionary content knowledge, religiosity and educational background of Slovene preschool and primary school pre-service teachers. *Eurasia Journal of Mathematics, Science and Technology Education*, *16*(7), 1–13. doi:10.29333/ejmste/7991

Torrance, E. P. (1979). *The search for satori and creativity*. Creative Education Foundation and Creative Synergetic Associates.

Torreblanca, M., Merino, G., & De Longhi, A. L. (2011). El enfoque histórico contextualizado como facilitador de la enseñanza de los mecanismos evolutivos. La utilidad de un enfoque histórico integrado a un encuadre CTS en la enseñanza de los mecanismos micro y macroevolutivos. *Revista De Educación En Biología*, *14*(1), 56–59.

Torsney, B. M., Ponnock, A. R., & Lombardi, D. (2017). The Role of Values in Preservice Teachers' Decision to Teach. *Teacher Educator*, *52*(1), 39–56. doi:10.1080/08878730.2016.1196986

Toth, E. E., Brem, S. K., & Erdos, G. (2009). "Virtual Inquiry": Teaching Molecular Aspects of Evolutionary Biology Through Computer-Based Inquiry. *Evolution (New York)*, *2*(4), 679–687. doi:10.100712052-009-0163-y

Treacy, P. (2021). A conceptual framework for integrating mathematics and science in the secondary classroom. *SN Social Sciences*, *1*(6), 1–17. doi:10.100743545-021-00166-x

Treacy, P., & O'Donoghue, J. (2014). Authentic integration: A model for integrating mathematics and science in the classroom. *International Journal of Mathematical Education in Science and Technology*, *45*(5), 703–718. doi:10.1080/0020739X.2013.868543

Trumbo, J. (2006). Making science visible: Visual literacy in science communication. In L. Pauwels (Ed.), *Visual culture of science: Re-thinking representational practices in knowledge building and science communication* (pp. 266–283). Dartmouth College Press, University Press of New England.

Truss, A., & Anderson, V. (2023). The navigational challenges of a blended learning approach to teaching in business and management. *International Journal of Management Education*, *21*(1), 100733. doi:10.1016/j.ijme.2022.100733

Tshuma, T., & Sanders, M. (2015). Textbooks as a possible influence on unscientific ideas about evolution. *Journal of Biological Education*, *49*(4), 354–369. doi:10.1080/00219266.2014.967274

Turkka, J., Haatainen, O., & Aksela, M. (2017). Integrating art into science education: A survey of science teachers' practices. *International Journal of Science Education*, *39*(10), 1403–1419. doi:10.1080/09500693.2017.1333656

Turner, F., & Rowland, T. (2011). The knowledge quartet as an organising framework for developing and deepening teachers' mathematics knowledge. In K. Ruthven & T. Rowland (Eds.), *Mathematical knowlege in teaching* (pp. 195–212). Springer. doi:10.1007/978-90-481-9766-8_12

Turşucu, S., Spandaw, J., Flipse, S., & de Vries, M. J. (2017). Teachers' beliefs about improving transfer of algebraic skills from mathematics into physics in senior pre-university education. *International Journal of Science Education*, *39*(5), 587–604. doi:10.1080/09500693.2017.1296981

Tytler, R., Mulligan, J., Prain, V., White, P., Xu, L., Kirk, M., Nielsen, C., & Speldewinde, C. (2021). An interdisciplinary approach to primary school mathematics and science learning. *International Journal of Science Education*, *43*(12), 1926–1949. doi:10.1080/09500693.2021.1946727

Tytler, R., Prain, V., & Hobbs, L. (2021). Rethinking Disciplinary Links in Interdisciplinary STEM Learning: A temporal model. *Research in Science Education*, *51*(S1, Suppl 1), S269–S287. doi:10.100711165-019-09872-2

Tytler, R., Williams, G., Hobbs, L., & Anderson, J. (2019). Challenges and opportunities for a STEM interdisciplinary agenda. In B. Doig, J. Williams, D. Swanson, R. B. Ferri, & P. Drake (Eds.), *Interdisciplinary Mathematics Education. The state of the art and beyond* (pp. 51–81). Springer Open. doi:10.1007/978-3-030-11066-6_5

United Nations. (2022). *Make the SDGS a Reality*. United Nations Department of Economic and Social Affairs, Sustainable Development. https://sdgs.un.org/#goal_section

Urich, J. A., & Sasse, E. A. (2011). An Ap"peel"ing Activity. *Mathematics Teacher*, *105*(3), 189–193. doi:10.5951/mathteacher.105.3.0189

Vallo, D., & Valovicova, L. (2019). Interdisciplinary relations of mathematics and physics in curriculum of conic sections. *AIP Conference Proceedings*, 1–7. doi:10.1063/1.5124779

Van Dijk, E., & Reydon, T. (2010). A conceptual analysis of evolutionary theory for teacher education. *Science and Education*, *19*(6-8), 655–677. doi:10.100711191-009-9190-x

Van Dooren, W., Verschaffel, L., & Onghena, P. (2003). Pre-service teachers' preferred strategies for solving arithmetic and algebra word problems. *Journal of Mathematics Teacher Education*, *6*(1), 27–52. doi:10.1023/A:1022109006658

Vázquez, A., & Manassero, M. A. (2008). El declive de las actitudes hacia la ciencia de los estudiantes: Un indicador inquietante para la educación científica [The decline of students' attitudes towards science: a disturbing indicator for science education]. *Revista Eureka*, *5*(3), 274–292. doi:10.25267/Rev_Eureka_ensen_divulg_cienc.2008.v5.i3.03

Vázquez, A., & Manassero, M. A. (2011). The decline of girls' and boys' attitudes toward science in compulsory education. *Ciência & Educação (Bauru)*, *17*(2), 249–268. doi:10.1590/S1516-73132011000200001

Venville, G., Sheffield, R., Rennie, L., & Wallace, J. (2008). The writing on the classroom wall: The effect of school context on learning in integrated, community-based science projects. *Journal of Research in Science Teaching*, *45*, 857–880. doi:10.1002/tea.20245

Venville, G., Wallace, J., Rennie, L., & Malone, J. (2002). Curriculum Integration: Eroding the high ground of science as a school subject. *Studies in Science Education*, *37*(1), 43–83. doi:10.1080/03057260208560177

Verspoor, A., & Wu, K. (1990). *Textbooks and educational development*. Educational and Employment Division, Population and Human Resources Department, The World Bank.

Vicente-Romero, R. (2016). *Trabajar ideas alternativas en el proceso de enseñanza-aprendizaje de la evolución a través de juegos: "Survival & Evolution"* [tesis de maestría, Universidad de Zaragoza]. Repositorio institucional https://zaguan.unizar.es/record/58753

Vieira, F. (2009). Em contracorrente: O valor da indagação da pedagogia na universidade. *Educação Sociedade & Culturas*, *28*, 107–126. https://www.fpce.up.pt/ciie/revistaesc/ESC28/28_flavia.pdf

Viiri, J., & Savinainen, A. (2008). Teaching-learning sequences: A comparison of learning demand analysis and educational reconstruction. *Latin-American Journal of Physics Education*, *2*(2), 80–86.

Vondrasek, J. R., Janis, A., & Taylor, D. R. (2004). Evolution Kills: A Web Resource for Instructors of Evolutionary Biology. *Bioscene: Journal of College Biology Teaching*, *30*(2), 3–8.

Wade-Jaimes, K., Demir, K., & Qureshi, A. (2018). Modeling strategies enhanced by metacognitive tools in high school physics to support student conceptual trajectories and understanding of electricity. *Science Education*, *102*(4), 711–743. doi:10.1002ce.21444

Wagensberg, J. (2006). *CosmoCaixa: The total museum through conversation between architects and museologists* (No. 069.2: 5 COS).

Wallin, A. (2008). One year after teaching - How consistent are students in using the scientific theory of biological evolution by natural selection? In M. Hammann, M. Reiss, C. Boulter & S. D. Tunnicliffe (Eds.), *Biology in Context – Learning and teaching for the twenty-first century. Proceedings of the VI Conference of European Researchers in Didactic of Biology (ERIDOB)* (pp. 52-63). University of London.

Wang, H., Moore, T., Roehrig, G., & Park, M. (2011). STEM Integration: Teacher Perceptions and Practice. *Journal of Pre-College Engineering Education Research*, *1*(2), Article 2. doi:10.5703/1288284314636

Watson, J. D., & Crick, F. (1953). Molecular Structure of Nucleic Acids: A Structure for Deoxyribose Nucleic Acid. *Nature*, *171*(4356), 737–738. doi:10.1038/171737a0 PMID:13054692

Weber, B. W. (2019, November). Climbing Bloom's Taxonomy With Jupyter Notebooks: Experiences in Mechanical Engineering. In *ASME International Mechanical Engineering Congress and Exposition* (*Vol. 59421*, p. V005T07A022). American Society of Mechanical Engineers. 10.1115/IMECE2019-10615

Weinberg, A., & McMeeking, L. (2017). Toward meaningful interdisciplinary education: High school teachers' views of mathematics and science integration: Meaningful interdisciplinary education. *School Science and Mathematics*, *117*(5), 204–213. doi:10.1111sm.12224

West, S. S., Vasquez-Mireles, S., & Coker, C. (2006). Mathematics and/or science education: Separate or integrate. *Journal of Mathematical Sciences and Mathematics Education*, *1*(2), 11–18.

West, S., Vásquez-Mireles, S., & Coker, C. (2006). Math and/or science education: Separate or integrate? *Journal of Mathematical Science and Mathematics Education*, *1*(2), 11–18.

White, D., & Delaney, S. (2021). Full STEAM Ahead, but Who Has the Map for Integration? A PRISMA Systematic Review on the Incorporation of Interdisciplinary Learning into Schools. *International Journal on Math. Science and Technology Education*, *9*(2), 9–32.

Widjaja, W., Hubber, P., & Aranda, G. (2019). Potential and challenges in integrating science and mathematics in the classroom through real-world problems: A case of implementing an interdisciplinary approach to STEM. In Y. S. Hsu & Y. F. Yeh (Eds.), *Asia-Pacific STEM teaching practices* (pp. 157–171). Springer. doi:10.1007/978-981-15-0768-7_10

Wieman, C. E. (2014). Large-scale comparison of science teaching methods sends clear message. *Proceedings of the National Academy of Sciences of the United States of America*, *111*(23), 8319–8320. doi:10.1073/pnas.1407304111 PMID:24853505

Wiles, J., & Bondi, J. (2007). *Curriculum development. A guide to practice* (7th ed.). Pearson Education.

Wiley, E. O. (2010). Why Trees Are Important. *Evo Edu Outreach*, *3*(4), 499–505. doi:10.100712052-010-0279-0

Williams, J., Roth, W.-M., Swanson, D., Doig, B., Groves, S., Omuvwie, M., . . . Mousoulides, N. (2016). *Interdisciplinary Mathematics Education.* Springer Open. Retrieved from https://link.springer.com/book/10.1007/978-3-030-11066-6

Williams, J. (2011). STEM Education: Proceed with caution. *Design and Technology Education. International Journal (Toronto, Ont.)*, *16*(1), 26–35.

Winarno, N., Rusdiana, D., Riandi, R., Susilowati, E., & Afifah, R. (2020). Implementation of integrated science curriculum: A critical review of the literature. *Journal for the Education of the Gifted*, *8*(2), 795–817.

Winarno, N., Widodo, A., Rusdiana, D., Rochintaniawati, D., & Afifah, R. M. A. (2019). Pre-service Science Teachers' Conceptual Understanding of Integrated Science Subject: A Case Study. *Journal of Physics*, *1204*(1), 012104.

Wong, V. J. (2018). *The relationship between school science and mathematics education* [Unpublished doctoral dissertation]. King's College London. King's Research Portal: https://kclpure.kcl.ac.uk/portal/

Wong, V. (2017). Variation in graphing practices between mathematics and science: Implications for science teaching. *The School Science Review*, *98*(365), 109–115.

Wong, V. (2018). *The relationship between school science and mathematics education.* King's College London.

Wong, V. (2019). Authenticity, transition and mathematical competence: An exploration of the values and ideology underpinning an increase in the amount of mathematics in the science curriculum in England. *International Journal of Science Education*, *41*(13), 1805–1826. doi:10.1080/09500693.2019.1641249

Wong, V., & Dillon, J. (2019). 'Voodoo maths', asymmetric dependency and maths blame: Why collaboration between school science and mathematics teachers is so rare. *International Journal of Science Education*, *41*(6), 782–802. doi:10.1080/09500693.2019.1579945

Wong, V., & Dillon, J. (2020). Crossing the boundaries: Collaborations between mathematics and science departments in English secondary (high) schools. *Research in Science & Technological Education, 38*(4), 396–416. doi:10.1080/02635143.2019.1636024

Wong, V., Dillon, J., & King, H. (2016). STEM in England: Meanings and motivations in the policy arena. *International Journal of Science Education, 38*(15), 2346–2366. doi:10.1080/09500693.2016.1242818

Woodland, R. H. (2016). Evaluating PK–12 professional learning communities: An improvement science perspective. *The American Journal of Evaluation, 37*(4), 505–521. doi:10.1177/1098214016634203

Woodland, R., Lee, M. K., & Randall, J. (2013). A validation study of the teacher collaboration assessment survey. *Educational Research and Evaluation, 19*(5), 442–460. doi:10.1080/13803611.2013.795118

Woodland, R., & Mazur, R. (2015). Beyond hammers versus hugs: Leveraging educator evaluation and professional learning communities into job-embedded professional development. *NASSP Bulletin, 99*(1), 5–25. doi:10.1177/0192636515571934

World Economic Forum. (2016). *New vision for education: Fostering social and emotional learning through technology.* https://www3.weforum.org/docs/WEF_New_Vision_for_Education.pdf

Wrensch, D. L., & Young, S. S. Y. (1975). Effects of quality of resource and fertilization status on some fitness traits in the two-spotted spider mite, *Tetranychus urticae* Koch. *Oecologia, 18*(4), 259–267. doi:10.1007/BF00345850 PMID:28308917

Xu, L., Prain, V., & Speldewinde, C. (2021). Challenges in designing and assessing student interdisciplinary learning of optics using a representation construction approach. *International Journal of Science Education, 43*(6), 844–867. doi:10.1080/09500693.2021.1889070

Yakman, G. (2012). Recognizing the A in STEM Education. *Middle Ground Magazine, 16*(1), 15–16.

Yates, T., & Marek, E. (2015). A Study Identifying Biological Evolution-Related Misconceptions Held by Prebiology High School Students. *Creative Education, 6*(8), 811–834. doi:10.4236/ce.2015.68085

Yeboah, R., Abonyi, U. K., & Luguterah, A. W. (2019). Making primary school science education more practical through appropriate interactive instructional resources: A case study of Ghana. *Cogent Education, 6*(1), 1–14. doi:10.1080/2331186X.2019.1611033

Yeo, C., & Lee, H. (2016). Developing and Assessing a Learning Progression for the Ecosystem. *Journal of The Korean Association For Science Education, 36*(1), 29–43. doi:10.14697/jkase.2016.36.1.0029

Yetilmezsoy, K., & Mungan, C. E. (2018). MATLAB time-based simulations of projectile motion, pendulum oscillation, and water discharge. *European Journal of Physics, 39*, 065803.

Yew, E. H. J., & Gohb, K. (2016). Problem-Based Learning: An Overview of its Process and Impact on Learning. *Health Profession Education, 2*(2), 75–79. doi:10.1016/j.hpe.2016.01.004

Yıldırım, A., & Şimşek, H. (2016). *Sosyal bilimlerde nitel araştırma yöntemleri* [Qualitative research methods in the social sciences]. Seçkin.

Yin, R. K. (1989). *Case study research: Design and methods.* Sage.

Yoho, R. A., & Vanmali, B. H. (2016). Controversy in Biology Classrooms-Citizen Science Approaches to Evolution and Applications to Climate Change Discussions. *Journal of Microbiology & Biology Education, 17*(1), 110–114. doi:10.1128/jmbe.v17i1.1026 PMID:27047604

You, H. S. (2017). Why teach science with an interdisciplinary approach: History, trends, and conceptual frameworks. *Journal of Education and Learning*, *6*(4), 66–77. doi:10.5539/jel.v6n4p66

Young, M. (2011). The return to subjects: A sociological perspective on the UK Coalition government's approach to the 14–19 curriculum. *Curriculum Journal*, *22*(2), 265–278. doi:10.1080/09585176.2011.574994

Young, M. (2014). The progressive case for a subject-based curriculum. In M. Young & D. Lambert (Eds.), *R. C., & M. Roberts, Knowledge and the future school: Curriculum and social justice* (pp. 89–110). Bloomsbury Academic.

Zabel, J., & Gropengiesser, H. (2011). Learning progress in evolution theory: Climbing a ladder or roaming a landscape? *Journal of Biological Education*, *45*(3), 143–149. doi:10.1080/00219266.2011.586714

Zhang, D., Orrill, C., & Campbell, T. (2015). Using the mixture Rasch model to explore knowledge resources students invoke in mathematic and science assessments. *School Science and Mathematics*, *115*(7), 356–365. doi:10.1111sm.12135

Zollman, A. (2012). Learning for STEM Literacy: STEM Literacy for Learning. *School Science and Mathematics*, *112*(1), 12–19. doi:10.1111/j.1949-8594.2012.00101.x

Zubrowski, B. (2002). Integrating science into Design Technology projects: Using a standard model in the design process. *Journal of Technology Education*, *13*(2), 48–67. doi:10.21061/jte.v13i2.a.4

Zudaire, I., & Napal-Fraile, M. (2020). Exploring the Conceptual Challenges of Integrating Epigenetics in Secondary-Level Science Teaching. *Research in Science Education*. Advance online publication. doi:10.100711165-019-09899-5

About the Contributors

Bento Cavadas holds a degree in Biology/Geology Teaching from the Faculty of Sciences of the University of Porto and a PhD in the field of Educational Sciences from the Faculty of Education of the University of Salamanca, with the thesis "The evolution of Natural Science textbooks in secondary education in Portugal 1836- 2005". Currently is a professor in the Polytechnic Institute of Santarém/ School of Education, in which co-coordinates the project CreativeLab_Sci&Math®. Is a researcher at CeiED, from University Lusófona. Author of textbooks of natural sciences, biology and geology. The main research interests are in the field of science education, teacher education, science textbooks and integration of science and mathematics.

Neusa Branco holds a degree in Teaching of Mathematics, from the Faculdade de Ciências of the Universidade de Lisboa, a master's degree in Education, in the specialization in Didactics of Mathematics, from the Faculdade de Ciências of the Universidade de Lisboa and a PhD in Education, specialization in Didactics of Mathematics by the Instituto de Educação of the Universidade de Lisboa. She is a Professor at the Instituto Politécnico de Santarém-Escola Superior de Educação de Santarém, where she is coordinator of the Master in Teaching of the 1st Cycle of Basic Education and Mathematics and Science of 2nd Cycle of Basic Education and is sub-coordinator of the Master in Pre-school Education and Teaching of the 1st cycle of basic education. She was President of the Technical-Scientific Council, from 2018 to 2022. She collaborates in research projects in the area of Education and in international cooperation projects. The main research interests are: Teaching and learning of Algebra; Didactics of Mathematics; Initial teacher training; Mathematical connections; Interdisciplinarity between science and mathematics; Curricular integration of ICT and computacional thinking.

* * *

Nazir Amir is a Master Teacher at the Academy of Singapore Teachers, Ministry of Education, Singapore. A passionate teacher who actively carries out classroom research, he mentors teachers to design enjoyable lessons to foster intrinsic motivation, engagement and creativity amongst students. Apart from working with Singapore teachers, he has also been invited to present his work at international platforms, and has published numerous research articles in international peer-reviewed journals and book chapters so that the wider education community may benefit from his classroom research. He is a recipient of the Singapore President's Award for Teachers, the U.S. Fulbright Distinguished Award in Teaching, and a top finalist in the Global Teacher Prize.

Sevim Bezen received her MS in Physics Education from Hacettepe University in 2014. She has earned doctoral degrees in Physics Education from Hacettepe University in 2019. In 2014 she was hired as the research assistant by the Physics Education Department at Hacettepe University. Her field of study includes physics education, teacher training, curricula, metaphor, concept maps, conceptual understanding, qualitative, quantitative, action research, and mixed research methods. She has published numerous national and international articles in the field of education. She writes and presents widely on issues of physics education.

Rosário Cardoso is teacher of the 520 disciplinary group – Biology and Geology. Since September 2018 Maria do Rosário has been appointed to the MHNC-UP. She has experience in the implementation of curricular flexibility programmes and promotion of school success, having worked as coordinator of projects of Promotion of Health Education and Experimental Science Teaching. Currently performing pedagogical functions within the MHNC-UP's educational services department, producing scientific and didactic content for the Museum's Educational Offer for Schools, she is the liaison between the educational and the scientific communities, namely through Ciência Viva at School organisation.

Violeta Meneses Carvalho completed the Integrated Master in Biomedical Engineering in 2020 by the University of Minho and her area of expertise is Biomaterials, Rehabilitation, and Biomechanics. Currently, she is an invited assistant at the Production and Systems Department teaching statistics and numerical methods. At the same time, she is a PhD student in the Mechanical Engineering Doctoral program at the University of Minho focused on the development and optimization of innovative nanoparticles and a preclinical organ-on-a-chip platform for nanomedicine.

Marisa Correia is Adjunct Professor at School of Education/Polytechnic Institute of Santarém. Has a Post-doctorate in Education from the Institute of Education - University of Lisbon (IE-UL), within the framework of the GoSTEM Project, funded by the FCT—Portuguese Foundation for Science and Technology. Has a degree in Chemistry and Physics, a Master in Education from the Faculty of Sciences - University of Lisbon, a PhD in Science Education and post-graduation in Education and Digital Technologies from IE-UL. CV: https://www.cienciavitae.pt/F319-BEDD-099B.c61ba884-53ef-4fd7-94bf-c178df8637a7

İpek Derman graduated from Hacettepe University, Department of Biology Education in 2009. She received her MS in 2013 and earned a doctoral degree in 2019 in the Division of Curriculum and Instruction from Hacettepe University. She earned doctoral degrees in the same Division from Hacettepe University in 2019. In 2011 she was hired as a research assistant by the Department of Educational Sciences at Hacettepe University. Her field of study includes curriculum development, teacher training, learning theories, teaching methods, science education, higher education, qualitative, quantitative, and mixed research methods. She has published numerous national and international articles in the field of education. She writes and presents extensively on curriculum and instruction areas.

María Rocío Esteban is the Associate Professor of Science Education at the University of Extremadura, Spain, holding a Ph.D. in Plant Biology, a Master's degree, and a Bachelor's degree in Biology. Her research focus is on utilizing active learning approaches to effectively transfer fundamental concepts of science to students.

Ana C. M. Ferreira (Ph.D.) is an Assistant Professor at the Faculty of Engineering and Technologies at Lusíada Norte University (Vila Nova de Famalicão Campus) and a researcher at ALGORITMI and MEtRICs centers, both research units from the School of Engineering, University of Minho. She is a Post-Doc Researcher and collaborator member at ALGORITMI Centre, within the Industrial Engineering and Management (IEM) research line, Ergonomics & Human Factors Group (EHF). Additionally, she is also an integrated member of MEtRICs - Mechanical Engineering and Resource Sustainability Center.

Maria João Fonseca, with a background in animal biology and biology teaching and a PhD in science education and communication, is currently Head of Communication at the Natural History and Science Museum of the University of Porto and Director of the Ciência Viva Center Hall of Biodiversity. She is in charge of general administration and of coordinating and promoting science communication and outreach activities. She has been involved in the design and evaluation of formal and informal educational activities, teacher training, and outreach initiatives for all sorts of audiences. Also, she is currently part of the team responsible for the reconstruction of the Natural History and Science Museum of the University of Porto.

Cristina Guerra Ph.D. in Physics (Condensed Matter Physics), University of Lisbon (2001); Master in Physics (Condensed Matter Physics), University of Lisbon (1995); Bachelor of science in Chemistry, University of Lisbon (1991). Associate Professor at Lusofona University since 2007, teaches subjects in the area of physics and programming for engineering courses. Since 2019, has been the Director of the Faculty of Engineering at Lusofona University, performing between 2010 and 2019 the position of deputy director of the same faculty.

Fulden Guler-Nalbantoglu graduated from the science education department at Middle East Technical University in 2012. She received her master's degree in Curriculum and Instruction program from the same university in 2015. She is currently a Ph.D. candidate in the science education field. Her research interests are pedagogical content knowledge, STEM education, and teacher education. She has been working as a full-time research assistant in the science education department at Ege University since 2013.

José María Marcos-Merino is professor at the University of Extremadura. He is a doctor in Science Education and has collaborated with international centers such as the Leibniz Institute for Science and Mathematics Education (Germany) and the University of Santarém (Portugal). He is currently working on teacher training of future teachers and his research focuses on teaching and learning, cognitive and affective aspects of science.

Pedro Carmona Marques holds a PhD in Leaders for Technical Industries - Engineering Design & Advanced Manufacturing from IST/MIT (2012), a Master in Engineering Physics and a Degree in Industrial Engineering. Currently he is an Associate Professor at Lusófona University and an Invited Adjunct Professor at ISL – Polytechnic of Lisbon. Main areas of work and R&D focus are on Industrial Management, Innovation and STEM.

María Martín-Peciña graduated in Biology by the University of Salamanca (Spain) in 2013 and received a MS in Genetics and Evolution (2014) and a MS in Secondary Education Teaching (2021) by University of Granada (Spain). From this last institution, she obtained her PhD in Fundamental and

Systems Biology in 2021 that was awarded as Outstanding Cum Laude and received an International Mention enyoing different research stays at the Katholieke Universiteit of Leuven (Belgium). Since then, she has been dedicated to research in Science Education at the University of Jaén (Spain) focusing on erroneous mental model about Evolutionary Biology, Biological Evolution and sustainability and STEM sustainable education. She works on European projects aimed at teachers professional development as well as at implementation of teaching strategies based on research evidence to improve the teaching-learning processes of STEM disciplines through open schooling, sustainability, critical-thinking and cross-curricular approaches.

Guadalupe Martínez Borreguero is a full professor. Lines of research: Professional development of science and mathematics teachers; Enhancing STEM competences in Extremadura students. Design, application and validation of metacognitive and emotional regulation strategies for teachers in training; Promotion of sustainable development in preservice teachers through integrated teaching in STEM; Emotions in science teaching and learning as a tool for a scientific education that promotes integral sustainability.

Maria Martins is an Adjunct Professor at School of Education/Polytechnic Institute of Santarem. Specialist in Elementary Teacher Education, Master in Mathematics for Teaching by Coimbra University Faculty of Science and Technology. Degree in Mathematics, Educational Training Branch by the same university. Research interests: Mathematics Education; Preservice Teachers Education; Mathematics in the Early Years; Mathematics and Literature for Children; STEAM; Mathematical Games and Manipulatives; ICT in Mathematics Education. CV: https://www.cienciavitae.pt//pt/0F1B-1E5E-AAB9.a054c7b4-89b1-44e7-940f-42803bcea1ac

Milagros Mateos Núñez is Assistant Professor Doctor. Lines of research: Professional development of science and mathematics teachers; Enhancing STEM competencies in Extremadura students. Design, application and validation of metacognitive and emotional regulation strategies for teachers in training; Promotion of sustainable development in preservice teachers through integrated teaching in STEM.

Nuno Melo is a professor in the field of Physical and Natural Sciences at the Escola Superior de Educação de Lisboa. Holds a PhD in Biochemistry and a postgraduate specialization degree in Education – Health Education. He has developed research work in the area of education for sustainable development and didactics of natural sciences in basic education.

Nelson Mestrinho holds a degree in Mathematics (branch of Mathematical Analysis) from the University of Évora (1996) and a Master's degree in Applied Mathematics in 2002 from the same University. He also has a specialisation in Teacher Training for Basic Education (1st and 2nd Cycles) from the Polytechnic Institutes of Santarém, Lisbon and Leiria. He is an Adjunct Professor at the Polytechnic Institute of Santarém. He works in the area(s) of Exact Sciences with emphasis on Mathematics and Social Sciences with emphasis on Education Sciences.

Aurora Muñoz-Losa is professor at the University of Extremadura. She is a PhD in Theoretical and Computational Chemistry and has worked in international centers such as the University of Pisa (Italy), CEA-Saclay (France), and the University of Vienna (Austria). She is currently working on teacher training of future teachers and his research focuses on the teaching and learning of science at different educational stages.

João Nogueira has a degree in Physics from the Faculty of Sciences of the University of Lisbon, Master in Environmental Engineering from the Instituto Superior Técnico, he holds a PhD in Education from the Institute of Education of the Lusophone University of Humanities and Technologies, with the thesis "Computational programming and controlled learning in Physics: an exploratory study with high school students". As a supplementary training, he did the school part of the Master in Gravitation and High Energies at the Faculty of Sciences of the University of Lisbon. He is an assistant professor at the Faculty of Engineering of ULHT, an institution in which he has taught, since 1997, subjects in the areas of Physics, Mathematics and Programming. He was, for three years, Professor of Physics (Trainee Assistant) at the University of Beira Interior, also having a two-year experience in Basic and Secondary Education, in which he taught Physics, Chemistry and Mathematics. Since 2010, he has been coordinator of the scientific area of Physics at the Faculty of Engineering at ULHT. Between 2003 and 2013, he was an assistant researcher in the MARETEC group of the Environment and Energy section of the Mechanical Engineering Department of Instituto Superior Técnico. During this period, I work in the area of oceanographic modeling, as well as in biological modeling. Also in this group, he participated in several international projects, having written book chapters, as well as publications and oral publications in international and national criteria. He is currently an PhD Integrated Researcher, Lusófona University, CeiED (Interdisciplinary Research Centre for Education and Development), continuing to publish articles in international journals, book chapters, as well as presentations at national and international conferences and participation in national and international projects. Currently, he works in the field of science teaching, focusing on the use of programming and computational environments in the teaching / learning processes of Physics, Chemistry and Mathematics. With these methodologies, it seeks to enhance the meaningful learning of Sciences, based on interdisciplinarity and integration of concepts.

Jesús Ochoa de Alda is an associate professor at the Teacher Training School (Cáceres, Spain). He has followed an extended and diverse carrier that involves stages as a molecular biologist at the King's College (London), Institut Pasteur (Paris), and Ecole normale supérieure (Paris), responsibilities as dean of the Faculty of Biology, and deputy director of the Teacher Training School among others. His research interest focus on the role of emotions in learning and fundamental aspects of life sciences.

José Alexandre da Silva Pinto was born in Paredes, Portugal, in 1968. Holds a BSc in Mathematics and Science Teaching from the Higher Education School of – Polytechnic Porto (ESE/IPP), an MSc in Science Teaching Supervision from the University of Minho, and a Ph.D. in Didactics of Science and Technology from the University of Trás-os-Montes and Alto Douro (UTAD). Teaches in the Department of Mathematics, Sciences and Technology of ESE/IPP since 1995, mainly Didactics of Science Teaching and Internships Supervision. He has several national and international publications.

Nora Rodilla-Martín has a Bachelor's degree in Primary Education. Her research interest focus on active learning methodologies

Cristina S. Rodrigues is an Assistant Professor at the University of Minho, where teaches and is responsible for courses in applied statistics at undergraduate and postgraduate studies in engineering. She holds a PhD in Industrial and Systems Engineering (2008), specialization in Numerical and Statistical Methods from the University of Minho, a Msc in Business Sciences (2002) from a School of Economics and Management of the University of Porto, a Graduation in Engineering and Industrial Management (1994) from University of Aveiro. Her current research interests are in industrial engineering and quality management, entrepreneurship, engineering education, research by questionnaire, experimental design and multivariate statistics analysis.

Margarida Rodrigues is PhD in Mathematics Education from the University of Lisbon (Portugal). Coordinator Professor at the Escola Superior de Educação of Instituto Politécnico de Lisboa, Portugal. Integrated member of the Unidade de Investigação e Desenvolvimento em Educação e Formação (UIDEF) at Instituto de Educação, Universidade de Lisboa, Portugal. President of the Scientific Technical Council and Coordinator of Master's Degree in the 1st cycle of Primary Education and Mathematics and Natural Sciences in the 2nd cycle of Primary Education. Member of diverse research projects in mathematics education and environmental education.

Nelson Rodrigues is an invited professor and researcher at the University of Minho.

Xana Sá Pinto is a Portuguese biologist, passionate by biodiversity and evolutionary processes. Until recently, her research career mostly focused on the study of evolutionary processes driving divergence and speciation in marine realm. During her PhD and post-doc research period she participated and organised in several education and outreach projects related with evolution, which made her become increasingly interested and engaged with science education. In 2015 she joined the research centre in didactics and technologies in teacher training at the University of Aveiro (CIDTFF.UA). She is now researching on the effects of teaching science at the light of evolution to elementary school students and on the impacts of evolution literacy on individuals' attitudes related with social problems. She is a founder member of EvoKE - a European network that aims to promote transdisciplinary collaborations to foster science literacy about evolution. She is the leader of the Formal Education working group (WG2) of EuroScitizen Cost Action.

Pedro Sarreira holds a BSc in Physics and a MSc in Biophysics (University of Lisbon, Faculty of Sciences). He completed the curricular part of the PhD in Education (specialty Didactics of Sciences). He has a vast experience in training preservice teachers of the first levels (1st to 6th year of schooling), teaching elementary sciences (physics, astronomy, chemistry, environmental education) and its didactics, and supervising their traineeships. His research interests are in the field of science education and environmental education.

Maria João Silva is Coordinator Professor with Aggregation and Tenure in Escola Superior de Educação, Politécnico de Lisboa (IPL, ESELx), since 2015; Professor at the in Escola Superior de Educação, Politécnico do Porto, from 1987 to 2015; Aggregation in Education (Universidade Lusófona), Degree and PhD in Environmental Engineering (Universidade Nova de Lisboa), Master's and Post-doctorate degree in Information Systems and Technologies in Education (Coimbra University). Coordinator of the Interdisciplinary Center for Educational Studies (CIED, IPL, ESELx), since 2021. Coordinator and member of diverse research projects in environmental education. Coordinator of the Eco-Sensors4Health

Project (from 12/10/2017 to 11/3/2019), funded by Portugal2020 and FCT. She has several publications on environmental education.

Inês Maria Baptista Teixeira completed the Integrated Master in Mechanical Engineering in 2021/08/04 by Universidade of Minho. She is Invited Assistant in University of Minho - Campus de Azurém and attends the Doctoral Program in Industrial and Systems Engineering by the University of Minho since 2022/01/14.

Senhorinha de Fátima Capela Fortunas Teixeira is Associate Professor in University of Minho, Researcher in University of Minho Centro ALGORITMI and Member of the Coordinating Committee of the Degree Program Integrated Master in Mechanical Engineering (MIEM) in University of Minho. From 2011 to 2015, she was Department Director, and from 2016 to 2021, Erasmus Academic Coordinator in Production and Systems Area in University of Minho. She completed the Bachelor in Systems Engineering and Informatics in 1984 by University of Minho and Doctor in Chemical Engineering in 1989 by University of Birmingham School of Chemical Engineering.

Nuno Teles, since 2008, has participated in several pedagogical projects at the University of Porto, having a diversified experience in education, teaching, promoting, and communicating sciences, and being responsible for developing activities of the educational service in MHNC-UP. Student at the Ph.D. in science teaching and dissemination since 2020. He is interested in research areas there are centered on formal and non-formal education with the interest of promoting science and education in museums.

Joana Torres has been passionate about nature since she was a child. She graduated in Biology in 2008 and since then has been working in the area of communication, promotion and scientific dissemination, collaborating with institutions such as the University of Porto, the Fundação para o Desenvolvimento Social do Porto, the Serralves Foundation and CIBIO-InBIO, where he was part of the communication team. She was also scientific advisor in the study "The evolution of science in Portugal (1987-2016)". Collaborating with Museu de História Natural e da Ciência da Universidade do Porto since 2019, she performs functions of communication advisory, idealization, organization and dissemination of activities and events.

Victoria Wong is a senior lecturer in STEM (science) education at the University of Exeter, UK. Her research interests include the relationship between school science and mathematics education, students' use of mathematics in science and how science education policy is made and enacted. After studying for a degree in chemistry, Victoria qualified as a teacher and taught in England, Spain and New Zealand. She has also worked as an independent science education consultant and as a teacher educator at King's College London and the University of Oxford.

Index

F

G

H

I

S

T

U

V

W

Ensure Quality Research is Introduced to the Academic Community

Become an Evaluator for IGI Global Authored Book Projects

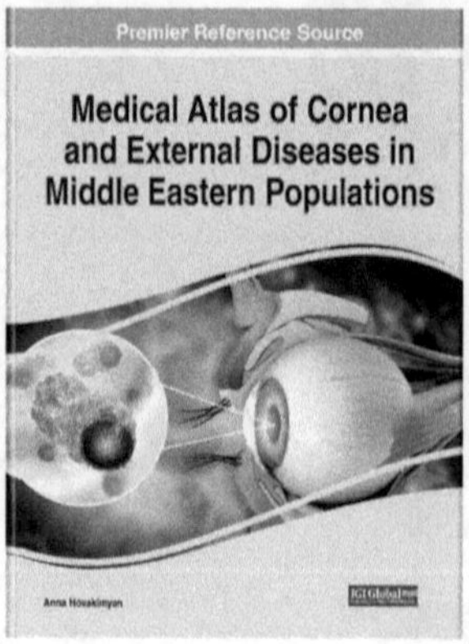

The overall success of an authored book project is dependent on quality and timely manuscript evaluations.

Applications and Inquiries may be sent to:
development@igi-global.com

Applicants must have a doctorate (or equivalent degree) as well as publishing, research, and reviewing experience. Authored Book Evaluators are appointed for one-year terms and are expected to complete at least three evaluations per term. Upon successful completion of this term, evaluators can be considered for an additional term.

If you have a colleague that may be interested in this opportunity, we encourage you to share this information with them.

www.ingramcontent.com/pod-product-compliance
Lightning Source LLC
LaVergne TN
LVHW081250011026
849162LV00006BA/1293
* 9 7 8 1 6 6 8 4 5 7 6 5 8 *